Birkhäuser

Trends in Mathematics

Trends in Mathematics is a series devoted to the publication of volumes arising from conferences and lecture series focusing on a particular topic from any area of mathematics. Its aim is to make current developments available to the community as rapidly as possible without compromise to quality and to archive these for reference.

Proposals for volumes can be submitted using the Online Book Project Submission Form at our website www.birkhauser-science.com.

Material submitted for publication must be screened and prepared as follows:

All contributions should undergo a reviewing process similar to that carried out by journals and be checked for correct use of language which, as a rule, is English. Articles without proofs, or which do not contain any significantly new results, should be rejected. High quality survey papers, however, are welcome.

We expect the organizers to deliver manuscripts in a form that is essentially ready for direct reproduction. Any version of $\TeX$ is acceptable , but the entire collection of files must be in one particular dialect of $\TeX$ and unified according to simple instructions available from Birkhäuser.

Furthermore, in order to guarantee the timely appearance of the proceedings it is essential that the final version of the entire material be submitted no later than one year after the conference.

More information about this series at http://www.springer.com/series/4961

Galina Filipuk • Yoshishige Haraoka •
Sławomir Michalik
Editors

Analytic, Algebraic and Geometric Aspects of Differential Equations

Będlewo, Poland, September 2015

Birkhäuser

Editors
Galina Filipuk
Faculty of Mathematics, Informatics and Mechanics
University of Warsaw
Warsaw, Poland

Yoshishige Haraoka
Kumamoto University
Kumamoto, Japan

Sławomir Michalik
Faculty of Mathematics and Natural Sciences
Cardinal Stefan Wyszynski University
Warsaw, Poland

ISSN 2297-0215 ISSN 2297-024X (electronic)
Trends in Mathematics
ISBN 978-3-319-52841-0 ISBN 978-3-319-52842-7 (eBook)
DOI 10.1007/978-3-319-52842-7

Library of Congress Control Number: 2017939936

Mathematics Subject Classification (2010): 34Mxx, 35R03, 17B22, 33C52

Printed on acid-free paper

This book is published under the trade name Birkhäuser, www.birkhauser-science.com
The registered company is Springer International Publishing AG
The registered company address is: Gewerbestrasse 11, 6330 Cham, Switzerland

Preface

Differential equations play a substantial role in various research fields in mathematics and its applications, and as such they play a substantial role in a variety of aspects. Experience shows that finding and understanding the relationships between these aspects lead to new developments to the theory of differential equations and also to other research fields. With this in mind, we organized the school and conference *Analytic, Algebraic and Geometric Aspects of Differential Equations*. The overall goal of the school and the conference was to bring together leading experts in the theory of differential and difference equations in the complex domain from different countries, to tackle and find approaches to various open problems, present recent research results, learn new methods in the related areas (which is invaluable especially for younger researchers), and identify new topics for future research.

During the school and the conference the participants discussed various aspects of analysis of differential equations concerning irregular singular points, and summability of formal solutions, and topics from representation theory, mathematical physics, theory of singularities, differential Galois theory, difference algebra, theory of integrable systems, theory of hypergeometric functions and Painlevé transcendental functions, theory of orthogonal polynomials, and so on.

In this volume we collected all lecture notes from the school, where five lectures on exact WKB analysis, sub-Riemannian geometry, summability, holonomic systems and Dunkl theory were delivered, and also several research papers corresponding to the talks at the conference. We hope that this volume will be useful for studying the particular topics as well for understanding the variety of relationships among diverse research fields in mathematics.

The school and conference *Analytic, Algebraic and Geometric Aspects of Differential Equations* (bcc.impan.pl/15AAGA) was held from September 6 to September 19, 2015, at the Mathematical Research and Conference Center in Będlewo, Poland. The conference was co-financed by the Institute of Mathematics of the Polish Academy of Sciences (IMPAN) and the Warsaw Center of Mathematics and Computer Science. The Organizing Committee consisted of Galina Filipuk, Yoshishige Haraoka, Grzegorz Łysik, and Sławomir Michalik. The Scientific

Committee included Werner Balser, Moulay Barkatou, Stefan Hilger, Masatake Miyake, Hidetoshi Tahara, Masafumi Yoshino, and Henryk Żołądek.

The editors would like to express their deep gratitude to Alexander Vasiliev for his invaluable help with publishing this volume, and to Grzegorz Łysik, who was the head of the Organizing Committee of the school and the conference.

Warsaw, Poland — Galina Filipuk
Kumamoto, Japan — Yoshishige Haraoka
Warsaw, Poland — Sławomir Michalik

Contents

Part I Lecture Notes

An Introduction to Dunkl Theory and Its Analytic Aspects 3
Jean-Philippe Anker

Holonomic Systems 59
Yoshishige Haraoka

Sub-Riemannian Geometry and Hypoelliptic Operators 89
Irina Markina

Asymptotic Analysis and Summability of Formal Power Series 199
Javier Sanz

WKB Analysis and Stokes Geometry of Differential Equations 263
Yoshitsugu Takei

Part II Research Papers

Transcendental Meromorphic Solutions of P_{34} and Small Targets 307
Ewa Ciechanowicz and Galina Filipuk

Towards the Convergence of Generalized Power Series Solutions of Algebraic ODEs 325
Renat R. Gontsov and Irina V. Goryuchkina

Connection Problem for Regular Holonomic Systems in Several Variables 337
Yoshishige Haraoka

On k-Summability of Formal Solutions for Certain Higher Order Partial Differential Operators with Polynomial Coefficients 351
Kunio Ichinobe

On Stokes Phenomena for the Alternate Discrete PI Equation 369
Nalini Joshi and Yoshitsugu Takei

Flat Structures and Algebraic Solutions to Painlevé VI Equation 383
Mitsuo Kato, Toshiyuki Mano, and Jiro Sekiguchi

Relation of Semi-Classical Orthogonal Polynomials to General Schlesinger Systems via Twistor Theory 399
Hironobu Kimura

Some Notes on the Multi-Level Gevrey Solutions of Singularly Perturbed Linear Partial Differential Equations 415
Alberto Lastra and Stéphane Malek

Reducibility of Hypergeometric Equations 429
Toshio Oshima

Parametric Borel Summability of Partial Differential Equations of Irregular Singular Type 455
Masafumi Yoshino

Part I
Lecture Notes

An Introduction to Dunkl Theory and Its Analytic Aspects

Jean-Philippe Anker

Abstract The aims of these lecture notes are twofold. On the one hand, they are meant as an introduction to rational and trigonometric Dunkl theory, starting with the historical examples of special functions associated with radial Fourier analysis on rank one symmetric spaces. On the other hand, we have tried to give an up-to-date account of the main results and problems, with an emphasis on analytic aspects.

Keywords Dunkl theory • Special functions associated with root systems • Spherical Fourier analysis

Mathematics Subject Classification (2000). Primary 33C67; Secondary 05E05, 20F55, 22E30, 33C80, 33D67, 39A70, 42B10, 43A32, 43A90

1 Introduction

Dunkl theory is a far reaching generalization of Fourier analysis and special function theory related to root systems. During the sixties and seventies, it became gradually clear that radial Fourier analysis on rank one symmetric spaces was closely connected with certain classes of special functions in one variable:

- Bessel functions in connection with radial Fourier analysis on Euclidean spaces,
- Jacobi polynomials in connection with radial Fourier analysis on spheres,
- Jacobi functions (i.e. the Gauss hypergeometric function ${}_2F_1$) in connection with radial Fourier analysis on hyperbolic spaces.

See [51] for a survey. During the eighties, several attempts were made, mainly by the Dutch school (Koornwinder, Heckman, Opdam), to extend these results in higher rank (i.e. in several variables), until the discovery of Dunkl operators in the rational

J.-P. Anker (✉)
Fédération Denis Poisson (FR 2964), Laboratoire MAPMO (UMR 7349), Université d'Orléans & CNRS, Bâtiment de Mathématiques, B.P. 6759, 45067 Orléans Cedex 2, France
e-mail: anker@univ-orleans.fr

G. Filipuk et al. (eds.), *Analytic, Algebraic and Geometric Aspects of Differential Equations*, Trends in Mathematics, DOI 10.1007/978-3-319-52842-7_1

case and Cherednik operators in the trigonometric case. Together with q-special functions introduced by Macdonald, this has led to a beautiful theory, developed by several authors which encompasses in a unified way harmonic analysis on all Riemannian symmetric spaces and spherical functions thereon:

- generalized Bessel functions on flat symmetric spaces, and their asymmetric version, known as the Dunkl kernel,
- Heckman–Opdam hypergeometric functions on Riemannian symmetric spaces of compact or noncompact type, and their asymmetric version, due to Opdam,
- Macdonald polynomials on affine buildings.

Beside Fourier analysis and special functions, this theory has also deep and fruitful interactions with

- algebra (double affine Hecke algebras),
- mathematical physics (Calogero-Moser-Sutherland models, quantum many body problems),
- probability theory (Feller processes with jumps).

There are already several surveys about Dunkl theory available in the literature:

- [67] (see also [29]) about rational Dunkl theory (state of the art in 2002),
- [64] about trigonometric Dunkl theory (state of the art in 1998),
- [31] about integrable systems related to Dunkl theory,
- [54] and [17] about q-Dunkl theory and affine Hecke algebras,
- [42] about probabilistic aspects of Dunkl theory (state of the art in 2006).

These lectures are intended to give an overview of some analytic aspects of Dunkl theory. The topics are indicated in italic in Fig. 1, where we have tried to summarize relations between several theories of special functions, which were alluded to above,

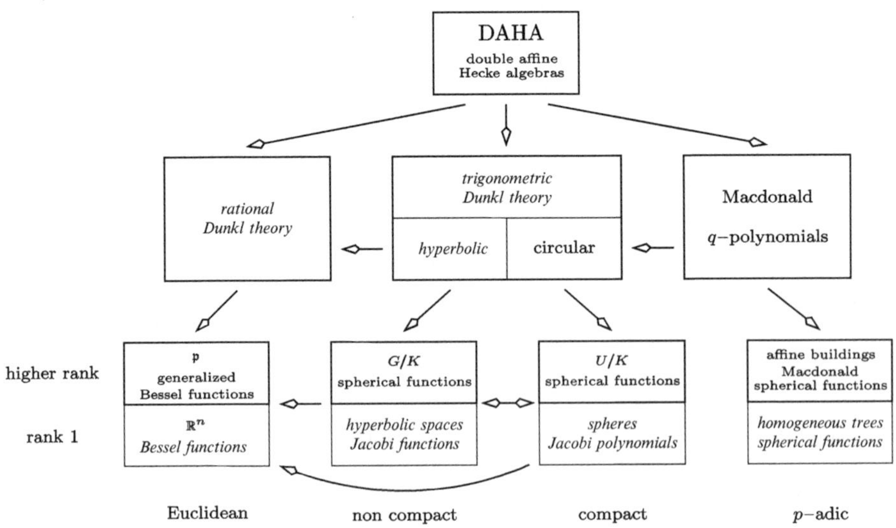

Fig. 1 Relation between various special function theories

and where arrows mean limits. Let us describe the content of our notes. In Sect. 2, we consider several geometric settings (Euclidean spaces, spheres, hyperbolic spaces, homogeneous trees, ...) where radial Fourier analysis is available and can be applied successfully, for instance to study evolutions equations (heat equation, wave equation, Schrödinger equation, ...). Section 3 is devoted to the rational Dunkl theory and Sect. 4 to the trigonometric Dunkl theory. In both cases, we first review the basics and next address some important analytic issues. We conclude with an appendix about root systems and with a comprehensive bibliography. For lack of time and competence, we haven't touched upon other aforementioned aspects of Dunkl theory, for which we refer to the bibliography.

2 Spherical Fourier Analysis in Rank 1

Cosine Transform

Let us start with an elementary example. Within the framework of even functions on the real line $\mathbb{R}$, the Fourier transform is given by

$$\hat{f}(\lambda) = \int_{\mathbb{R}} dx f(x) \cos \lambda x$$

and the inverse Fourier transform by

$$f(x) = \tfrac{1}{2\pi} \int_{\mathbb{R}} d\lambda \hat{f}(\lambda) \cos \lambda x .$$

The cosine functions $\varphi_\lambda(x) = \cos \lambda x$ ($\lambda \in \mathbb{C}$) occurring in these expressions can be characterized in various ways. Let us mention

- *Power series expansion*:

$$\varphi_\lambda(x) = \sum\nolimits_{\ell=0}^{+\infty} \tfrac{(-1)^\ell}{(2\ell)!} (\lambda x)^{2\ell} \qquad \forall\, \lambda, x \in \mathbb{C} .$$

- *Differential equation*: the functions $\varphi = \varphi_\lambda$ are the smooth eigenfunctions of $(\frac{\partial}{\partial x})^2$, which are even and normalized by $\varphi(0) = 1$.
- *Functional equation*: the functions $\varphi = \varphi_\lambda$ are the nonzero continuous functions on $\mathbb{R}$ which satisfy

$$\frac{\varphi(x+y) + \varphi(x-y)}{2} = \varphi(x)\, \varphi(y) \qquad \forall\, x, y \in \mathbb{R}.$$

Hankel Transform on Euclidean Spaces

The Fourier transform on $\mathbb{R}^n$ and its inverse are given by

$$\hat{f}(\xi) = \int_{\mathbb{R}^n} dx f(x)\, e^{-i\langle \xi, x\rangle} \tag{1}$$

and

$$f(x) = (2\pi)^{-n} \int_{\mathbb{R}^n} d\lambda\, \hat{f}(\xi)\, e^{i\langle \xi, x\rangle} \tag{2}$$

Notice that the Fourier transform of a radial function $f = f(r)$ on $\mathbb{R}^n$ is again a radial function $\hat{f} = \hat{f}(\lambda)$. In this case, (1) and (2) become

$$\hat{f}(\lambda) = \tfrac{2\pi^{\frac{n}{2}}}{\Gamma(\frac{n}{2})} \int_0^{+\infty} dr\, r^{n-1} f(r)\, j_{\frac{n-2}{2}}(i\lambda r) \tag{3}$$

and

$$f(r) = \frac{1}{2^{n-1}\pi^{\frac{n}{2}}\,\Gamma(\frac{n}{2})} \int_0^{+\infty} d\lambda\, \lambda^{n-1} \hat{f}(\lambda)\, j_{\frac{n-2}{2}}(i\lambda r)\,. \tag{4}$$

Instead of the exponential function or the cosine function, (3) and (4) involve now the modified Bessel function $j_{\frac{n-2}{2}}$, which can be characterized again in various ways:

- *Relation with classical special functions and power series expansion.*

$$\begin{aligned} j_{\frac{n-2}{2}}(z) &= \Gamma(\tfrac{n}{2})\, \big(\tfrac{iz}{2}\big)^{\frac{2-n}{2}} J_{\frac{n-2}{2}}(iz) \\ &= \sum\nolimits_{\ell=0}^{+\infty} \tfrac{\Gamma(\frac{n}{2})}{\ell!\,\Gamma(\frac{n}{2}+\ell)} \big(\tfrac{z}{2}\big)^{2\ell} \\ &= {}_0\mathrm{F}_1(\tfrac{n}{2};\tfrac{z^2}{4}) = e^{-z}\,{}_1\mathrm{F}_1(\tfrac{n-1}{2};n-1;2z) \qquad \forall\, z\in\mathbb{C}, \end{aligned}$$

where J_ν denotes the classical Bessel function of the first kind and

$${}_p\mathrm{F}_q(a_1,\dots,a_p;b_1,\dots,b_q;z) = \sum\nolimits_{\ell=0}^{+\infty} \tfrac{(a_1)_\ell \dots (a_p)_\ell}{(b_1)_\ell \dots (b_q)_\ell} \tfrac{z^\ell}{\ell!}$$

the generalized hypergeometric function.

- *Differential equations*. The function $\varphi_\lambda(r) = j_{\frac{n-2}{2}}(i\lambda r)$ is the unique smooth solution to the differential equation

$$\left(\tfrac{\partial}{\partial r}\right)^2 \varphi_\lambda + \tfrac{n-1}{r}\left(\tfrac{\partial}{\partial r}\right)\varphi_\lambda + \lambda^2 \varphi_\lambda = 0\,,$$

which is normalized by $\varphi_\lambda(0) = 1$. Equivalently, the function

$$x \longmapsto \varphi_\lambda(|x|) = j_{\frac{n-2}{2}}(i\lambda\,|x|) \tag{5}$$

is the unique smooth radial normalized eigenfunction of the Euclidean Laplacian

$$\Delta_{\mathbb{R}^n} = \sum\nolimits_{j=1}^{n}\left(\tfrac{\partial}{\partial x_j}\right)^2 = \left(\tfrac{\partial}{\partial r}\right)^2 + \tfrac{n-1}{r}\left(\tfrac{\partial}{\partial r}\right) + \tfrac{1}{r^2}\,\Delta_{\mathbb{S}^{n-1}}$$

corresponding to the eigenvalue $-\lambda^2$.

Remark 2.1 The function (5) is a matrix coefficient of a continuous unitary representation of the Euclidean motion group $\mathbb{R}^n \rtimes \mathrm{O}(n)$.

The function (5) is a spherical average of plane waves. Specifically,

$$\varphi_\lambda(|x|) = \int_{\mathrm{O}(n)} dk\; e^{i\lambda\langle u,k.x\rangle} = \frac{\Gamma(\frac{n}{2})}{2\pi^{\frac{n}{2}}}\int_{\mathbb{S}^{n-1}} dv\; e^{i\lambda\langle v,x\rangle}\,,$$

where u is any unit vector in $\mathbb{R}^n$. Hence the integral representation

$$\begin{aligned}\varphi_\lambda(r) &= \frac{\Gamma(\frac{n}{2})}{\sqrt{\pi}\;\Gamma(\frac{n-1}{2})}\int_0^\pi d\theta\,(\sin\theta)^{n-2}\,e^{i\lambda r\cos\theta}\\ &= \frac{2\,\Gamma(\frac{n}{2})}{\sqrt{\pi}\;\Gamma(\frac{n-1}{2})}\,r^{2-n}\int_0^r ds\,(r^2-s^2)^{\frac{n-3}{2}}\,\cos\lambda s\,.\end{aligned} \tag{6}$$

Spherical Fourier Analysis on Real Spheres

Real spheres

$$\mathbb{S}^n = \{\,x = (x_0, x_1, \dots, x_n) \in \mathbb{R}^{1+n} \mid |x|^2 = x_0^2 + \dots + x_n^2 = 1\,\}$$

of dimension $n \geq 2$ are the simplest examples of Riemannian symmetric spaces of compact type. They are simply connected Riemannian manifolds, with constant positive sectional curvature. The Riemannian structure on $\mathbb{S}^n$ is induced by the Euclidean metric in $\mathbb{R}^{1+n}$, restricted to the tangent bundle of $\mathbb{S}^n$, and the Laplacian on $\mathbb{S}^n$ is given by $\Delta f = \tilde{\Delta}\tilde{f}\,\big|_{\mathbb{S}^n}$, where $\tilde{\Delta} = \sum\nolimits_{j=0}^{n}\left(\tfrac{\partial}{\partial x_j}\right)^2$ denotes the Euclidean

Laplacian in $\mathbb{R}^{1+n}$ and $\tilde{f}(x)=f\big(\frac{x}{|x|}\big)$ the homogeneous extension of f to $\mathbb{R}^{1+n}\smallsetminus\{0\}$. In spherical coordinates

$$\begin{cases} x_0 &= \cos\theta_1\,,\\ x_1 &= \sin\theta_1\,\cos\theta_2\,,\\ &\vdots\\ x_{n-1} &= \sin\theta_1\,\sin\theta_2\,\dots\,\sin\theta_{n-1}\,\cos\theta_n\,,\\ x_n &= \sin\theta_1\,\sin\theta_2\,\dots\,\sin\theta_{n-1}\,\sin\theta_n\,, \end{cases}$$

the Riemannian metric, the Riemannian volume and the Laplacian read respectively

$$ds^2 = \sum\nolimits_{j=1}^{n}(\sin\theta_1)^2\dots(\sin\theta_{j-1})^2\,(d\theta_j)^2\,,$$

$$d\,\mathrm{vol} = (\sin\theta_1)^{n-1}\dots(\sin\theta_{n-1})d\theta_1\dots d\theta_n$$

and

$$\Delta = \sum\nolimits_{j=1}^{n}\tfrac{1}{(\sin\theta_1)^2\dots(\sin\theta_{j-1})^2}\left\{\left(\tfrac{\partial}{\partial\theta_j}\right)^2 + (n-j)\,(\cot\theta_j)\,\tfrac{\partial}{\partial\theta_j}\right\}.$$

Let $G=\mathrm{O}(n+1)$ be the isometry group of $\mathbb{S}^n$ and let $K\approx\mathrm{O}(n)$ be the stabilizer of $e_0=(1,0,\dots,0)$. Then $\mathbb{S}^n$ can be realized as the homogeneous space G/K (see Fig. 2). As usual, we identify right-K-invariant functions on G with functions on $\mathbb{S}^n$, and bi-K-invariant functions on G with radial functions on $\mathbb{S}^n$ i.e. functions on $\mathbb{S}^n$ which depend only on $x_0=\cos\theta_1$. For such functions,

$$\int_{\mathbb{S}^n} d\,\mathrm{vol}\,f \;=\; 2\,\frac{\pi^{\frac{n}{2}}}{\Gamma(\frac{n}{2})}\int_0^{\pi} d\theta_1\,(\sin\theta_1)^{n-1}f(\cos\theta_1)$$

and

$$\Delta f = \tfrac{\partial^2 f}{\partial\theta_1^2} + (n-1)\,(\cot\theta_1)\,\tfrac{\partial f}{\partial\theta_1}\,.$$

The spherical functions on $\mathbb{S}^n$ are the smooth normalized radial eigenfunctions of the Laplacian on $\mathbb{S}^n$. Specifically,

$$\begin{cases} \Delta\,\varphi_\ell = -\ell\,(\ell+n-1)\,\varphi_\ell\,,\\ \varphi_\ell(e_0) = 1\,, \end{cases}$$

where $\ell\in\mathbb{N}$. They can be expressed in terms of classical special functions, namely

$$\varphi_\ell(x_0) = \tfrac{\ell\,!\,(n-2)\,!}{(\ell+n-2)\,!}\,C_\ell^{(\frac{n-1}{2})}(x_0) = \tfrac{\ell\,!}{(\frac{n}{2})_\ell}\,P_\ell^{(\frac{n}{2}-1,\frac{n}{2}-1)}(x_0)$$

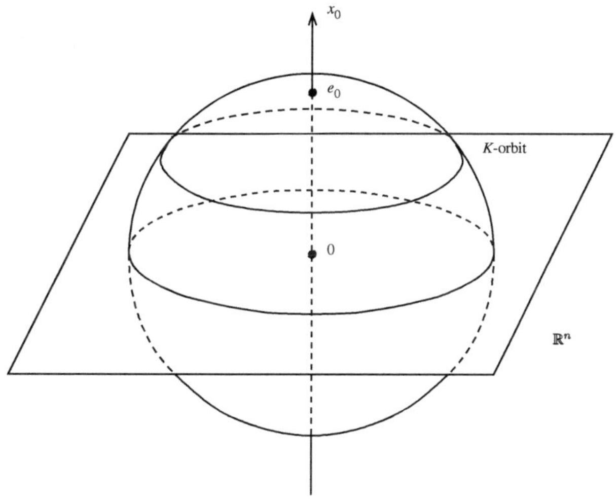

Fig. 2 Real sphere $\mathbb{S}^n$

or

$$\varphi_\ell(\cos\theta_1) = {}_2\mathrm{F}_1\big(-\ell, \ell+n-1; \tfrac{n}{2}; \sin^2\theta_1\big),$$

where $C_\ell^{(\lambda)}$ are the Gegenbauer or ultraspherical polynomials, $P_\ell^{(\alpha,\beta)}$ the Jacobi polynomials and ${}_2\mathrm{F}_1$ the Gauss hypergeometric function.

Remark 2.2 We have emphasized the characterization of spherical functions on $\mathbb{S}^n$ by a differential equation. Here are other characterizations:

- The spherical functions are the continuous bi-K-invariant functions φ on G which satisfy the functional equation

$$\int_K dk\ \varphi(xky) = \varphi(x)\,\varphi(y) \qquad \forall\ x,y\in G. \tag{7}$$

- The spherical functions are the continuous bi-K-invariant functions φ on G such that

$$f \longmapsto \int_G dx f(x)\,\varphi(x) \tag{8}$$

defines a character of the (commutative) convolution algebra $C_c(K\backslash G/K)$.

- The spherical functions are the matrix coefficients

$$\varphi(x) = \langle \pi(x)v, v\rangle,$$

where π is a continuous unitary representation of G, which has nonzero K-fixed vectors and which is irreducible, and v is a K-fixed vector, which is normalized by $|v| = 1$.

- Integral representation:

$$\begin{aligned}
&\varphi_\ell(\cos\theta_1) \\
&= \frac{\Gamma(\frac{n}{2})}{\sqrt{\pi}\,\Gamma(\frac{n-1}{2})} \int_0^\pi d\theta_2\,(\sin\theta_2)^{n-2}\big[\cos\theta_1 + i\,(\sin\theta_1)(\cos\theta_2)\big]^\ell \\
&= \frac{\Gamma(\frac{n}{2})}{\sqrt{\pi}\,\Gamma(\frac{n-1}{2})}\,(\sin\theta_1)^{2-n} \int_{-\sin\theta_1}^{\sin\theta_1} ds\,(\sin^2\theta_1 - s^2)^{\frac{n-3}{2}}\,(\cos\theta_1 + i\,s)^\ell\,.
\end{aligned} \tag{9}$$

The spherical Fourier expansion of radial functions on $\mathbb{S}^n$ reads

$$f(x) = \sum\nolimits_{\ell\in\mathbb{N}} d_\ell\,\langle f, \varphi_\ell\rangle\,\varphi_\ell(x)\,,$$

where

$$d_\ell = \tfrac{n\,(n+2\ell-1)\,(n+\ell-2)!}{n!\,\ell!}$$

and

$$\begin{aligned}
\langle f, \varphi_\ell\rangle &= \frac{\Gamma(\frac{n+1}{2})}{2\pi^{\frac{n+1}{2}}} \int_{\mathbb{S}^n} dx f(x)\,\varphi_\ell(x) \\
&= \frac{\Gamma(\frac{n+1}{2})}{\sqrt{\pi}\,\Gamma(\frac{n}{2})} \int_0^\pi d\theta_1\,(\sin\theta_1)^{n-1} f(\cos\theta_1)\,\varphi_\ell(\cos\theta_1)\,.
\end{aligned}$$

Spherical Fourier Analysis on Real Hyperbolic Spaces

Real hyperbolic spaces $\mathbb{H}^n$ are the simplest examples of Riemannian symmetric spaces of noncompact type. They are simply connected Riemannian manifolds, with constant negative sectional curvature. Let us recall the following three models of $\mathbb{H}^n$.

- **Model 1: Hyperboloid (Fig. 3)**

In this model, $\mathbb{H}^n$ consists of the hyperboloid sheed

$$\{\, x = (x_0, x_1, \dots, x_n) \in \mathbb{R}^{1+n} = \mathbb{R}\times\mathbb{R}^n \mid L(x,x) = -1,\ x_0 \ge 1 \,\}$$

defined by the Lorentz quadratic form $L(x,x) = -x_0^2 + x_1^2 + \ldots + x_n^2$. The Riemannian structure is given by the metric $ds^2 = L(dx, dx)$, restricted to the tangent bundle of

$\mathbb{H}^n$, and the Laplacian by

$$\Delta f = L\big(\tfrac{\partial}{\partial x},\tfrac{\partial}{\partial x}\big)\tilde{f}\,\big|_{\mathbb{H}^n},$$

where $\tilde{f}(x) = f\big(\frac{x}{\sqrt{-L(x,x)}}\big)$ denotes the homogeneous extension of f to the light cone $\{\,x\in\mathbb{R}^{1+n}\,|\,L(x,x)<0,\ x_0>0\,\}$.

- **Model 2: Upper half-space (Fig. 4)**

In this model, $\mathbb{H}^n$ consists of the upper half-space $\mathbb{R}^n_+ = \{y \in \mathbb{R}^n\,|\,y_n > 0\}$ equipped with the Riemannian metric $ds^2 = y_n^{-2}\,|dy|^2$. The volume is given by $d\mathrm{vol} = y_n^{-n}\,dy_1 \dots dy_n$ and the Laplacian by

$$\Delta = y_n^2 \sum\nolimits_{j=1}^{n} \tfrac{\partial^2}{\partial y_j^2} - (n-2)\,y_n\,\tfrac{\partial}{\partial y_n}\,.$$

- **Model 3: Ball (Fig. 5)**

In this model, $\mathbb{H}^n$ consists of the unit ball $\mathbb{B}^n = \{\,z\in\mathbb{R}^n\,|\,|z|<1\,\}$. The Riemannian metric is given by $ds^2 = \big(\frac{1-|z|^2}{2}\big)^{-2}\,|dz|^2$, the volume by $d\mathrm{vol} = \big(\frac{1-|z|^2}{2}\big)^{-n}$

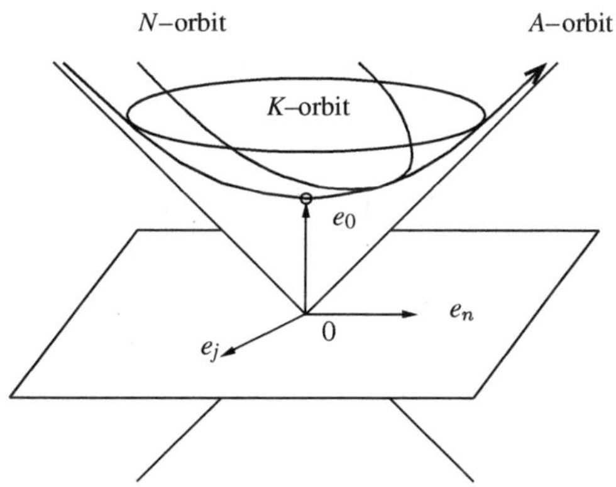

Fig. 3 Hyperboloid model of $\mathbb{H}^n$

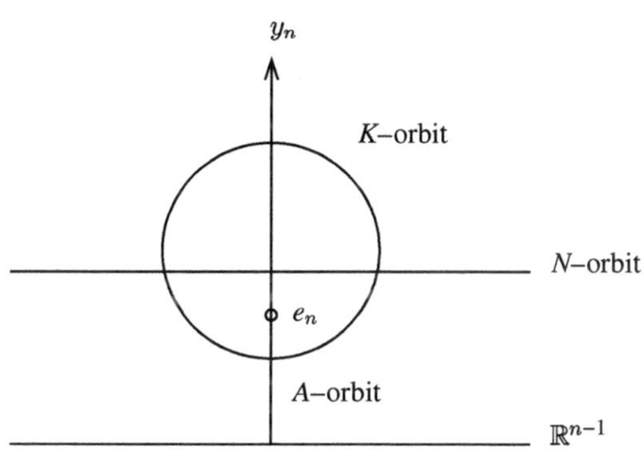

Fig. 4 Upper half-space model of $\mathbb{H}^n$

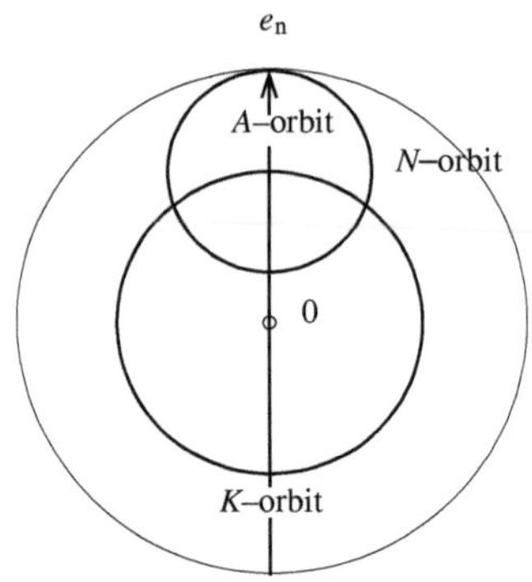

Fig. 5 Ball model of $\mathbb{H}^n$

$dz_1 \dots dz_n$, the distance to the origin by $r = 2\,\operatorname{artanh}|z| = \log\frac{1+|z|}{1-|z|}$, and the Laplacian by

$$\Delta = \left(\tfrac{1-|z|^2}{2}\right)^2 \sum\nolimits_{j=1}^{n} \tfrac{\partial^2}{\partial z_j^2} + (n-2)\,\tfrac{1-|z|^2}{2} \sum\nolimits_{j=1}^{n} z_j \tfrac{\partial}{\partial z_j}\,.$$

Remark 2.3 Model 1 and Model 3 are mapped onto each other by the stereographic projection with respect to $-e_0$, while Model 2 and Model 3 are mapped onto each other by the inversion with respect to the sphere $S(-e_n,\sqrt{2})$. This leads to the following formulae

$$\begin{cases} x_0 = \frac{1+|y|^2}{2y_n} = \frac{1+|z|^2}{1-|z|^2} & \\ x_j = \frac{y_j}{y_n} = \frac{2z_j}{1-|z|^2} & (j=1,\dots,n-1) \\ x_n = \frac{1-|y|^2}{2y_n} = \frac{2z_n}{1-|z|^2} & \end{cases}$$

$$\begin{cases} y_j = \frac{x_j}{x_0+x_n} = \frac{2z_j}{1+|z|^2+2z_n} & (j=1,\dots,n-1) \\ y_n = \frac{1}{x_0+x_n} = \frac{1-|z|^2}{1+|z|^2+2z_n} & \end{cases}$$

$$\begin{cases} z_j = \frac{x_j}{1+x_0} = \frac{2y_j}{1+|y|^2+2y_n} & (j=1,\dots,n-1) \\ z_n = \frac{x_n}{1+x_0} = \frac{1-|y|^2}{1+|y|^2+2y_n}\,. & \end{cases}$$

Let G be the isometry group of $\mathbb{H}^n$ and let K be the stabilizer of a base point in $\mathbb{H}^n$. Then $\mathbb{H}^n$ can be realized as the homogeneous space G/K. In Model 1, G is made up of two among the four connected components of the Lorentz group $\mathrm{O}(1,n)$, and the stabilizer of e_0 is $K = \mathrm{O}(n)$. Consider the subgroup $A \approx \mathbb{R}$ in G consisting of

- the matrices

$$a_r = \begin{pmatrix} \cosh r & 0 & \sinh r \\ 0 & I & 0 \\ \sinh r & 0 & \cosh r \end{pmatrix} \qquad (r \in \mathbb{R})$$

  in Model 1,
- the dilations $a_r : y \longmapsto e^{-r} y$ in Model 2,

and the subgroup $N \approx \mathbb{R}^{n-1}$ consisting of horizontal translations $n_\upsilon : y \longmapsto y + \upsilon$ $(\upsilon \in \mathbb{R}^{n-1})$ in Model 2. Then we have

- the Cartan decomposition $G = K\overline{A^+}K$, which corresponds to polar coordinates in Model 3,
- the Iwasawa decomposition $G = NAK$, which corresponds to Cartesian coordinates in Model 2.

We shall denote by $a_{r(g)}$ and $a_{h(g)}$ the $\overline{A^+}$ and A components of $g \in G$ in the Cartan and Iwasawa decompositions.

Remark 2.4 In small dimensions, $\mathbb{H}^2 \approx \mathrm{SL}(2,\mathbb{R})/\mathrm{SO}(2)$ and $\mathbb{H}^3 \approx \mathrm{SL}(2,\mathbb{C})/\mathrm{SU}(2)$.

As usual, we identify right-K-invariant functions on G with functions on $\mathbb{H}^n$, and bi-K-invariant functions on G with radial functions on $\mathbb{H}^n$ i.e. functions on $\mathbb{H}^n$ which depend only on the distance r to the origin. For radial functions $f = f(r)$,

$$\int_{\mathbb{H}^n} d\,\mathrm{vol}\, f = 2\,\frac{\pi^{\frac{n}{2}}}{\Gamma(\frac{n}{2})} \int_0^{+\infty} dr\,(\sinh r)^{n-1} f(r)$$

and

$$\Delta f = \tfrac{\partial^2 f}{\partial r^2} + (n-1)(\coth r)\,\tfrac{\partial f}{\partial r}\,.$$

The spherical functions φ_λ are the smooth normalized radial eigenfunctions of the Laplacian on $\mathbb{H}^n$. Specifically,

$$\begin{cases} \Delta\,\varphi_\lambda = -\left\{\lambda^2 + \rho^2\right\}\varphi_\lambda\,, \\ \varphi_\lambda(0) = 1\,, \end{cases}$$

where $\rho = \frac{n-1}{2}$.

Remark 2.5 The spherical functions on $\mathbb{H}^n$ can be characterized again in several other ways. Notably,

- Differential equation: the function $\varphi_\lambda(r)$ is the unique smooth solution to the differential equation

$$\big(\tfrac{\partial}{\partial r}\big)^2\varphi_\lambda+(n-1)(\coth r)\,\big(\tfrac{\partial}{\partial r}\big)\,\varphi_\lambda+(\lambda^2+\rho^2)\,\varphi_\lambda=0\,,$$

which is normalized by $\varphi_\lambda(0)=1$.
- Relation with classical special functions:

$$\varphi_\lambda(r)=\varphi_\lambda^{\frac{n-2}{2},-\frac12}(r)={}_2\mathrm{F}_1\big(\tfrac{\rho+i\lambda}{2},\tfrac{\rho-i\lambda}{2};\tfrac{n}{2};-\sinh^2 r\big)\,,$$

where $\varphi_\lambda^{\alpha,\beta}$ are the Jacobi functions and ${}_2\mathrm{F}_1$ is the Gauss hypergeometric function.
- Same functional equations as (7) and (8).
- The spherical functions are the matrix coefficients

$$\varphi_\lambda(x)=\tfrac{\Gamma(\frac{n}{2})}{2\pi^{\frac{n}{2}}}\,\langle\pi_\lambda(x)1,1\rangle\,,$$

of the spherical principal series representations of G on $L^2(\mathbb{S}^{n-1})$.
- According to the Harish–Chandra formula

$$\varphi_\lambda(x)=\int_K dk\;e^{(\rho-i\lambda)\,h(kx)}\,,$$

the function φ_λ is a spherical average of horocyclic waves. Let us make this integral representation more explicit:

$$\begin{aligned}\varphi_\lambda(r)&=\tfrac{\Gamma(\frac{n}{2})}{2\pi^{\frac{n}{2}}}\int_{\mathbb{S}^{n-1}}dv\;\langle\cosh r-(\sinh r)\,v,e_n\rangle^{i\lambda-\rho}\\&=\frac{\Gamma(\frac{n}{2})}{\sqrt{\pi}\;\Gamma(\frac{n-1}{2})}\int_0^\pi d\theta\,(\sin\theta)^{n-2}\big[\cosh r-(\sinh r)(\cos\theta)\big]^{i\lambda-\rho}\\&=\frac{2^{\frac{n-1}{2}}\,\Gamma(\frac{n}{2})}{\sqrt{\pi}\;\Gamma(\frac{n-1}{2})}\,(\sinh r)^{2-n}\int_0^r ds\,(\cosh r-\cosh s)^{\frac{n-3}{2}}\,\cos\lambda s\,.\end{aligned}\tag{10}$$

Remark 2.6 The asymptotic behavior of the spherical functions is given by the Harish–Chandra expansion

$$\varphi_\lambda(r)=\mathbf{c}(\lambda)\,\Phi_\lambda(r)+\mathbf{c}(-\lambda)\,\Phi_{-\lambda}(r)\,,$$

where

$$\mathbf{c}(\lambda)=\tfrac{\Gamma(2\rho)}{\Gamma(\rho)}\,\tfrac{\Gamma(i\lambda)}{\Gamma(i\lambda+\rho)}$$

and

$$\begin{aligned}\Phi_\lambda(r) &= (2\cosh r)^{i\lambda-\rho}\,{}_2\mathrm{F}_1\big(\tfrac{\rho-i\lambda}{2},\tfrac{\rho+1-i\lambda}{2}\,;1-i\lambda\,;\cosh^{-2}r\big)\\ &= e^{(i\lambda-\rho)r}\sum\nolimits_{\ell=0}^{+\infty}\Gamma_\ell(\lambda)\,e^{-2\ell r},\end{aligned}$$

with $\Gamma_0 \equiv 1$.

The spherical Fourier transform (or Harish–Chandra transform) of radial functions on $\mathbb{H}^n$ is defined by

$$\mathcal{H}f(\lambda) = \int_{\mathbb{H}^n} dx\, f(x)\,\varphi_\lambda(x) = \frac{\Gamma(\frac{n}{2})}{2\pi^{\frac{n}{2}}}\int_0^{+\infty} dr\,(\sinh r)^{n-1} f(r)\,\varphi_\lambda(r) \tag{11}$$

and the inversion formula reads

$$f(x) = 2^{n-3}\,\pi^{-\frac{n}{2}-1}\,\Gamma(\tfrac{n}{2})\int_0^{+\infty} d\lambda\;|\mathbf{c}(\lambda)|^{-2}\,\mathcal{H}f(\lambda)\,\varphi_\lambda(x)\,. \tag{12}$$

Remark 2.7

- The Plancherel density reads

$$|\mathbf{c}(\lambda)|^{-2} = \frac{\pi}{2^{2n-4}\,\Gamma(\frac{n}{2})^2}\prod\nolimits_{j=0}^{\frac{n-3}{2}}(\lambda^2+j^2)$$

  in odd dimension, and

$$|\mathbf{c}(\lambda)|^{-2} = \frac{\pi}{2^{2n-4}\,\Gamma(\frac{n}{2})^2}\,\lambda\tanh\pi\lambda\prod\nolimits_{j=0}^{\frac{n}{2}-1}\big[\lambda^2+(j+\tfrac{1}{2})^2\big]$$

  in even dimension. Notice the different behaviors

$$|\mathbf{c}(\lambda)|^{-2} \sim \frac{\pi}{2^{2n-4}\,\Gamma(\frac{n}{2})^2}\,|\lambda|^{n-1}$$

  at infinity, and

$$|\mathbf{c}(\lambda)|^{-2} \sim \frac{\pi\,\Gamma(\frac{n-1}{2})^2}{2^{2n-4}\,\Gamma(\frac{n}{2})^2}\,\lambda^2$$

  at the origin.
- Observe that (11) and (12) are not symmetric, unlike (1) and (2), or (3) and (4).

The spherical Fourier transform (11), which is somewhat abstract, can be bypassed by considering the Abel transform, which is essentially the horocyclic Radon transform restricted to radial functions. Specifically,

$$\begin{aligned} \mathcal{A}f(r) &= e^{-\rho r} \int_N dn\, f(na_r) \\ &= \frac{(2\pi)^{\frac{n-1}{2}}}{\Gamma(\frac{n-1}{2})} \int_{|r|}^{+\infty} ds\ \sinh s\,(\cosh s - \cosh r)^{\frac{n-3}{2}} f(s)\,. \end{aligned}$$

Then the following commutative diagram holds, let say in the Schwartz space setting:

$$\begin{array}{ccc} & \mathcal{S}_{\text{even}}(\mathbb{R}) & \\ \mathcal{H} \nearrow & & \nwarrow \mathcal{F} \\ \mathcal{S}_{\text{rad}}(\mathbb{H}^n) & \underset{\mathcal{A}}{\longrightarrow} & \mathcal{S}_{\text{even}}(\mathbb{R}) \end{array}$$

Here $\mathcal{S}_{\text{rad}}(\mathbb{H}^n)$ denotes the L^2 radial Schwartz space on $\mathbb{H}^n$, which can be identified with $(\cosh r)^{-\rho}\, \mathcal{S}_{\text{even}}(\mathbb{R})$, $\mathcal{F}$ the Euclidean Fourier transform on $\mathbb{R}$ and each arrow is an isomorphism. Thus the inversion of the spherical Fourier transform $\mathcal{H}$ boils down to the inversion of the Abel transform $\mathcal{A}$. In odd dimension,

$$\mathcal{A}^{-1}g(r) = (2\pi)^{-\frac{n-1}{2}} \left(-\tfrac{1}{\sinh r}\tfrac{\partial}{\partial r}\right)^{\frac{n-1}{2}} g(r)$$

while, in even dimension,

$$\mathcal{A}^{-1}g(r) = \frac{1}{2^{\frac{n-1}{2}}\pi^{\frac{n}{2}}} \int_{|r|}^{+\infty} \frac{ds}{\sqrt{\cosh s - \cosh r}} \left(-\tfrac{\partial}{\partial s}\right) \left(-\tfrac{1}{\sinh s}\tfrac{\partial}{\partial s}\right)^{\frac{n}{2}-1} g(s)\,.$$

Consider finally the transform

$$\mathcal{A}^*g(x) = \int_K dk\ e^{\rho\, h(kx)}\, g(h(kx))\,,$$

which is dual to the Abel transform, i.e.,

$$\int_{\mathbb{H}^n} dx\, f(x)\, \mathcal{A}^*g(x) = \int_{-\infty}^{+\infty} dr\ \mathcal{A}f(r)\, g(r)\,,$$

and which is an isomorphism between $C^\infty_{\text{even}}(\mathbb{R})$ and $C^\infty_{\text{rad}}(\mathbb{H}^n)$. It is given explicitly by

$$\mathcal{A}^*g(r) = \frac{2^{\frac{n-1}{2}}\,\Gamma(\frac{n}{2})}{\sqrt{\pi}\,\Gamma(\frac{n-1}{2})} (\sinh r)^{2-n} \int_0^r ds\,(\cosh r - \cosh s)^{\frac{n-3}{2}}\, g(s)$$

and its inverse by

$$(\mathcal{A}^*)^{-1}f(r) = \frac{\sqrt{\pi}}{2^{\frac{n-1}{2}}\,\Gamma(\frac{n}{2})}\,\frac{\partial}{\partial r}\Big(\frac{1}{\sinh r}\,\frac{\partial}{\sinh r}\Big)^{\frac{n-3}{2}}\{(\sinh r)^{n-2}f(r)\}$$

in odd dimension and by

$$(\mathcal{A}^*)^{-1}f(r) = \frac{1}{2^{\frac{n-1}{2}}\,(\frac{n}{2}-1)!}\,\frac{\partial}{\partial r}\Big(\frac{1}{\sinh r}\,\frac{\partial}{\partial r}\Big)^{\frac{n}{2}-1}\int_0^r \frac{ds}{\sqrt{\cosh r-\cosh s}}\,(\sinh s)^{n-1}f(s)$$

in even dimension.

Remark 2.8 Notice that the spherical function $\varphi_\lambda(r)$ is the dual Abel transform of the cosine function $\cos\lambda s$.

Applications Spherical Fourier analysis is an efficient tool for solving invariant PDEs on $\mathbb{H}^n$. Here are some examples of evolution equations.

- The heat equation

$$\begin{cases} \partial_t u(x,t) = \Delta_x u(x,t) \\ u(x,0) = f(x) \end{cases}$$

can be solved explicitly by means of the inverse Abel transform. Specifically,

$$u(x,t) = f * h_t(x)\,,$$

where the heat kernel is given by

$$h_t(r) = \frac{1}{2^{\frac{n+1}{2}}\,\pi^{\frac{n}{2}}}\,t^{-\frac{1}{2}}\,e^{-\rho^2 t}\Big(-\frac{1}{\sinh r}\,\frac{\partial}{\partial r}\Big)^{\frac{n-1}{2}}\,e^{-\frac{r^2}{4t}}$$

in odd dimension and by

$$h_t(r) = (2\pi)^{-\frac{n+1}{2}}\,t^{-\frac{1}{2}}\,e^{-\rho^2 t}\int_{|r|}^{+\infty}\frac{ds}{\sqrt{\cosh s-\cosh r}}\,\Big(-\frac{\partial}{\partial s}\Big)\Big(-\frac{1}{\sinh s}\,\frac{\partial}{\partial s}\Big)^{\frac{n}{2}-1}\,e^{-\frac{s^2}{4t}}$$

in even dimension. Moreover, the following global estimate holds:

$$h_t(r) \asymp e^{-\rho^2 t}\,e^{-\rho r}\,e^{-\frac{r^2}{4t}} \times \begin{cases} t^{-\frac{3}{2}}\,(1+r) & \text{if } t \ge 1+r, \\ t^{-\frac{n}{2}}\,(1+r)^{\frac{n-1}{2}} & \text{if } 0<t\le 1+r. \end{cases} \tag{13}$$

- Similarly for the Schrödinger equation

$$\begin{cases} i\,\partial_t u(x,t) = \Delta_x u(x,t), \\ u(x,0) = f(x). \end{cases}$$

In this case $u(x,t) = f * h_{-it}(x)$, where

$$h_{-it}(r) = \frac{1}{2^{\frac{n+1}{2}}\,\pi^{\frac{n}{2}}}\; e^{i\frac{\pi}{4}\operatorname{sign}(t)}\, |t|^{-\frac{1}{2}}\, e^{-i\rho^2 t} \Big(-\frac{1}{\sinh r}\frac{\partial}{\partial r}\Big)^{\frac{n-1}{2}} e^{-i\frac{r^2}{4t}}$$

in odd dimension,

$$\begin{aligned} h_{-it}(r) &= (2\pi)^{-\frac{n+1}{2}}\, e^{i\frac{\pi}{4}\operatorname{sign}(t)}\, |t|^{-\frac{1}{2}}\, e^{-i\rho^2 t} \\ &\quad \times \int_{|r|}^{+\infty} \frac{ds}{\sqrt{\cosh s - \cosh r}} \Big(-\frac{\partial}{\partial s}\Big)\Big(-\frac{1}{\sinh s}\frac{\partial}{\partial s}\Big)^{\frac{n}{2}-1} e^{-i\frac{s^2}{4t}} \end{aligned}$$

in even dimension, and

$$|h_{-it}(r)| \lesssim e^{-\rho r} \times \begin{cases} |t|^{-\frac{3}{2}}\,(1+r) & \text{if } |t| \ge 1+r \\ |t|^{-\frac{n}{2}}\,(1+r)^{\frac{n-1}{2}} & \text{if } 0 < |t| \le 1+r \end{cases}$$

in all dimensions.

- The shifted wave equation

$$\begin{cases} \partial_t^2 u(x,t) = \big(\Delta_x + \rho^2\big)\, u(x,t) \\ u(x,0) = f(x),\ \partial_t|_{t=0}\, u(x,t) = g(x) \end{cases}$$

can be solved explicitly by means of the inverse dual Abel transform. Specifically,

$$\begin{aligned} u(t,x) &= \frac{1}{2^{\frac{n+1}{2}}\,\pi^{\frac{n-1}{2}}}\, \frac{\partial}{\partial t} \Big(\frac{1}{\sinh t}\frac{\partial}{\partial t}\Big)^{\frac{n-3}{2}} \Big\{\frac{1}{\sinh t}\int_{S(x,|t|)} dy\, f(y)\Big\} \\ &\quad + \frac{1}{2^{\frac{n+1}{2}}\,\pi^{\frac{n-1}{2}}} \Big(\frac{1}{\sinh t}\frac{\partial}{\partial t}\Big)^{\frac{n-3}{2}} \Big\{\frac{1}{\sinh t}\int_{S(x,|t|)} dy\, g(y)\Big\} \end{aligned}$$

in odd dimension and

$$\begin{aligned} u(t,x) &= \frac{1}{2^{\frac{n+1}{2}}\,\pi^{\frac{n}{2}}}\, \frac{\partial}{\partial |t|} \Big(\frac{\partial}{\partial(\cosh t)}\Big)^{\frac{n}{2}-1} \int_{B(x,|t|)} dy\, \frac{f(y)}{\sqrt{\cosh t - \cosh d(y,x)}} \\ &\quad + \frac{1}{2^{\frac{n+1}{2}}\,\pi^{\frac{n}{2}}}\, \operatorname{sign}(t) \Big(\frac{\partial}{\partial(\cosh t)}\Big)^{\frac{n}{2}-1} \int_{B(x,|t|)} dy\, \frac{g(y)}{\sqrt{\cosh t - \cosh d(y,x)}} \end{aligned}$$

in even dimension.

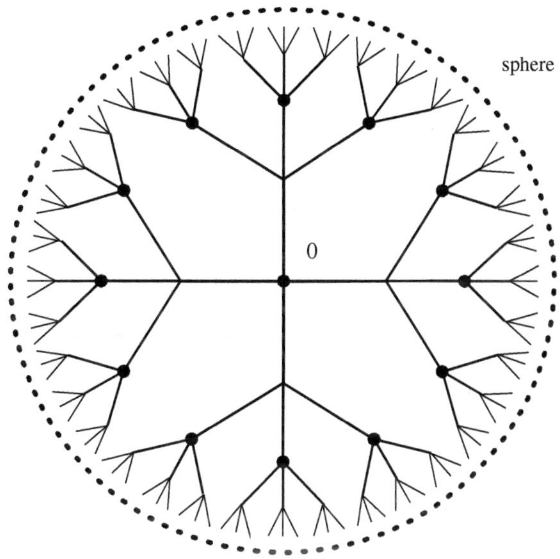

Fig. 6 Ball picture of $\mathbb{T}_3$

Spherical Fourier Analysis on Homogeneous Trees

A homogeneous tree is a connected graph, with no loop and with the same number of edges at each vertex. Let us denote by $\mathbb{T}_q$ the set of vertices of the homogeneous tree with $q+1>2$ edges (Fig. 6). It is equipped with the counting measure and with the geodesic distance, given by the number of edges between two points. The volume of any sphere $S(x,r)$ of radius $r\in\mathbb{N}$ is given by

$$\delta(r) = \begin{cases} 1 & \text{if } r=0, \\ (q+1)\,q^{r-1} & \text{if } r\in\mathbb{N}^*. \end{cases}$$

Once we have chosen an origin $0\in\mathbb{T}_q$ and an oriented geodesic $\omega:\mathbb{Z}\to\mathbb{T}_q$ through 0, let us denote by $|x|\in\mathbb{N}$ the distance of a vertex $x\in\mathbb{T}_q$ to the origin and by $h(x)\in\mathbb{Z}$ its horocyclic height (see Fig. 7). Let G be the isometry group of $\mathbb{T}_q$ and let K be the stabilizer of 0. Then G is a locally compact group, K is a compact open subgroup, and $\mathbb{T}_q \approx G/K$.

Remark 2.9 If q is a prime number, then $\mathbb{T}_q \approx \mathrm{PGL}(2,\mathbb{Q}_q)/\,\mathrm{PGL}(2,\mathbb{Z}_q)$.

The combinatorial Laplacian Δ on $\mathbb{T}_q$ is defined by $\Delta f = Af - f$, where A denotes the average operator

$$Af(x) = \tfrac{1}{q+1}\sum\nolimits_{y\in S(x,1)} f(y)\,.$$

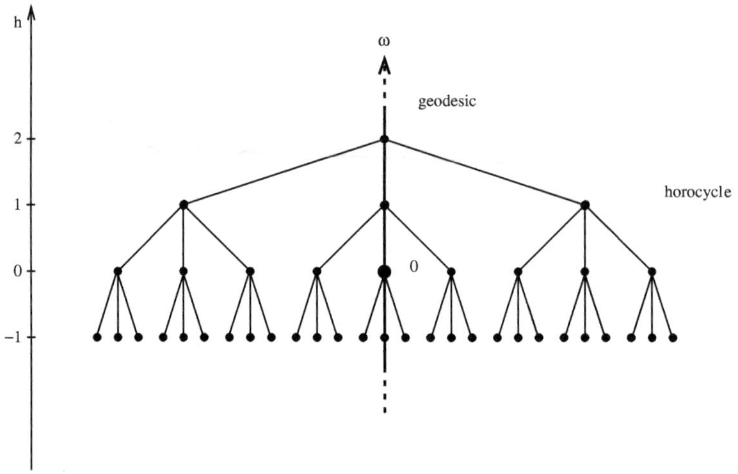

Fig. 7 Upper half-space picture of $\mathbb{T}_3$

Remark 2.10 Notice that f is harmonic, i.e., $\Delta f = 0$ if and only if f has the mean value property, i.e., $f = Af$.

The spherical functions φ_λ on $\mathbb{T}_q$ are the normalized radial eigenfunctions of Δ, or equivalently A. Specifically,

$$\begin{cases} A\,\varphi_\lambda = \gamma(\lambda)\,\varphi_\lambda\,, \\ \varphi_\lambda(0) = 1\,, \end{cases}$$

where $\gamma(\lambda) = \frac{q^{i\lambda}+q^{-i\lambda}}{q^{1/2}+q^{-1/2}}$.

Remark 2.11 The spherical functions on $\mathbb{T}_q$ can be characterized again in several other ways. Notably,

- Explicit expression:

$$\varphi_\lambda(r) = \begin{cases} \mathbf{c}(\lambda)\,q^{(-1/2+i\lambda)r} + \mathbf{c}(-\lambda)\,q^{(-1/2-i\lambda)r} & \text{if } \lambda \in \mathbb{C} \smallsetminus \frac{\tau}{2}\mathbb{Z}, \\ (-1)^{jr}\left(1 + \frac{q^{1/2}-q^{-1/2}}{q^{1/2}+q^{-1/2}}\,r\right) q^{-r/2} & \text{if } \lambda = \frac{\tau}{2}j, \end{cases} \tag{14}$$

  where $\mathbf{c}(\lambda) = \frac{1}{q^{1/2}+q^{-1/2}}\,\frac{q^{1/2+i\lambda}-q^{-1/2-i\lambda}}{q^{i\lambda}-q^{-i\lambda}}$ and $\tau = \frac{2\pi}{\log q}$. Notice that (14) is even and τ-periodic in λ.
- Same functional equations as (7) and (8).
- The spherical functions are the bi-K-invariant matrix coefficients of the spherical principal series representations of G on $L^2(\partial\,\mathbb{T}_q)$.

The spherical Fourier transform of radial functions on $\mathbb{T}_q$ is defined by

$$\mathcal{H}f(\lambda) = \sum\nolimits_{x\in\mathbb{T}_q} f(x)\,\varphi_\lambda(x) = f(0) + \tfrac{q^{1/2}+q^{-1/2}}{q^{1/2}} \sum\nolimits_{r=1}^{+\infty} q^r f(r)\,\varphi_\lambda(r) \tag{15}$$

and the inversion formula reads

$$f(r) = \tfrac{q^{1/2}}{q^{1/2}+q^{-1/2}}\,\tfrac{1}{\tau} \int_0^{\tau/2} d\lambda\, |\mathbf{c}(\lambda)|^{-2}\, \mathcal{H}f(\lambda)\,\varphi_\lambda(r)\,. \tag{16}$$

Consider the Abel transform

$$\begin{aligned}\mathcal{A}f(h) &= q^{\frac{h}{2}} \sum\nolimits_{\substack{x\in\mathbb{T}_q\\ h(x)=h}} f(|x|)\\ &= q^{\frac{|h|}{2}} f(|h|) + \tfrac{q-1}{q} \sum\nolimits_{j=1}^{+\infty} q^{\frac{|h|}{2}+j} f(|h|+2j)\,,\end{aligned}$$

which is essentially the horocyclic Radon transform restricted to radial functions. Then the following commutative diagram holds, let say in the Schwartz space setting:

$$\begin{array}{ccc} & C^\infty_{\mathrm{even}}(\mathbb{R}/\tau\mathbb{Z}) & \\ \mathcal{H}\nearrow & & \nwarrow\mathcal{F} \\ \mathcal{S}_{\mathrm{rad}}(\mathbb{T}_q) & \underset{\mathcal{A}}{\longrightarrow} & \mathcal{S}_{\mathrm{even}}(\mathbb{Z}) \end{array}$$

Here $\mathcal{S}_{\mathrm{even}}(\mathbb{Z})$ denotes the space of even functions on $\mathbb{Z}$ such that

$$\sup\nolimits_{r\in\mathbb{N}^*} r^m\,|f(r)| < +\infty \qquad \forall\; m\in\mathbb{N},$$

$\mathcal{S}_{\mathrm{rad}}(\mathbb{T}_q)$ the space of radial functions on $\mathbb{T}_q$, whose radial part coincides with $q^{-\frac{r}{2}}\mathcal{S}(\mathbb{N})$,

$$\mathcal{F}f(\lambda) = \sum\nolimits_{h\in\mathbb{Z}} q^{i\lambda h} f(h)$$

is a variant of the Fourier transform on $\mathbb{Z}$, and each arrow is an isomorphism. The inverse Abel transform is given by

$$\mathcal{A}^{-1}f(r) = \sum\nolimits_{j=0}^{+\infty} q^{-\frac{r}{2}-j} \big\{f(r+2j) - f(r+2j+2)\big\}\,.$$

Consider finally the transform

$$\mathcal{A}^*g(r) = \tfrac{1}{\delta(r)} \sum\nolimits_{\substack{x\in\mathbb{T}_q\\ |x|=r}} q^{\frac{h(x)}{2}} g\big(h(x)\big)\,,$$

which is dual to the Abel transform, i.e.,

$$\sum\nolimits_{x\in\mathbb{T}_q} f(x)\,\mathcal{A}^*g(|x|) = \sum\nolimits_{h\in\mathbb{Z}} \mathcal{A}f(h)\,g(h)\,,$$

and which is an isomorphism between the space of all even functions on $\mathbb{Z}$ and the space of all radial functions on $\mathbb{T}_q$. It is given explicitly by $\mathcal{A}^*g\,(0) = 0$, and

$$\mathcal{A}^*g\,(r) = \tfrac{2q}{q+1}\,q^{-\frac{r}{2}}f(r) + \tfrac{q-1}{q+1}\,q^{-\frac{r}{2}} \sum\nolimits_{\substack{-r<h<r\\ h \text{ has same parity as } r}} f(h)$$

if $r\in\mathbb{N}^*$. Its inverse is given by $(\mathcal{A}^*)^{-1}f\,(0) = f\,(0)$,

$$\begin{aligned}(\mathcal{A}^*)^{-1}f(h) &= \tfrac{q^{1/2}+q^{-1/2}}{2}\,q^{\frac{h-1}{2}}f(h)\\ &\quad - \tfrac{q-q^{-1}}{2}\,q^{-\frac{h}{2}} \sum\nolimits_{0<r\text{ odd}<h} q^r f(r)\end{aligned}$$

if $h\in\mathbb{N}$ is odd, and

$$\begin{aligned}(\mathcal{A}^*)^{-1}f(h) &= \tfrac{q^{1/2}+q^{-1/2}}{2}\,q^{\frac{h-1}{2}}f(h) - \tfrac{q^{1/2}-q^{-1/2}}{2}\,q^{-\frac{h-1}{2}}f(0)\\ &\quad - \tfrac{q-q^{-1}}{2}\,q^{-\frac{h}{2}} \sum\nolimits_{0<r\text{ even}<h} q^r f(r)\end{aligned}$$

if $h\in\mathbb{N}^*$ is even.

Applications Let us use spherical Fourier analysis to study discrete evolution equations on $\mathbb{T}_q$, as we did in the differential setting on $\mathbb{H}^n$.

- Consider the heat equation with discrete time $t\in\mathbb{N}$

$$\begin{cases} u(x,t+1) - u(x,t) = \Delta_x u(x,t)\\ u(x,0) = f(x)\end{cases}$$

  or, equivalently, the simple random walk

$$u(x,t) = A^t f(x) = f * h_t(x)\,.$$

  Its density $h_t(x)$ vanishes unless $|x|\le t$ have the same parity. In this case,

$$h_t(x) \asymp \frac{1+|x|}{(1+t)\sqrt{1+t-|x|}}\,\gamma_0^t\,q^{-\frac{|x|}{2}}\,e^{-t\,\psi\left(\frac{1+|x|}{1+t}\right)}\,, \tag{17}$$

  where $\psi(z) = \frac{1+z}{2}\log(1+z) - \frac{1-z}{2}\log(1-z)$ and $\gamma_0 = \gamma(0) = \frac{2}{q^{1/2}+q^{-1/2}} < 1$ is the spectral radius of A.

- The shifted wave equation with discrete time $t \in \mathbb{Z}$

$$\begin{cases} \gamma_0\, \Delta_t^{\mathbb{Z}} u(x,t) = \big(\Delta_x^{\mathbb{T}_q} + 1 - \gamma_0\big)\, u(x,t) \\ u(x,0) = f(x), \;\; \{u(x,1) - u(x,-1)\}/2 = g(x) \end{cases}$$

can be solved explicitly by using the inverse dual Abel transform. Specifically,

$$u(x,t) = C_t f(x) + S_t\, g(x)\,,$$

where

$$\begin{cases} C_t = \frac{M_{|t|} - M_{|t|-2}}{2} \\ S_t = \operatorname{sign}(t)\, M_{|t|-1} \end{cases}$$

and

$$M_t f(x) = q^{-\frac{t}{2}} \sum\nolimits_{\substack{d(y,x) \le t \\ t - d(y,x) \text{ even}}} f(y)$$

if $t \ge 0$, while $M_t = 0$ if $t < 0$.

Comments, References and Further Results

- Our main reference for classical special functions is [25].
- The spherical Fourier analysis presented in this section takes place on homogeneous spaces G/K associated with Gelfand pairs (G,K).
- The sphere $\mathbb{S}^n$ and the hyperbolic space $\mathbb{H}^n$ are dual symmetric spaces. By letting their curvature tend to 0, the Euclidean space $\mathbb{R}^n$ is obtained as a limit. At the level of spherical functions, the duality is reflected by the relation

$$\varphi_\ell^{\mathbb{S}^n}(\cos\theta_1) = \varphi_\lambda^{\mathbb{H}^n}(r)$$

between (9) and (10), when we specialize $\lambda = \pm i\,(\rho + \ell)$ $(\ell \in \mathbb{N})$ and take $r = \pm i\,\theta_1$. And (6) is a limit of (10) and (9):

$$\varphi_\lambda^{\mathbb{R}^n}(r) = \lim\nolimits_{\varepsilon \to 0} \varphi_{\frac{\lambda}{\varepsilon}}^{\mathbb{H}^n}(\varepsilon r) = \lim\nolimits_{\ell \to +\infty} \varphi_\ell^{\mathbb{S}^n}(\cos\tfrac{\lambda r}{\ell})\,.$$

Homogeneous trees may be considered as discrete analogs of hyperbolic spaces. This may be justified by the structural similarity between $\mathbb{H}^2 \approx \mathrm{PSL}(2,\mathbb{R}) / \mathrm{PSO}(2)$ and $\mathbb{T}_q \approx \mathrm{PGL}(2,\mathbb{Q}_q) / \mathrm{PGL}(2,\mathbb{Z}_q)$ when q is a prime number. At the analytic level, an actual relation is provided by the meta-theory developed by Cherednik [17], which includes as limit cases spherical Fourier analysis on $\mathbb{T}_q$ and on $\mathbb{H}^n$, as well as on $\mathbb{R}^n$ or on $\mathbb{S}^n$.

- The material in subsection "Spherical Fourier Analysis on Real Hyperbolic Spaces" generalizes to the class of Riemannian symmetric spaces of noncompact type and of rank 1, which consist of all hyperbolic spaces

$$\mathbb{H}^n = \mathrm{H}^n(\mathbb{R}),\ \mathrm{H}^n(\mathbb{C}),\ \mathrm{H}^n(\mathbb{H}),\ \mathrm{H}^2(\mathbb{O}),$$

and further to the class of Damek–Ricci spaces. One obtains this way a group theoretical interpretation of Jacobi functions $\varphi_\lambda^{\alpha,\beta}$ for infinitely many discrete parameters

$$\alpha = \tfrac{m_1+m_2-1}{2},\ \beta = \tfrac{m_2-1}{2}.$$

Our main references are [9, 33, 51, 75] and [12].

- Our main references for subsection "Spherical Fourier Analysis on Homogeneous Trees" are [19, 34], and [12]. Evolution equations (heat, Schrödinger, wave) with continuous time were also considered on homogeneous trees (see [20, 57, 82] and [49]).
- Spherical Fourier analysis generalizes to higher rank (see [41, 47] for Riemannian symmetric spaces and [53, 55, 65] for affine buildings).

Classification

Type	Constant curvature	Rank 1	General case
Euclidean	$\mathbb{R}^n$		$\mathfrak{p} \rtimes K/K$
Compact	$\mathbb{S}^n = \mathrm{S}(\mathbb{R}^{n+1})$	$\mathrm{S}(\mathbb{F}^n)$	U/K
Non compact	$\mathbb{H}^n = \mathrm{H}^n(\mathbb{R})$	$\mathrm{H}^n(\mathbb{F})$	G/K
p-adic		Homogeneous trees	Affine buildings

Notation

- $\mathfrak{g}_{\mathbb{C}}$ is a complex semisimple Lie algebra,
- $\mathfrak{g}$ is a noncompact real form of $\mathfrak{g}_{\mathbb{C}}$ and $\mathfrak{g} = \mathfrak{k} \oplus \mathfrak{p}$ a Cartan decomposition of $\mathfrak{g}$,
- $\mathfrak{u} = \mathfrak{k} \oplus i\mathfrak{p}$ is the compact dual form of $\mathfrak{g}$,
- $\mathfrak{a}$ is a Cartan subspace of $\mathfrak{p}$,
- $\mathfrak{g} = \mathfrak{a} \oplus \mathfrak{m} \oplus \big(\bigoplus_{\alpha\in\mathcal{R}} \mathfrak{g}_\alpha\big)$ is the root space decomposition of $(\mathfrak{g},\mathfrak{a})$,
- $\mathcal{R}^+$ is a positive root subsystem and $\mathfrak{a}^+$ the corresponding positive Weyl chamber in $\mathfrak{a}$,
- $\mathfrak{n} = \bigoplus_{\alpha\in\mathcal{R}^+} \mathfrak{g}_\alpha$ is the corresponding nilpotent Lie subalgebra,
- $m_\alpha = \dim \mathfrak{g}_\alpha$ is the multiplicity of the root α,
- $\rho = \frac{1}{2}\sum_{\alpha\in\mathcal{R}^+} m_\alpha$,

- $G_{\mathbb{C}}$ is a complex Lie group with finite center and Lie algebra $\mathfrak{g}_{\mathbb{C}}$,
- G, U, K and N are the Lie subgroups of $G_{\mathbb{C}}$ corresponding to the Lie subalgebras $\mathfrak{g}$, $\mathfrak{u}$, $\mathfrak{k}$ and $\mathfrak{n}$.

Special Functions

- Bessel functions on $\mathfrak{p}$:

$$\varphi_\lambda^{\mathfrak{p}}(x) = \int_K dk\, e^{i\langle\lambda,(\mathrm{Ad}\,k)x\rangle} \qquad \forall\,\lambda\in\mathfrak{p}_{\mathbb{C}},\ \forall\,x\in\mathfrak{p}_{\mathbb{C}}. \tag{18}$$

- Spherical functions on G/K:

$$\varphi_\lambda^{G}(x) = \int_K dk\, e^{\langle i\lambda+\rho,\,\mathfrak{a}(kx)\rangle} \qquad \forall\,\lambda\in\mathfrak{a}_{\mathbb{C}},\ \forall\,x\in G, \tag{19}$$

  where $\mathfrak{a}(y)$ denotes the $\mathfrak{a}$-component of $y\in G$ in the Iwasawa decomposition $G=N(\exp\mathfrak{a})K$.
- Spherical functions on affine buildings are classical Macdonald polynomials (i.e. Hall–Littlewood polynomials for the type $\tilde{A}$).

- The global heat kernel estimate (13) generalizes as follows to G/K:

$$\begin{aligned} h_t(x) \asymp\, & t^{-\frac{n}{2}} \Big\{ \prod\nolimits_{\alpha\in R^+} \big(1+\langle\alpha,x^+\rangle\big)\big(1+t+\langle\alpha,x^+\rangle\big)^{\frac{m_\alpha+m_{2\alpha}}{2}-1} \Big\} \\ & \times\, e^{-|\rho|^2 t-\langle\rho,x^+\rangle-\frac{|x^+|^2}{4t}}, \end{aligned} \tag{20}$$

  where R^+ is the set of positive indivisible roots in $\mathcal{R}$ and x^+ denotes the $\overline{\mathfrak{a}^+}$-component of $x\in G$ in the Cartan decomposition $G=K(\exp\overline{\mathfrak{a}^+})K$ (see [8] and the references therein).
- Random walks are harder to analyze on affine buildings. A global estimate similar to (17) and (20) was established in [10] for the simple random walk on affine buildings of type $\tilde{\mathrm{A}}_2$. In general, the main asymptotics of random walks were obtained in [91].

Epilogue

This section outlines spherical Fourier analysis around 1980 (except for the later applications to evolution equations). In the 1980s, Heckman and Opdam addressed the following problem (which goes back to Koornwinder for the root system BC_2): for any root system R, construct a continuous family of special functions on $\mathfrak{a}$ generalizing spherical functions on the corresponding symmetric spaces G/K, as Jacobi functions (or equivalently the Gauss hypergeometric function) generalize spherical functions on hyperbolic spaces. This problem was solved during the 1990s, mainly by Cherednik, Dunkl, Heckman, Macdonald, and Opdam, and has actually given rise to a large theory of special functions associated to root system, which is nowadays often referred to as Dunkl theory.

Remark 2.12 A different generalization of hypergeometric functions to Grassmannians was developed by Aomoto and Gelfand at the end of the twentieth century.

3 Rational Dunkl Theory

Rational Dunkl theory originates from the seminal paper [26]. This theory of special functions in several variables encompasses

- Euclidean Fourier analysis (which corresponds to the multiplicity $k=0$),
- classical Bessel functions in dimension 1,
- generalized Bessel functions associated with Riemannian symmetric spaces of Euclidean type (which correspond to a discrete set of multiplicities k).

In this subsection, we use [67] (or alternately [42, pp. 1–69]) as our primary reference and quote only later works. Our notation goes as follows (see the Appendix for more details):

- $\mathfrak{a}$ is a Euclidean vector space of dimension n, which we identify with its dual space,
- R is a root system, which is reduced but not necessarily crystallographic,
- W is the associated reflection group,
- $\mathfrak{a}^+$ is a positive Weyl chamber in $\mathfrak{a}$, $\overline{\mathfrak{a}^+}$ its closure, R^+ the corresponding positive root subsystem, and S the subset of simple roots,
- $\overline{\mathfrak{a}_+}$ denotes the closed cone generated by R^+, which is the dual cone of $\overline{\mathfrak{a}^+}$,
- for every $x\in\mathfrak{a}$, x^+ denotes the element of the orbit Wx which lies in $\overline{\mathfrak{a}^+}$,
- w_0 denotes the longest element in W, which interchanges $\mathfrak{a}^+$ and $-\mathfrak{a}^+$, respectively R^+ and $R^-=-R^+$,
- k is a multiplicity, which will remain implicit in most formulae and which will be assumed to be nonnegative after a while,
- $\gamma=\sum_{\alpha\in R^+}k_\alpha$,
- $\delta(x)=\prod_{\alpha\in R^+}|\langle\alpha,x\rangle|^{2k_\alpha}$ is the reference density in the case $k\geq 0$.

Dunkl Operators

Definition 3.1 The rational Dunkl operators, which are often simply called Dunkl operators, are the differential-difference operators defined by

$$D_\xi f(x) = \partial_\xi f(x) + \sum\nolimits_{\alpha\in R^+} k_\alpha \tfrac{\langle\alpha,\xi\rangle}{\langle\alpha,x\rangle} \{f(x) - f(r_\alpha x)\} \tag{21}$$

for every $\xi \in \mathfrak{a}$.

Remark 3.2

- Notice that Dunkl operators D_ξ reduce to partial derivatives ∂_ξ when $k = 0$.
- The choice of R^+ plays no role in Definition 3.1, as

$$\sum\nolimits_{\alpha\in R^+} k_\alpha \tfrac{\langle\alpha,\xi\rangle}{\langle\alpha,x\rangle} \{f(x) - f(r_\alpha x)\} = \sum\nolimits_{\alpha\in R} \tfrac{k_\alpha}{2} \tfrac{\langle\alpha,\xi\rangle}{\langle\alpha,x\rangle} \{f(x) - f(r_\alpha x)\}.$$

- Dividing by $\langle\alpha, x\rangle$ produces no actual singularity in (21), as

$$\begin{aligned}\tfrac{f(x)-f(r_\alpha x)}{\langle\alpha,x\rangle} &= -\tfrac{1}{\langle\alpha,x\rangle} \int_0^1 dt\, \tfrac{\partial}{\partial t} f(x - t\langle\alpha^\vee, x\rangle\,\alpha) \\ &= \tfrac{2}{|\alpha|^2} \int_0^1 dt\, \partial_\alpha f(x - t\langle\alpha^\vee, x\rangle\,\alpha).\end{aligned}$$

Commutativity is a remarkable property of Dunkl operators.

Theorem 3.3 *Fix a multiplicity k. Then*

$$D_\xi \circ D_\eta = D_\eta \circ D_\xi \qquad \forall\, \xi, \eta \in \mathfrak{a}.$$

This result leads to the notion of Dunkl operators D_p, for every polynomial $p \in \mathcal{P}(\mathfrak{a})$, and of their symmetric part $\tilde{D}_p$ on symmetric (i.e. W-invariant) functions.

Examples

- The Dunkl Laplacian is given by

$$\Delta f(x) = \sum\nolimits_{j=1}^n D_j^2 f(x) = \overbrace{\sum\nolimits_{j=1}^n \partial_j^2 f(x) + \sum\nolimits_{\alpha\in R^+} \tfrac{2k_\alpha}{\langle\alpha,x\rangle} \partial_\alpha f(x)}^{\text{differential part } \tilde{\Delta} f(x)}$$
$$\underbrace{-\sum\nolimits_{\alpha\in R^+} \tfrac{k_\alpha |\alpha|^2}{\langle\alpha,x\rangle^2} \{f(x) - f(r_\alpha x)\}}_{\text{difference part}},$$

where D_j, respectively ∂_j denote the Dunkl operators, respectively the partial derivatives with respect to an orthonormal basis of $\mathfrak{a}$.

- In dimension 1, the Dunkl operator is given by

$$Df(x) = \left(\tfrac{\partial}{\partial x}\right) f(x) + \tfrac{k}{x} \left\{ f(x) - f(-x) \right\}$$

and the Dunkl Laplacian by

$$Lf(x) = \left(\tfrac{\partial}{\partial x}\right)^2 f(x) + \tfrac{2k}{x} \left(\tfrac{\partial}{\partial x}\right) f(x) - \tfrac{k}{x^2} \left\{ f(x) - f(-x) \right\}.$$

Here are some other properties of Dunkl operators.

Proposition 3.4

- *The Dunkl operators map the following function spaces into themselves:*

$$\mathcal{P}(\mathfrak{a}),\ \mathcal{C}^\infty(\mathfrak{a}),\ \mathcal{C}_c^\infty(\mathfrak{a}),\ \mathcal{S}(\mathfrak{a}),\ \dots$$

More precisely, the Dunkl operators D_ξ, with $\xi \in \mathfrak{a}$, are homogeneous operators of degree -1 on polynomials.

- W-equivariance: *For every $w \in W$ and $p \in \mathcal{P}(\mathfrak{a})$, we have*

$$w \circ D_p \circ w^{-1} = D_{wp}\,.$$

Hence $\tilde{D}_{pq} = \tilde{D}_p \circ \tilde{D}_q$, for all symmetric (i.e. W-invariant) polynomials $p, q \in \mathcal{P}(\mathfrak{a})^W$.

- Skew-adjointness: *Assume that $k \geq 0$. Then, for every $\xi \in \mathfrak{a}$,*

$$\int_{\mathfrak{a}} dx\, \delta(x)\, D_\xi f(x)\, g(x) = -\int_{\mathfrak{a}} dx\, \delta(x) f(x)\, D_\xi g(x)\,.$$

Dunkl Kernel

Theorem 3.5 *For generic multiplicities k and for every $\lambda \in \mathfrak{a}_{\mathbb{C}}$, the system*

$$\begin{cases} D_\xi E_\lambda = \langle \lambda, \xi \rangle\, E_\lambda & \forall\ \xi \in \mathfrak{a}, \\ E_\lambda(0) = 1, \end{cases}$$

has a unique smooth solution on $\mathfrak{a}$, which is called the Dunkl kernel.

Remark 3.6 In this statement, generic means that k belongs to a dense open subset K_{reg} of K, whose complement is a countable union of algebraic sets. The set K_{reg} is known explicitly and it contains in particular $\{\, k \in K \mid \operatorname{Re} k \geq 0 \,\}$.

Definition 3.7 The generalized Bessel function is the average

$$J_\lambda(x) = \tfrac{1}{|W|} \sum\nolimits_{w \in W} E_\lambda(wx) = \tfrac{1}{|W|} \sum\nolimits_{w \in W} E_{w\lambda}(x)\,. \tag{22}$$

Remark 3.8

- Mind the possible formal confusion between (22) and the classical Bessel function of the first kind J_ν.
- Conversely, the Dunkl kernel $E_\lambda(x)$ can be recovered by applying to the generalized Bessel function $J_\lambda(x)$ a linear differential operator in x whose coefficients are rational functions of λ (see [62, proposition 6.8.(4)]).
- In dimension 1, K_{reg} is the complement of $-\mathbb{N} - \frac{1}{2}$ in $\mathbb{C}$. The generalized Bessel function (22) reduces to the modified Bessel function encountered in section "Hankel Transform on Euclidean Spaces":

  $$J_\lambda(x) = j_{k-\frac{1}{2}}(\lambda x)\,.$$

  The Dunkl kernel is a combination of two such functions:

  $$E_\lambda(x) = \underbrace{j_{k-\frac{1}{2}}(\lambda x)}_{\text{even}} + \underbrace{\tfrac{\lambda x}{2k+1}\, j_{k+\frac{1}{2}}(\lambda x)}_{\text{odd}}$$

  It can be also expressed in terms of the confluent hypergeometric function:

  $$\begin{aligned} E_\lambda(x) &= \tfrac{\Gamma(k+\frac{1}{2})}{\sqrt{\pi}\,\Gamma(k)} \int_{-1}^{+1} du\, (1-u)^{k-1} (1+u)^k\, e^{\lambda x u} \\ &= e^{\lambda x} \underbrace{\tfrac{\Gamma(2k+1)}{\Gamma(k)\,\Gamma(k+1)} \int_0^1 dv\, v^{k-1} (1-v)^k\, e^{-2\lambda x v}}_{{}_1\mathrm{F}_1(k;2k+1;-2\lambda x)}\,. \end{aligned}$$

- When $k = 0$, the Dunkl kernel $E_\lambda(x)$ reduces to the exponential $e^{\langle \lambda, x \rangle}$ and the generalized Bessel function $J_\lambda(x)$ to

  $$\operatorname{Cosh}_\lambda(y) = \tfrac{1}{|W|} \sum\nolimits_{w \in W} e^{\langle w\lambda, y \rangle}\,. \tag{23}$$

- As far as we know, the non-symmetric Dunkl kernel had not occurred previously in special functions, group theory or geometric analysis.
- Bessel functions (18) on Riemannian symmetric spaces of Euclidean type $\mathfrak{p} \rtimes K/K$ are special cases of (22), corresponding to crystallographic root systems and to certain discrete sets of multiplicities. More precisely, if

  - $\mathfrak{g} = \mathfrak{p} \oplus \mathfrak{k}$ is the associated semisimple Lie algebra,
  - $\mathfrak{a}$ is a Cartan subspace of $\mathfrak{p}$,

- $\mathcal{R}$ is the root system of $(\mathfrak{g}, \mathfrak{a})$,
- m_α is the multiplicity of $\alpha \in \mathcal{R}$,

and

- R is the subsystem of indivisible roots in $\mathcal{R}$,
- $k_\alpha = \frac{m_\alpha + m_{2\alpha}}{2} \quad \forall\, \alpha \in R$,

then

$$\varphi_\lambda^{\mathfrak{p}}(x) = J_{i\lambda}(x) \qquad \forall\, \lambda \in \mathfrak{a}_{\mathbb{C}},\ \forall\, x \in \mathfrak{a}.$$

In next proposition, we collect some properties of the Dunkl kernel.

Proposition 3.9

- Regularity: *$E_\lambda(x)$ extends to a holomorphic function in $\lambda \in \mathfrak{a}_{\mathbb{C}}$, $x \in \mathfrak{a}_{\mathbb{C}}$ and $k \in K_{\text{reg}}$.*
- Symmetries:

$$\begin{cases} E_\lambda(x) = E_x(\lambda), & \\ E_{w\lambda}(wx) = E_\lambda(x) & \forall\, w \in W, \\ E_\lambda(tx) = E_{t\lambda}(x) & \forall\, t \in \mathbb{C}, \\ \overline{E_\lambda(x)} = E_{\overline{\lambda}}(\overline{x}) & \textit{when } k \geq 0. \end{cases}$$

- Positivity: *Assume that $k \geq 0$. Then,*

$$0 < E_\lambda(x) \leq e^{\langle \lambda^+, x^+ \rangle} \qquad \forall\, \lambda \in \mathfrak{a},\ \forall\, x \in \mathfrak{a}.$$

- Global estimate: *Assume that $k \geq 0$. Then, for every $\xi_1, \ldots, \xi_N \in \mathfrak{a}$,*

$$\left| \partial_{\xi_1} \ldots \partial_{\xi_N} E_\lambda(x) \right| \leq |\xi_1| \ldots |\xi_N|\, |\lambda|^N\, e^{\langle (\operatorname{Re}\lambda)^+, (\operatorname{Re} x)^+ \rangle} \qquad \forall\, \lambda \in \mathfrak{a}_{\mathbb{C}},\ \forall\, x \in \mathfrak{a}_{\mathbb{C}}.$$

Dunkl Transform

From now on, we assume that $k \geq 0$.

Definition 3.10 The Dunkl transform is defined by

$$\mathcal{H}f(\lambda) = \int_{\mathfrak{a}} dx\, \delta(x) f(x)\, E_{-i\lambda}(x)\,. \tag{24}$$

In next theorem, we collect the main properties of the Dunkl transform.

Theorem 3.11

- *The Dunkl transform is an automorphism of the Schwartz space $\mathcal{S}(\mathfrak{a})$.*
- *We have*

$$\begin{cases} \mathcal{H}(D_\xi f)(\lambda) = i\,\langle \xi, \lambda\rangle\, \mathcal{H}f(\lambda) & \forall\ \xi \in \mathfrak{a}, \\ \mathcal{H}(\langle \xi, .\rangle f) = i\,D_\xi\, \mathcal{H}f & \forall\ \xi \in \mathfrak{a}, \\ \mathcal{H}(wf)(\lambda) = \mathcal{H}f(w\lambda) & \forall\ w \in W, \\ \mathcal{H}\big[f(t\,.)\big](\lambda) = |t|^{-n-2\gamma}\,(\mathcal{H}f)(t^{-1}\lambda) & \forall\ t \in \mathbb{R}^*. \end{cases}$$

- Inversion formula:

$$f(x) = c^{-2} \int_{\mathfrak{a}} d\lambda\ \delta(\lambda)\, \mathcal{H}f(\lambda)\, E_{i\lambda}(x)\,, \tag{25}$$

  where

$$c = \int_{\mathfrak{a}} dx\ \delta(x)\, e^{-\frac{|x|^2}{2}} \tag{26}$$

  is the so-called Mehta–Macdonald integral.
- Plancherel identity: *The Dunkl transform extends to an isometric automorphism of $L^2(\mathfrak{a}, \delta(x)\,dx)$, up to a positive constant. Specifically,*

$$\int_{\mathfrak{a}} d\lambda\ \delta(\lambda)\, |\mathcal{H}f(\lambda)|^2 = c^2 \int_{\mathfrak{a}} dx\ \delta(x)\, |f(x)|^2.$$

- Riemann–Lebesgue Lemma: *The Dunkl transform maps $L^1(\mathfrak{a}, \delta(x)\,dx)$ into $\mathcal{C}_0(\mathfrak{a})$.*
- Paley–Wiener Theorem: *The Dunkl transform is an isomorphism between $\mathcal{C}_c^\infty(\mathfrak{a})$ and the Paley–Wiener space $\mathcal{PW}(\mathfrak{a}_\mathbb{C})$, which consists of all holomorphic functions $h : \mathfrak{a}_\mathbb{C} \longrightarrow \mathbb{C}$ such that*

$$\exists\ R > 0,\ \forall\ N \in \mathbb{N},\ \sup\nolimits_{\lambda \in \mathfrak{a}_\mathbb{C}} (1+|\lambda|)^N e^{-R\,|\operatorname{Im}\lambda|}\, |h(\lambda)| < +\infty\,. \tag{27}$$

  More precisely, the support of $f \in \mathcal{C}_c^\infty(\mathfrak{a})$ is contained in the closed ball $\overline{B(0,R)}$ if and only if $h = \mathcal{H}f$ satisfies (27).

Remark 3.12

- Notice that (24) and (25) are symmetric, as the Euclidean Fourier transform (1) and its inverse (2), or the Hankel transform (3) and its inverse (4).
- In the W-invariant case, $E_{\pm i\lambda}(x)$ is replaced by $J_{\pm i\lambda}(x)$ in (24) and (25).
- The Dunkl transform of a radial function is again a radial function.
- The following sharper version of the Paley–Wiener Theorem was proved in [6], as a consequence of the corresponding result in the trigonometric setting (see Theorem 4.9) and thus under the assumption that R is crystallographic. Given

a W-invariant convex compact neighborhood C of the origin in $\mathfrak{a}$, consider the gauge $\chi(\lambda)=\max_{x\in C}\langle\lambda,x\rangle$. Then the support of $f\in\mathcal{C}_c^\infty(\mathfrak{a})$ is contained in C if and only if its Dunkl transform $h=\mathcal{H}f$ satisfies the condition

$$\forall\, N\in\mathbb{N},\ \sup_{\lambda\in\mathfrak{a}_{\mathbb{C}}}(1+|\lambda|)^N e^{-\chi(\operatorname{Im}\lambda)}\,|h(\lambda)|<+\infty\,. \tag{28}$$

Problem Extend the latter result to the non-crystallographic case.

Heat Kernel

The heat equation

$$\begin{cases}\partial_t u(x,t)=\Delta_x u(x,t)\\ u(x,0)=f(x)\end{cases}$$

can be solved via the Dunkl transform (under suitable assumptions). This way, one obtains

$$u(x,t)=\int_{\mathfrak{a}} dy\;\delta(y) f(y)\, h_t(x,y)\,,$$

where the heat kernel is given by

$$h_t(x,y)=c^{-2}\int_{\mathfrak{a}} d\lambda\;\delta(\lambda)\,e^{-t|\lambda|^2}E_{i\lambda}(x)\,E_{-i\lambda}(y)\qquad\forall\,t>0\,,\ \forall\,x,y\in\mathfrak{a}\,.$$

In next proposition, we collect some properties of the heat kernel established by Rösler.

Proposition 3.13

- *$h_t(x,y)$ is an smooth symmetric probability density. More precisely,*
  - *$h_t(x,y)$ is an analytic function in $(t,x,y)\in(0,+\infty)\times\mathfrak{a}\times\mathfrak{a}$,*
  - *$h_t(x,y)=h_t(y,x)$,*
  - *$h_t(x,y)>0$ and $\int_{\mathfrak{a}} dy\,\delta(y)\,h_t(x,y)=1$.*
- Semigroup property:

$$h_{s+t}(x,y)=\int_{\mathfrak{a}} dz\;\delta(z)\,h_s(x,z)\,h_t(z,y)\,.$$

- Expression by means of the Dunkl kernel:

$$h_t(x,y) = c^{-1}\,(2t)^{-\frac{n}{2}-\gamma}\,e^{-\frac{|x|^2}{4t}-\frac{|y|^2}{4t}}\,E_{\frac{x}{\sqrt{2t}}}\big(\tfrac{y}{\sqrt{2t}}\big) \qquad \forall\,t>0,\ \forall\,x,y\in\mathfrak{a}\,. \tag{29}$$

- Upper estimate:

$$h_t(x,y) \le c^{-1}\,(2t)^{-\frac{n}{2}-\gamma}\,\max{}_{w\in W}\,e^{-\frac{|wx-y|^2}{4t}}\,.$$

Remark 3.14

- In [13], the following sharp heat kernel estimates were obtained in dimension 1 (and also in the product case):

$$h_t(x,y) \asymp \begin{cases} t^{-k-\frac{1}{2}}\,e^{-\frac{x^2+y^2}{4t}} & \text{if } |xy|\le t,\\ t^{-\frac{1}{2}}\,(xy)^{-k}\,e^{-\frac{(x-y)^2}{4t}} & \text{if } xy\ge t,\\ t^{\frac{1}{2}}\,(-xy)^{-k-1}\,e^{-\frac{(x+y)^2}{4t}} & \text{if } -xy\ge t. \end{cases} \tag{30}$$

  Notice the lack of Gaussian behavior when $-xy\ge t$, in particular when $y=-x$ tends to infinity faster then $\sqrt{t}$.
- The Dunkl Laplacian is the infinitesimal generator of a Feller Markov process on $\mathfrak{a}$, which has remarkable features (Brownian motion with jumps) and which has drawn a lot of attention in the 2000s. We refer to [42] and [24] for probabilistic aspects of Dunkl theory.

Problem Prove in general heat kernel estimates similar to (30).

Intertwining Operator and (Dual) Abel Transform

Consider the Abel transform

$$\mathcal{A} = \mathcal{F}^{-1}\circ\mathcal{H}\,,$$

which is obtained by composing the Dunkl transform $\mathcal{H}$ with the inverse Euclidean Fourier transform $\mathcal{F}^{-1}$ on $\mathfrak{a}$, and the dual Abel transform $\mathcal{A}^*$, which satisfies

$$\int_{\mathfrak{a}} dx\;\delta(x)f(x)\,\mathcal{A}^*g(x) = \int_{\mathfrak{a}} dy\;\mathcal{A}f(y)\,g(y)\,.$$

Theorem 3.15

- *The dual Abel transform $\mathcal{A}^*$ coincides with the intertwining operator $\mathcal{V}$ defined on polynomials by Dunkl and extended to smooth functions by Trimèche.*

- Intertwining property: *for every* $\xi \in \mathfrak{a}$,

$$\mathcal{A} \circ D_\xi = \partial_\xi \circ \mathcal{A} \qquad \textit{and} \qquad \mathcal{V} \circ \partial_\xi = D_\xi \circ \mathcal{V}\,. \tag{31}$$

- Symmetries:

$$\mathcal{A}(wf) = w(\mathcal{A}f) \qquad \textit{and} \qquad \mathcal{V}(wg) = w(\mathcal{V}g) \qquad \forall\, w\in W,$$
$$\mathcal{A}\big[f(t\,.)\big](y) = |t|^{-2\gamma}(\mathcal{A}f)(ty) \textit{ and } \mathcal{V}\big[g(t\,.)\big](x) = (\mathcal{V}g)(tx) \;\forall\, t\in\mathbb{R}^*.$$

- *For every* $x \in \mathfrak{a}$, *there is a unique Borel probability measure* μ_x *on* $\mathfrak{a}$ *such that*

$$\mathcal{V}g(x) = \int_{\mathfrak{a}} d\mu_x(y)\, g(y)\,. \tag{32}$$

  The support of μ_x *is contained in the convex hull of* Wx. *Moreover, if* $k > 0$, *the support of* μ_x *is* W*-invariant and contains* Wx.
- $\mathcal{A}$ *is an automorphism of the spaces* $\mathcal{C}_c^\infty(\mathfrak{a})$ *and* $\mathcal{S}(\mathfrak{a})$, *while* $\mathcal{V}$ *is an automorphism of* $\mathcal{C}^\infty(\mathfrak{a})$, *with*

$$|\mathcal{V}g(x)| \le \max\nolimits_{y\in\mathrm{co}(Wx)} |g(y)| \qquad \forall\, x\in\mathfrak{a}.$$

The following integral representations, which follow from (22), (31) and (32), generalize (6) and (18) in the present setting.

Corollary 3.16 *For every* $\lambda \in \mathfrak{a}_{\mathbb{C}}$, *we have*

$$E_\lambda(x) = \int_{\mathfrak{a}} d\mu_x(y)\, e^{\langle\lambda,y\rangle}$$

and

$$J_\lambda(x) = \int_{\mathfrak{a}} d\mu_x^W(y)\, \operatorname{Cosh}_\lambda(y)\,,$$

where $\operatorname{Cosh}_\lambda$ *is defined in* (23) *and*

$$\mu_x^W = \tfrac{1}{|W|} \sum\nolimits_{w\in W} \mu_{wx}\,.$$

Remark 3.17

- The first three items in Theorem 3.15 hold for all multiplicities $k \in K_{\mathrm{reg}}$.
- The following symmetries hold :

$$\begin{cases} d\mu_{wx}(wy) = d\mu_x(y) & \forall\, w\in W,\\ d\mu_{tx}(ty) = d\mu_x(y) & \forall\, t\in\mathbb{R}^*. \end{cases}$$

- In [22], it is conjectured that the measure μ_x is absolutely continuous with respect to the Lebesgue measure under the following two assumptions:
  - x is regular (which means that $\langle \alpha, x \rangle \neq 0$, for every $\alpha \in R$),
  - $\mathfrak{a}$ is spanned by the roots α with multiplicity $k_\alpha > 0$.
- These conjectures hold in dimension 1 (hence in the product case), where

$$d\mu_x(y) = \frac{\Gamma(k+\frac{1}{2})}{\sqrt{\pi}\,\Gamma(k)}\,(|x| + \operatorname{sign}(x)\,y)\,(x^2 - y^2)^{k-1}\,\mathbb{1}_{(-|x|,+|x|)}(y)\,dy$$

  if $x \neq 0$, while μ_0 is the Dirac measure at the origin.

Generalized Translations, Convolution and Product Formula

Definition 3.18

- The generalized convolution corresponds, via the Dunkl transform, to pointwise multiplication:

$$(f * g)(x) = c^{-2} \int_{\mathfrak{a}} d\lambda\; \delta(\lambda)\, \mathcal{H}f(\lambda)\, \mathcal{H}g(\lambda)\, E_{i\lambda}(x)\,. \tag{33}$$

- The generalized translations are defined by

$$(\tau_y f)(x) = c^{-2} \int_{\mathfrak{a}} d\lambda\; \delta(\lambda)\, \mathcal{H}f(\lambda)\, E_{i\lambda}(x)\, E_{i\lambda}(y) = (\tau_x f)(y)\,. \tag{34}$$

The key objects here are the tempered distributions

$$f \longmapsto \langle\, \nu_{x,y}\,, f \,\rangle\,, \tag{35}$$

which are defined by (34) and which enter the product formula

$$E_\lambda(x)\, E_\lambda(y) = \langle\, \nu_{x,y}\,, E_\lambda \,\rangle\,. \tag{36}$$

Remark 3.19

- When $k = 0$, then (33) reduces to the usual convolution on $\mathfrak{a}$, (34) to $(\tau_y f)(x) = f(x+y)$, and $\nu_{x,y} = \delta_{x+y}$.
- In the W- invariant case, (33) becomes

$$(f * g)(x) = c^{-2} \int_{\mathfrak{a}} d\lambda\; \delta(\lambda)\, \mathcal{H}f(\lambda)\, \mathcal{H}g(\lambda)\, J_{i\lambda}(x)$$

and (36)

$$J_\lambda(x)\, J_\lambda(y) = \langle\, \nu^W_{x,y}\,, J_\lambda \,\rangle\,, \tag{37}$$

where

$$\nu^W_{x,y} = \frac{1}{|W|} \sum\nolimits_{w\in W} \nu_{wx,wy}\,.$$

Lemma 3.20 *The distributions* (35) *are compactly supported.*

- *Specifically,* $\nu_{x,y}$ *is supported in the spherical shell*

$$\big\{ z\in\mathfrak{a} \;\big|\; \big||x|-|y|\big| \le |z| \le |x|+|y| \,\big\}\,.$$

- *Assume that W is crystallographic. Then* $\nu_{x,y}$ *is actually supported in*

$$\big\{ z\in\mathfrak{a} \;\big|\; z^+ \preccurlyeq x^+ + y^+,\; z^+ \succcurlyeq y^+ + w_0 x^+ \text{ and } x^+ + w_0 y^+ \big\}\,, \tag{38}$$

where $\preccurlyeq$ *denotes the partial order on* $\mathfrak{a}$ *associated with the cone* $\overline{\mathfrak{a}_+}$ *(Fig. 8).*

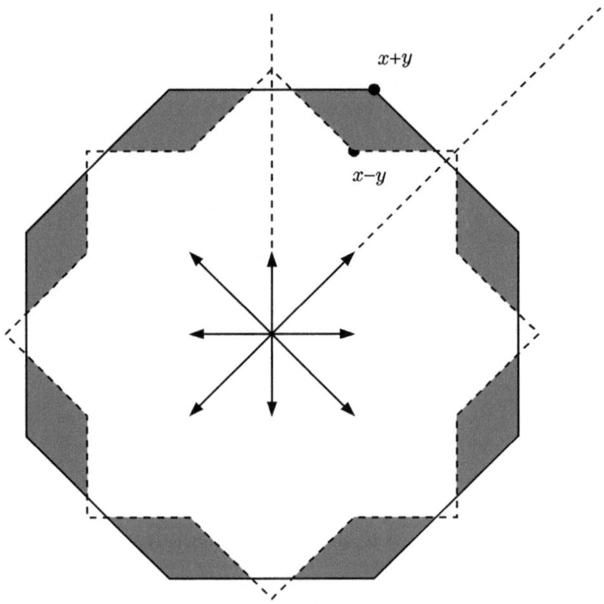

Fig. 8 Picture of the set (38) for the root system B_2

Example In dimension 1, $\nu_{x,y}$ is a bounded signed measure. Specifically,

$$d\nu_{x,y}(z) = \begin{cases} \nu(x,y,z)\,|z|^{2k}\,dz & \text{if } x,y\in\mathbb{R}^*, \\ d\delta_y(z) & \text{if } x=0\,, \\ d\delta_x(z) & \text{if } y=0\,, \end{cases}$$

where

$$\nu(x,y,z) = \frac{\Gamma(k+\frac12)}{\sqrt{\pi}\,\Gamma(k)}\,\frac{(z+x+y)(z+x-y)(z-x+y)}{2xyz}\,\frac{\{|z|^2-(|x|-|y|)^2\}^{k-1}\{(|x|+|y|)^2-|z|^2\}^{k-1}}{\{2\,|x|\,|y|\,|z|\}^{2k-1}}$$

if $x,y,z\in\mathbb{R}^*$ satisfy the triangular inequality $\big||x|-|y|\big|<|z|<|x|+|y|$ and

$$\nu(x,y,z) = 0$$

otherwise. Moreover, let

$$M = \sup_{x,y\in\mathbb{R}} \int_{\mathbb{R}} d|\nu_{x,y}|(z) = \sqrt{2}\,\frac{\left[\Gamma(k+\frac12)\right]^2}{\Gamma(k+\frac14)\,\Gamma(k+\frac34)}\,.$$

Then $M\geq 1$ and $M\nearrow\sqrt{2}$ as $k\nearrow+\infty$.

In general there is a lack information about (35) and the following facts are conjectured [68].

Problems

(a) The distributions $\nu_{x,y}$ are bounded signed Borel measures.
(b) They are uniformly bounded in x and y.
(c) The measures $\nu^W_{x,y}$ are positive.

If $\nu^{(W)}_{x,y}$ is a measure, notice that it is normalized by

$$\int_{\mathfrak{a}} d\nu^{(W)}_{x,y}(z) = 1\,.$$

These problems and especially (b) are important for harmonic analysis. They imply indeed the following facts, for the reference measure $\delta(x)\,dx$ on $\mathfrak{a}$.

Problems

(d) The generalized translations (34) are uniformly bounded on L^1 and hence on L^p, for every $1\leq p\leq\infty$.
(e) *Young's inequality:* For all $1\leq p,q,r\leq\infty$ satisfying $\frac1p+\frac1q-\frac1r=1$, there exists a constant $C\geq 0$ such that

$$\|f*g\|_{L^r} \leq C\,\|f\|_{L^p}\,\|g\|_{L^q}\,. \tag{39}$$

Beside the trivial L^2 setting and the one-dimensional case (hence the product case), here are two more situations where these problems have been solved.

- **Radial case [68].** Translations of radial functions are positive. Specifically, for radial functions $f(z) = {'\!f}(|z|)$ and nonzero $y \in \mathfrak{a}$, we have

$$(\tau_y f)(x) = \int_{\mathfrak{a}} d\mu_{\frac{y}{|y|}}(z)\, {'\!f}\big(\sqrt{|x|^2 + |y|^2 + 2\,\langle x, z\rangle\, |y|}\,\big)\,. \tag{40}$$

Hence (39) holds if f or g is radial.

- **Symmetric space case.** Assume that the multiplicity k corresponds to a Riemannian symmetric space of Euclidean type. Then $\nu^W_{x,y}$ is a positive measure and (39) holds for W-invariant functions.

Comments, References and Further Results

- The computation of the integral (26) has a long history. A closed form was conjectured by Mehta for the root systems of type A and by Macdonald for general root systems. For the four infinite families of classical root systems, it can be actually deduced from an earlier integral formula of Selberg. In general, the Mehta–Macdonald formula was proved by Opdam, first for crystallographic root systems [61] and next for all root systems [62]. His proof was simplified by Etingof [32], who removed in particular the computer-assisted calculations used in the last cases.
- We have not discussed the shift operators, which move the multiplicity k by integers and which have proven useful in the W-invariant setting (see [62]).
- The following asymptotics hold for the Dunkl kernel, under the assumption that $k \geq 0$ (see [22]): there exists $\upsilon : W \longrightarrow \mathbb{C}$ such that, for every $w \in W$ and for every $\lambda, x \in \mathfrak{a}^+$,

$$\lim\nolimits_{t\to+\infty} (it)^{\gamma}\, e^{-it\langle\lambda, wx\rangle}\, E_{it\lambda}(wx) = \upsilon(w)\, \delta(\lambda)^{-\frac{1}{2}}\, \delta(x)^{-\frac{1}{2}}\,. \tag{41}$$

If $w = \mathrm{I}$, we have $\upsilon(\mathrm{I}) = (2\pi)^{-\frac{n}{2}}\, c$, where c is defined by (26), and (41) holds more generally when $\tilde{t} = it$ tends to infinity in the half complex space $\{\tilde{t} \in \mathbb{C} \mid \operatorname{Re}\tilde{t} \geq 0\}$. If $w \neq \mathrm{I}$, the Dunkl kernel is expected to have a different asymptotic behavior, when $\operatorname{Re}\tilde{t}$ becomes positive. In dimension 1, we have indeed

$$\lim_{\substack{\tilde{t}\to+\infty \\ |\arg\tilde{t}|\leq\frac{\pi}{2}-\varepsilon}} \tilde{t}^{\,k+1} e^{-\tilde{t}\lambda x} E_{\tilde{t}\lambda}(-x) = \frac{2^{k-1}\, k\, \Gamma(k+\frac{1}{2})}{\sqrt{\pi}}\, \lambda^{-k-1}\, x^{-k-1}\,,$$

for any $0 < \varepsilon \leq \frac{\pi}{2}$. This discrepancy plays an important role in [13].
- In the fourth item of Theorem 3.15, the sharper results about the support of μ_x when $k > 0$ were obtained in [38].

- Specific information is available in the W-invariant setting for the root systems A_n. In this case, an integral recurrence formula over n was obtained in [1] and [77] for J_λ, by taking rational limits of corresponding formulae in the trigonometric case (see the 11th item in subsection "Comments, References and Further Results"). Moreover, an explicit expression of μ_x^W is deduced in [77]. As a consequence, the support of μ_x is shown to be equal to the convex hull of Wx, when $k>0$.
- The asymmetric setting is harder. Beyond the one-dimensional case (and the product case), explicit expressions of the measure μ_x are presently available in some two-dimensional cases. For the root system A_2, two closely related expressions were obtained, first in [27] and recently in [2]. For the root system B_2, a complicated formula was obtained in [28] and a simpler one recently in [3]. The case of dihedral root systems $I_2(m)$ is currently investigated (see [18, 23] and the references therein).
- In Lemma 3.20, the sharper result in the crystallographic case was obtained in [6].
- An explicit product formula was obtained in [69] for generalized Bessel functions associated with root systems of type B and for three one-dimensional families of multiplicities (which are two-dimensional in this case). The method consists in computing a product formula in the symmetric space case, which corresponds to a discrete set of multiplicities k, and in extending it holomorphically in k. The resulting measure lives in a matrix cone, which projects continuously onto $\overline{\mathfrak{a}^+}$, and its image $|W|\,\nu_{x,y}^W$ is a probability measure if $k\geq 0$.
- Potential theory in the rational Dunkl setting has been studied in [35–38, 40, 43, 56, 58] (see also [66] and the references therein).
- Many current works deal with generalizations of results in Euclidean harmonic analysis to the rational Dunkl setting. Among others, let us mention
  - [84] about the Hardy–Littlewood and the Poisson maximal functions,
  - [4] and [5] about singular integrals and Calderon–Zygmund theory,
  - [13] and [30] about the Hardy space H^1.
- A further interesting deformation of Euclidean Fourier analysis, encompassing rational Dunkl theory and the Laguerre semigroup, was introduced and studied in [16].

4 Trigonometric Dunkl Theory

Trigonometric Dunkl theory was developed in the symmetric case by Heckman and Opdam in the 1980s, and in the non-symmetric case by Opdam and Cherednik in the 1990s. This theory of special functions in several variables encompasses

- Euclidean Fourier analysis (which corresponds to the multiplicity $k=0$),
- Jacobi functions in dimension 1,

- spherical functions associated with Riemannian symmetric spaces of noncompact type (which correspond to a discrete set of multiplicities k).

In this section, we use [64] as our primary reference and quote only later works. We resume the notation of Sect. 3, with some modifications:

- the root system R is now assumed to be crystallographic but not necessarily reduced,
- $\tilde{R}$ denotes the subsystem of non-multipliable roots,
- the reference density in the case $k \geq 0$ is now $\delta(x) = \prod_{\alpha \in R^+} \left| 2 \sinh \frac{\langle \alpha, x \rangle}{2} \right|^{2k_\alpha}$,

and some addenda:

- Q denotes the root lattice, $Q^\vee$ the coroot lattice and P the weight lattice,
- $\rho = \sum_{\alpha \in \mathbb{R}^+} \frac{k_\alpha}{2} \alpha$.

Cherednik Operators

Definition 4.1 The trigonometric Dunkl operators, which are often called Cherednik operators, are the differential-difference operators defined by

$$D_\xi f(x) = \partial_\xi f(x) + \sum_{\alpha \in R^+} k_\alpha \frac{\langle \alpha, \xi \rangle}{1 - e^{-\langle \alpha, x \rangle}} \{ f(x) - f(r_\alpha x) \} - \langle \rho, \xi \rangle f(x) \tag{42}$$

for every $\xi \in \mathfrak{a}$.

Notice that the counterpart of Remark 3.2 holds in the present setting. In next theorem, we collect properties of Cherednik operators. The main one is again commutativity, which leads to Cherednik operators D_p, for every polynomial $p \in \mathcal{P}(\mathfrak{a})$, and to their symmetric parts $\tilde{D}_p$ on W-invariant functions.

Theorem 4.2

- *For any fixed multiplicity k, the Cherednik operators* (42) *commute pairwise.*
- *The Cherednik operators map the following function spaces into themselves*:

$$\mathbb{C}[e^P],\ \mathcal{P}(\mathfrak{a}),\ \mathcal{C}^\infty(\mathfrak{a}),\ \mathcal{C}_c^\infty(\mathfrak{a}),\ \mathcal{S}^2(\mathfrak{a}) = \left(\operatorname{Cosh}_\rho\right)^{-1} \mathcal{S}(\mathfrak{a}),\ \ldots$$

where $\mathbb{C}[e^P]$ *denotes the algebra of polynomials in* e^λ $(\lambda \in P)$ *and* Cosh_ρ *is defined in* (23).
- W-equivariance: *For every* $w \in W$ *and* $\xi \in \mathfrak{a}$, *we have*

$$\left(w \circ D_\xi \circ w^{-1}\right) f(x) = D_{w\xi} f(x) + \sum_{\alpha \in R^+ \cap wR^-} k_\alpha \langle \alpha, w\xi \rangle f(r_\alpha x) .$$

Hence $\tilde{D}_{pq} = \tilde{D}_p \circ \tilde{D}_q$ *for all symmetric polynomials* $p, q \in \mathcal{P}(\mathfrak{a})^W$.

- Adjointness: *Assume that $k \geq 0$. Then, for every $\xi \in \mathfrak{a}$,*

$$\int_{\mathfrak{a}} dx\, \delta(x)\, D_\xi f(x)\; g(-x) = \int_{\mathfrak{a}} dx\, \delta(x) f(x)\, D_\xi g(-x)\,.$$

Examples

- The Heckman–Opdam Laplacian is given by

$$\begin{aligned}\Delta f(x) &= \sum\nolimits_{j=1}^{n} D_j^2 f(x) \\ &= \overbrace{\sum\nolimits_{j=1}^{n} \partial_j^2 f(x) + \sum\nolimits_{\alpha\in R^+} k_\alpha \coth\tfrac{\langle\alpha,x\rangle}{2}\, \partial_\alpha f(x) + |\rho|^2 f(x)}^{\text{differential part } \tilde{\Delta} f(x)} \\ &\quad \underbrace{- \sum\nolimits_{\alpha\in R^+} k_\alpha \frac{|\alpha|^2}{4\sinh^2\frac{\langle\alpha,x\rangle}{2}} \big\{ f(x) - f(r_\alpha x) \big\}}_{\text{difference part}},\end{aligned}$$

where D_j, respectively ∂_j denote the Cherednik operators, respectively the partial derivatives with respect to an orthonormal basis of $\mathfrak{a}$.
- In dimension 1, the Cherednik operator is given by

$$\begin{aligned}Df(x) &= \big(\tfrac{\partial}{\partial x}\big) f(x) + \big\{\tfrac{k_1}{1-e^{-x}} + \tfrac{2k_2}{1-e^{-2x}}\big\} \big\{ f(x) - f(-x)\big\} - \rho f(x) \\ &= \big(\tfrac{\partial}{\partial x}\big) f(x) + \big\{\tfrac{k_1}{2}\coth\tfrac{x}{2} + k_2 \coth x\big\} \big\{ f(x) - f(-x)\big\} - \rho f(-x)\end{aligned}$$

and the Heckman–Opdam Laplacian by

$$\begin{aligned}\Delta f(x) &= \big(\tfrac{\partial}{\partial x}\big)^2 f(x) + \big\{k_1 \coth\tfrac{x}{2} + 2\,k_2 \coth x\big\} \big(\tfrac{\partial}{\partial x}\big) f(x) + \rho^2 f(x) \\ &\quad - \Big\{\frac{k_1}{4\sinh^2\frac{x}{2}} + \frac{k_2}{\sinh^2 x}\Big\} \big\{ f(x) - f(-x)\big\},\end{aligned}$$

where $\rho = \frac{k_1}{2} + k_2$.

Hypergeometric Functions

Theorem 4.3 *Assume that $k \geq 0$. Then, for every $\lambda \in \mathfrak{a}_{\mathbb{C}}$, the system*

$$\begin{cases} D_\xi G_\lambda = \langle \lambda, \xi \rangle\, G_\lambda & \forall\, \xi \in \mathfrak{a}, \\ G_\lambda(0) = 1. \end{cases}$$

has a unique smooth solution on $\mathfrak{a}$, which is called the Opdam hypergeometric function.

Definition 4.4 The Heckman–Opdam hypergeometric function is the average

$$F_\lambda(x) = \frac{1}{|W|} \sum\nolimits_{w\in W} G_\lambda(wx)\,. \tag{43}$$

Remark 4.5

- Conversely, $G_\lambda(x)$ can be recovered by applying to $F_\lambda(x)$ a linear differential operator in x whose coefficients are rational functions of λ.
- The expression $G_\lambda(x)$ extends to a holomorphic function of $\lambda \in \mathfrak{a}_{\mathbb{C}}$, $x \in \mathfrak{a} + i\,U$ and $k \in V$, where U is a W-invariant open neighborhood of 0 in $\mathfrak{a}$ and V is a W-invariant open neighborhood of $\{k \in K \mid k \geq 0\}$.
- The Heckman–Opdam hypergeometric function (43) is characterized by the system

$$\begin{cases} \tilde{D}_p\, F_\lambda = p(\lambda)\, F_\lambda & \forall\, p \in \mathcal{P}(\mathfrak{a})^W, \\ F_\lambda(0) = 1. \end{cases}$$

- In dimension 1, the Heckman–Opdam hypergeometric function reduces to the Gauss hypergeometric function ${}_2\mathrm{F}_1$ or, equivalently, to the Jacobi functions $\varphi_\lambda^{\alpha,\beta}$:

$$F_\lambda(x) = {}_2\mathrm{F}_1(\rho+\lambda, \rho-\lambda; k_1+k_2+\tfrac{1}{2}; -\sinh^2\tfrac{x}{2}) = \varphi_{i2\lambda}^{k_1+k_2-\frac{1}{2}, k_2-\frac{1}{2}}(\tfrac{x}{2})\,,$$

and the Opdam hypergeometric function to a combination of two such functions:

$$G_\lambda(x) = \varphi_{i2\lambda}^{k_1+k_2-\frac{1}{2}, k_2-\frac{1}{2}}(\tfrac{x}{2}) + \frac{\rho+\lambda}{2k_1+2k_2+1}\,(\sinh x)\, \varphi_{i2\lambda}^{k_1+k_2+\frac{1}{2}, k_2+\frac{1}{2}}(\tfrac{x}{2})\,.$$

- When $k=0$, $G_\lambda(x)$ reduces to the exponential $e^{\langle\lambda,x\rangle}$ and $F_\lambda(x)$ to the function $\operatorname{Cosh}_\lambda(x)$.
- The functions $G_{-\rho}$ and $F_{-\rho}$ are equal to 1.
- Spherical functions φ_λ^G on Riemannian symmetric space G/K of noncompact type are Heckman–Opdam hypergeometric functions. Specifically, if

$$\begin{cases} \mathcal{R} \text{ is the root system of } (\mathfrak{g}, \mathfrak{a}), \\ m_\alpha = \dim \mathfrak{g}_\alpha\,, \end{cases}$$

set

$$\begin{cases} R = 2\,\mathcal{R}, \\ k_{2\alpha} = \tfrac{1}{2} m_\alpha\,. \end{cases}$$

Then

$$\varphi_\lambda^G(\exp x) = F_{i\frac{\lambda}{2}}(2x)\,.$$

We collect in the next two propositions asymptotics and estimates of the hypergeometric functions.

Proposition 4.6 *The following Harish–Chandra type expansions hold*:

$$F_\lambda(x) = \sum\nolimits_{w\in W} \mathbf{c}(w\lambda)\, \Phi_{w\lambda}(x)\,,$$

$$G_\lambda(x) = \frac{1}{\prod_{\alpha\in\tilde{R}^+}(\langle\lambda,\alpha^\vee\rangle) - \frac{1}{2}k_{\alpha/2} - k_\alpha)} \sum\nolimits_{w\in W} \mathbf{c}(w\lambda)\, \Psi_{w,\lambda}(x)\,.$$

Here

$$\mathbf{c}(\lambda) = c_0 \prod\nolimits_{\alpha\in R^+} \frac{\Gamma(\langle\lambda,\alpha^\vee\rangle + \frac{1}{2}k_{\alpha/2})}{\Gamma(\langle\lambda,\alpha^\vee\rangle + \frac{1}{2}k_{\alpha/2} + k_\alpha)}\,,$$

where c_0 is a positive constant such that $\mathbf{c}(\rho) = 1$, and

$$\Phi_\lambda(x) = \sum\nolimits_{\ell\in Q^+} \Gamma_\ell(\lambda)\, e^{\langle\lambda-\rho-\ell,x\rangle}\,,$$

$$\Psi_{w,\lambda}(x) = \sum\nolimits_{\ell\in Q^+} \Gamma_\ell(w,\lambda)\, e^{\langle w\lambda-\rho-\ell,x\rangle}$$

are converging series, for generic $\lambda\in\mathfrak{a}_{\mathbb{C}}$ and for every $x\in\mathfrak{a}^+$.

Proposition 4.7 *Assume that $k\geq 0$.*

- *All functions G_λ with $\lambda\in\mathfrak{a}$ are strictly positive.*
- *The ground function G_0 has the following behavior:*

$$G_0(x) \asymp \Big\{ \prod\nolimits_{\substack{\alpha\in\tilde{R}^+ \\ \langle\alpha,x\rangle\geq 0}} \big(1+\langle\alpha,x\rangle\big) \Big\}\, e^{-\langle\rho,x^+\rangle} \qquad \forall\, x\in\mathfrak{a}.$$

In particular,

$$G_0(x) \asymp \Big\{ \prod\nolimits_{\alpha\in\tilde{R}^+} \big(1+\langle\alpha,x\rangle\big) \Big\}\, e^{-\langle\rho,x^+\rangle}$$

if $x\in\overline{\mathfrak{a}^+}$, while

$$G_0(x) \asymp e^{-\langle\rho,x^+\rangle}$$

if $x\in-\overline{\mathfrak{a}^+}$.

- *For every* $\lambda\in\mathfrak{a}_{\mathbb{C}}$, $\mu\in\mathfrak{a}$ *and* $x\in\mathfrak{a}$, *we have*

$$|G_{\lambda+\mu}(x)|\le e^{\langle(\operatorname{Re}\lambda)^+,x^+\rangle}\,G_\mu(x)\,.$$

In particular, the following estimates hold, for every $\lambda\in\mathfrak{a}_{\mathbb{C}}$ *and* $x\in\mathfrak{a}$:

$$|G_\lambda(x)|\le G_{\operatorname{Re}\lambda}(x)\le G_0(x)\,e^{\langle(\operatorname{Re}\lambda)^+,x^+\rangle}\,.$$

Cherednik Transform

From now on, we assume that $k\ge 0$.

Definition 4.8 The Cherednik transform is defined by

$$\mathcal{H}f(\lambda)=\int_{\mathfrak{a}}dx\,\delta(x)f(x)\,G_{i\lambda}(-x)\,. \tag{44}$$

In next theorem, we collect the main properties of the Cherednik transform.

Theorem 4.9

- *The Cherednik transform is an isomorphism between the* L^2 *Schwartz space*

$$\mathcal{S}^2(\mathfrak{a})=\big(\operatorname{Cosh}_\rho\big)^{-1}\,\mathcal{S}(\mathfrak{a})$$

and the Euclidean Schwartz space $\mathcal{S}(\mathfrak{a})$.
- Paley–Wiener Theorem: *The Cherednik transform is an isomorphism between* $\mathcal{C}_c^\infty(\mathfrak{a})$ *and the Paley–Wiener space* $\mathcal{PW}(\mathfrak{a}_{\mathbb{C}})$. *More precisely, let* C *be a* W-*invariant convex compact neighborhood of the origin in* $\mathfrak{a}$ *and let* $\chi(\lambda)=\max_{x\in C}\langle\lambda,x\rangle$ *be the associated gauge. Then the support of* $f\in\mathcal{C}_c^\infty(\mathfrak{a})$ *is contained in* C *if and only if* $h=\mathcal{H}f$ *satisfies* (28).
- Inversion formula:

$$f(x)=c_1\int_{\mathfrak{a}}d\lambda\,\tilde{\delta}(\lambda)\,\mathcal{H}f(\lambda)\,G_{i\lambda}(x)\,, \tag{45}$$

where

$$\begin{aligned}\tilde{\delta}(\lambda)&=\frac{c_0^2}{|\mathbf{c}(i\lambda)|^2}\prod\nolimits_{\alpha\in R^+}\frac{-i\langle\lambda,\alpha^\vee\rangle+\frac12k_{\alpha/2}+k_\alpha}{-i\langle\lambda,\alpha^\vee\rangle+\frac12k_{\alpha/2}}\\&=\prod\nolimits_{\alpha\in R^+}\frac{\Gamma(i\langle\lambda,\alpha^\vee\rangle+\frac12k_{\alpha/2}+k_\alpha)}{\Gamma(i\langle\lambda,\alpha^\vee\rangle+\frac12k_{\alpha/2})}\,\frac{\Gamma(-i\langle\lambda,\alpha^\vee\rangle+\frac12k_{\alpha/2}+k_\alpha+1)}{\Gamma(-i\langle\lambda,\alpha^\vee\rangle+\frac12k_{\alpha/2}+1)}\end{aligned} \tag{46}$$

and c_1 *is a positive constant.*

Remark 4.10

- In the W-invariant case, the Cherednik transform (44) reduces to

$$\mathcal{H}f(\lambda) = \int_{\mathfrak{a}} dx\, \delta(x) f(x)\, F_{i\lambda}(-x) \tag{47}$$

and its inverse (45) to

$$f(x) = c_1\, c_0^2 \int_{\mathfrak{a}} d\lambda\, |\mathbf{c}(i\lambda)|^{-2}\, \mathcal{H}f(\lambda)\, F_{i\lambda}(x)\,. \tag{48}$$

It is an isomorphism between $\mathcal{S}^2(\mathfrak{a})^W = \big(\operatorname{Cosh}_\rho\big)^{-1} \mathcal{S}(\mathfrak{a})^W$ and $\mathcal{S}(\mathfrak{a})^W$, which extends to an isometric isomorphism, up to a positive constant, between $L^2(\mathfrak{a}, \delta(x)\,dx)^W$ and $L^2(\mathfrak{a}, |\mathbf{c}(i\lambda)|^{-2} d\lambda)^W$.
- Formulae (47) and (48) are not symmetric, as the spherical Fourier transform (11) and its inverse (12) on hyperbolic spaces $\mathbb{H}^n$, or their counterparts (15) and (16) on homogeneous trees $\mathbb{T}_q$. The asymmetry is even greater between (44) and (45), where the density (46) is complex-valued.
- There is no straightforward Plancherel identity for the full Cherednik transform (44). Opdam has defined in [63] a vector-valued transform leading to a Plancherel identity in the non-W-invariant case.

Rational Limit

Rational Dunk theory (in the crystallographic case) is a suitable limit of trigonometric Dunk theory, as Hankel analysis on $\mathbb{R}^n$ is a limit of spherical Fourier analysis on $\mathbb{H}^n$. More precisely, assume that the root system R is both crystallographic and reduced. Then,

- the Dunkl kernel is the following limit of Opdam hypergeometric functions:

$$E_\lambda(x) = \lim\nolimits_{\varepsilon\to 0} G_{\varepsilon^{-1}\lambda}(\varepsilon x)\,,$$

- the Dunkl transform $\mathcal{H}_{\mathrm{rat}}$ is a limit case of the Cherednik transform $\mathcal{H}_{\mathrm{trig}}$:

$$(\mathcal{H}_{\mathrm{rat}} f)(\lambda) = \lim\nolimits_{\varepsilon\to 0}\, \varepsilon^{-n-2\gamma} \big\{\mathcal{H}_{\mathrm{trig}}[f(\varepsilon^{-1}.)]\big\}(\varepsilon^{-1}\lambda)\,,$$

- likewise for the inversion formulae (25) and (45):

$$(\mathcal{H}_{\mathrm{rat}}^{-1} f)(x) = \text{const.}\ \lim\nolimits_{\varepsilon\to 0}\, \varepsilon^{n+2\gamma} \big\{\mathcal{H}_{\mathrm{trig}}^{-1}[f(\varepsilon\,.)]\big\}(\varepsilon x)\,.$$

Intertwining Operator and (Dual) Abel Transform

In the trigonometric setting, consider again the Abel transform

$$\mathcal{A} = \mathcal{F}^{-1} \circ \mathcal{H},$$

which is obtained by composing the Cherednik transform $\mathcal{H}$ with the inverse Euclidean Fourier transform $\mathcal{F}^{-1}$ on $\mathfrak{a}$, and the dual Abel transform $\mathcal{A}^*$, which satisfies

$$\int_{\mathfrak{a}} dx\, \delta(x) f(x)\, \mathcal{A}^* g(x) = \int_{\mathfrak{a}} dy\, \mathcal{A}f(y)\, g(y)\,.$$

Remark 4.11 In [85], $\mathcal{A}^* = \mathcal{V}$ is called the trigonometric Dunkl intertwining operator and $\mathcal{A} = \mathcal{V}^*$ the dual operator.

Proposition 4.12

- *For every* $\xi \in \mathfrak{a}$*,*

$$\mathcal{A} \circ D_\xi = \partial_\xi \circ \mathcal{A} \qquad \textit{and} \qquad \mathcal{V} \circ \partial_\xi = D_\xi \circ \mathcal{V}\,.$$

- *For every* $x \in \mathfrak{a}$*, there is a unique tempered distribution* μ_x *on* $\mathfrak{a}$ *such that*

$$\mathcal{V}g(x) = \langle\, \mu_x, g\,\rangle\,.$$

Moreover, the support of μ_x *is contained in the convex hull of* Wx*.*

Corollary 4.13 *For every* $\lambda \in \mathfrak{a}_{\mathbb{C}}$*, we have*

$$G_\lambda(x) = \langle\, \mu_x, e^\lambda \,\rangle \quad \textit{and} \quad F_\lambda(x) = \langle\, \mu_x^W, \operatorname{Cosh}_\lambda \,\rangle\,, \tag{49}$$

where $\operatorname{Cosh}_\lambda$ *is defined in* (23) *and* $\mu_x^W = \frac{1}{|W|} \sum_{w \in W} \mu_{wx}$.

Remark 4.14

- The distribution μ_x is most likely a probability measure, as in the rational setting.
- This is true in dimension 1 (hence in the product case), where

$$d\mu_x(y) = \begin{cases} d\delta_x(y) & \text{if } x=0 \text{ or if } k_1=k_2=0. \\ \mu(x,y)\,dy & \text{otherwise.} \end{cases}$$

As far as it is concerned, the density $\mu(x,y)$ vanishes unless $|y| < |x|$. In the generic case, where $k_1 > 0$ and $k_2 > 0$, it is given explicitly by

$$\begin{aligned}\mu(x,y) &= 2^{k_1+k_2-2}\,\frac{\Gamma(k_1+k_2+\frac{1}{2})}{\sqrt{\pi}\,\Gamma(k_1)\,\Gamma(k_2)}\,|\sinh\tfrac{x}{2}|^{-2k_1}\,|\sinh x|^{-2k_2}\\ &\quad\times\int_{|y|}^{|x|} dz\,(\sinh\tfrac{z}{2})\,(\cosh\tfrac{z}{2}-\cosh\tfrac{y}{2})^{k_1-1}\,(\cosh x-\cosh z)^{k_2-1}\\ &\qquad\times(\operatorname{sign}x)\left\{e^{\frac{x}{2}}\,(2\cosh\tfrac{x}{2})-e^{-\frac{y}{2}}\,(2\cosh\tfrac{z}{2})\right\}.\end{aligned} \tag{50}$$

In the limit case, where $k_1 = 0$ and $k_2 > 0$,

$$\mu(x,y) = 2^{k_2-1}\,\frac{\Gamma(k_2+\frac{1}{2})}{\sqrt{\pi}\,\Gamma(k_2)}\,|\sinh x|^{-2k_2}\,(\cosh x-\cosh y)^{k_2-1}\,(\operatorname{sign}x)\,(e^{x}-e^{-y})\,. \tag{51}$$

In the other limit case, where $k_1 > 0$ and $k_2 = 0$, the density is half of (51), with k_2, x, y replaced respectively by $k_1, \frac{x}{2}, \frac{y}{2}$.

Generalized Translations, Convolution and Product Formula

Definition 4.15

- The generalized convolution corresponds, via the Cherednik transform, to pointwise multiplication:

$$(f*g)(x) = c_1\int_{\mathfrak{a}} d\lambda\,\tilde{\delta}(\lambda)\,\mathcal{H}f(\lambda)\,\mathcal{H}g(\lambda)\,G_{i\lambda}(x)\,.$$

- The generalized translations are defined by

$$(\tau_y f)(x) = c_1\int_{\mathfrak{a}} d\lambda\,\tilde{\delta}(\lambda)\,\mathcal{H}f(\lambda)\,G_{i\lambda}(x)\,G_{i\lambda}(y)\,. \tag{52}$$

The key objects are again the tempered distributions

$$f \longmapsto \langle\,\nu_{x,y}\,,f\,\rangle$$

defined by (52) and their averages

$$\nu_{x,y}^{W} = \frac{1}{|W|}\sum\nolimits_{w\in W}\nu_{wx,wy}\,,$$

which enter the product formulae

$$G_\lambda(x)\,G_\lambda(y) = \langle\,\nu_{x,y}, G_\lambda\,\rangle$$

and

$$F_\lambda(x)\,F_\lambda(y) = \langle\, \nu^W_{x,y}, F_\lambda \,\rangle\,. \tag{53}$$

Example In dimension 1, the distributions $\nu_{x,y}$ are signed measures, which are uniformly bounded in x and y. Explicitly [11],

$$d\nu_{x,y}(z) = \begin{cases} \nu(x,y,z)\,dz & \text{if } x,y\in\mathbb{R}^*,\\ d\delta_y(z) & \text{if } x=0,\\ d\delta_x(z) & \text{if } y=0, \end{cases}$$

where the density $\nu(x,y,z)$ is given by the following formulae, when $x,y,z\in\mathbb{R}^*$ satisfy the triangular inequality

$$\big|\,|x|-|y|\,\big| < |z| < |x|+|y|\,,$$

and vanishes otherwise.

- Assume that $k_1>0$ and $k_2>0$. Then

$$\begin{aligned} \nu(x,y,z) &= 2^{k_1-2}\,\tfrac{\Gamma(k_1+k_2+\frac12)}{\sqrt{\pi}\,\Gamma(k_1)\,\Gamma(k_2)}\,\operatorname{sign}(xyz)\,|\sinh\tfrac{x}{2}\sinh\tfrac{y}{2}|^{-2k_1-2k_2}(\cosh\tfrac{z}{2})^{2k_2}\\ &\times\int_0^\pi d\chi\,(\sin\chi)^{2k_2-1}\\ &\quad\times\Big[\cosh\tfrac{x}{2}\cosh\tfrac{y}{2}\cosh\tfrac{z}{2}\cos\chi-\tfrac{1+\cosh x+\cosh y+\cosh z}{4}\Big]_+^{k_1-1}\\ &\quad\times\Big[\sinh\tfrac{x+y+z}{2}-2\cosh\tfrac{x}{2}\cosh\tfrac{y}{2}\sinh\tfrac{z}{2}\\ &\qquad+\tfrac{k_1+2k_2}{k_2}\cosh\tfrac{x}{2}\cosh\tfrac{y}{2}\cosh\tfrac{z}{2}(\sin\chi)^2\\ &\qquad+\tfrac{\sinh z-\sinh x-\sinh y}{2}\cos\chi\Big]\,. \end{aligned}$$

- Assume that $k_1=0$ and $k_2>0$. Then

$$\begin{aligned} \nu(x,y,z) &= 2^{2k_2-1}\,\tfrac{\Gamma(k_2+\frac12)}{\sqrt{\pi}\,\Gamma(k_2)}\,\operatorname{sign}(xyz)\,|(\sinh x)(\sinh y)|^{-2k_2}\\ &\times\Big[\sinh\tfrac{x+y+z}{2}\sinh\tfrac{-x+y+z}{2}\sinh\tfrac{x-y+z}{2}\sinh\tfrac{x+y-z}{2}\Big]^{k_2}\\ &\times\Big[\sinh\tfrac{x+y-z}{2}\Big]^{-1}e^{\frac{x+y-z}{2}}\,. \end{aligned}$$

- In the other limit case, where $k_1>0$ and $k_2=0$, the density is again half of the previous one, with k_2, x, y replaced respectively by k_1, $\frac{x}{2}$, $\frac{y}{2}$.

In higher dimension, we have the trigonometric counterparts of Problems (a), (b), (c), (d), (e) in section "Generalized Translations, Convolution and Product Formula" but fewer results than in the rational case. In particular, there is no formula like (40) for radial functions. A new property is the following Kunze–Stein phenomenon, which is typical of the semisimple setting and which was proved in [11] (see also [14]) and [86].

Proposition 4.16 *Let* $1 \leq p < 2$. *Then there exists a constant* $C > 0$ *such that*

$$\| f * g \|_{L^2} \leq C \, \| f \|_{L^p} \, \| g \|_{L^2} \,,$$

for every $f \in L^p(\mathfrak{a}, \delta(x)\,dx)$ *and* $g \in L^2(\mathfrak{a}, \delta(x)\,dx)$.

Comments, References and Further Results

- The joint action of the Cherednik operators D_p, with $p \in \mathcal{P}(\mathfrak{a})$, and of the Weyl group W may look intricate. It corresponds actually to a faithful representation of a graded affine Hecke algebra [63].
- Heckman [45] considered initially the following trigonometric version

$$'D_\xi f(x) = \partial_\xi f(x) + \sum\nolimits_{\alpha \in R^+} \tfrac{k_\alpha}{2} \, \langle \alpha, \xi \rangle \, \coth \tfrac{\langle \alpha, x \rangle}{2} \, \big\{ f(x) - f(r_\alpha x) \big\}$$

  of rational Dunkl operators, which are closely connected to (42):

$$D_\xi f(x) = {}'D_\xi f(x) - \sum\nolimits_{\alpha \in R^+} \tfrac{k_\alpha}{2} \, \langle \alpha, \xi \rangle f(r_\alpha x) \,.$$

  These operators are W-equivariant:

$$w \circ {}'D_\xi \circ w^{-1} = {}'D_{w\xi} \,,$$

  and skew-invariant:

$$\int_{\mathfrak{a}} dx \, \delta(x) \, ({}'D_\xi f)(x) \, g(x) = - \int_{\mathfrak{a}} dx \, \delta(x) f(x) \, ({}'D_\xi g)(x) \,,$$

  but they don't commute:

$$[{}'D_\xi, {}'D_\eta] f(x) = \sum\nolimits_{\alpha, \beta \in R^+} \tfrac{k_\alpha k_\beta}{4} \, \big\{ \langle \alpha, \xi \rangle \, \langle \beta, \eta \rangle - \langle \beta, \xi \rangle \, \langle \alpha, \eta \rangle \big\} f(r_\alpha r_\beta x) \,.$$

- The hypergeometric functions $x \longmapsto G_\lambda(x)$ and $x \longmapsto F_\lambda(x)$ extend holomorphically to a tube $\mathfrak{a} + iU$ in $\mathfrak{a}_{\mathbb{C}}$. The optimal width for F_λ was investigated in [52].

- Proposition 4.6 was obtained in [63]. The asymptotic behavior of F_λ was fully determined in [59]. This paper contains in particular a proof of the estimate

$$F_\lambda(x) \asymp \Big\{ \prod\nolimits_{\substack{\alpha\in\tilde{R}^+\\ \langle\alpha,\lambda\rangle\neq 0}} \big(1+\langle\alpha,x\rangle\big) \Big\}\, e^{\langle\lambda-\rho,x\rangle} \qquad \forall\,\lambda,x\in\overline{\mathfrak{a}^+},$$

  which was stated in [80] (see also [78]), and the following generalization of a celebrated result of Helgason & Johnson in the symmetric space case:

  F_λ is bounded if and only if λ belongs to the convex hull of $W\rho$.

- The sharp estimates in Proposition 4.7 were obtained in [80] (see also [78]) and [74].
- As in the rational case, we have not discussed the shift operators, which move the multiplicity k by integers and which have proven useful in the W-invariant setting (see [46, Part I, Chap. 3], [64, Sect. 5]).
- Rational limits in subsection "Rational Limit" have a long prehistory. In the Dunkl setting, they have been used for instance in [1, 6, 15, 21, 71, 77], …(seemingly first and independently in preprint versions of [15] and [21]).
- There are other interesting limits between special functions occurring in Dunkl theory. For instance, in [74] and [73], Heckman–Opdam hypergeometric functions associated with the root system A_{n-1} are obtained as limits of Heckman–Opdam hypergeometric functions associated with the root system BC_n, when some multiplicities tend to infinity. See [72] for a similar result about generalized Bessel functions.
- The expressions (49) are substitutes for the integral representations (10) and (19). A different integral representation of F_λ is established in [83].
- Formula (50) was obtained in [7] and used there to prove the positivity of μ_x when $k_1>0$ and $k_2>0$. A more complicated expression was obtained previously in [39] and in [14]. It was used in [39] to disprove mistakenly the positivity of μ_x. Another approach, which consists in proving the positivity of a heat type kernel, was followed in [87–90]. But, as pointed out in [7] (see Remark 3.4), the same flaw occurs in [87–89].
- It is natural to look for recurrence formulae over n for the five families of classical crystallographic root systems A_n, B_n, C_n, BC_n, D_n (see the Appendix). In the case of A_n, an integral recurrence formula for F_λ (or for Jack polynomials) was discovered independently by several authors (see for instance [44, 60, 76]). An explicit expression of μ_x^W is deduced in [76] and [77], first for $x\in\mathfrak{a}^+$ and next for any $x\in\overline{\mathfrak{a}^+}$. In particular, if $k>0$, then μ_x^W is a probability measure, whose support is equal to the convex hull of Wx and which is absolutely continuous with respect to the Lebesgue measure, except for $x=0$ where $\mu_x^W=\delta_0$.
- As in the rational case (see the eighth item in subsection "Comments, References and Further Results"), an explicit product formula was obtained in [70] and [92] for Heckman–Opdam hypergeometric functions associated with root systems of type BC and for certain continuous families of multiplicities.

- Probabilistic aspects of trigonometric Dunkl theory were studied in [80] and [79] (see also [78]). Regarding the heat kernel $h_t(x,y)$, the estimate (20) was shown to hold for $h_t(x,0)$ and some asymptotics were obtained for $h_t(x,y)$. But there is no trigonometric counterpart of the expression (29), neither precise information like (30) about the full behavior of $h_t(x,y)$.
- The bounded harmonic functions for the Heckman–Opdam Laplacian were determined in [81].

Acknowledgements Work partially supported by the regional project MADACA (Marches Aléatoires et processus de Dunkl–Approches Combinatoires et Algébriques, www.fdpoisson.fr/madaca). We thank the organizers of the school and conference AAGADE 2015 (*Analytic, Algebraic and Geometric Aspects of Differential Equations*, Mathematical Research and Conference Center, Będlewo, Poland, September 2015) for their invitation and the warm welcome at Pałac Będlewo. We also thank Béchir Amri, Nizar Demni, Léonard Gallardo, Chaabane Rejeb, Margit Rösler, Simon Ruijsenaars, Patrice Sawyer and Michael Voit for helpful comments and discussions.

Appendix: Root Systems

In this appendix, we collect some information about root systems and reflection groups. More details can be found in classical textbooks such as [48] or [50].

Definition A.1 Let $\mathfrak{a} \approx \mathbb{R}^n$ be a Euclidean space.

- A (crystallographic) root system in $\mathfrak{a}$ is a finite set R of nonzero vectors satisfying the following conditions:
  - (a) for every $\alpha \in R$, the reflection $r_\alpha(x) = x - 2\,\frac{\langle \alpha, x\rangle}{|\alpha|^2}\,\alpha$ maps R onto itself,
  - (b) $2\,\frac{\langle \alpha, \beta\rangle}{|\alpha|^2} \in \mathbb{Z}$ for all $\alpha, \beta \in R$.
- A root system R is reducible if it can be splitted into two orthogonal root systems, and irreducible otherwise.

Remark A.2

- Unless specified, we shall assume that R spans $\mathfrak{a}$.
- $\alpha^\vee = \frac{2}{|\alpha|^2}\,\alpha$ denotes the coroot corresponding to a root α. If R is a root system, then $R^\vee$ is again a root system.
- Most root systems are reduced, which means that
  - (c) the roots proportional to any root α are reduced to $\pm\alpha$.
    Otherwise the only possible alignment of roots is

$$-2\alpha, -\alpha, +\alpha, +2\alpha\,.$$

  A root α is called
  - ∘ indivisible if $\frac{\alpha}{2}$ is not a root,

  ◦ non-multipliable if 2α is not a root.

- We shall also consider non-crystallographic reduced root systems R, which satisfy (a) and (c), but not necessarily (b).

Definition A.3

- The connected components of

$$\{\, x\in\mathfrak{a} \mid \langle\alpha,x\rangle\neq 0 \;\; \forall\,\alpha\in R \,\}$$

  are called Weyl chambers. We choose any of them, which is called positive and denoted by $\mathfrak{a}^+$. R^+ denotes the set of roots which are positive on $\mathfrak{a}^+$.
- The Weyl or Coxeter group W associated with R is the finite subgroup of the orthogonal group $\mathrm{O}(\mathfrak{a})$ generated by the root reflections $\{\, r_\alpha \mid \alpha\in R \,\}$.

Remark A.4

- The group W acts simply transitively on the set of Weyl chambers.
- The longest element w_0 in W interchanges $\mathfrak{a}^+$ and $-\mathfrak{a}^+$.
- Every $x\in\mathfrak{a}$ belongs to the W-orbit of a single $x^+\in\overline{\mathfrak{a}^+}$.

There are six classical families of irreducible root systems:

- A_n $(n\geq 1)$:
  $\mathfrak{a}=\{\, x\in\mathbb{R}^{n+1} \mid x_0+x_1+\ldots+x_n=0 \,\}$
  $R=\{\, e_i-e_j \mid 0\leq i\neq j\leq n \,\}$
  $\mathfrak{a}^+=\{\, x\in\mathfrak{a} \mid x_0>x_1>\ldots>x_n \,\}$
  $W=\mathrm{S}_{n+1}$
- B_n $(n\geq 2)$:
  $\mathfrak{a}=\mathbb{R}^n$
  $R=\{\, \pm e_i \mid 1\leq i\leq n \,\}\cup\{\, \pm e_i\pm e_j \mid 1\leq i<j\leq n \,\}$
  $\mathfrak{a}^+=\{\, x\in\mathbb{R}^n \mid x_1>\ldots>x_n>0 \,\}$
  $W=\{\pm 1\}^n\rtimes\mathrm{S}_n$
- C_n $(n\geq 2)$:
  $\mathfrak{a}=\mathbb{R}^n$
  $R=\{\, \pm 2\,e_i \mid 1\leq i\leq n \,\}\cup\{\, \pm e_i\pm e_j \mid 1\leq i<j\leq n \,\}$
  $\mathfrak{a}^+=\{\, x\in\mathbb{R}^n \mid x_1>\ldots>x_n>0 \,\}$
  $W=\{\pm 1\}^n\rtimes\mathrm{S}_n$
- BC_n $(n\geq 1)$:
  $\mathfrak{a}=\mathbb{R}^n$
  $R=\{\, \pm e_i, \pm 2\,e_i \mid 1\leq i\leq n \,\}\cup\{\, \pm e_i\pm e_j \mid 1\leq i<j\leq n \,\}$
  $\mathfrak{a}^+=\{\, x\in\mathbb{R}^n \mid x_1>\ldots>x_n>0 \,\}$
  $W=\{\pm 1\}^n\rtimes\mathrm{S}_n$
- D_n $(n\geq 3)$:
  $\mathfrak{a}=\mathbb{R}^n$
  $R=\{\, \pm e_i\pm e_j \mid 1\leq i<j\leq n \,\}$
  $\mathfrak{a}^+=\{\, x\in\mathbb{R}^n \mid x_1>\ldots>|x_n| \,\}$
  $W=\{\, \varepsilon\in\{\pm 1\}^n \mid \varepsilon_1\ldots\varepsilon_n=1 \,\}\rtimes\mathrm{S}_n$
- $\mathrm{I}_2(m)$ $(m\geq 3)$:
  $\mathfrak{a}=\mathbb{C}$
  $R=\{\, e^{i\pi\frac{j}{m}} \mid 0\leq j<2m \,\}$
  $\mathfrak{a}^+=\{\, z\in\mathbb{C}^* \mid \big(\tfrac{1}{2}-\tfrac{1}{m}\big)\,\pi<\arg z<\tfrac{\pi}{2} \,\}$
  $W=(\mathbb{Z}/m\mathbb{Z})\rtimes(\mathbb{Z}/2\mathbb{Z})$ (dihedral group)

The full list of irreducible root systems (crystallographic or reduced) includes in addition a finite number of exceptional cases:

$$\mathrm{E}_6,\ \mathrm{E}_7,\ \mathrm{E}_8,\ \mathrm{F}_4,\ \mathrm{G}_2,\ \mathrm{H}_3,\ \mathrm{H}_4.$$

Remark A.5 In the list above,

- the non crystallographic root systems are

$$\mathrm{H}_3, \mathrm{H}_4 \text{ and } \mathrm{I}_2(m) \text{ with } \begin{cases} m = 5, \\ m \geq 7, \end{cases}$$

- all root systems are reduced, with the exception of BC_n,
- there are some redundancies in low dimension:

$$\begin{cases} \mathrm{A}_1 \times \mathrm{A}_1 \approx \mathrm{D}_2 \approx \mathrm{I}_2(2) & \\ \mathrm{B}_2 \approx \mathrm{C}_2 \approx \mathrm{I}_2(4) & \text{(up to the root length)} \\ \mathrm{A}_2 \approx \mathrm{I}_2(3) & \\ \mathrm{G}_2 \approx \mathrm{I}_2(6) & \text{(up to the root length)} \end{cases}$$

The 2-dimensional root systems (crystallographic or reduced) are depicted in Fig. 9.

Definition A.6 A multiplicity is a W-invariant function $k : R \longrightarrow \mathbb{C}$.

Remark A.7

- In Dunkl theory, one assumes most of the time that $k \geq 0$.
- Assume that R is crystallographic and irreducible. Then two roots belong to the same W-orbit if and only if they have the same length. Thus k takes at most three values. In the non crystallographic case, there are one or two W-orbits in R. Specifically, by resuming the classification of root systems, k takes
  - 1 value in the following cases :

$$\mathrm{A}_n,\ \mathrm{D}_n,\ \mathrm{E}_6,\ \mathrm{E}_7,\ \mathrm{E}_8,\ \mathrm{H}_3,\ \mathrm{H}_4,\ \mathrm{I}_2(m) \text{ with } m \text{ odd,}$$

  - 2 values in the following cases :

$$\mathrm{B}_n,\ \mathrm{C}_n,\ \mathrm{F}_4,\ \mathrm{G}_2,\ \mathrm{I}_2(m) \text{ with } m \text{ even,}$$

  - 3 values in the case of BC_n.

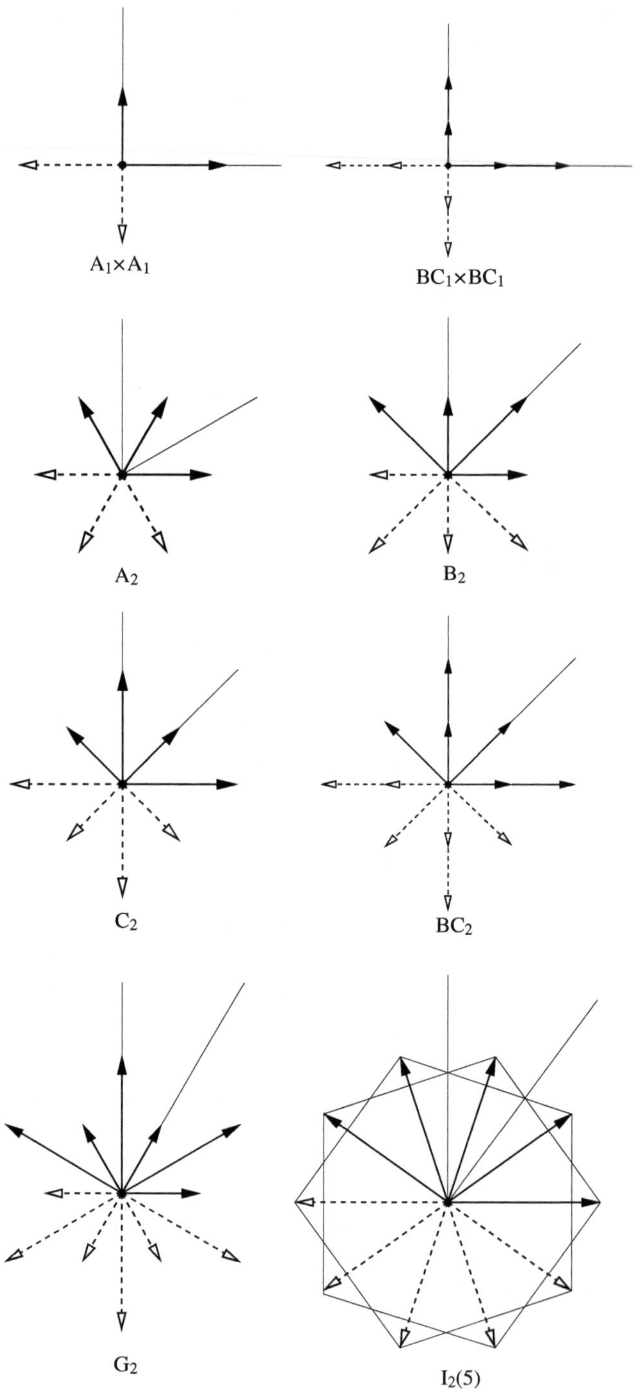

Fig. 9 Two-dimensional root systems

References

1. B. Amri, Note on Bessel functions of type A_{N-1}. Integral Transf. Spec. Funct. **25**(6), 448–461 (2014)
2. B. Amri, On the integral representations for Dunkl kernels of type A_2. J. Lie Theory **26**(4), 1163–1175 (2016)
3. B. Amri, N. Demni, Laplace–type integral representations of the generalized Bessel function and of the Dunkl kernel of type B (2017). Preprint [arXiv:1611.05696]
4. B. Amri, M. Sifi, Riesz transforms for Dunkl transform. Ann. Math. Blaise Pascal **19**(1), 247–262 (2012)
5. B. Amri M. Sifi, Singular integral operators in Dunkl setting. J. Lie Theory **22**(3), 723–739 (2012)
6. B. Amri, J.P. Anker, M. Sifi, Three results in Dunkl theory. Colloq. Math. **118**(1), 299–312 (2010)
7. J.P. Anker, An elementary proof of the positivity of the intertwining operator in one-dimensional trigonometric Dunkl theory. Proc. Am. Math. Soc. (2017). doi:https://doi.org/10.1090/proc/13679
8. J.P. Anker P. Ostellari, The heat kernel on symmetric spaces, in *Lie Groups and Symmetric Spaces: In Memory of F.I. Karpelevich*, ed. by S.G. Gindikin, pp. 27–46, American Mathematical Society Translation (2), vol. 210 (American Mathematical Society, Providence, RI, 2004)
9. J.P. Anker, E. Damek, C. Yacoub, Spherical analysis on harmonic *AN* groups. Ann. Scuola Norm. Sup. Pisa (4) **23**(4), 643–679 (1996)
10. J.P. Anker, B. Schapira, B. Trojan, Heat kernel and Green function estimates on affine buildings of type $\widetilde{A}_r$ (2017). Preprint [arXiv0612385]
11. J.P. Anker, F. Ayadi M. Sifi, Opdam's hypergeometric functions (product formula and convolution structure in dimension 1). Adv. Pure Appl. Math. **3**(1), 11–44 (2012)
12. J.P. Anker, P. Martinot, E. Pedon, A.G. Setti, The shifted wave equation on Damek–Ricci spaces and on homogeneous trees, in *Trends in Harmonic Analysis* (XXXV Convegno di Analisi Armonica, Roma, 2011), ed. by M.A. Picardello. Springer INdAM Series, vol. 3 (Springer, Milan, 2013), pp. 1–25
13. J.P. Anker, N. Ben Salem, J. Dziubański N. Hamda, The Hardy space H^1 in the rational Dunkl setting. Constr. Approx. **42**(1), 93–128 (2015)
14. F. Ayadi, Analyse harmonique et équation de Schrödinger associées au laplacien de Dunkl trigonométrique. Ph.D. Thesis, Université d'Orléans Université de Tunis El Manar (2011). https//tel.archives-ouvertes.fr/tel–00664822
15. S. Ben Saïd, B. Ørsted, Bessel functions for root systems via the trigonometric setting, Int. Math. Res. Not. **9**, 551–585 (2005)
16. S. Ben Saïd, T. Kobayashi B. Ørsted, Laguerre semigroup and Dunkl operators. Compos. Math. **148**(4), 1265–1336 (2012)
17. I. Cherednik, *Double Affine Hecke Algebras*. London Mathematical Society Lecture Notes Series, vol. 319 (Cambridge University Press, Cambridge, 2005), 434 pp.
18. D. Constales, H. De Bie, P. Lian, Explicit formulas for the Dunkl dihedral kernel and the (κ, a)–generalized Fourier kernel (2017). Preprint [arXiv:1610.00098]
19. M.G. Cowling, S. Meda, A.G. Setti, An overview of harmonic analysis on the group of isometries of a homogeneous tree. Expo. Math. **16**, 385–423 (1998)
20. M.G. Cowling, S. Meda A.G. Setti, Estimates for functions of the Laplace operator on homogeneous trees, Trans. Am. Math. Soc. **352**(9), 4271–4293 (2000)
21. M.F.E. de Jeu, Paley–Wiener theorems for the Dunkl transform. Trans. Am. Math. Soc. **358**(10), 4225–4250 (2006)
22. M. de Jeu, M. Rösler, Asymptotic analysis for the Dunkl kernel. J. Approx. Theory **119**, 110–126 (2002)
23. L. Deléaval, N. Demni, H. Youssfi, Dunkl kernel associated with dihedral groups. J. Math. Anal. Appl. **432**, 928–944 (2015)

24. N. Demni, Dunkl operators (an overview). Updated lecture notes (2015). CIMPA Spring School Analytical and probabilistic aspects of Dunkl theory (Monastir, April 13–25, 2009). http://perso.univ--rennes1.fr/nizar.demni/Sitenizar/Travaux_files/SurveyDunkl.pdf
25. DLMF, Digital library of mathematical functions. NIST (National Institute of Standards and Technology). http://dlmf.nist.gov/
26. C.F. Dunkl, Differential–difference operators associated to reflection groups. Trans. Am. Math. Soc. **311**, 167–183 (1989)
27. C.F. Dunkl, Intertwining operators associated to the group S_3, Trans. Am. Math. Soc. **347**(9), 3347–3374 (1995)
28. C.F. Dunkl, An intertwining operator for the group B_2. Glasgow Math. J. **49**, 291–319 (2007)
29. C.F. Dunkl, Y. Xu, *Orthogonal Polynomials of Several Variables*. Encyclopedia of Mathematics and its Applications, vol. 81 (Cambridge University Press, Cambridge, 2001)
30. J. Dziubański, Riesz transforms characterizations of Hardy spaces H^1 for the rational Dunkl setting and multidimensional Bessel operators. J. Geom. Anal. **26**(4), 2639–2663 (2016)
31. P. Etingof, *Calogero–Moser Systems and Representation Theory*. Zurich Lectures in Advances Mathematics, vol. 4 (European Mathematical Society, Zürich, 2007)
32. P. Etingof, A uniform proof of the Macdonald–Mehta–Opdam identity for finite Coxeter groups. Math. Res. Lett. **17**(2), 275–282 (2010)
33. J. Faraut, Analyse harmonique sur les paires de Guelfand et les espaces hyperboliques, in *Analyse Harmonique*, ed. by J.L. Clerc, J. Faraut, M. Rais, R. Takahashi (Cours CIMPA, Nice, 1982), pp. 315–446
34. A. Figà–Talamanca, M.A. Picardello, *Harmonic Analysis on Free Groups*. Lecture Notes in Pure and Applied Mathematics, vol. 87 (Marcel Dekker, New York, 1983)
35. L. Gallardo, L. Godefroy, Propriété de Liouville et équation de Poisson pour le laplacien généralisé de Dunkl. C. R. Math. Acad. Sci. Paris **337**(10), 639–644 (2003)
36. L. Gallardo, C. Rejeb, A new mean value property for harmonic functions relative to the Dunkl–Laplacian operator and applications. Trans. Am. Math. Soc. **368**, 3727–3753 (2016)
37. L. Gallardo, C. Rejeb, Newtonian potentials and subharmonic functions associated to the Dunkl–Laplace operator (2017). Preprint [hal–01368871]
38. L. Gallardo, C. Rejeb, Support properties of the intertwining and the mean value operators in Dunkl's analysis (2017). Preprint [hal–01331693]
39. L. Gallardo, K. Trimèche, Positivity of the Jacobi–Cherednik intertwining operator and its dual, Adv. Pure Appl. Math. **1**(2), 163–194 (2010)
40. L. Gallardo, C. Rejeb, M. Sifi, Dunkl–Riesz potentials of Radon measures. Preprint (2015)
41. R. Gangolli, V.S. Varadarajan, *Harmonic Analysis of Spherical Functions on Real Reductive Groups* (Springer, Berlin, 1988)
42. P. Graczyk, M. Rösler, M. Yor. (eds.), *Harmonic and Stochastic Analysis of Dunkl Processes*. Travaux en Cours 71 (Hermann, Paris, 2008)
43. P. Graczyk, T. Luks, M. Rösler, On the Green function and Poisson integrals of the Dunkl Laplacian (2017). Preprint [arXiv:1607.08746]
44. M. Hallnäs, S.N.M. Ruijsenaars, A recursive construction of joint eigenfunctions for the hyperbolic nonrelativistic Calogero–Moser Hamiltonians. Int. Math. Res. Not. **20**, 10278–10313 (2015)
45. G.J. Heckman, An elementary approach to the hypergeometric shift operators of Opdam, Invent. Math. **103**(2), 341–350 (1991)
46. G.J. Heckman, H. Schlichtkrull, *Harmonic Analysis and Special Functions on Symmetric Spaces*. Perspectives in Mathematics, vol. 16 (Academic Press, San Diego, 1994)
47. S. Helgason, *Groups and Geometric Analysis: Integral Geometry, Invariant Differential Operators, and Spherical Functions*. Pure and Applied Mathematics, vol. 113 (Academic, Orlando, 1984); Mathematical Surveys Monographs, vol. 83 (American Mathematical Society, Providence, RI, 2000)
48. J.E. Humphreys, *Reflection Groups and Coxeter Groups*. Cambridge Studies in Advances Mathematics, vol. 29 (Cambridge University Press, Cambridge, 1990)

49. A. Jamal Eddine, Schrödinger equation on homogeneous trees. J. Lie Theory **23**, 779–794 (2013)
50. R. Kane, *Reflection Groups and Invariant Theory*. CMS Books in Mathematics, vol. 5 (Springer, New York, 2001)
51. T.H. Koornwinder, Jacobi functions and analysis on noncompact semisimple Lie groups, in *Special Functions: Group Theoretical Aspects and Applications*, ed. by R.A. Askey, T.H. Koornwinder, W. Schempp (Reidel, Dordrecht, 1984), pp. 1–84
52. B. Krötz, E.M. Opdam, Analysis on the crown domain. Geom. Funct. Anal. **18**(4), 1326–1421 (2008)
53. I.G. Macdonald, Spherical functions on a group of p-adic type. Publications Ramanujan Institute 2, Centre for Advances Study in Mathematics, University of Madras, Madras, 1971
54. I.G. Macdonald, *Affine Hecke Algebras and Orthogonal Polynomials*. Cambridge Tracts in Mathematics, vol. 157 (Cambridge University Press, Cambridge, 2003)
55. A.M. Mantero, A. Zappa, Macdonald formula for spherical functions on affine buildings. Ann. Fac. Sci. Toulouse Math. (6) **20**(4), 669–758 (2011)
56. M. Maslouhi, E.H. Youssfi, Harmonic functions associated to Dunkl operators. Monatsh. Math. **152**(4), 337–345 (2007)
57. G. Medolla, A.G. Setti, The wave equation on homogeneous trees. Ann. Mat. Pura Appl. (4) **176**, 1–27 (1999)
58. H. Mejjaoli, K. Trimèche, On a mean value property associated with the Dunkl Laplacian operator and applications. Integral Transf. Spec. Funct. **12**(3), 279–302 (2001)
59. E.K. Narayanan, A. Pasquale, S. Pusti, Asymptotics of Harish–Chandra expansions, bounded hypergeometric functions associated with root systems, and applications. Adv. Math. **252**, 227–259 (2014)
60. A. Okounkov, G. Olshanski, Shifted Jack polynomials, binomial formula, and applications. Math. Res. Lett. **4**(1), 69–78 (1997)
61. E.M. Opdam, Some applications of hypergeometric shift operators. Invent. Math. **98**(1), 1–18 (1989)
62. E.M. Opdam, Dunkl operators, Bessel functions and the discriminant of a finite Coxeter group. Compos. Math. **85**, 333–373 (1993)
63. E.M. Opdam, Harmonic analysis for certain representations of graded Hecke algebras. Acta Math. **175**, 75–121 (1995)
64. E.M. Opdam, *Lecture Notes on Dunkl Operators for Real and Complex Reflection Groups*. Mathematical Society of Japan Memoirs, vol. 8 (Mathematical Society Japan, Tokyo, 2000)
65. J. Parkinson, Spherical harmonic analysis on affine buildings. Math. Z. **253**(3), 571–606 (2006)
66. C. Rejeb, Fonctions harmoniques et sous–harmoniques associées à des systèmes de racines. Ph.D. Thesis, Université François–Rabelais de Tours & Université de Tunis El Manar (2015). httpstel.archives-ouvertes.fr/tel–01291741
67. M. Rösler, Dunkl operators (theory and applications), in *Orthogonal Polynomials and Special Functions* (*Leuven, 2002*), ed, by E. Koelink, W. Van Assche. Lecture Notes in Mathematics, vol. 1817 (Springer, Berlin, 2003), pp. 93–135
68. M. Rösler, A positive radial product formula for the Dunkl kernel. Trans. Am. Math. Soc. **355**, 2413–2438 (2003)
69. M. Rösler, Bessel convolutions on matrix cones. Compos. Math. **143**, 749–779 (2007)
70. M. Rösler, Positive convolution structure for a class of Heckman–Opdam hypergeometric functions of type BC. J. Funct. Anal. **258**, 2779–2800 (2010)
71. M. Rösler, M. Voit, Positivity of Dunkl's intertwining operator via the trigonometric setting. Int. Math. Res. Not. **63**, 3379–3389 (2004)
72. M. Rösler, M. Voit, A limit relation for Dunkl–Bessel functions of type A and B. Symmetry Integrability Geom. Methods Appl. **4**, 9 pp. (2008). Paper 083
73. M. Rösler, M. Voit, Integral representation and uniform limits for some Heckman–Opdam hypergeometric functions of type BC. Trans. Am. Math. **368**, 6005–6032 (2016)
74. M. Rösler, T.H. Koornwinder, M. Voit, Limit transition between hypergeometric functions of type BC and type A. Compos. Math. **149**, 1381–1400 (2013)

75. F. Rouvière, Espaces de Damek-Ricci, géométrie et analyse, in Analyse sur les groupes de Lie et théorie des représentations (Kénitra, 1999), pp. 45–100. Sémin. Congr. 7, Soc. Math. France, Paris, 2003
76. P. Sawyer, Spherical functions on symmetric cones. Trans. Am. Math. Soc. **349**(9), 3569–3584 (1997)
77. P. Sawyer, A Laplace–type representation of the generalized spherical functions associated to the root systems of type A (2017). Preprint [arXiv:1608.03567]
78. B. Schapira, Etude analytique et probabiliste de laplaciens associés à des systèmes de racines (laplacien hypergéométrique de Heckman–Opdam et laplacien combinatoire sur les immeubles affines). Ph.D. Thesis, Université d'Orléans (2006). https://tel.archives-ouvertes.fr/tel-00115557
79. B. Schapira, The Heckman–Opdam Markov processes. Probab. Theory Relat. Fields **138**(3–4), 495–519 (2007)
80. B. Schapira, Contributions to the hypergeometric function theory of Heckman and Opdam (sharp estimates, Schwartz space, heat kernel). Geom. Funct. Anal. **18**(1), 222–250 (2008)
81. B. Schapira, Bounded harmonic functions for the Heckman–Opdam Laplacian. Int. Math. Res. Not. **17**, 3149–3159 (2009)
82. A.G. Setti, L^p and operator norm estimates for the complex time heat operator on homogeneous trees. Trans. Am. Math. Soc. **350**(2), 743–768 (1998)
83. Y. Sun, A new integral formula for Heckman-Opdam hypergeometric functions. Adv. Math. **289**, 1157–1204 (2016)
84. S. Thangavelu, Y. Xu, Convolution operator and maximal function for the Dunkl transform. J. Anal. Math. **97**, 25–55 (2005)
85. K. Trimèche, The trigonometric Dunkl intertwining operator and its dual associated with the Cherednik operators and the Heckman–Opdam theory. Adv. Pure Appl. Math. **1**(3), 293–323 (2010)
86. K. Trimèche, Hypergeometric convolution structure on L^p-spaces and applications for the Heckman-Opdam theory. Preprint (2012)
87. K. Trimèche, Positivity of the transmutation operators associated with a Cherednik type operator on the real line. Adv. Pure Appl. Math. **3**, 361–376 (2012)
88. K. Trimèche, Positivity of the transmutation operators and absolute continuity of their representing measures for a root system on $\mathbb{R}^d$. Int. J. Appl. Math. **28**(4), 427–453 (2015)
89. K. Trimèche, The positivity of the transmutation operators associated with the Cherednik operators attached to the root system of type A_2. Adv. Pure Appl. Math. **6**(2), 125–134 (2015)
90. K. Trimèche, The positivity of the transmutation operators associated to the Cherednik operators for the root system BC_2. Math. J. Okayama Univ. **58**, 183–198 (2016)
91. B. Trojan, Heat kernel and Green function estimates on affine buildings (2017). Preprint [arXiv:1310.2288]
92. M. Voit, Product formulas for a two-parameter family of Heckman–Opdam hypergeometric functions of type BC. J. Lie Theory **25**(1), 9–36 (2015)

Holonomic Systems

Yoshishige Haraoka

To the memory of my mother Reiko and my father Kenkichi

Abstract New understanding of Fuchsian ordinary differential equations due to Katz and Oshima is presented. We extend it to regular holonomic systems, and proceed to global analysis by using the extended notions. Problem of constructing regular holonomic systems is also discussed.

Keywords Middle convolution • Monodromy • Regular holonomic system • Rigid local system

Mathematics Subject Classification (2000). Primary 34M35; Secondary 35N99

1 Introduction

Holonomic systems are natural extensions of ordinary differential equations, and appear in physics, representation theory, theory of automorphic functions, and so on. The deformation theory of linear differential equations is in a sense a study of holonomic systems. Thus holonomic systems concern various branches of physics and mathematics, and then seem to be a substantial object. There are many similarities between holonomic systems and ordinary differential equations, and also several differences. Recently, in the theory of Fuchsian ordinary differential equations, a big progress is caused by Katz [12] and Oshima [14]. These results can be applied to the study of holonomic systems, and will bring a new development.

In this lecture, we first explain the new understanding of Fuchsian ordinary differential equations given by Katz and Oshima. Fuchsian ordinary differential equations are classified by using the spectral types, and in each class equations are connected by two operations—addition and middle convolution. Analytic properties of solutions are transmitted by these operations. Next we apply these results to

Y. Haraoka (✉)
Department of Mathematics, Kumamoto University, Kumamoto 860-8555, Japan
e-mail: haraoka@kumamoto-u.ac.jp

G. Filipuk et al. (eds.), *Analytic, Algebraic and Geometric Aspects of Differential Equations*, Trends in Mathematics, DOI 10.1007/978-3-319-52842-7_2

the study of regular holonomic systems. We can define the spectral type and the middle convolution in this case similarly as in ODE case. These notions will become powerful tools. We use them to construct regular holonomic systems. We also focus on the difference between Fuchsian ordinary differential equations and regular holonomic systems. For the global analysis of regular holonomic systems, geometry of the singular locus plays a decisive role. We explain how to compute the fundamental group of the complement of the singular locus, and how the topology determines the analytic nature of solutions.

2 Basics of the Theory of Linear ODE in the Complex Domain

In this section, we briefly review basic results for linear ordinary differential equations in the complex domain bearing the extension to holonomic case in mind.

Let D be a domain in $\mathbb{C}$ (or in $\mathbb{P}^1 = \mathbb{C} \cup \{\infty\}$). We consider a system of differential equations

$$\frac{dY}{dx} = A(x)Y \tag{1}$$

of the first order, where $Y = {}^t(y_1, y_2, \ldots, y_n)$ is a vector of unknowns and $A = (a_{jk}(x))$ an $n \times n$-matrix with entries $a_{jk}(x)$ holomorphic in D. Let a be any point in D.

The following is the most fundamental theorem.

Theorem 2.1 *For any vector $Y_0 \in \mathbb{C}^n$, there exists a unique solution $Y(x)$ of (1) satisfying the initial condition*

$$Y(a) = Y_0,$$

and $Y(x)$ is holomorphic in any disc centered at a contained in D.

To prove the theorem, we first show that there exists a unique formal solution at a, and then prove the convergence of the formal solution by using a majorant series method.

The following is also a fundamental result, which is almost a direct consequence of the above theorem.

Theorem 2.2 *Let $Y(x)$ be the solution in the previous theorem. For any curve γ in D with starting point a, the solution $Y(x)$ can be analytically continued along γ.*

We roughly sketch the proof. We can take a chain of discs each of which is contained in D with center on the curve γ, and which cover γ. We may assume that each disc contains the center of the next disc. Let $B(a; r)$ be the first disc, where the solution $Y(x)$ converges. The center a_1 of the next disc $B(a_1; r_1)$ is contained in $B(a;r)$. Thanks to Theorem 2.1, there exists a unique solution $Y_1(x)$ in $B(a_1;r_1)$ satisfying the initial condition $Y_1(a_1) = Y(a_1)$. By the uniqueness of the solution at $x = a_1$, we have $Y_1(x) = Y(x)$ on a disc contained in $B(a;r) \cap B(a_1;r_1)$ centered at a_1, which implies that $Y(x)$ can be continued analytically to the next disc $B(a_1;r_1)$. Continuing this process, we come to the conclusion.

Theorem 2.2 implies that any solution of (1) is defined on the universal covering $\tilde{D}$ of D. In other words, any solution of (1) is a multi-valued function on D. Let U be a simply connected subdomain of D or the universal covering $\tilde{D}$. By the above theorems, we can show that the set of solutions on U makes a linear space of dimension n. A basis of the linear space is called a *fundamental system of solutions*, and the matrix consisting of a basis is called a *fundamental matrix solution* (FMS in short).

Exercise 1 Show that, if a tuple of solutions of (1) is linearly independent at a point, then it is linearly independent at any point of a common domain of definition.

Regular Singularity

Definition Let $f(x)$ be a (multi-valued) holomorphic function in $0 < |x-a| < r$ which is not holomorphic at $x = a$. $f(x)$ is said to be *regular singular* at $x = a$ if there exists a positive constant N such that, for any $\theta_1 < \theta_2$,

$$|x-a|^N |f(x)| \to 0 \quad (x \to a,\ \theta_1 < \arg(x-a) < \theta_2)$$

holds.

For example, $f(x) = \log(x-a)$ and $f(x) = (x-a)^\lambda$ with $\lambda \in \mathbb{C} \setminus \mathbb{Z}_{\geq 0}$ are regular singular at $x = a$.

Lemma 2.3 *If $f(x)$ is regular singular at $x = a$ and single valued in a neighborhood of $x = a$, then $x = a$ is a pole of $f(x)$.*

Exercise 2 Show this lemma.

Now we consider Eq. (1), where $A(x)$ is meromorphic in some domain $\hat{D}$. A singular point of $A(x)$ is called a singular point of Eq. (1).

Definition A singular point $x = a$ of Eq. (1) is called a *regular singular point* if any solution of (1) is at most regular singular at $x = a$.

For a scalar higher order differential equation, there is a simple criterion for regular singularity (Fuchs's theorem). We have only to look at the order of the poles of the coefficients. There is no such simple criterion for the system (1) of the first order, however, if $x = a$ is a simple pole of $A(x)$, then $x = a$ is a regular singular point of Eq. (1).

Let $x = a$ be a simple pole of $A(x)$, and

$$A(x) = \frac{A_{-1}}{x-a} + \sum_{m=0}^{\infty} A_m (x-a)^m$$

the Laurent expansion. A square matrix is said to be *non-resonant* if difference of any two distinct eigenvalues is not an integer.

Theorem 2.4 *Assume that the residue matrix A_{-1} is non-resonant. Then there exists a unique fundamental matrix solution of (1) of the form*

$$\mathcal{Y}(x) = F(x)(x-a)^{A_{-1}},$$

where

$$F(x) = I + \sum_{m=1}^{\infty} F_m (x-a)^m$$

is a convergent series.

Corollary 2.5 *Under the same assumption as Theorem 2.4, there exists a fundamental matrix solution of (1) of the form*

$$\tilde{\mathcal{Y}}(x) = \tilde{F}(x)(x-a)^J,$$

where J is the Jordan canonical form of A_{-1},

$$\tilde{F}(x) = P + \sum_{m=1}^{\infty} \tilde{F}_m (x-a)^m$$

a convergent series, and P a matrix satisfying $P^{-1}A_{-1}P = J$. The coefficients $\tilde{F}_m$ are uniquely determined by P.

To prove Theorem 2.4, we show that F_m's are uniquely determined, and then that the series $F(x)$ is convergent. We can do this in a similar manner as the proof of Theorem 2.1. The series $\tilde{F}(x)$ in Corollary 2.5 is obtained by $\tilde{F}(x) = F(x)P$.

Monodromy Representations

Let b be any point in D, and $\mathcal{Y}(x)$ a fundamental matrix solution of (1) on a simply connected neighborhood of b in D. For any loop γ in D with base point b, there exists $M_\gamma \in \mathrm{GL}(n, \mathbb{C})$ such that

$$\gamma_* \mathcal{Y}(x) = \mathcal{Y}(x) M_\gamma$$

holds, where γ_* denotes the analytic continuation along γ. It is shown that M_γ depends only on the homotopy class $[\gamma] \in \pi_1(D, b)$. Then we have a map

$$\begin{array}{rccc} \rho : & \pi_1(D, b) & \to & \mathrm{GL}(n, \mathbb{C}) \\ & \gamma & \mapsto & M_\gamma \end{array}$$

where we identified γ with $[\gamma]$. It is easily shown that ρ is an anti-homomorphism. We call ρ the *monodromy representation* of (1) with respect to $\mathcal{Y}(x)$. The image of ρ is said to be a *monodromy group*.

Two anti-homomorphisms

$$\rho, \rho' : \pi_1(D, b) \to \mathrm{GL}(n, \mathbb{C})$$

are said to be equivalent, if there exists $C \in \mathrm{GL}(n, \mathbb{C})$ such that $\rho(\gamma) = C\rho'(\gamma)C^{-1}$ holds for any $\gamma \in \pi_1(D, b)$. It is easy to show that, if we take another FMS, then we get an equivalent monodromy representation.

Now we consider the domain

$$D = \mathbb{P}^1 \setminus \{a_0, a_1, \ldots, a_p\},$$

where $a_0, a_1, \ldots, a_p$ are distinct points in $\mathbb{P}^1$. Take $b \in D$. For each a_j, we take a circle K with center a_j such that the other a_k's are in the outside of K. We choose a point c on K as a base point, and give K a positive direction. Take any path L in D from b to c. We call the loop LKL^{-1} or the element in $\pi_1(D, b)$ represented by LKL^{-1} a $(+1)$-*loop* (or *monodromy*) for a_j. By the definition, we see that any two $(+1)$-loops for a_j are conjugate in $\pi_1(D, b)$. Then, for any $(+1)$-loop γ for a_j, the conjugacy class $[\rho(\gamma)]$ is uniquely determined by a_j. We call the conjugacy class $[\rho(\gamma)]$ the *local monodromy* at a_j.

3 Fuchsian Ordinary Differential Equations on $\mathbb{P}^1$

We consider the case where $A(x)$ is a rational function in x, and let $\{a_0, a_1, \ldots, a_p\}$ be the set of singular points of (1). In this case, Eq. (1) is said to be *Fuchsian* if any singular point a_j is a regular singular point. Without loss of generality we may assume that $a_0 = \infty$.

A typical example of Fuchsian differential equation is given by

$$\frac{dY}{dx} = \left(\sum_{j=1}^{p} \frac{A_j}{x - a_j} \right) Y, \tag{2}$$

where $A_1, A_2, \ldots, A_p$ are constant $n \times n$-matrices. We set

$$A_0 = -\sum_{j=1}^{p} A_j.$$

In this note we call (2) a *Fuchsian system*. A Fuchsian system (2) is called non-resonant if every A_j $(0 \le j \le p)$ is non-resonant, and semi-simple if every A_j is semi-simple.

Theorem 3.1 *We consider two Fuchsian equations (1) and*

$$\frac{dZ}{dx} = B(x)Z \tag{3}$$

of the same size and with the same set of singular points. Let ρ and ρ' be monodromy representations of (1) and (3), respectively. Two monodromy representations ρ and ρ' are equivalent if and only if there exists $P(x) \in \mathrm{GL}(n, \mathbb{C}(x))$ such that the gauge transformation $Y = P(x)Z$ sends (1) to (3).

Proof Suppose that ρ and ρ' are equivalent. We can choose FMSs $\mathcal{Y}(x)$ and $\mathcal{Z}(x)$ of (1) and (3), respectively, so that $\rho = \rho'$. Then, if we define $P(x) = \mathcal{Y}(x)\mathcal{Z}(x)^{-1}$, it is single valued, and the entries of $P(x)$ has at most regular singularity. By Lemma 2.3, we have $P(x) \in \mathrm{M}(n, \mathbb{C}(x))$, and hence $P(x) \in \mathrm{GL}(n, \mathbb{C}(x))$.

The converse assertion is easy. □

The assertion of this theorem is a fundamental piece of the equivalence of categories between Fuchsian ordinary differential equations and monodromy representations. There are many other relations between Fuchsian equations and monodromy representations.

Theorem 3.2 *The monodromy representation ρ of (1) is reducible if and only if there exists $P(x) \in \mathrm{GL}(n, \mathbb{C}(x))$ such that the gauge transformation $Y = P(x)Z$ sends (1) to (3) with a coefficient of block triangular form*

$$B(x) = \left(\begin{array}{c|c} * & * \\ \hline O & * \end{array} \right).$$

Exercise 3 Prove Theorem 3.2.

Theorem 3.3 *The monodromy group of (1) is finite if and only if any solution of (1) is an algebraic function.*

To prove the theorem, we note that the number of the values of any entry of any solution at any point is at most the order of the monodromy group. We can prove the assertion by considering elementary symmetric polynomials of the branches.

These theorems show that it substantial to study monodromy representations for the study of Fuchsian ordinary differential equations.

Rigidity of Monodromy and Spectral Types

Note that we are considering the domain $D = \mathbb{P}^1 \setminus \{a_0, a_1, \ldots, a_p\}$ and $b \in D$. Then we have a presentation of the fundamental group given by

$$\pi_1(D, b) = \langle \gamma_0, \gamma_1, \ldots, \gamma_p \mid \gamma_0\gamma_1 \cdots \gamma_p = 1 \rangle, \tag{4}$$

where each γ_j is a $(+1)$-loop for a_j. For $0 \le j \le p$, we set

$$\rho(\gamma_j) = M_j.$$

By the presentation, ρ is determined by the tuple $(M_0, M_1, \ldots, M_p)$. Then we denote $\rho = (M_0, M_1, \ldots, M_p)$. Also by the presentation, we have

$$\begin{cases} M_p \cdots M_1 M_0 = I, \\ [M_j] \text{ is the local monodromy at } x = a_j \ (0 \le j \le p). \end{cases} \tag{5}$$

Note that, if we can apply Theorem 2.4 at a regular singular point $x = a_j$, we have

$$[M_j] = [e^{2\pi i A_{-1}}],$$

where A_{-1} is the residue matrix of $A(x)$ at $x = a_j$. Even for general case, the local monodromy $[M_j]$ can be explicitly obtained from the differential equation (1).

Question Are the conditions (5) enough to determine the equivalence class $[\rho]$?

To consider this Question, we now look at several examples. We call ρ semi-simple if all local monodromies are semi-simple.

Example

(i) $n = 2, p = 2$, ρ: irreducible and semi-simple.
We can set

$$\rho = (A, B, C)$$

with

$$\begin{cases} CBA = I, \\ A \sim \begin{pmatrix} \alpha_1 & \\ & \alpha_2 \end{pmatrix}, \ B \sim \begin{pmatrix} \beta_1 & \\ & \beta_2 \end{pmatrix}, \ C \sim \begin{pmatrix} \gamma_1 & \\ & \gamma_2 \end{pmatrix}, \end{cases}$$

where $\alpha_1\alpha_2\beta_1\beta_2\gamma_1\gamma_2 = 1$ and $\alpha_1 \neq \alpha_2, \beta_1 \neq \beta_2, \gamma_1 \neq \gamma_2$. The last three conditions come from the irreducibility. Since $C = (BA)^{-1}$, it is enough to determine the equivalence class $[A, B]$. We have

$$[A, B] = \left[\begin{pmatrix} \alpha_1 & \\ & \alpha_2 \end{pmatrix}, \begin{pmatrix} a & b \\ c & d \end{pmatrix} \right].$$

By the irreducibility, we see $bc \neq 0$. Then we have

$$\left[\begin{pmatrix} \alpha_1 & \\ & \alpha_2 \end{pmatrix}, \begin{pmatrix} a & b \\ c & d \end{pmatrix} \right] = \left[\begin{pmatrix} \alpha_1 & \\ & \alpha_2 \end{pmatrix}, \begin{pmatrix} a & b' \\ 1 & d \end{pmatrix} \right],$$

where we set $b' = bc$. From

$$\begin{pmatrix} a & b' \\ 1 & d \end{pmatrix} \sim \begin{pmatrix} \beta_1 & \\ & \beta_2 \end{pmatrix}$$

we obtain

$$\begin{cases} a + d = \beta_1 + \beta_2, \\ ad - b' = \beta_1\beta_2. \end{cases}$$

From

$$BA = C^{-1} \sim \begin{pmatrix} \gamma_1^{-1} & \\ & \gamma_2^{-1} \end{pmatrix}$$

we obtain

$$\alpha_1 a + \alpha_2 d = \gamma_1^{-1} + \gamma_2^{-1}.$$

These three equations determine a, b', d uniquely. Thus the equivalence class $[\rho]$ is uniquely determined.

(ii) $n = 2, p > 2$, ρ: irreducible and semi-simple.
We set $\rho = (A, B, C_1, C_2, \dots)$. Similarly to (i), we may set

$$[A, B, C_1, C_2, \dots] = \left[\begin{pmatrix} \alpha_1 & \\ & \alpha_2 \end{pmatrix}, \begin{pmatrix} a & b' \\ 1 & d \end{pmatrix}, \begin{pmatrix} x_1 & y_1 \\ z_1 & w_1 \end{pmatrix}, \begin{pmatrix} x_2 & y_2 \\ z_2 & w_2 \end{pmatrix}, \dots \right],$$

where we have

$$\begin{pmatrix} x_j & y_j \\ z_j & w_j \end{pmatrix} \sim \begin{pmatrix} \gamma_{j1} & \\ & \gamma_{j2} \end{pmatrix}$$

for $j = 1, 2, \ldots$. Compared with the first case, we have $4(p-2)$ more unknowns, while the number of conditions increases by $2(p-2)$, which implies that the solution contains $2(p-2)$ parameters. Thus the equivalence class $[\rho]$ is not unique.

(iii) $n = 3, p = 2$, ρ: irreducible and semi-simple. For $\rho = (A, B, C)$ with $CBA = I$, we assume

$$A \sim \begin{pmatrix} \alpha & & \\ & \alpha & \\ & & \alpha' \end{pmatrix}$$

with $\alpha \neq \alpha'$ and B, C do not have multiple eigenvalues. We may set

$$[A, B] = \left[A, \begin{pmatrix} \beta_1 & & \\ & \beta_2 & \\ & & \beta_3 \end{pmatrix} \right].$$

Since $\operatorname{rank}(A - \alpha I) = 1$, we have

$$A = \alpha I + \begin{pmatrix} x \\ y \\ z \end{pmatrix} \begin{pmatrix} 1 & p & q \end{pmatrix},$$

and we may send (p, q) to $(1, 1)$ by a similar transformation which leaves diagonal matrices invariant. Now calculating the trace of A, we get

$$3\alpha + x + y + z = 2\alpha + \alpha'.$$

Then we can reduce the unknowns to y and z, while we have two conditions coming from the eigenvalues of $AB = C^{-1}$. Thus $[\rho]$ is uniquely determined in this case.

(iv) $n = 3, p = 2$, ρ: irreducible and semi-simple, and we assume that the local monodromies do not have multiple eigenvalues. After a similar normalization, A depends on four parameters, while we have still two conditions. Then in this case $[\rho]$ is not unique.

Looking at these examples, we notice that the multiplicities of the eigenvalues of the local monodromies play a substantial role. Then we define the following notion.

For a semi-simple matrix A, we define the spectral type by the multiplicities of the eigenvalues. Namely, if

$$A \sim \mathrm{diag}[\overbrace{\lambda_1, \dots, \lambda_1}^{m_1}, \overbrace{\lambda_2, \dots, \lambda_2}^{m_2}, \dots, \overbrace{\lambda_q, \dots, \lambda_q}^{m_q}]$$

with $\lambda_j \neq \lambda_k$ $(j \neq k)$, we set

$$A^\natural = (m_1, m_2, \dots .m_q)$$

and call it the *spectral type* of A. In general, we denote by $e_k(A, \lambda)$ the number of the Jordan cells in the Jordan canonical form of A of eigenvalue λ and of size $\geq k$. Let $\lambda_1, \lambda_2, \dots$ be the eigenvalues of A. Then the *spectral type* of A is defined by

$$A^\natural = ((e_1(A, \lambda_1), e_2(A, \lambda_1), \dots), (e_1(A, \lambda_2), e_2(A, \lambda_2), \dots), \dots).$$

For example, if

$$A \sim \begin{pmatrix} \lambda & 1 & & & & \\ & \lambda & 1 & & & \\ & & \lambda & & & \\ & & & \lambda & & \\ & & & & \mu & 1 \\ & & & & & \mu \end{pmatrix},$$

we have

$$A^\natural = ((211), (11)).$$

For a monodromy representation $\rho = (M_0, M_1, \dots, M_p)$, we define its *spectral type* by

$$\rho^\natural = (M_0^\natural, M_1^\natural, \dots, M_p^\natural).$$

Now we return to the Question. We have seen that the equivalence class $[\rho]$ is sometimes uniquely determined by the local monodromies and sometimes is not.

Definition A monodromy representation ρ is called *rigid* if its equivalence class $[\rho]$ is uniquely determined by the local monodromies.

There is a very simple criterion for the rigidity.

Definition For a monodromy representation $\rho = (M_0, M_1, \dots, M_p)$ of rank n, we define the *index of rigidity* ι by

$$\iota = (1-p)n^2 + \sum_{j=0}^{p} \dim Z(M_j),$$

where $Z(M)$ denotes the centralizer of M.

Theorem 3.4 (Katz [12]) *If ρ is irreducible, $\iota \le 2$ holds. In this case, ρ is rigid if and only if $\iota = 2$.*

Thus to know the rigidity, we have only to compute the index of rigidity. The dimension of the centralizer can be calculated as follows.

Proposition 3.5 *Let the spectral type of M be given by*

$$M^\natural = ((e_1^{(1)}, e_2^{(1)}, \dots), (e_1^{(2)}, e_2^{(2)}, \dots), \dots).$$

Then we have

$$\dim Z(M) = \sum_j \sum_k \left(e_k^{(j)}\right)^2.$$

Exercise 4 Prove Proposition 3.5.

Exercise 5 Show that the index of rigidity is an even number.

Hence, the rigidity is determined by the rank n, the number of the singular points $p+1$ and the spectral type $\rho^\natural$. We have a natural question. Does there exist a rigid monodromy representation for any $(n, p, \rho^\natural)$ with $\iota = 2$?

We give a table of $(n, \rho^\natural)$ with $\iota = 2$ for $n \le 4$:

$$\begin{aligned}
n = 2:\quad & ((11), (11), (11)),\\
n = 3:\quad & ((21), (111), (111)), ((21), (21), (21), (21)),\\
n = 4:\quad & ((31), (1^4), (1^4)), ((22), (211), (1^4)), ((211), (211), (211)),\\
& ((31), (31), (22), (211)), ((31), (22), (22), (22)),\\
& ((31), (31), (31), (1^4)), ((31), (31), (31), (31), (31)).
\end{aligned}$$

We shall see later that, in the above table, there is a spectral type which cannot be realized by any monodromy representation.

Notational Remark In the above table, each inner parenthesis in each $\rho^\natural$ represents a spectral type of a local monodromy, and in particular it represents a semi-simple one. However, thanks to Proposition 3.5, the index of rigidity does not change if we put parentheses further in the inner parenthesis to denote a non semi-simple spectral type: e.g. $(111), ((11)1)$ and $((111))$ give the same dimension of the centralizer. Then we can understand that the above table also gives non semi-simple spectral types.

As mentioned above, in the above table there is a spectral type which cannot be realized by any monodromy representation. Thus we notice another problem. It is Kostov [13] who formulated and solved the problem below.

Problem (Deligne-Simpson Problem, DSP for Short) For an irreducible representation ρ, the spectral type $\rho^\natural$ is called *irreducibly realizable*. DSP asks a numerical criterion for irreducibly realizable spectral types.

DSP is solved by Kostov and also by Crawley-Boevey [1]. We combine the solution by the latter with the operation middle convolution to get an algorithm for the irreducibly realizability. Now we explain the middle convolution.

Middle Convolution

Riemann-Liouville integral is defined by

$$(I_a^\lambda)(f)(x) = \frac{1}{\Gamma(\lambda)} \int_a^x f(t)(x-t)^{\lambda-1}\, dt,$$

where $\lambda \in \mathbb{C}$. It is regarded as a "complex time" differentiation, since we have

$$(I_a^{-n})(f)(x) = f^{(n)}(x)$$

for $n \in \mathbb{Z}_{\geq 0}$. (This equality can be shown by using the method of finite part of divergent integrals by Hadamard.)

We consider a Fuchsian system (2). Let $Y(x)$ be any solution of (2), and construct a differential equation satisfied by $(I_a^\lambda)(Y)(x)$. The obtained equation may be reducible. If so, take the irreducible part. It turns out that the resulting equation is again a Fuchsian system with the same set of singular points. We call this operation the *middle convolution* with parameter λ, and denote by mc_λ:

$$(2) \xrightarrow{mc_\lambda} \frac{dZ}{dx} = \left(\sum_{j=1}^{p} \frac{B_j}{x-a_j} \right) Z.$$

We also denote this correspondence by

$$mc_\lambda(A_0, A_1, \ldots, A_p) = (B_0, B_1, \ldots, B_p),$$

where $B_0 = -\sum_{j=1}^{p} B_j$. Explicit description of the middle convolution is given in [3].

Middle convolution mc_λ induces a transformation of the monodromies. For a monodromy representation ρ of the system (2), we get a monodromy representation ρ' of the system $mc_\lambda(2)$. We denote by MC_λ the operation $\rho \mapsto \rho'$.

Middle convolution possesses the following basic properties. We assume some generic condition on the system (2). Then we have

$$mc_0 = \text{id.},$$

$$mc_\lambda \circ mc_\mu = mc_{\lambda+\mu},$$

mc_λ keeps the index of rigidity,

mc_λ keeps the irreducibility,

mc_λ keeps the deformation equation.

(Refer to [3, 4, 8, 9, 12].) Corresponding properties also hold for MC_λ.

Now we introduce another operation. Let $\alpha = (\alpha_1, \alpha_2, \ldots, \alpha_p)$ be a point in $\mathbb{C}^p$. The operation

$$(A_1, A_2, \ldots, A_p) \mapsto (A_1 + \alpha_1 I, A_2 + \alpha_2 I, \ldots, A_p + \alpha_p I)$$

is called the *addition* with parameter α. This operation corresponds to the gauge transformation

$$Y(x) \mapsto \prod_{j=1}^{p} (x - a_j)^{\alpha_j} \cdot Y(x)$$

of the system (2).

The following is one of the main results in [12].

Theorem 3.6 (Katz [12]) *Any irreducible rigid Fuchsian system can be obtained from a differential equation of rank 1 by a finite iteration of additions and middle convolutions.*

A differential equation of rank 1 is given by

$$\frac{dy}{dx} = \left(\sum_{j=1}^{p} \frac{\alpha_j}{x - a_j} \right) y,$$

which is explicitly solved by

$$y(x) = \prod_{j=1}^{p} (x - a_j)^{\alpha_j}.$$

As a consequence of Theorem 3.6, many quantities for rigid equations can be computed by the iteration of additions and middle convolutions: Monodromy representations, connection coefficients, integral representations of solutions, power series expansions of solutions, irreducibility conditions, and so on [4, 10, 14].

We can describe the change of the spectral type by middle convolution. Assume that the Fuchsian system (2) is non-resonant. This is an essential assumption because, under this assumption, we have $M_j \sim e^{2\pi i A_j}$ and $M_j^\natural = A_j^\natural$. Moreover, for simplicity, we assume that the system (2) is semi-simple. Set

$$mc_\lambda(A_0, A_1, \ldots, A_p) = (B_0, B_1, \ldots, B_p).$$

For each j, we set

$$A_j^\natural = (m_1^{(j)}, m_2^{(j)}, \ldots, m_{q_j}^{(j)}).$$

Here $m_1^{(j)}$ denotes the multiplicity of the eigenvalue 0 (resp. λ) for $1 \le j \le p$ (resp. $j = 0$). Then, by this convention, $m_1^{(j)}$ can be 0. We set

$$g = \sum_{j=0}^{p} m_1^{(j)}.$$

Then, if $mc_\lambda(2)$ is also semi-simple, we obtain

$$B_j^\natural = \left(m_1^{(j)} - (g - (p-1)n), m_2^{(j)}, \ldots, m_{q_j}^{(j)}\right) \quad (0 \le j \le p),$$

where n denotes the rank of (2). Thus the rank of $mc_\lambda(2)$ becomes $pn - g$. We set $d = g - (p-1)n$.

Example

(i) $(A_0^\natural, A_1^\natural, A_2^\natural) = ((21), (111), (111))$. We have $g = 2 + 1 + 1 = 4$, and then $d = 4 - (2-1) \times 3 = 1$. Then the spectral type changes to

$$((11), (011), (011)) = ((11), (11), (11)).$$

(ii) $((31), (31), (31), (1111))$. We have $g = 10$ and $d = 2$. Then the spectral type changes to

$$((11), (11), (11), (-1\,111)),$$

which is impossible by the existence of -1. This implies that the spectral type $((31), (31), (31), (1111))$ is not realizable.

Algorithm for Irreducibly Realizable Spectral Types

We give an algorithm which gives a solution to DSP. Here we explain only in semi-simple case. For the general case, refer to [8].

We start from a spectral type

$$\left((m_1^{(0)}, m_2^{(0)}, \ldots), (m_1^{(1)}, m_2^{(1)}, \ldots), \ldots, (m_1^{(p)}, m_2^{(p)}, \ldots)\right)$$

with

$$\sum_k m_k^{(j)} = n \quad (0 \le j \le p).$$

Step 1. Calculate the index of rigidity ι.

$$\left[\begin{array}{lll} \iota > 2 & \to & \text{Not irreducibly realizable} \\ \iota \le 2 & \to & \left[\begin{array}{l} \iota = 0 \ \& \ \mathrm{g.c.d}_{j,k}\{m_k^{(j)}\} \ge 2 \to \ \text{Not irreducibly realizable} \\ \text{otherwise} \to \ \text{Go to Step 2} \end{array}\right. \end{array}\right.$$

Step 2. Assume that, for each j, $m_1^{(j)}$ is the maximum among $m_k^{(j)}$ $(k = 1, 2, \ldots)$.
Set

$$d = \sum_{j=0}^{p} m_1^{(j)} - (p-1)n.$$

$$\left[\begin{array}{lll} d \le 0 & \to & \text{Irreducibly realizable} \\ d > 0 & \to & \left[\begin{array}{l} m_1^{(j)} < d \ (\exists j) \to \ \text{Not irreducibly realizable} \\ m_1^{(j)} \ge d \ (\forall j) \to \ \text{Go to Step 3} \end{array}\right. \end{array}\right.$$

Step 3. Make a new spectral type

$$\left((m_1^{(0)} - d, m_2^{(0)}, \ldots), (m_1^{(1)} - d, m_2^{(1)}, \ldots), \ldots, (m_1^{(p)} - d, m_2^{(p)}, \ldots)\right)$$

with $n' = n - d$.

$$\left[\begin{array}{lll} n' = 1 & \to & \text{Irreducibly realizable \& Rigid} \\ n' > 1 & \to & \text{Go back to Step 2} \end{array}\right.$$

Thus the spectral type is irreducibly realizable if $d \leq 0$ in Step 2 or if $n' = 1$ in Step 3. You will see that the spectral type made in Step 3 is the result of middle convolution. If $n' = 1$, the corresponding equation is of rank 1, which evidently exists. In this case, our algorithm gives a proof of Theorem 3.6. Irreducibly realizability for the case $d \leq 0$ in Step 2 is a consequence of the result of Crawley-Boevey [1].

An irreducibly realizable spectral type is called *basic* if it is an output of this algorithm.

Theorem 3.7 (Oshima [14]) *For every value of ι (≤ 2), the number of the basic spectral types is finite.*

We give a list of basic spectral types for $\iota = 0$ and -2.

$\iota = 0$:

$$\begin{gathered}((11),(11),(11),(11)),\ ((111),(111),(111)),\\ ((22),(1^4),(1^4)),\ ((33),(222),(1^6))\end{gathered}$$

$\iota = -2$:

$$\begin{aligned}&((11),(11),(11),(11),(11)),\ ((21),(21),(111),(111)),\\ &((22),(22),(22),(211)),\ ((31),(22),(22),(1^4)),\\ &((211),(1^4),(1^4)),\ (32),(1^5),(1^5)),\\ &((221),(221),(1^5)),\ ((33),(2211),(1^6)),\\ &((222),(222),(2211)),\ ((44),(2^4),(2^3 11)),\\ &((44),(332),(1^8)),\ ((55),(3^3 1),(2^5)),\\ &((66),(4^3),(2^5 11))\end{aligned}$$

Since the above algorithm is based on the middle convolution (and addition), any irreducibly realizable spectral type is connected to a basic spectral type by a finite iteration of middle convolutions and additions. Then as in rigid case, many quantities of an irreducible Fuchsian system can be obtained from ones of the corresponding Fuchsian system with basic spectral type. Thus the study of Fuchsian differential equations is reduced to the study of basic ones.

4 Regular Holonomic Systems

In this section, we consider linear Pfaffian systems. A Pfaffian system is a useful normal form of holonomic systems. In the theory of holonomic systems, one usually consider the behavior of solutions in a regular domain together with the behavior at singular locus. In this lecture, we study only the behavior in a regular domain; namely, we consider only an "open part" of holonomic systems. For detailed proofs of the assertions in this section, please refer to [8]. For readers who are interested in purely algebraic descriptions, we recommend [2].

Let X be a domain in $\mathbb{C}^n$, and $x = (x_1, x_2, \ldots, x_n)$ a coordinate of $\mathbb{C}^n$. We consider a linear Pfaffian system

$$du = \left(\sum_{k=1}^{n} A_k(x)\, dx_k \right) u, \tag{6}$$

where $u = {}^t(u_1, u_2, \ldots, u_N)$ is a vector of unknowns and, for each k, $A_k(x)$ is an $N \times N$-matrix function holomorphic on X. Note that (6) is a collection of partial differential equations

$$\frac{\partial u}{\partial x_k} = A_k(x)u \quad (1 \leq k \leq n).$$

Definition *Integrability condition* (I.C.) means the system of differential equations

$$\frac{\partial A_k}{\partial x_l} + A_k A_l = \frac{\partial A_l}{\partial x_k} + A_l A_k \quad (1 \leq k, l \leq n).$$

Theorem 4.1 *Suppose that the Pfaffian system (6) satisfies the integrability condition. Then, for any $a \in X$ and $u_0 \in \mathbb{C}^N$, there exists a unique solution $u(x)$ satisfying*

$$u(a) = u_0.$$

$u(x)$ is holomorphic in any polydisc with center a which is contained in X.

Theorem 4.2 *Assume the integrability condition. Then any solution of (6) can be analytically continued along any curve in X.*

Theorems 4.1 and 4.2 can be shown in analogous ways as the proofs of Theorems 2.1 and 2.1, respectively. Thanks to Theorem 4.2, we can define the monodromy representation for (6) in a completely similar way as in one variable case.

Regular Singularity

Let U be a domain in $\mathbb{C}^n$, and $\varphi(x)$ a holomorphic function on U. We set

$$E = \{x \in U \mid \varphi(x) = 0\},$$
$$E^0 = \{x \in E \mid \operatorname{grad} \varphi(x) \neq 0\}.$$

We consider a Pfaffian system with logarithmic singularity along E:

$$du = (A(x)d\log\varphi(x) + \Omega_1)\,u,$$

where $A(x)$ is an $N\times N$-matrix function holomorphic on U and Ω_1 an $N\times N$-matrix 1-form holomorphic on U. We study this Pfaffian system locally in a neighborhood of a point $a \in E^0$. By a coordinate change, we can send E to $x_1 = 0$ locally and a to 0. Then we have

$$du = \left(\left(\frac{B_1(x')}{x_1} + C_1(x)\right)dx_1 + \sum_{k=2}^{n} A_k(x)\,dx_k\right)u, \tag{7}$$

where we set $x' = (x_2,\dots,x_n)$, and B_1, C_1, A_k are holomorphic at 0.

The following is a particular assertion of a basic fact which holds for regular holonomic systems in general.

Theorem 4.3 *Suppose that the Pfaffian system (7) satisfies the integrability condition. Then the Jordan canonical form of $B_1(x')$ does not depend on x'.*

Proof From the integrability condition, we obtain

$$\frac{\partial}{\partial x_k}\left(\frac{B_1}{x_1} + C_1\right) + \left(\frac{B_1}{x_1} + C_1\right)A_k = \frac{\partial A_k}{\partial x_1} + A_k\left(\frac{B_1}{x_1} + C_1\right)$$

for $k \geq 2$. We set $D_k(x') = A_k(0, x')$. Then comparing the coefficients of $1/x_1$ in both sides of the above equation, we get

$$\frac{\partial B_1}{\partial x_k} = [D_k, B_1] \quad (k \geq 2).$$

We also obtain

$$\frac{\partial D_k}{\partial x_l} + D_k D_l = \frac{\partial D_l}{\partial x_k} + D_l D_k \quad (2 \leq k, l \leq n)$$

from the integrability condition. The latter equalities are the integrability condition for the Pfaffian system

$$dp = \left(\sum_{k=2}^{n} D_k(x')\,dx_k\right)p. \tag{8}$$

Thanks to Theorem 4.1, we have a fundamental matrix solution $P(x')$ of the Pfaffian system. Now we have

$$\begin{aligned}\frac{\partial}{\partial x_k}(P^{-1}B_1P) &= -P^{-1}\frac{\partial P}{\partial x_k}P^{-1}B_1P + P^{-1}\frac{\partial B_1}{\partial x_k}P + P^{-1}B_1\frac{\partial P}{\partial x_k} \\ &= -P^{-1}D_kB_1P + P^{-1}[D_k, B_1]P + P^{-1}B_1D_kP \\ &= O.\end{aligned}$$

This completes the proof. □

Theorem 4.4 *Suppose that the Pfaffian system (7) satisfies the integrability condition, and that the Jordan canonical form J of $B_1(x')$ is non-resonant. Then we have a fundamental matrix solution*

$$\mathcal{U}(x) = F(x){x_1}^J,$$

where $F(x)$ is holomorphic and invertible at $x = 0$.

Proof We can take a fundamental matrix solution $P(x')$ of (8) such that $P^{-1}B_1P$ becomes the Jordan canonical form J. Look at the x_1-equation of the Pfaffian system (7):

$$\frac{\partial u}{\partial x_1} = \left(\frac{B_1(x')}{x_1} + C_1(x)\right)u.$$

Applying Corollary 2.5, we get a fundamental matrix solution

$$\mathcal{U}(x_1, x') = F(x_1, x'){x_1}^J,$$

where

$$F(x_1, x') = P(x') + \sum_{m=1}^{\infty} F_m(x'){x_1}^m.$$

F_m's are uniquely determined by P. We shall show that $\mathcal{U}(x_1, x')$ satisfies the Pfaffian system (7). Put this solution to the x_k-equation of (7):

$$\frac{\partial u}{\partial x_k} = A_k(x)u.$$

Then we have

$$\frac{\partial F}{\partial x_k}{x_1}^J = A_k(x)F{x_1}^J,$$

which induces

$$\frac{\partial F}{\partial x_k} = A_k(x)F. \tag{9}$$

To show Eq. (9), we consider the function

$$\varphi(x) = \frac{\partial F}{\partial x_k} {x_1}^J - A_k F {x_1}^J.$$

By the integrability condition, we see that φ satisfies the x_1-equation, and $\varphi(x){x_1}^{-J}$ takes 0 at $x = 0$. Then uniqueness of solution implies $\varphi = 0$. Thus $\mathcal{U}(x)$ becomes a solution of the Pfaffian system (7). □

Monodromy Representations

We proceed to the global analysis. Let $\varphi(x) = \varphi(x_1, x_2, \ldots, x_n)$ be a reduced polynomial, and

$$\varphi(x) = \prod_j \varphi_j(x)$$

its irreducible decomposition. We set

$$S = \{x \in \mathbb{C}^n \mid \varphi(x) = 0\}, \; S_j = \{x \in \mathbb{C}^n \mid \varphi_j(x) = 0\}.$$

We denote the hyperplane at infinity in $\mathbb{P}^n$ by H_∞, and set $\bar{S} = S \cup H_\infty$. As a domain of definition, we take

$$X = \mathbb{C}^n \setminus S = \mathbb{P}^n \setminus \bar{S}.$$

We consider a regular holonomic system singular along $\bar{S}$. Let

$$\rho : \pi_1(X, b) \to \mathrm{GL}(N, \mathbb{C})$$

be its monodromy representation. For each irreducible component S_j, take a point $p \in S_j$ which is a regular point of S (i.e. $\operatorname{grad}\varphi(p) \neq 0$). Take a complex line H passing through p in general position with respect to S. Let K be a $(+1)$-loop for p on H, and take a path L in X from b to the base point of K. We call the loop LKL^{-1} a $(+1)$-*loop* (or *monodromy*) for S_j. The following proposition is fundamental.

Proposition 4.5 *Take any irreducible component S_j. Any two $(+1)$-loops for S_j are conjugate in $\pi_1(X, b)$.*

Thanks to Proposition 4.5, for any $(+1)$-loop γ for S_j, $[\rho(\gamma)]$ is uniquely determined by S_j. We call it the *local monodromy* at S_j. Thus we notice that irreducible components S_j of S play a similar role as singular points in ordinary differential equations.

In the case of Fuchsian ordinary differential equations, we have fundamental conditions (5) for the generators of the monodromy. The first one comes from the presentation of the fundamental group. Then also in regular holonomic case, the condition coming from a presentation of the fundamental group will play a basic role. For the presentation, Zariski gave very effective ways. First, he showed that

$$\pi_1(\mathbb{P}^n \setminus \bar{S}) \simeq \pi_1(\mathbb{P}^{n-1} \setminus \bar{S}')$$

for $n > 2$, where $\mathbb{P}^{n-1}$ denotes a hyperplane in $\mathbb{P}^n$ in general position with respect to $\bar{S}$, and $\bar{S}' = \mathbb{P}^{n-1} \cap \bar{S}$. This result is called Zariski's hyperplane section theorem (cf. [6, 15]). Then we have only to study the case $n = 2$. Next he gave an effective way to get a presentation of $\pi_1(\mathbb{P}^2 \setminus \bar{S})$, which is called Zariski-van Kampen theorem (cf. [16]). Let S be a curve in $\mathbb{C}^2$ which is defined by a polynomial of total degree d. Take a line H which meets with S at d points. Then Zariski-van Kampen theorem asserts that

$$\pi_1(\mathbb{C}^2 \setminus S) = \langle \gamma_1, \dots, \gamma_d \mid \gamma_j^{g_k} = \gamma_j \ (1 \le j \le d,\ 1 \le k \le e) \rangle,$$

where $\gamma_1, \dots, \gamma_d$ are generators of $\pi_1(H \setminus (H \cap S))$. We shall not explain $\gamma_j^{g_k}$ here.

We give several examples of fundamental groups whose presentations are obtained by applying Zariski-van Kampen theorem.

Example

(i) $S = \{xy = 0\} \subset \mathbb{C}^2$.

$$\pi_1(\mathbb{C}^2 \setminus S) = \langle \gamma_1, \gamma_2 \mid \gamma_1\gamma_2 = \gamma_2\gamma_1 \rangle \simeq \mathbb{Z}^2.$$

(ii) $S = \{xy(x - y) = 0\} \subset \mathbb{C}^2$.

$$\pi_1(\mathbb{C}^2 \setminus S) = \langle \gamma_1, \gamma_2, \gamma_3 \mid \gamma_1\gamma_2\gamma_3 = \gamma_2\gamma_3\gamma_1 = \gamma_3\gamma_1\gamma_2 \rangle .$$

(iii) $S = \{x(x - 1)y(y - 1)(x - y) = 0\} \subset \mathbb{C}^2$.

$$\pi(\mathbb{C}^2 \setminus S) = \left\langle \gamma_1, \gamma_2, \gamma_3, \gamma_4, \gamma_5 \,\middle|\, \begin{array}{l} \gamma_1\gamma_2 = \gamma_2\gamma_1,\ \gamma_4\gamma_5 = \gamma_5\gamma_4, \\ \gamma_1\gamma_3\gamma_5 = \gamma_3\gamma_5\gamma_1 = \gamma_5\gamma_1\gamma_3, \\ \gamma_2\gamma_3\gamma_4 = \gamma_3\gamma_4\gamma_2 = \gamma_4\gamma_2\gamma_3 \end{array} \right\rangle.$$

We find that, compared with the one dimensional case (4), there appear various relations among the generators. This makes the global study of regular holonomic systems very different from ODE case, and also interesting.

As in ODE case, we can see global natures of solutions of regular holonomic systems by studying monodromy representations. The following theorem is fundamental.

Theorem 4.6 *Consider a holonomic system with logarithmic singularity along a hypersurface $\bar{S} \subset \mathbb{P}^n$. Let $f(x)$ be a rational function (i.e. a ratio of two polynomials with constant coefficients) in the entries of solutions of the holonomic system. If $f(x)$ is holomorphic and single valued on $\mathbb{P}^n \setminus \bar{S}$, it is a rational function in $x_1, x_2, \ldots, x_n$.*

To show the theorem, we first take a complex line which go through a regular point a of S. We apply Lemma 2.3 to show that a is at most a pole on the line. By varying the complex line, we see that $\varphi(x)^m f(x)$ is holomorphic in a neighborhood of a, where $\varphi(x)$ is a defining polynomial of S and m is some non-negative integer. Since the set of the singular points of S is of codimension at least 2, by Hartogs's theorem, we see that $\varphi(x)^m f(x)$ is globally holomorphic. Since H_∞ is also at most regular singular, it is at most a pole. Then $\varphi(x)^m f(x)$ is a polynomial, which implies the conclusion.

By using this theorem, we obtain several results. The first one is a similar assertion to ODE case.

Theorem 4.7 *Two regular holonomic systems of the same rank singular along a common hypersurface $\bar{S} \subset \mathbb{P}^n$ have isomorphic monodromy representations if and only if they can be transformed from one to the other by a gauge transformation with coefficients in rational functions in $\mathbb{C}^n$.*

The second one is a proper phenomenon for higher dimensional case.

Theorem 4.8 *If the fundamental group $\pi_1(\mathbb{P}^n \setminus \bar{S})$ is commutative, then, for any completely integrable linear Pfaffian system with logarithmic singularity along $\bar{S}$, all solutions are elementary.*

Proof For each j, let γ_j be a $(+1)$-loop for the irreducible component S_j. Since $\pi_1(\mathbb{P}^n \setminus \bar{S})$ is commutative, by the help of Zariski's hyperplane section theorem and Zariski-van Kampen theorem, we see that $\pi_1(\mathbb{P}^n \setminus \bar{S})$ is an abelian group generated by γ_j's. Take any FMS $\mathcal{U}(x)$, and consider the monodromy representation ρ with respect to $\mathcal{U}(x)$. Set $\rho(\gamma_j) = M_j$. Note that M_j's are commutative. We can take A_j such that $e^{2\pi i A_j} = M_j$ and that A_j's are commutative each other. Define $\Phi(x) = \prod_j \varphi_j(x)^{A_j}$. Then we see that $\mathcal{U}(x)\Phi(x)^{-1}$ is single valued, which implies that it is rational by Theorem 4.6. □

This assertion is first obtained by Gérard-Levelt [5]. Our proof can be regarded as a multiplicative version of their proof.

The rigidity of monodromy representations can be defined in a completely similar manner.

Definition A monodromy representation ρ is called *rigid* if its equivalence class $[\rho]$ is uniquely determined by the local monodromies. ρ is called *almost rigid* if there are only a finite number of equivalence classes which have the same local monodromies as ρ.

In general, there are many more relations among the generators of π_1 compared with one dimensional case. Then monodromy representations are prone to be (almost) rigid.

Example Appell's hypergeometric series F_1, F_2, F_3 and F_4 satisfy regular holonomic systems, and the holonomic systems can be transformed to linear Pfaffian systems with the same singular locus

$$\bar{S} = \{x_1(x_1 - 1)x_2(x_2 - 1)(x_1 - x_2) = 0\} \cup H_\infty.$$

We can show that the monodromy representations of these Pfaffian systems are almost rigid [11].

We explain F_1's case. We set $S_1 = \{x_1 = 0\}$, $S_2 = \{x_2 = 1\}$, $S_3 = \{x_1 = x_2\}$, $S_4 = \{x_1 = 1\}$ and $S_5 = \{x_2 = 0\}$. The Pfaffian system is of rank 3, and the spectral types of local monodromies at $S_1, S_2, S_3, S_4, S_5; H_\infty$ are $(21, 21, 21, 21, 21; 21)$. Let

$$\rho : \pi_1(\mathbb{P}^2 \setminus \bar{S}) \to \mathrm{GL}(3, \mathbb{C})$$

be an anti-homomorphism. We take the generators of the fundamental group in the previous example (iii), and set $\rho(\gamma_j) = M_j$ $(1 \le j \le 5)$. We assume that the spectral types of M_j are (21). Without loss of generality, we may assume

$$M_j \sim \begin{pmatrix} 1 & & \\ & 1 & \\ & & e_j \end{pmatrix}$$

with $e_j \neq 0, 1$ $(1 \le j \le 5)$. We look for an irreducible tuple $(M_1, M_2, \ldots, M_5)$ with the above Jordan canonical form satisfying $[M_1, M_2] = O$, $[M_4, M_5] = O$, $M_5M_3M_1 = M_3M_1M_5 = M_1M_5M_3$ and $M_4M_3M_2 = M_3M_2M_4 = M_2M_4M_3$. We find that such tuple exists only when $e_1e_2 = e_4e_5$ holds. Under this condition, there are two equivalence classes of irreducible ρ. Thus the monodromy representation of F_1 is almost rigid. Note that these two equivalence classes have distinct local monodromy at $x = \infty$ and $y = \infty$ in the compactification $\mathbb{P}^1 \times \mathbb{P}^1$ of $\mathbb{C}^2$. Namely, the monodromy representation of F_1 is rigid if we consider it in $\mathbb{P}^1 \times \mathbb{P}^1$.

For the details of this argument and for the results in the case of F_2, F_3, F_4, please refer to [11].

Middle Convolution

Middle convolution plays a substantial role in the Katz theory for Fuchsian ordinary differential equations. We expect it also important in the study of regular holonomic systems. In this section, we try to extend the definition of middle convolution to a class of Pfaffian systems with logarithmic singularities along hyperplane arrangements, and apply it to the study of constructing regular holonomic systems.

Let

$$H_j = \{x \in \mathbb{C} \mid h_j(x) = 0\} \quad (1 \leq j \leq g)$$

be hyperplanes in $\mathbb{C}^n$, where $h_j(x)$ is a polynomial of degree 1. Let $A_1, A_2, \ldots, A_g$ be constant $N \times N$-matrices. We consider a Pfaffian system

$$du = \left(\sum_{j=1}^{g} A_j d \log h_j \right) u. \tag{10}$$

We assume that the system (10) is completely integrable. This condition can be explicitly described [7].

We define the middle convolution of (10) in x_k-direction. Without loss of generality, we take $k = 1$. We set $J_1 = \{j \mid \partial h_j / \partial x_1 \neq 0\}$, and for $j \in J_1$, we write $h_j(x) = c_j(x_1 - a_j)$ with $c_j \in \mathbb{C}^{\times}$ and $a_j \in \mathbb{C}[x_2, \ldots, x_n]$. Then the x_1-equation of (10) is

$$\frac{\partial u}{\partial x_1} = \left(\sum_{j \in J_1} \frac{A_j}{x_1 - a_j} \right) u. \tag{11}$$

We take a middle convolution of (11) with parameter λ as an ODE, and denote it by $mc_{\lambda}(11)$.

Theorem 4.9 *The differential equation $mc_{\lambda}(11)$ can be prolonged to a Pfaffian system in $x_1, x_2, \ldots, x_n$, which is completely integrable.*

We call the prolonged Pfaffian system the middle convolution of (10) with parameter λ in x_1-direction, and denote it by $mc_{\lambda}^{x_1}(10)$. As we have explained in section "Middle Convolution" in section 3, solutions of the equation $mc_{\lambda}(11)$ can be expressed by Riemann-Liouville integrals of solutions of (11), which are functions in $x_1, x_2, \ldots, x_n$ and satisfy a completely integrable system. In this way, we can prove Theorem 4.9. For the detailed proof, please refer to [7]. In the same paper, we show the following fundamental properties of the middle convolution:

$$mc_0^{x_k} = \text{id.},$$

$$mc_{\lambda}^{x_k} \circ mc_{\mu}^{x_k} = mc_{\lambda+\mu}^{x_k},$$

$mc_{\lambda}^{x_k}$ keeps the irreducibility.

Example Let $A_1, A_2, \ldots, A_5$ be constant $N \times N$-matrices. We consider a completely integrable Pfaffian system

$$du = \left(A_1 \frac{dx}{x} + A_2 \frac{dy}{y-1} + A_3 \frac{d(x-y)}{x-y} + A_4 \frac{dx}{x-1} + A_5 \frac{dy}{y}\right) u.$$

Take $\lambda \in \mathbb{C}$. By making a Riemann-Liouville integral of a solution of the x-equation of this system with parameter λ, we find that the integral satisfies the Pfaffian system

$$dv = \left(B_1 \frac{dx}{x} + B_2 \frac{dy}{y-1} + B_3 \frac{d(x-y)}{x-y} + B_4 \frac{dx}{x-1} + B_5 \frac{dy}{y}\right) v$$

of rank $3N$, where $B_1, B_2, \ldots, B_5$ are defined by

$$B_1 = \begin{pmatrix} A_1 + \lambda I & A_3 & A_4 \\ O & O & O \\ O & O & O \end{pmatrix}, \quad B_3 = \begin{pmatrix} O & O & O \\ A_1 & A_3 + \lambda I & A_4 \\ O & O & O \end{pmatrix},$$

$$B_4 = \begin{pmatrix} O & O & O \\ O & O & O \\ A_1 & A_3 & A_4 + \lambda I \end{pmatrix},$$

$$B_2 = \begin{pmatrix} A_1 & O & O \\ O & A_2 + A_4 & -A_4 \\ O & -A_3 & A_2 + A_3 \end{pmatrix}, \quad B_5 = \begin{pmatrix} A_3 + A_5 & -A_3 & O \\ -A_1 & A_1 + A_5 & O \\ O & O & A_5 \end{pmatrix}.$$

It is shown that this system is also completely integrable. The system may be reducible, and in such case, we can get the maximal invariant subspace W with respect to the action of $B_1, B_2, \ldots, B_5$. Explicitly, we have $W = \mathcal{K} + \mathcal{L}$, where

$$\mathcal{K} = \{{}^t(v_1, v_3, v_4) \mid v_j \in \operatorname{Ker} A_j \ (j = 1, 3, 4)\},$$
$$\mathcal{L} = \operatorname{Ker}(B_1 + B_3 + B_4).$$

By taking the action on the quotient space $\mathbb{C}^{3N}/W$, we get an irreducible completely integrable Pfaffian system, which is the middle convolution with parameter λ in x-direction.

By combining Theorem 3.6 and middle convolution for Pfaffian systems, we have a way of constructing completely integrable Pfaffian systems from rigid Fuchsian ordinary differential equations.

Let (E) be a rigid Fuchsian ordinary differential equation with $(m+1)$ singular points ($m \geq 3$). We may assume that three singular points are normalized to $0, 1, \infty$, and denote the singular point by $y_1 = 0, y_2 = 1, y_3, \ldots, y_m, y_{m+1} = \infty$. Thanks to Theorem 3.6, we have a chain of additions and middle convolutions which connects (E) to a differential equation (F) of rank 1. Set $y_0 = x$, where x denotes the independent variable of the equation (E). We have a Pfaffian system

$$du = \left(\sum_{0\le j<k\le m} a_{jk} \frac{d(y_j - y_k)}{y_j - y_k} \right) u$$

of rank 1 such that the y_0-equation coincides with (F). We can take any value of $a_{jk} \in \mathbb{C}$ for $1 \le j < k \le m$, and for these any values, the system is completely integrable because it is of rank 1. Now apply the chain of additions and middle convolutions in the converse way, in which the middle convolutions are ones in y_0-direction. Then we obtain a completely integrable Pfaffian system whose y_0-equation coincides with (E).

In this way, we can prolong rigid equations with at least four singular points to completely integrable Pfaffian systems in the original independent variable together with the positions of singular points as new variables. In other words, we can deform (E) in this way.

Example In section "Rigidity of Monodromy and Spectral Types" we give a table of rigid spectral types. There we find three spectral types with four components:

$$((21),(21),(21),(21)),\ ((31),(31),(22),(211)),\ ((31),(22),(22),(22)).$$

We shall construct Pfaffian systems by using these spectral types.

We have the following chains of additions and middle convolutions connecting the above spectral types to the spectral type $((1),(1),(1),(1))$ of rank 1:

$$((\underline{3}1),(\underline{3}1),(\underline{2}2),(2\underline{1}1)) \xrightarrow{d=3+3+2+1-2\times 4=1} ((21),(21),(12),(21)),$$

$$((\underline{3}1),(\underline{2}2),(\underline{2}2),(\underline{2}2)) \xrightarrow{d=3+2+2+2-2\times 4=1} ((21),(12),(12),(12)),$$

$$((\underline{2}1),(\underline{2}1),(\underline{2}1),(\underline{2}1)) \xrightarrow{d=2+2+2+2-2\times 3=2} ((1),(1),(1),(1)).$$

According to the manner explained just before, we can construct completely integrable Pfaffian systems from a rank 1 system by using these chains. First we take a Pfaffian system

$$du = \left(a_1 \frac{dx}{x} + a_2 \frac{dy}{y-1} + a_3 \frac{d(x-y)}{x-y} + a_4 \frac{dx}{x-1} + a_5 \frac{dy}{y} \right) u \tag{12}$$

of rank 1, where $a_1, a_2, \ldots, a_5 \in \mathbb{C}$ are almost arbitrarily chosen. Apply a middle convolution mc_λ^x to the system (12). Then we get

$$dv = \left(A_1 \frac{dx}{x} + A_2 \frac{dy}{y-1} + A_3 \frac{d(x-y)}{x-y} + A_4 \frac{dx}{x-1} + A_5 \frac{dy}{y} \right) v, \tag{13}$$

where

$$A_1 = \begin{pmatrix} a_1+\lambda & a_3 & a_4 \\ 0 & 0 & 0 \\ 0 & 0 & 0 \end{pmatrix},\ A_3 = \begin{pmatrix} 0 & 0 & 0 \\ a_1 & a_3+\lambda & a_4 \\ 0 & 0 & 0 \end{pmatrix},\ A_4 = \begin{pmatrix} 0 & 0 & 0 \\ 0 & 0 & 0 \\ a_1 & a_3 & a_4+\lambda \end{pmatrix},$$

$$A_2 = \begin{pmatrix} a_1 & 0 & 0 \\ 0 & a_2+a_4 & -a_4 \\ 0 & -a_3 & a_2+a_3 \end{pmatrix},\ A_5 = \begin{pmatrix} a_3+a_5 & -a_3 & 0 \\ -a_1 & a_1+a_5 & 0 \\ 0 & 0 & a_5 \end{pmatrix}.$$

The spectral type of the x-equation of (13) is $((21),(21),(21),(21))$. The spectral type of (13) as a Pfaffian system will be given at the end of this example.

At this moment, we apply addition and middle convolution to the system (13) in two ways. As the first case, we apply the addition

$$A_1' = A_1 - (a_1+\lambda)I,\ A_j' = A_j \quad (2 \le j \le 5)$$

to the system (13), and then apply middle convolution mc_μ^x. Since $\dim \operatorname{Ker} A_1' = 1, \dim \operatorname{Ker} A_3' = \dim \operatorname{Ker} A_4' = 2$, we have $\dim W = 5$, where $W = \mathcal{K} + \mathcal{L}$ is the maximal invariant subspace. Then we obtain a completely integrable Pfaffian system

$$dw = \left(B_1 \frac{dx}{x} + B_2 \frac{dy}{y-1} + B_3 \frac{d(x-y)}{x-y} + B_4 \frac{dx}{x-1} + B_5 \frac{dy}{y} \right) w \qquad (14)$$

of rank $3 \times 3 - 5 = 4$. The spectral type of the x-equation of (14) is $[(22),(31),(31),(211)]$.

As the second case, we apply the addition

$$A_1'' = A_1 - (a_1+\lambda)I,\ A_4'' = A_4 - (a_4+\lambda)I,\ A_j'' = A_j \quad (j = 2,3,5),$$

to the system (13), and then apply the middle convolution $mc_{\lambda-a_3}^x$. In this case, we have $\dim \operatorname{Ker} A_1'' = \dim \operatorname{Ker} A_4'' = 1, \dim \operatorname{Ker} A_3'' = 2$ and $\dim \mathcal{L} = 1$, which induces $\dim W = 1+1+2+1 = 5$. Then we obtain a completely integrable Pfaffian system

$$dz = \left(C_1 \frac{dx}{x} + C_2 \frac{dy}{y-1} + C_3 \frac{d(x-y)}{x-y} + C_4 \frac{dx}{x-1} + C_5 \frac{dy}{y} \right) z \qquad (15)$$

of rank $3 \times 3 - 5 = 4$.

We give a table of spectral types of the above systems as Pfaffian systems.

System	$x=0$	$y=1$	$x=y$	$x=1$	$y=0$	H_∞	$x=\infty$	$y=\infty$
(12)	(1)	(1)	(1)	(1)	(1)	(1)	(1)	(1)
(13)	(21)	(21)	(21)	(21)	(21)	(21)	(21)	(21)
(14)	(22)	(22)	(31)	(31)	(31)	(31)	(211)	(211)
(15)	(22)	(22)	(31)	(22)	(22)	(211)	(22)	(22)

We see that the system (13) is essentially a Pfaffian system corresponding to a system of partial differential equations satisfied by Appell's hypergeometric series F_1, (14) is that by F_2 and F_3, and (15) is essentially a Pfaffian system corresponding to a pull-back of a system of partial differential equations satisfied by F_4.

Exercise 6 Obtain explicit forms of the matrices B_j, C_j $(1 \le j \le 5)$ in (14) and (15).

Thus, we can obtain a completely integrable Pfaffian system whose one dimensional section is a rigid Fuchsian equation. However, for a completely integrable Pfaffian system, its one dimensional sections can be non-rigid. In such case, we understand that the Pfaffian system is obtained from an algebraic solution of the deformation equation of the non-rigid one dimensional section.

Acknowledgements This work is supported by the JSPS grants-in-aid for scientific research B, No. 15H03628.

References

1. W. Crawley-Boevey, On matrices in prescribed conjugacy classes with no common invariant subspace and sum zero. Duke Math. J. **118**, 339–352 (2003)
2. P. Deligne, *Équations différentielles à points singuliers réguliers*. Lecture Notes in Mathematics, vol. 163 (Springer, Berlin, 1970)
3. M. Dettweiler, S. Reiter, An algorithm of Katz and its application to the inverse Galois problem. J. Symb. Comput. **30**, 761–798 (2000)
4. M. Dettweiler, S. Reiter, Middle convolution of Fuchsian systems and the construction of rigid differential systems. J. Algebra **318**, 1–24 (2007)
5. R. Gérard, A.H.M. Levelt, Étude d'une classe particulière de systèmes de Pfaff du type de Fuchs sur l'espace projectif complexe. J. Math. pures et appl. **51**, 189–217 (1972)
6. H.A. Hamm, D.T. Lê, Un théorème de Zariski du type de Lefschetz. Ann. Sci. École Norm. Sup. **6**, 317–366 (1973)
7. Y. Haraoka, Middle convolution for completely integrable systems with logarithmic singularities along hyperplane arrangements. Adv. Stud. Pure Math. **62**, 109–136 (2012)
8. Y. Haraoka, *Linear differential equations in the complex domain.* (Sugaku Shobo, Tokyo, 2015) (in Japanese)
9. Y. Haraoka, G. Filipuk, Middle convolution and deformation for Fuchsian systems. J. Lond. Math. Soc. **76**, 438–450 (2007)
10. Y. Haraoka, S. Hamaguchi, Topological theory for Selberg type integral associated with rigid Fuchsian systems. Math. Ann. **353**, 1239–1271 (2012)

11. Y. Haraoka, T. Kikukawa, Rigidity of monodromies for Appell's hypergeometric functions. Opuscula Math. **35**, 567–594 (2015)
12. N.M. Katz, *Rigid Local Systems* (Princeton University Press, Princeton, NJ, 1996)
13. V.P. Kostov, On the Deligne-Simpson problem. Proc. Steklov Inst. Math. **238**, 148–185 (2002)
14. T. Oshima, *Fractional Calculus of Weyl Algebra and Fuchsian Differential Equations*. MSJ Memoirs, vol. 28 (Mathematical Society of Japan, Tokyo, 2012)
15. I. Shimada, Fundamental groups of complements to hypersurfaces. RIMS Kôkyûroku **1033**, 27–33 (1998)
16. E.R. van Kampen, On the fundamental group of an algebraic curve. Am. J. Math. **55**, 255–260 (1933)

Sub-Riemannian Geometry and Hypoelliptic Operators

Irina Markina

Abstract In this course we carefully define the notion of a non-holonomic manifold, which is a manifold with a certain non-integrable smooth sub-bundle of the tangent bundle, also called a distribution. We define such concepts as horizontal distributions, bracket generating condition for distributions, a sub-Riemannian structure, hypoelliptic and subelliptic operators and others. We connect the geometry of a manifold, defined by non-holonomic constraints, with properties of hypoelliptic operators on that manifold.

Keywords CR manifolds • Differential complexes • Hörmander condition • Hypoelliptic operators • Sub-Riemannian geometry • $\overline{\partial}_b$-Neumann problem

Mathematics Subject Classification (2000). Primary 35H10, 35H20, 53C17; Secondary 32W10, 32V05, 35N15

1 Introduction

These lecture notes were prepared for the school "Analytic, Algebraic and Geometric Aspects of Differential Equations" that took place in Będlewo, Poland, September 6–12, 2015. The presented material is well known in the literature, except for the last part, which contains recent results. More than 140 years ago, in 1867, Eugenio Beltrami [15] introduced the Laplace operator for a Riemannian metric, which is also referred to as the Laplace-Beltrami operator. The Laplace operator is a core example of partial differential operators possessing the marvellous regularity property, called the hypoellipticity: the distribution solutions are actually C^{∞} smooth. In the last two centuries the geometric analysis on smooth manifolds, with imposed non-holonomic constraints was intensively developed. Geometrically, non-holonomic constraints are described by the presence of the distinguished sub-bundle of the tangent bundle of the manifold. If the manifold is endowed with

I. Markina (✉)
Department of Mathematics, University of Bergen, Bergen, Norway
e-mail: irina.markina@uib.no

G. Filipuk et al. (eds.), *Analytic, Algebraic and Geometric Aspects of Differential Equations*, Trends in Mathematics, DOI 10.1007/978-3-319-52842-7_3

a Riemannian metric, measuring the sections of the sub-bundle, then the non-holonomic manifold with the metric received the name sub-Riemannian manifold. The theory of regularity of solutions to elliptic and parabolic equations, defined by differential operators, related to the sub-bundle, attracted a lot of attention. It turned out that the hypoellipticity property is closely related to the geometry of the non-holonomic manifold.

The main aim of the notes is to connect the sub-Riemannian geometry with completely non-holonomic distributions and the hypoellipticity property of the related differential operators. After the introduction of sub-Riemannian manifolds and necessary analytic tools we sketch the proof of the Hörmander theorem and show the appearance of both notions in the analysis and PDE in several complex variables. The last part of the notes is the extension of some presented ideas for the construction of solutions of a boundary value problem in domains of the space of quaternions.

2 Main Definitions

Smooth Manifolds, Vector Fields, Tangent Map

It is supposed that the reader is familiar with the notion of smooth or C^∞ manifolds [37, 84, 92]. We set up main definitions and notations. A smooth manifold is a Hausdorff, second countable topological space, where a smooth complete atlas is defined. We write M for a smooth manifold, or rather M^n if we want to emphasize the dimension n of the manifold. Let $C^\infty(M)$ denote the space of smooth real valued functions defined on M.

The tangent space at a point $q \in M$ is denoted by T_qM. Recall that any element $v_q \in T_qM$ is a function $v_q\colon C^\infty(M) \to \mathbb{R}$ satisfying two properties

1. $\mathbb{R}$-linearity: $v_q(af + bg) = av_q(f) + bv_q(g)$,
2. Leibnizian property: $v_q(fg) = v_q(f)g(q) + f(q)v_q(g)$

for all $a, b \in \mathbb{R}$, $f, g \in C^\infty(M)$, $q \in M$. The space T_qM, $q \in M$, is a real vector space and therefore v_q is called *tangent vector*.

The notion of a tangent vector v_q is the generalisation of the derivative of C^∞ smooth functions along the direction v_q. If the chart $\big(U, \varphi = (x^1, \dots, x^n)\big)\colon U \subset M$, $\varphi\colon U \to \mathbb{R}^n$, is chosen, then the standard notation for the basis for T_qM at $q \in M$ is $\left(\frac{\partial}{\partial x^1}, \dots, \frac{\partial}{\partial x^n}\right)$ or shortly $(\partial_1, \dots, \partial_n)$. Any vector $v_q \in T_qM$ will be written in coordinates as $v_q = \sum_{j=1}^n v_q^j \partial_j$.

The dual space to T_qM is denoted by T_q^*M and the pairing is written as $\langle .\,, . \rangle_q$, where we usually omit the subscript "q". Recall that a pairing between the tangent and the co-tangent space is a map $\langle .\,, . \rangle\colon T_qM \times T_q^*M \to \mathbb{R}$, which is bi-linear, non-degenerate, and smoothly varying with respect to $q \in M$. It is non-degenerate in the sense that if $\langle v, \lambda \rangle = 0$ for all $v \in T_qM$, then $\lambda \equiv 0$. The dual basis to $(\partial_1, \dots, \partial_n)$

with respect to the pairing is denoted by $(dx^1, \dots, dx^n)$ and, by definition, it satisfies $\langle dx^i, \partial_j \rangle = \delta_{ij}$, where δ_{ij} is the Kronecker symbol. Then any co-vector $\lambda \in T_q^*M$ is written in coordinates as $\lambda = \sum_{k=1}^n \lambda_k dx^k$. The elements of T_q^*M are usually called *co-vectors*.

The tangent and co-tangent bundles are denoted by TM and T^*M, respectively. Both vector bundles are C^∞ smooth manifolds [37, 84]. The reader can find more information about fiber bundles, for instance, in [59]. The notations

$$\begin{aligned} \mathrm{pr}_M\colon\ TM &\to M \\ (q, v) &\to q \end{aligned} \qquad \text{and} \qquad \begin{aligned} \mathrm{pr}_M^*\colon\ T^*M &\to M \\ (q, \lambda) &\to q \end{aligned}$$

will be fixed for the canonical projections from the tangent and co-tangent bundles to the underlying manifold.

A *vector field* X on a manifold M is a function that assigns to each point $q \in M$ a tangent vector $X_q \in T_qM$. If $f \in C^\infty(M)$, then Xf denotes a real valued function on M given by

$$(Xf)(q) = X_q f, \qquad \text{for all} \quad q \in M.$$

A vector field X is called smooth if for any $f \in C^\infty(M)$ the function $Xf\colon M \to \mathbb{R}$ belongs to $C^\infty(M)$. If $\big(U, \varphi = (x^1, \dots, x^n)\big)$ is a coordinate chart, then any vector field X can be written in terms of coordinates as $X_q = \sum_{j=1}^n X^j(q)\partial_j$. Then the C^∞ smoothness condition of the vector field X in a neighbourhood $U \subset M$ is equivalent to the requirement that all functions $X^j\colon U \to \mathbb{R}$, $j = 1, \dots, n$, are of class $C^\infty(U)$. If the functions X^j, $j = 1, \dots, n$, are analytic in U, then the corresponding vector field X is called an *analytic* vector field. Thus, the vector fields are the differential operators of the first order.

Another way to define a vector field X is to use the definition of a local section. Namely, a vector field X is a smooth map $X\colon U \to TM$, such that $\mathrm{pr}_M \circ X = \mathrm{id}_U$ for any open set $U \subset M$. The section is global if U can be taken as the entire M. We write $\mathrm{Vect}\, M$ ($\mathrm{Vect}\, U$) for the collection of smooth vector fields or smooth sections, defined on M (U, $U \subset M$). Algebraically, $\mathrm{Vect}\, M$ is a module over the ring $C^\infty(M)$ and a vector space over the field $\mathbb{R}$ (or $\mathbb{C}$ if the manifold M is modelled over $\mathbb{C}^n$). Moreover, a multiplication of two vector fields can be defined. The multiplication $[\cdot\,,\cdot]\colon \mathrm{Vect}\, M \times \mathrm{Vect}\, M \to \mathrm{Vect}\, M$, which received the name *commutator* or the *Lie product* is defined by

$$[X, Y]f = X(Yf) - Y(Xf). \tag{1}$$

The Lie product satisfies three axioms given in the following definition.

Definition 2.1 A Lie algebra over $\mathbb{R}$ ($\mathbb{C}$) is a real (complex) vector space V together with an operation $[\cdot\,,\cdot]: V \times V \to V$ (called the bracket, commutator, or Lie product) satisfying the following three axioms:

1. skew symmetry: $[X, Y] = -[Y, X]$,
2. bi-linearity: $[aX + bY, Z] = a[X, Z] + b[Y, Z]$, $a, b \in \mathbb{R}$ ($\mathbb{C}$) (and the same with respect to the second term in $[.\,,.]$),
3. Jacobi identity: $[[X, Y], Z] + [[Z, X], Y] + [[Y, Z], X] = 0$

for any $X, Y, Z \in V$.

The set of smooth vector fields $\operatorname{Vect} M$, considered as a real vector space endowed with the Lie multiplication, forms a Lie algebra.

Definition 2.2 Let M and N be two smooth manifolds and $F: M \to N$ be a map. The map F is smooth if for any $q \in M$ and for any local charts (U, φ) of $q \in M$ and (V, ψ) of $F(q) \in N$, the composition $\psi \circ F \circ \varphi^{-1}: \varphi(U) \to \psi(V)$ is a smooth map in the sense of smoothness defined in the Euclidean space $\mathbb{R}^n$.

Definition 2.3 Let $F: M \to N$ be a smooth map. The differential of F at $q \in M$ is the linear map $d_qF: T_qM \to T_{F(q)}N$ defined by

$$\big(d_qF(X_q)\big)f := X_q(f \circ F)$$

for any $f \in C^\infty(N)$ and $X_q \in T_qM$.

If the local charts $\big(U, \varphi = (x^1, \dots, x^m)\big)$ of $q \in M$ and $\big(V, \psi = (y^1, \dots, y^n)\big)$ of $F(q) \in N$ are chosen, then

$$d_qF(\partial_{x^j}) = \sum_{k=1}^{n} \frac{\partial}{\partial x^j}\Big(y^k\big(F(q)\big)\Big)\partial_{y^k}|_{F(q)}, \quad j = 1, \dots, m.$$

The matrix $\left\{\frac{\partial}{\partial x^j}\Big(y^k\big(F(q)\big)\Big)\right\}_{k,j}$ is called the *Jacobi matrix* of the map F with respect to the given coordinate charts.

Distributions and Non-holonomic Constraints

Definition 2.4 A sub-bundle D of the tangent bundle TM is called *smooth* on M, if for any $q \in M$ there is a neighborhood $U(q)$ and smooth linearly independent vector fields $X_1, \dots, X_k$, such that $D_x = \operatorname{span}\{X_1, \dots, X_k\}|_x$, for all $x \in U(q)$.

A sub-bundle D is called *analytic* if the vector fields $X_1, \dots, X_k$ in Definition 2.4 are analytic. From now on, we will work only with smooth sub-bundles and smooth manifolds and therefore we omit the word "smooth". The notion of a sub-bundle naturally leads to the following question. When does a sub-bundle D of TM define a submanifold N inside of the original manifold M? The answer was given by Frobenius [44].

Definition 2.5 A smooth sub-bundle D on M is called involutive or integrable if $[X, Y]$ is a smooth section of D for any choice of smooth sections X and Y of D.

Definition 2.6 A smooth submanifold N of a manifold M is the integral manifold of a sub-bundle D of TM if for any point $q \in N$ there is an open neighbourhood $U(q) \subset N$ such that $T_xN = D_x$ for any $x \in U(q)$.

Theorem 2.7 ([44, 113]) *A submanifold N of a manifold M is the integral manifold of a sub-bundle D of TM, if and only if, D is involutive.*

In this case a foliation of the manifold M by integral manifolds N passing through different points $q \in M$ is produced. Somehow, one cannot leave a chosen leaf N of the foliation while being tangent to the sub-bundle D.

A smooth or, more generally, absolutely continuous curve $c: I \to M$, $I \subset \mathbb{R}$, can be considered as a smooth map between two manifolds. In this case the image of the tangent vector $\frac{d}{dr} \in T_tI$ under the tangent map $d_tc: T_tI \to T_{c(t)}M$ is denoted by $\dot{c}(t)$, i.e. $d_tc\left(\frac{d}{dr}\right) = \dot{c}(t)$, and is called the *velocity vector* of the curve c at $t \in I$.

Definition 2.8 We say that an absolutely continuous curve $c: I \to M$ is tangent to the sub-bundle D of TM (or *the curve c is horizontal*) if the tangent vector $\dot{c}(t)$ (where it exists) belongs to the vector space $D_{c(t)}$ for any $t \in I$, whenever $\dot{c}(t)$ is defined.

If a sub-bundle D of TM is involutive, then starting from a point $q \in M$ and being tangent to D one can reach only the points inside the integral manifold N passing through q. If the sub-bundle D is not involutive, then under some conditions we can reach any point on the original manifold M. To describe those conditions we introduce the notion of a bracket generating sub-bundle. Let $\Gamma(D)$ be all smooth sections $X: M \to D$ of the sub-bundle D of TM. Define

$$\text{Lie}(D) = \left\{D^1 = \Gamma(D), \dots, D^{k+1} = D^k + [D, D^k], \dots\right\}.$$

If

$$\text{Lie}_q(D) = T_qM \quad \text{for every} \quad q \in M,$$

then we say that the sub-bundle D is *bracket generating*, or *non-integrable*, or *completely non-holonomic*. A sub-bundle D is said to be *k-step bracket generating* if $D^j = \{0\}$ for all $j > k$. We say that a sub-bundle D is *strongly bracket generating* if $D^{k+1} = D^k + [X, D^k]$ for all non-vanishing $X \in \Gamma(D)$. If a sub-bundle D is bracket generating and the rank of D^k does not depend on the point $q \in M$ for any k, then the sub-bundle D is called *regular*.

Example Consider vector fields in $\mathbb{R}^4$ written in coordinates (x, y, z, w):

$$X_1 = \frac{\partial}{\partial x}, \quad X_2 = \frac{\partial}{\partial y}, \quad X_3 = \frac{\partial}{\partial z} + x^k \frac{\partial}{\partial w}.$$

The sub-bundle $D = \text{span}\{X_1, X_2, X_3\}$ is smooth $(k+1)$-step bracket generating but not strongly bracket generating.

Now we are ready to formulate a sufficient condition for the connectivity problem. This condition was independently proved by Rashevskiĭ [89] and Chow [31].

Theorem 2.9 ([31, 89]) *If a manifold M is topologically connected and if a sub-bundle D on M is bracket generating, then any two points on M can be connected by a piecewise smooth curve tangent to D.*

Necessary and sufficient conditions for the connectivity problem in the case of C^∞ manifolds and C^∞ smooth sub-bundles can be found in [99] and references therein.

Example In the following example we show that the Chow-Rashevskiĭ condition is not necessary for connectivity. Let $M = \mathbb{R}^2$, $X_1 = \frac{\partial}{\partial x}$, $X_2 = \phi(x)\frac{\partial}{\partial y}$, where the C^∞ function ϕ satisfies

$$\begin{aligned} \phi(x) &> 0, \quad \text{if } x > 0, \\ \phi(x) &= 0, \quad \text{if } x \leq 0. \end{aligned}$$

It is clear that one can not move vertically in the left half-plane, but one can move horizontally to the right half-plane, displace arbitrarily in the right half-plane and proceed to the left half-plane, see Fig. 1. Thus, one can connect any points in the plane being tangent to the sub-bundle $D = \text{span}\{X_1, X_2\}$, but the vector fields definitely do not span the entire plane at points $q = (x, y)$ with $x \leq 0$.

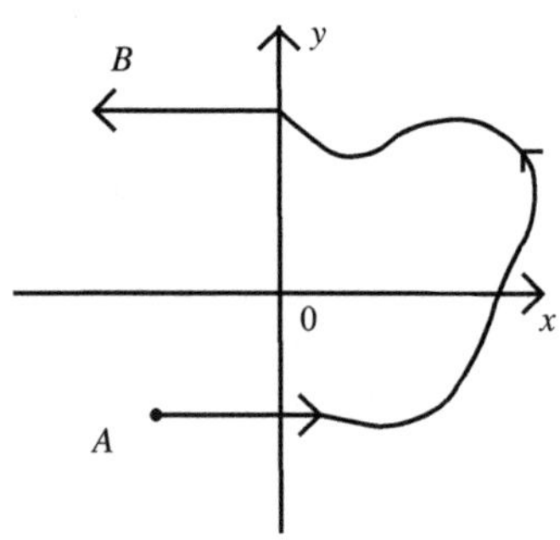

Fig. 1 $(\mathbb{R}^2, D)$ is horizontally connected, but D is not bracket generating

Example Another example of a bracket generating sub-bundle is the Grušin sub-bundle spanned by vector fields

$$X_1 = \frac{\partial}{\partial x}, \quad X_2 = x\frac{\partial}{\partial y},$$

in $\mathbb{R}^2$ studied by Baouendi in his PhD thesis in the early 1970s, and later by numerous authors, see for instance, [6, 28, 38, 45, 51, 103]. The sub-bundle is not smooth, since along the line $x = 0$ it degenerates from two dimensional sub-bundle to one dimensional.

Example Historically the integrability condition was given in terms of one-forms, not in terms of vector fields. Let M be a manifold of dimension n and we want to describe a sub-bundle D of TM of rank k, $k < n$. To achieve this we need to find $n-k$ one-forms $\Theta_1, \dots, \Theta_{n-k}$, such that the sub-bundle D belongs to their common kernel. The forms $\Theta_j, j = 1, \dots, n-k$, are called annihilators of D. It is equivalent to solving the system

$$\begin{cases} \Theta_1(x^1, \dots, x^n) = 0 \\ \dots\dots\dots\dots\dots \\ \Theta_{n-k}(x^1, \dots, x^n) = 0, \end{cases}$$

that received the name Pfaffian system. This system is integrable if the one-forms $\Theta_1, \dots, \Theta_{n-k}$ are exact forms:

$$\begin{cases} \Theta_1(x^1, \dots, x^n) = d\theta_1(x^1, \dots, x^n) = 0 \\ \dots\dots\dots\dots\dots \\ \Theta_{n-k}(x^1, \dots, x^n) = d\theta_{n-k}(x^1, \dots, x^n) = 0. \end{cases} \tag{2}$$

After integrating the latter system we get $n-k$ functions describing a k-dimensional integral submanifold of M, defined by the integrable system (2) or by the involutive sub-bundle D.

The Chow-Rashevskiĭ Theorem 2.9 for an analytic co-rank one sub-bundle D, or for one Pfaffian equation was proved by C. Carathéodory. The result states the following. *Let M be a connected manifold endowed with an analytic co-rank one sub-bundle D. If there exist two points A and $B \in M$ that cannot be connected by a horizontal curve, then the sub-bundle D is integrable.* Or, formulating the negation of the above statement: *If for any points $A, B \in M$ there is a horizontal curve connecting these points, then the sub-bundle D is non-integrable* (*completely non-holonomic, bracket generating*).

C. Carathéodory developed this theory due to the question posted by M. Born to derive the second law of thermodynamics and the existence of the entropy function. Translating the problem into the geometric language we work with a manifold M

that is the set of all possible thermodynamical states of some isolated system. The admissible or horizontal curves are adiabatic curves, such curves that correspond to slow processes in time and such that during these processes no heat Θ is exchanged. C. Carathéodory wrote the condition of an adiabatic process as a Pfaffian equation $\Theta = 0$ on M. It was known at that moment from works by S. Carnot, J.P. Joule and others, that there are thermodynamical states $A, B \in M$, which cannot be connected by an adiabatic process (by a horizontal curve). Carathéodory's theorem states in this case that the sub-bundle defined by the Pfaffian equation $\Theta = 0$ is integrable, which leads to the existence of two functions T (temperature) and S (entropy) that locally satisfy the relation $\Theta = TdS$. This proves the existence of the entropy function S, as well as proving that the adiabatic process remains in a leaf (hypersurface) of the state space M corresponding to the entropy function. The entropy function S tends not to decrease, being constant or increasing, according to the second law of thermodynamics.

Due to the names of S. Carnot and C. Carathéodory involved in this discovery, M. Gromov called the geometry of manifolds with non-integrable sub-bundles as the *Carnot-Carathéodory geometry*.

Exercises Decide whether the following sub-bundles $D = \operatorname{span}\{X_1, X_2\}$ in $\mathbb{R}^3$ are bracket generating and regular.

1. Heisenberg sub-bundle: $X_1 = \frac{\partial}{\partial x}, X_2 = \frac{\partial}{\partial y} + x\frac{\partial}{\partial z}$.
2. Martinet sub-bundle: $X_1 = \frac{\partial}{\partial x}, X_2 = \frac{\partial}{\partial y} + x^2\frac{\partial}{\partial z}$.

Riemannian and Sub-Riemannian Manifolds

One of the first works that gave the name *sub-Riemannian geometry* and established the subject as a part of geometry was the paper of Strichartz [97], published in 1986. We recall some notions from the Riemannian geometry and compare basic definitions in the Riemannian and sub-Riemannian settings.

Definition 2.10 A Riemannian metric is a map $g\colon T_qM \times T_qM \to \mathbb{R}$, which is symmetric, bilinear, positively definite for any $q \in M$, and smoothly varying with respect to q.

If the coordinate chart $\big(U, \varphi = (x^1, \dots, x^n)\big)$ is chosen and $(\partial_1, \dots, \partial_n)$ is the local basis for T_qM, $q \in U$, then $g_{ij} = g(\partial_i, \partial_j)$ is the associated matrix to the metric g. Smoothness of g means that the matrix $g_{ij}(q) = g_{ij}(x^1, \dots, x^n)$ is a smooth function of $(x^1, \dots, x^n)$ in $\varphi(U)$.

The couple (M, g) is called a Riemannian manifold. It would be more correct to say that the triplet (M, TM, g) is called a Riemannian manifold.

Definition 2.11 The distance $d(q_0, q_1)$ between two points $q_0, q_1 \in M$ related to the Riemannian metric g is defined by the equality

$$d(q_0, q_1) = \inf\left\{\int_0^1 \Big(g\big(\dot{c}(t), \dot{c}(t)\big)\Big)^{1/2} dt\right\},$$

where the infimum is taken over all curves $c: [0, 1] \to M$ differentiable almost everywhere in $[0, 1]$, and such that $c(0) = q_0$, $c(1) = q_1$.

We are now ready to define a sub-Riemannian manifold. Let M be a smooth manifold and let D be a smooth sub-bundle of the tangent bundle TM.

Definition 2.12 A map $g_D: D_q \times D_q \to \mathbb{R}$ which is symmetric, bilinear, positively definite for any $q \in M$ and smoothly varying with respect to q is called a sub-Riemannian metric.

Definition 2.13 The couple (D, g_D) is called a sub-Riemannian structure and the triplet (M, D, g_D) is called a sub-Riemannian manifold.

If $D = TM$, then Definition 2.13 is reduced to the definition of a Riemannian manifold. In this sense the sub-Riemannian geometry is a generalization of the Riemannian geometry. The distance function related to a sub-Riemannian metric g_D is defined by

$$d_{c-c}(q_0, q_1) = \inf\left\{\int_0^1 \Big(g_D\big(\dot{c}(t), \dot{c}(t)\big)\Big)^{1/2} dt\right\},$$

where the infimum is taken over all horizontal curves $c: [0, 1] \to M$ differentiable almost everywhere in $[0, 1]$ and such that $c(0) = q_0$, $c(1) = q_1$. Thus, we have added the horizontality condition, $\dot{c}(t) \in D_{c(t)}$, in the definition of the sub-Riemannian metric. The set of admissible curves is smaller, therefore, the d_{c-c}-distance is, in general, bigger than the Riemannian distance if both metrics are defined on the manifold and coincide on D_q, $q \in M$. Theorem 2.9 guarantees that the set of horizontal curves is not empty and therefore, the function d_{c-c} takes only finite values. The distance d_{c-c} is called the Carnot-Carathéodory distance due to the impact by S. Carnot and C. Carathéodory described in the last example of the previous section. Let us suppose that a Riemannian metric g and a sub-Riemannian metric g_D are defined on a smooth manifold M, and the Riemannian distance d and the Carnot-Carathéodory distance d_{c-c} on M are produced by them, respectively. As a result, two metric spaces (M, d) and (M, d_{c-c}) and two topological spaces (M, τ_d) and (M, τ_{c-c}) are defined, where the topology τ_d is generated by open balls in the d-metric and τ_{c-c} is generated by d_{c-c}-balls. It is established that the topological spaces (M, τ_d) and (M, τ_{c-c}) are equivalent, but the metric spaces (M, d) and (M, d_{c-c}) are not in general Lipschitz equivalent, see [76, p. 27], [14, 46, 50, 80]. The example in section "Heisenberg Sub-Riemannian Manifold as a Lie Group" shows non-equivalence of the metric spaces (M, d) and (M, d_{c-c}) in a particular case.

Riemannian and Sub-Riemannian Gradient

At the end of the subsection we would like to say some words about the gradient vector field in sub-Riemannian geometry. Let us recall that the gradient on the Riemannian manifold (M, g) is a vector field "grad" such that it is detected by its action on smooth functions by

$$g(\operatorname{grad} f, X) = Xf, \quad \text{for any} \quad X \in \operatorname{Vect} M \quad \text{and} \quad f \in C^{\infty}(M).$$

If a coordinate chart is chosen, then the gradient can be written as

$$\operatorname{grad} f = \sum_{i,j=1}^{N} g^{ij} \frac{\partial f}{\partial x^i} \partial_j, \tag{3}$$

where $\{g^{ij}\}_{i,j=1}^{n}$ is the inverse matrix to $g_{ij} = g(\partial_i, \partial_j)$, $i, j = 1, \ldots, n$. A reader can find more details about differential operators on Riemannian manifolds in [84].

In the case of a sub-Riemannian manifold (M, D, g_D) the definition is analogous. A sub-Riemannian gradient grad_D is a horizontal vector field, such that

$$g_D(\operatorname{grad}_D f, X) = Xf, \text{ for any smooth section } X \text{ of } D \text{ and } f \in C^{\infty}(M).$$

Short Introduction to Lie Groups

A Lie group is an object that nicely combines algebraic, geometric, and analytic properties. Namely, a Lie group $\mathbb{G}$ is a pair (M, ρ), where

1. M is a C^{∞} smooth finite dimensional manifold,
2. the map $\rho: M \times M \to M$ satisfies the axioms of the group product,
3. the map ρ is compatible with the smooth manifold structure in the sense that the map $\rho: M \times M \to M$ is C^{∞}-smooth.

As usual in mathematics, we will write only $\mathbb{G}$ instead of (M, ρ) to denote the Lie group and the underlying manifold M.

Recall, that a Lie algebra is a pair $(V, [\cdot\,,\cdot])$, where V is a vector space over the fields $\mathbb{R}$ or $\mathbb{C}$ and $[\cdot\,,\cdot]$ is the Lie product introduced in Definition 2.1.

There is a close relation between Lie groups and Lie algebras. To define the Lie algebra $\mathfrak{g}$ of a Lie group $\mathbb{G}$ we consider special vector fields on $\mathbb{G}$. We call the mappings

$$l_{\tau}(q) := \rho(\tau, q) = \tau q, \quad \tau \in \mathbb{G} \text{ fixed}, \quad q \in \mathbb{G} \text{ is arbitrary},$$

the *left translation* on $\mathbb{G}$ and

$$r_\tau(q) := \rho(q,\tau) = q\tau, \quad \tau \in \mathbb{G} \text{ fixed}, \quad q \in \mathbb{G} \text{ is arbitrary},$$

the *right translation* on $\mathbb{G}$. Since the group multiplication and the inversion are smooth, the maps $l_\tau, r_\tau \colon \mathbb{G} \to \mathbb{G}$ are smooth diffeomorphisms of $\mathbb{G}$. The differentials $d_q l_\tau \colon T_q\mathbb{G} \to T_{l_\tau(q)}\mathbb{G}$ and $d_q r_\tau \colon T_q\mathbb{G} \to T_{r_\tau(q)}\mathbb{G}$ are linear maps of the respective tangent spaces.

Definition 2.14 A vector field X on $\mathbb{G}$ satisfying

$$d_q l_\tau(X_q) = X_{l_\tau(q)} = X_{\tau q} \quad \Big(d_q r_\tau(X_q) = X_{r_\tau(q)} = X_{q\tau}\Big), \quad \forall\ \tau, q \in \mathbb{G}$$

is called left- (right-) invariant vector field.

The set of left invariant vector fields considered as a vector space over the field $\mathbb{R}$ with the Lie product defined by the commutator of vector fields (1) forms a real Lie algebra $\mathcal{L}$. Of course, one needs to verify that the commutator of left invariant vector fields is a left invariant vector field. Since any left invariant vector field is defined by its value at the identity of the group $e \in \mathbb{G}$, there is an isomorphism ι between the vector space $T_e\mathbb{G}$ and $\mathcal{L}$ defined by

$$\mathcal{L} \ni X \mapsto X_e \in T_e\mathbb{G}, \qquad T_e\mathbb{G} \ni v_e \mapsto dl(v_e) \in \mathcal{L}.$$

This isomorphism $\iota \colon T_e\mathbb{G} \to \mathcal{L}$ can be extended to an isomorphism of Lie algebras if we define Lie brackets in $T_e\mathbb{G}$ as $[X_e, Y_e] := [X, Y]_e$. The Lie algebra $(T_e\mathbb{G}, [\cdot\,,\cdot])$ is usually denoted by $\mathfrak{g}$ and is called the Lie algebra of the Lie group $\mathbb{G}$. The Lie algebra $\mathcal{R}$ of right invariant vector fields is isomorphic to $\mathfrak{g}$ if we set $\mathcal{R} \ni [X, Y] \leftrightarrow -[X, Y]_e \in \mathfrak{g}$.

The next step is to find a map between a given Lie group $\mathbb{G}$ and its Lie algebra $\mathfrak{g}$. The answer is given in terms of the exponential map $\exp \colon \mathfrak{g} \to \mathbb{G}$, that uses properties of solutions of ordinary differential equations [39].

Let $\mathbb{G}$ be a Lie group, $\mathfrak{g}$ be its Lie algebra, and let $X \in \mathfrak{g}$ be an arbitrary left invariant vector field. Then the theory of ordinary differential equations guarantees that the solution of the Cauchy problem

$$\begin{cases} \frac{dc_{X_e}(t)}{dt} = X(c_{X_e}(t)) \\ c_{X_e}(0) = e \end{cases}$$

is unique, possesses the properties of a one parameter subgroup of $\mathbb{G}$, and $\dot{c}_{X_e}(0) = X_e$, [39].

Definition 2.15 The map $\mathfrak{g} \ni X \to c_{X_e}(1) \in \mathbb{G}$ is called the group exponential map and is denoted by exp. Thus

$$\begin{aligned} \exp\colon \mathfrak{g} &\to \quad \mathbb{G} \\ X &\mapsto c_{X_e}(1). \end{aligned}$$

We will call the curve $c_{X_e}(t)$, $t \in \mathbb{R}$, the *exponential curve* and it is customary to use also the notation $\exp(tX_e)$ instead of $c_{X_e}(t)$. The main properties of the exponential map are listed in the following theorem:

Theorem 2.16 ([65, 113]) *Let X belong to the Lie algebra $\mathfrak{g}$ of a Lie group $\mathbb{G}$. Then the following properties hold.*

1. *The exponential curve* $\exp(tX_e) = c_{X_e}(t)$ *for each $t \in \mathbb{R}$ satisfies*

$$\frac{d}{dt}\Big|_{t=0} \exp(tX_e) = \frac{d}{dt}\Big|_{t=0} c_{X_e}(t) = \dot{c}_{X_e}(0) = X_e.$$

2. $\big(\exp(t_1+t_2)X_e\big) = \big(\exp(t_1X_e)\big)\big(\exp(t_2X_e)\big)$, *for all $t_1, t_2 \in \mathbb{R}$.*
3. $\exp(-tX_e) = \big(\exp(tX_e)\big)^{-1}$ *for each $t \in \mathbb{R}$.*
4. *The map* $\exp\colon \mathfrak{g} \to \mathbb{G}$ *is a C^∞ map between two manifolds.*
5. *The differential at the zero vector of the exponential map* $d_0 \exp\colon T_0\mathfrak{g} \to T_e\mathbb{G}$ *is the identity map $\mathfrak{g} \to \mathfrak{g}$, where we identify elements of $\mathfrak{g}$ with $T_0\mathfrak{g}$ for the domain of definition and $\mathfrak{g}$ with $T_e\mathbb{G}$ for the target space. An important corollary is that* exp *gives a diffeomorphism between a neighbourhood of $0 \in \mathfrak{g}$ and a neighborhood of $e \in \mathbb{G}$.*
6. *The left translation of c_{X_e} by $\tau \in \mathbb{G}$ given by $\tilde{c} = l_\tau(c_{X_e}(t)) = \tau c_{X_e}(t)$ is the unique integral curve of the left invariant vector field $\widetilde{X}$ ($\widetilde{X}_e = X_e$) such that it starts at the point $\tau = \tilde{c}(0)$. As a particular consequence, left invariant vector fields are always complete.*
7. *In the neighbourhoods of $0 \in \mathfrak{g}$ and $e \in \mathbb{G}$, where* exp *is a diffeomorphism, the inverse map is defined and is called the logarithm. It expresses the product of two exponents through the Baker-Campbell-Hausdorff formula [90], whose first terms are given as follows*

$$\begin{aligned} &\exp(X)\exp(Y) = \\ &\quad \exp\Big(X + Y + \frac{1}{2}[X,Y] + \frac{1}{12}[X,[X,Y]] - \frac{1}{12}[Y,[X,Y]] + \dots\Big). \end{aligned} \tag{4}$$

Let us assume now that the Lie algebra $\mathfrak{g}$ of a Lie group $\mathbb{G}$ is endowed with an inner product $(\cdot\,,\cdot)$. Then, by making use of left translations we can define a metric g on the group. Namely, let $v_q, w_q \in T_q\mathbb{G}$, then $dl_{q^{-1}}(v_q), dl_{q^{-1}}(w_q) \in T_e\mathbb{G}$. We define

$$g(v_q, w_q) := \big(dl_{q^{-1}}(v_q), dl_{q^{-1}}(w_q)\big) \quad \text{for any} \quad q \in \mathbb{G}. \tag{5}$$

Using right translations we can also define a metric.

Conversely, if there is a Riemannian metric g defined on a Lie group $\mathbb{G}$ considered as a smooth manifold, then it is compatible with the Lie structure if it is left (or right) invariant under the action of the group on itself.

Definition 2.17 A Riemannian metric g on a group $\mathbb{G}$ is called left invariant, if

$$g(X_q, Y_q) = g(d_q l_\tau(X_q), d_q l_\tau(Y_q)) = g(X_{\tau q}, Y_{\tau q})$$

for the left invariant vector fields $X, Y \in \mathfrak{g}$.

Exercises

1. Show that $(M, \cdot)$, where M is the set of (3×3) upper triangular real matrices

$$\begin{pmatrix} 1 & x & t \\ 0 & 1 & y \\ 0 & 0 & 1 \end{pmatrix}, \qquad x, y, t \in \mathbb{R}$$

   and "$\cdot$" stands for the matrix product, is a Lie group. Find the Lie algebra.
2. Show that if X, Y are left invariant (right invariant) vector fields on $\mathbb{G}$, then the commutator $[X, Y]$ is also a left (right) invariant vector field.
3. Show that the metric from (5) is a left invariant metric on the group.

Heisenberg Group

We start from the simplest example of a sub-Riemannian manifold that is called the Heisenberg group.

The Heisenberg Sub-Riemannian Manifold

Consider the smooth manifold $\mathbb{R}^3$ with coordinates $q = (x, y, t)$. Then the tangent and cotangent spaces are $T_q\mathbb{R}^3 = \text{span}\{\partial_x, \partial_y, \partial_t\}$, $T_q^*\mathbb{R}^3 = \text{span}\{dx, dy, dt\}$, and both are isomorphic to $\mathbb{R}^3$. We define the smooth 2-dimensional sub-bundle D of $T_q\mathbb{R}^3$ as the span of two vector fields

$$X = \partial_x - \frac{1}{2}y\partial_t, \quad Y = \partial_y + \frac{1}{2}x\partial_t. \tag{6}$$

Let us find a Riemannian metric g in coordinates (x, y, t) making X, Y and $T = [X, Y] = \partial_t$ orthonormal. Then we have $g(X, X) = g(X, Y) = g(T, T) = 1$ and other values vanish. We express the basis $(\partial_x, \partial_y, \partial_t)$ in the form

$$\partial_x = X + \frac{1}{2}yT, \quad \partial_y = Y - \frac{1}{2}xT, \quad \partial_t = T.$$

By making use of the bi-linearity of g, we obtain

$$g_{11} = g(\partial_x, \partial_x) = 1 + \frac{y^2}{4}, \quad g_{12} = g(\partial_x, \partial_y) = -\frac{xy}{4}, \quad g_{13} = g(\partial_x, \partial_t) = \frac{y}{2} \quad \ldots.$$

The matrix $\{g_{ij}\}$ takes the form

$$g_{ij} = \begin{pmatrix} 1 + \frac{y^2}{4} & -\frac{xy}{4} & \frac{y}{2} \\ -\frac{xy}{4} & 1 + \frac{x^2}{4} & -\frac{x}{2} \\ \frac{y}{2} & -\frac{x}{2} & 1 \end{pmatrix}. \tag{7}$$

Notice that $\det g = 1$. It implies that the volume form in $(\mathbb{R}^3, g)$ is given by the standard Lebesgue measure: $dx \wedge dy \wedge dt$.

The sub-bundle D is bracket generating of step 2 since $[X, Y] = \partial_t := T$ and $T_q\mathbb{R}^3 = \text{span}\{X, Y, T\}$. The dual basis to X, Y, T is

$$dx, \; dy, \; \omega = dt - \frac{1}{2}xdy + \frac{1}{2}ydx.$$

(Verify it!) The form ω is the annihilator of the sub-bundle D. We can also define the sub-bundle D as $D = \ker(\omega) = \{v = (x, y, z) \in \mathbb{R}^3 \mid \omega(v) = 0\}$. Define the sub-Riemannian metric g_D as the restriction of the metric g on the planes D_q for all $q \in \mathbb{R}^3$. Then $(\mathbb{R}^3, D, g_D)$ is the Heisenberg sub-Riemannian manifold.

Heisenberg Sub-Riemannian Manifold as a Lie Group

Let us consider the following non-commutative group on the smooth manifold $\mathbb{R}^3$. Define the product for $\tau = (x, y, t)$ and $q = (x_1, y_1, t_1)$ by

$$\tau q = (x, y, t)(x_1, y_1, t_1) = (x + x_1, y + y_1, t + t_1 + \frac{1}{2}(xy_1 - x_1y)). \tag{8}$$

As a motivation for this law one can consider the product of (4×4) real matrices

$$\begin{pmatrix} 1 & x & y & t \\ 0 & 1 & 0 & \frac{y}{2} \\ 0 & 0 & 1 & -\frac{x}{2} \\ 0 & 0 & 0 & 1 \end{pmatrix} \cdot \begin{pmatrix} 1 & x_1 & y_1 & t_1 \\ 0 & 1 & 0 & \frac{y_1}{2} \\ 0 & 0 & 1 & -\frac{x_1}{2} \\ 0 & 0 & 0 & 1 \end{pmatrix}$$

$$= \begin{pmatrix} 1 & x + x_1 & y + y_1 & t + t_1 + \frac{1}{2}(xy_1 - x_1y) \\ 0 & 1 & 0 & \frac{y+y_1}{2} \\ 0 & 0 & 1 & -\frac{x+x_1}{2} \\ 0 & 0 & 0 & 1 \end{pmatrix}$$

that leads to formula (8). It is an easy exercise to verify that the product (8) satisfies the group axioms. The identity e of the obtained group has coordinates $(0, 0, 0)$ with respect to this multiplication and the inverse element to (x, y, t) is $(-x, -y, -t)$. The pair, consisting of the smooth manifold $\mathbb{R}^3$ and the introduced group law, is called the *Heisenberg group* and is denoted by $\mathbb{H}^1$. This group law defines the left translation: $l_\tau(q) = \tau q$. The left translation l_τ by $\tau = (x, y, z)$ has the differential dl_τ written in coordinates (x, y, t) as

$$dl_\tau = \begin{pmatrix} 1 & 0 & 0 \\ 0 & 1 & 0 \\ -\frac{1}{2}y & \frac{1}{2}x & 1 \end{pmatrix}.$$

The action of $d_e l_\tau$ on the basis $(\partial_x, \partial_y, \partial_t) = (X_e, Y_e, T_e)$ at e, gives the basis (X, Y, T) at τ. We conclude that the basis (X, Y, T) is the basis of left invariant vector fields on the group $\mathbb{H}^1$. They form the Heisenberg algebra $\mathfrak{h}^1$ which is by definition a 3-dimensional Lie algebra with only one non-trivial commutator: $[X, Y] = T$ and all other commutators vanish. We use the identification of the Lie algebra of left-invariant vector fields with $T_e\mathbb{H}^1$. The exponential map is a global diffeomorphism [42] in this case. The coordinates on the group $\mathbb{H}^1$ can be given by

$$\mathbb{H}^1 \ni q = (x, y, t) = \exp(xX + yY + tT), \quad xX + yY + tT \in \mathfrak{h}^1.$$

These coordinates are called *of the first kind*. The inverse map of the exponential restores the group multiplication law from the commutation relations of the Heisenberg algebra in the following way. Let $V = xX + yY + tT \in \mathfrak{h}^1$, $V_1 = x_1X + y_1Y + t_1T \in \mathfrak{h}^1$ and $\tau = \exp(V)$, $q = \exp(V_1)$, then by the Baker-Campbell-Hausdorff formula (4) (BCH-formula for short) we obtain

$$\begin{aligned} \tau q &= \exp(V)\exp(V_1) = \exp(V + V_1 + \frac{1}{2}[V, V_1] + \dots) \\ &= \exp\big((x + x_1)X + (y + y_1)Y + (t + t_1)T + \frac{1}{2}(xy_1 - x_1y)T\big) \\ &= \big(x + x_1, y + y_1, t + t_1 + \frac{1}{2}(xy_1 - x_1y)\big), \end{aligned}$$

that coincides with (8).

There is a norm $\|\cdot\|_{\mathbb{H}^1}$ on the group $\mathbb{H}^1$ which is a direct analogue of the Euclidean norm in $\mathbb{R}^3$. It is defined by

$$\|\tau\|_{\mathbb{H}^1} = \Big((x^2 + y^2)^2 + t^2\Big)^{1/4}.$$

If we stretch the basis vectors X and Y of the Heisenberg algebra by a number $s > 0$, then the bi-linearity of the commutator implies $[sX, sY] = s^2T$. Making use of the

BCH-formula we get the dilatation δ_s on the group:

$$\delta_s(\tau) = \delta_s(x, y, t) = (sx, sy, s^2 t). \tag{9}$$

This dilatation, which is called the *homogeneous dilatation* is compatible with the norm in the sense that the norm becomes a homogeneous of order one function:

$$\|\delta_s(\tau)\|_{\mathbb{H}^1} = \|(sx, sy, s^2 t)\|_{\mathbb{H}^1} = s\|\tau\|_{\mathbb{H}^1}.$$

Compare this situation with the Euclidean norm and the usual dilatation in $\mathbb{R}^3$!

The Heisenberg distance function $d_{\mathbb{H}^1}$ is $d_{\mathbb{H}^1}(\tau, q) = \|\tau^{-1}q\|_{\mathbb{H}^1}$. The Heisenberg distance $d_{\mathbb{H}^1}$ and the Carnot-Carathéodory distance d_{c-c} are equivalent, since both are homogeneous functions.

Example Let us show that the Heisenberg distance and the Euclidean distance d_E are not Lipschitz equivalent, even locally in $\mathbb{R}^3$, see the definition of Lipschitz equivalent in Exercise 2 at the end of section "Heisenberg Group and Isoperimetric Problem". Take two points $e = (0, 0, 0)$ and $q = (0, 0, t)$. Then

$$d_{\mathbb{H}^1}(e, q) = \sqrt{|t|}, \quad d_E(e, q) = |t|,$$

which shows non-equivalence of the distance functions. This also proves that the metric spaces $(\mathbb{R}^3, d_E)$ and $(\mathbb{R}^3, d_{\mathbb{H}^1})$ are not equivalent. But the topological spaces $(\mathbb{R}^3, \tau_E)$ and $(\mathbb{R}^3, \tau_{\mathbb{H}^1})$ are equivalent since any Heisenberg ball contains an Euclidean ball and vice versa.

The metric with the matrix (7) is a left invariant metric on $\mathbb{H}^1$. The sub-bundle D with $D_\tau = \operatorname{span}\{X_\tau, Y_\tau\}$, where X, Y are defined in (6), itself can be called left invariant since it is completely defined by $X(e) = \partial_x$, $Y(e) = \partial_y$, and $D_\tau = dl_\tau D_e$.

The differential operator

$$\mathcal{L}_0 = X^2 + Y^2 = \frac{\partial^2}{\partial x^2} + \frac{\partial^2}{\partial y^2} + \frac{1}{4}(x^2 + y^2)\frac{\partial^2}{\partial t^2} - \Big(y\frac{\partial}{\partial x} - x\frac{\partial}{\partial y}\Big)\frac{\partial}{\partial t} \tag{10}$$

is called sub-Laplacian. It is an analogue of the Laplace-Beltrami operator $\Delta = \frac{\partial^2}{\partial x^2} + \frac{\partial^2}{\partial y^2} + \frac{\partial^2}{\partial t^2}$ in $\mathbb{R}^3$ with respect to the Euclidean metric. We will show in section "Fundamental Solution for $\Box_b$" that the homogeneous function $N(\tau) = \big((x^2 + y^2)^2 + 16t^2\big)^{1/4}$ for $\tau = (x, y, t) \in \mathbb{H}^1$ is connected to the fundamental solution $E(\tau)$ to the subelliptic operator (10) as follows $E(\tau) = \frac{c(Q)}{N(\tau)^{Q-2}}$. The constant $c(Q) < 0$ can be calculated explicitly and $Q = 4$ is the Hausdorff dimension of the metric space $(\mathbb{H}^1, d_{\mathbb{H}^1})$, see [41].

Let us present the exact formulas for the gradient on $\mathbb{H}^1$. In order to use formula (3), we calculate the inverse matrix to (7):

$$g^{ij} = \begin{pmatrix} 1 & 0 & -\frac{y}{2} \\ 0 & 1 & \frac{x}{2} \\ -\frac{y}{2} & \frac{x}{2} & 1 + \frac{x^2+y^2}{4} \end{pmatrix}.$$

Then

$$g^{ij}\begin{pmatrix} \partial_x f \\ \partial_y f \\ \partial_t f \end{pmatrix} = \begin{pmatrix} \partial_x f - \frac{y}{2}\partial_t f \\ \partial_y f + \frac{x}{2}\partial_t f \\ \frac{x\partial_y f - y\partial_x f}{2} + \left(1 + \frac{x^2+y^2}{4}\right)\partial_t f \end{pmatrix} = \begin{pmatrix} Xf \\ Yf \\ \frac{x}{2}Yf - \frac{y}{2}Xf + Tf \end{pmatrix}.$$

Thus

$$\text{grad} f = g^{ij}\begin{pmatrix} \partial_x f \\ \partial_y f \\ \partial_t f \end{pmatrix} \cdot \begin{pmatrix} \partial_x \\ \partial_y \\ \partial_t \end{pmatrix} = (Xf)\,X + (Yf)\,Y + (Tf)\,T.$$

The horizontal gradient "grad_D" is the projection of "grad" onto the space $D_q = \text{span}\{X_q, Y_q\}$ at each point $q \in \mathbb{H}^1$ and it is written as $\text{grad}_D f = (Xf, Yf)$ in the left invariant basis X, Y for D.

Heisenberg Group and Isoperimetric Problem

Let us recall the ancient story of Dido, or Elissa in the Greek version, the founder and the first Queen of Carthage (in modern-day Tunisia). She was daughter of the king of Tyre and after the life-threatening intrigues of her brother Pygmalion she had to leave her land. Eventually Elissa and her followers arrived at the coast of North Africa where Elissa asked the local inhabitants for a small piece of land for a temporary refuge until she could continue her journey. She was allowed to have only as much land as could be encompassed by an oxhide. Elissa cut the oxhide into thin strips so that she had enough to encircle an entire nearby hill. According to this legend, Elissa was the first person who solved the isoperimetric problem of enclosing the maximum area within a boundary of a fixed length.

The dual problem is to find a minimal length curve enclosing the fixed area. Let us formulate this problem mathematically. Introduce the coordinates (x, y) on the plane $\mathbb{R}^2$ and let $c(s) = (x(s), (y(s)),\ s \in I$, be a closed curve in $\mathbb{R}^2$ that encloses a bounded domain Ω. Then the area A of Ω can be calculated as $\int_\Omega dA = \int_c \frac{1}{2}(xdy - ydx)$ by the Stokes theorem. Here the area form $dA = dx \wedge dy$ is the differential of the one form $\frac{1}{2}(xdy - ydx)$. The variational problem with constraint is formulated as follows: *Find a closed curve* $c\colon I \to \mathbb{R}^2$ *of minimal length* $\int_c \sqrt{\dot{x}^2(s) + \dot{y}^2(s)}ds$, *such that the area* $A = \frac{1}{2}\int_c (xdy - ydx)$ *enclosed by this curve is fixed.* Let us introduce the third coordinate t that will reflect the change of the area swept by the curve $c(s) = (x(s), y(s))$, $s \in I$, i.e.,

$$\dot{t}(s) = \frac{1}{2}\big(x(s)\dot{y}(s) - y(s)\dot{x}(s)\big) \quad \text{for all} \quad s \in I. \tag{11}$$

We associate the family of curves $\gamma\colon I \to \mathbb{R}^3$, $\gamma(s) = \big(x(s), y(s), t(s)\big)$ to a single planar curve $c(s) = (x(s), y(s))$, $s \in I$, in such a way that they obey the constraint (11). Integrating condition (11), we get

$$t - t_0 = \frac{1}{2}\int_c \big(x(s)\dot{y}(s) - y(s)\dot{x}(s)\big)\, ds,$$

which means that the area enclosed by the planar curve c and the straight line connecting the end of c with the origin, is equal to the change of the vertical coordinate of γ (here we assumed $t_0 = 0$).

Another desirable condition is *to find a Riemannian metric* g *in* $\mathbb{R}^3$ *such that the length of* $\gamma\colon I \to \mathbb{R}^3$ *is equal to the length of the planar curve* c. In order to satisfy it, we find a sub-bundle D of planes in $\mathbb{R}^3$ such that γ will be tangent to D and the length of the vector $\dot{c}(s) = (\dot{x}(s), \dot{y}(s))$ in $\mathbb{R}^2$ coincides with the length of the vector $\dot{\gamma}(s) = (\dot{x}(s), \dot{y}(s), \dot{t}(s)) \in D_{\gamma(s)} \subset \mathbb{R}^3$. In this case we only need the restriction g_D of the Riemannian metric g to planes $D_{\gamma(s)}$ that will be the sub-Riemannian metric. Thus the sub-bundle D has to be annihilated by the form dual to the additional velocity coordinate of the spatial curve γ. So

$$D(x, y, t) = \ker(\omega) = \ker(dt - \frac{1}{2}(xdy - ydx)),$$

and the sub-Riemannian metric g_D is just the Euclidean metric on D making the basis for D orthonormal. The reader may recognize the Heisenberg manifold in the space $(\mathbb{R}^3, D, g_D)$ described in the first part of section "Heisenberg Group". More information about the relation between the isoperimetric problems and the Heisenberg groups the reader can find in [6, 24].

Exercises

1. Show that the Carnot-Carathéodory distance function on the Heisenberg sub-Riemannian manifold is homogeneous with respect to the dilatation (9).
2. Show that any two homogeneous with respect to the dilatation (9) distance functions d_1 and d_2 are equivalent on the Heisenberg group; that is, there are constants $C, \widetilde{C} > 0$, such that

$$Cd_1(\tau, q) \le d_2(\tau, q) \le \widetilde{C}d_1(\tau, q), \quad \tau, q \in \mathbb{H}^1.$$

Action of the Heisenberg Group on the Siegel Upper Half Space

Let $\mathbb{C}^{n+1}$ be the $(n+1)$-dimensional complex space. We use the notation $z = (z', z_{n+1})$, where $z' = (z_1, \ldots, z_n) \in \mathbb{C}^n$. The set

$$\mathcal{U}_n = \{(z_1, \ldots, z_{n+1}) \in \mathbb{C}^{n+1} : \text{Im}\,(z_{n+1}) > |z'|^2 = \sum_{l=1}^{n} |z_l|^2\}$$

defines the Siegel upper half space in $\mathbb{C}^{n+1}$. Let $B_{\mathbb{C}}$ denote the unit ball in $\mathbb{C}^{n+1}$:

$$B_{\mathbb{C}} = \{(w_1, \ldots, w_{n+1}) \in \mathbb{C}^{n+1} : \sum_{l=1}^{n+1} |w_l|^2 < 1\}.$$

Then the Cayley transformation

$$w_{n+1} = \frac{i - z_{n+1}}{i + z_{n+1}}, \qquad w_l = \frac{2iz_l}{i + z_{n+1}}, \qquad l = 1, \ldots, n,$$

and its inverse

$$z_{n+1} = i\frac{1 - w_{n+1}}{1 + w_{n+1}}, \qquad z_l = \frac{w_l}{1 + w_{n+1}}, \qquad l = 1, \ldots, n,$$

show that the unit ball $B_{\mathbb{C}}$ and the Siegel upper half space $\mathcal{U}_n$ are biholomorphically equivalent.

Dilatation Let $p = (z', z) \in \mathcal{U}_n$. For every positive number s we define a *dilatation* $\delta_s(p)$ by

$$\delta_s(p) = \delta_s(z', z) = (sz', s^2 z).$$

The non-isotropy of the dilatation comes from the definition of $\mathcal{U}_n$.

Rotation For every unitary linear transformation from $U(n, \mathbb{C})$ we define the *rotation* $\text{Rot}(p)$ on $\mathcal{U}_n$ by

$$\text{Rot}(p) = \text{Rot}(z', z) = (U(z'), z).$$

Both the dilatation and the rotation are extended to mappings on the boundary $\partial\mathcal{U}_n$.

Translation We introduce the n-dimensional analogue $\mathbb{H}^n$ of the Heisenberg group $\mathbb{H}^1$. Topologically the Heisenberg group is the space $\mathbb{H}^n = \mathbb{C}^n \times \mathbb{R}$, endowed with the Lie group product

$$[z, t][z', t'] = [z + z', t + t' + 2\text{Im} \sum_{k=1}^{n} z_k \bar{z}'_k]. \tag{12}$$

We change the constant $\frac{1}{2}$ to 2 in the last term of the multiplication law (12) in comparison to (8) for the convenience of calculations in sections "$\Box_b$ Operator on Heisenberg Group" and "Fundamental Solution for $\Box_b$". To every element $[z,t] \in \mathbb{H}^n$ we associate the following affine self-map of $\mathcal{U}_n$, that is the action on the left of the group $\mathbb{H}^n$ on $\mathcal{U}_n$. Notice, that we write $[z,t]$ for the element of the Heisenberg group and $p = (q', q)$ for the point in the Siegel upper half space $\mathcal{U}_n$.

$$[z,t].(q',q) \mapsto (q' + z, q + t + i|z|^2 + 2i\bar{z}q'). \tag{13}$$

This mapping preserves the level sets, given by the function $r(p) = \operatorname{Im}(q) - |q'|^2$. In fact, since $|q' + z|^2 = |q'|^2 + |z|^2 + 2\operatorname{Re}(\bar{z}q_1)$, we obtain

$$\operatorname{Im}(q + i|z|^2 + 2i\bar{z}q') - |q' + z|^2 = \operatorname{Im}(q) - |q'|^2.$$

Hence, the transformation (13) maps $\mathcal{U}_n$ onto itself and preserves the boundary $\partial\mathcal{U}_n$. Moreover, the constructed map is holomorphic in the domain $\mathcal{U}_n$.

One will only have to check that the mapping (13) defines an action of the group $\mathbb{H}^n$ on the space $\mathcal{U}_n$:

$$[w,s]\Big([z,t].(q',q)\Big) = \Big([w,s][z,t]\Big)(q',q).$$

Thus, (13) presents a realisation of $\mathbb{H}^n$ as a group of affine holomorphic bijections of $\mathcal{U}_n$. We can identify the elements of $\mathbb{H}^n$ with the boundary $\partial\mathcal{U}_n$ via its action at the origin $[z,t].(0,0) \mapsto (z, t + i|z|^2)$. Thus, $\mathbb{H}^n \ni [z,t]$ is identified with $(z, t + i|z|^2) \in \partial\mathcal{U}_n$. We may use the following coordinates (q', t, r) on $\mathcal{U}_n$:

$$\mathcal{U}_n \ni (q',q) = (q',t,r), \quad r = r(q',q) = \operatorname{Im}(q) - |q'|^2, \quad t = \operatorname{Re} q.$$

If $\operatorname{Im}(q) = |q'|^2$, then we get the coordinate representation of the boundary $\partial\mathcal{U}_n$

$$\partial\mathcal{U}_n \ni (q',q) = (q',t), \quad t = \operatorname{Re} q, \quad r = r(q',q) = 0.$$

In Sect. 5 we give the quaternion analogue of the Heisenberg group and its relation to quaternion Siegel upper half space and partial differential equations.

Carnot Groups

The following example includes connected simply connected Lie groups $\mathbb{G}$ whose Lie algebras are the direct sum of their subspaces

$$\mathfrak{g} = V_1 \oplus V_2 \oplus \dots V_m,$$

such that $[V_1, V_k] = V_{k+1}$, $k = 1, 2, \dots, m-1$, and $[V_1, V_m] = 0$. Since the commutators have finite length, the algebras and the groups are nilpotent. Define the sub-bundle D by left translations of V_1 and endow $\mathbb{G}$ by a left invariant metric $g_D: D_q \times D_q \to D_q$, $q \in \mathbb{G}$. The sub-bundle will be bracket generating since the space V_1 generates the entire Lie algebra. The sub-Riemannian manifold $(\mathbb{G}, D, g_D)$ received the name *Carnot groups* in literature.

2-Step Carnot Groups

2-step Carnot groups $\mathbb{G}$ have the Lie algebras $\mathfrak{g}$ satisfying

$$\mathfrak{g} = V_1 \oplus V_2, \qquad [V_1, V_1] = V_2, \quad [V_1, V_2] = [V_2, V_2] = \{0\}.$$

The underlying manifold is $\mathbb{R}^{\alpha+\beta}$, $\alpha = \dim V_1$, $\beta = \dim V_2$. The group multiplication law can be written by making use of a $\mathbb{R}^{\beta}$-valued skew symmetric form $\Omega: \mathbb{R}^{\alpha} \times \mathbb{R}^{\alpha} \to \mathbb{R}^{\beta}$. Namely, if we write $(v_1, v_2), (v_1', v_2') \in V_1 \oplus V_2$ for the Lie algebra elements, then

$$\left[(v_1, v_2), (v_1', v_2')\right] = (0, \Omega(v_1, v_1')).$$

If we use the coordinates of the first kind and write $\tau = (x, t) \in \mathbb{G}$, $q = (x_1, t_1) \in \mathbb{G}$, then

$$\tau q = (x, t)(x_1, t_1) := (x + x_1, t + t_1 + \frac{1}{2}\Omega(x, x_1))$$

by the BCH-formula (4).

$\mathbb{H}$-Type Groups

$\mathbb{H}$(eisenberg)-type groups are natural generalisations of the Heisenberg group. First of all, they are 2-step Carnot groups and their Lie algebras $\mathfrak{h}$ possess an inner product $(.,.)_{\mathfrak{h}}$ making the decomposition $V_1 \oplus V_2$ orthogonal. Moreover, by making use of the inner product and the given commutator we define a linear map $J: V_2 \to \mathrm{End}(V_1)$ by

$$(J_t v_1, v_2)_{\mathfrak{h}} = (t, [v_1, v_2])_{\mathfrak{h}}.$$

Under the above condition we say that $\mathbb{G}$ is an $\mathbb{H}$-type algebra if the map J satisfies the condition $J_t^2 v = -(t, t)_{\mathfrak{h}} \,\mathrm{Id}_{V_1}$ for any $t \in V_2$ and any $v \in V_1$. The $\mathbb{H}$-type groups were introduced by Kaplan [63] in the study of fundamental solutions of the associated sub-Laplace operator. Equivalent definitions of $\mathbb{H}$-type groups and their geometric properties can be found in [34, 64, 72, 73].

EXERCISE Show that the Heisenberg group is of $\mathbb{H}$-type.

Engel Group

The Engel group is an example of a 3-step Carnot group. The underlying manifold is $\mathbb{R}^4$. We use coordinates $q = (x, y, z, w)$. The Lie group multiplication law is given by the BCH-formula for a nilpotent group of step 3:

$$\exp(F_1)\exp(F_2) = \exp(F_1 + F_2 + \frac{1}{2}[F_1, F_2] + \frac{1}{12}[F_1, [F_1, F_2]] - \frac{1}{12}[F_2, [F_1, F_2]]). \tag{14}$$

The Lie algebra for the Engel group has to satisfy the relations

$$[X, Y] = Z, \quad [X, Z] = aW, \quad [Y, Z] = bW, \quad a, b \in \mathbb{R}.$$

For example, if we choose a slight modification of the Heisenberg vector fields

$$X = \partial_x - \frac{1}{2}y\partial_z + z\partial_w, \quad Y = \partial_y + \frac{1}{2}x\partial_z - z\partial_w,$$

then we get

$$[X, Y] = \partial_z := Z, \quad [X, Z] = -\partial_w = W, \quad [Y, Z] = \partial_w = -W. \tag{15}$$

If we write $F_i = x_iX + y_iY + z_iZ + w_iW$, $i = 1, 2$ then the BCH-formula (14) and the commutation relations (15) lead to the group law

$$(x_1, y_1, z_1, w_1)(x_2, y_2, z_2, w_2) = \big(x_1 + x_2,\ y_1 + y_2,\ z_1 + z_2 + \frac{1}{2}(x_1y_2 - x_2y_1),$$
$$w_1 + w_2 - \frac{1}{2}(x_1z_2 + x_2z_1) + \frac{1}{2}(y_1z_2 - y_2z_1) + \frac{1}{12}(y_1 - y_2)(x_1y_2 - x_2y_1)\big).$$

Another coordinate representation of the Engel group can be found in [36].

3 Hypoellipticity for Linear Equations

Functional Spaces and Differential Operators

Let Ω be an open connected set (domain) in the space $\mathbb{R}^n$. We use the notation $\alpha = (\alpha_1, \dots, \alpha_n)$, $\alpha_k \in \mathbb{N}$, for multi-index and $|\alpha| = \sum_k \alpha_k$ for its order. If f is a real function defined in Ω, then $\partial^\alpha f$ denotes its derivative of order $|\alpha|$ if it exists.

The vector space $C^m(\Omega)$, $m \in \mathbb{N}$, consists of all functions $f: \Omega \to \mathbb{R}$ which, together with all their derivatives $\partial^\alpha f$ of order $|\alpha| \le m$ are continuous in Ω. We write $C^\infty(\Omega) = \cap_{m=0}^\infty C^m(\Omega)$. The space $C_0^\infty(\Omega)$ is the subspace of $C^\infty(\Omega)$ of functions, having a compact support in the domain Ω.

The vector space $L^p(\Omega)$, $p > 0$, is the set of equivalence classes of measurable real or complex valued functions in Ω having the finite norm $\|u\|_{L^p} = \int_\Omega |u(x)|^p \, dx$. A class of equivalence consists of functions that coincide almost everywhere in Ω. The spaces $L^p(\Omega)$ are Banach spaces for $p \ge 1$ and for $p = 2$ it is a Hilbert space endowed with the inner product

$$(f_1, f_2)_{L^2(\Omega)} = \int_\Omega f_1(x) \bar{f}_2(x) \, dx.$$

The space $C_0^\infty(\Omega)$ is dense in $L^p(\Omega)$ in the topology defined by the norm $\| \cdot \|_{L^p}$.

The space of test functions $\mathcal{D}(\Omega)$ is the set C_0^∞ endowed with the following topology: a sequence φ_k converges to φ as $k \to \infty$ in $\mathcal{D}(\Omega)$ if

1. there is a compact $K \subset \Omega$ such that the support of $\varphi_k - \varphi$ belongs to K for every k;
2. $\lim_{k\to\infty} \partial^\alpha \varphi_k(x) = \partial^\alpha \varphi(x)$ uniformly on the compact K for each multi-index α.

With this definition of the convergence the space $\mathcal{D}(\Omega)$ is a complete locally convex topological vector space, where a set is compact if and only if it is closed and bounded. The dual space $\mathcal{D}'(\Omega)$ of $\mathcal{D}(\Omega)$ (space of linear functionals on $\mathcal{D}(\Omega)$) is called the space of (Schwartz) distributions on Ω. Functions $u \in L^1_{\text{loc}}(\Omega)$ can be identified with distributions $T_u(\varphi) = \int_\Omega u(x)\varphi(x) \, dx$.

The space of test functions $\mathcal{G}(\Omega)$ (rapidly decreasing functions) consists of functions $\varphi \in C^\infty(\Omega)$ having the following property: for any $N = 1, 2, , \ldots$ there is a constant M_N such that

$$|\varphi(x)| \le M_N |x|^{-N} \quad \text{as} \quad x \to \infty.$$

The space $\mathcal{G}(\Omega)$ can be endowed with countable family of seminorms

$$\sup_{x\in\Omega} |x^\beta \partial^\alpha \varphi(x)| < \infty, \quad \alpha, \beta \quad \text{are multi-indices,}$$

and thus it is a Fréchet space. The dual space $\mathcal{G}'(\Omega)$ of $\mathcal{G}(\Omega)$ is called the space of tempered distributions on Ω.

The last space of test functions $\mathcal{E}(\Omega)$ is the space $C^\infty(\Omega)$, endowed with the topology induced by semi-norms

$$\|\varphi\|_k = \sum_{|\alpha|\le k} \sup_{K\subset\Omega} |\partial^\alpha \varphi| < \infty, \quad k = 1, 2, \ldots,$$

where the supremum is taken over all compacts K in the domain Ω. The dual space $\mathcal{E}'(\Omega)$ to the Fréchet space $\mathcal{E}(\Omega)$ is the space of distributions having compact support.

Observe that since $\mathcal{D}(\Omega) \subset \mathcal{G}(\Omega) \subset \mathcal{E}(\Omega)$ as topological spaces and sets, we obtain that $\mathcal{E}'(\Omega) \subset \mathcal{G}'(\Omega) \subset \mathcal{D}'(\Omega)$.

The space of distributions is endowed with the weak* topology and the sequence $\{u_k\}$ from $\mathcal{D}'(\Omega)$ converges to $u \in \mathcal{D}'(\Omega)$ in the weak* topology if it converges pointwise

$$u_k(\varphi) \to u(\varphi) \quad \text{as} \quad k \to \infty \quad \text{for all} \quad \varphi \in \mathcal{D}(\Omega).$$

Analogously, the weak* topology is defined for the spaces $\mathcal{G}(\Omega)$ and $\mathcal{E}(\Omega)$. More information about distributions can be found in [1, 98].

If $f \in L^p(\mathbb{R}^n)$, $p \geq 1$, then the Fourier transform $F(f) = \widehat{f}$ of f is defined by

$$\widehat{f}(\xi) = \int_{\mathbb{R}^n} e^{-i(x,\xi)} f(x)\, dx, \quad \xi \in \mathbb{R}^n.$$

Here $(x, \xi) = \sum_{k=1}^n x_k \xi_k$ is the Euclidean inner product in $\mathbb{R}^n$. If $\widehat{f} \in L^p(\mathbb{R}^n)$, $p \geq 1$, then we can define the inverse Fourier transform by

$$f(x) = (2\pi)^{-n} \int_{\mathbb{R}^n} e^{i(x,\xi)} f(\xi)\, d\xi, \quad x \in \mathbb{R}^n.$$

The space $\mathcal{G}(\mathbb{R}^n)$ is closed under the Fourier transform and moreover the map

$$\begin{aligned} F \colon \mathcal{G}(\mathbb{R}^n) &\to \mathcal{G}(\mathbb{R}^n) \\ \varphi &\mapsto \widehat{\varphi} \end{aligned}$$

is continuous. The Fourier transform $\widehat{u}$ of a distribution $u \in \mathcal{G}'(\mathbb{R}^n)$ is defined by

$$\widehat{u}(\varphi) = u(\widehat{\varphi}), \quad \varphi \in \mathcal{G}(\mathbb{R}^n).$$

The map $F \colon \mathcal{G}'(\mathbb{R}^n) \to \mathcal{G}'(\mathbb{R}^n)$ is an isomorphism of vector spaces with weak* topology. The map F restricted to functions from to $L^2(\mathbb{R}^n)$ is an isometry, according to the Parseval equality.

Let Ω be an open set in $\mathbb{R}^n$. For $s \in \mathbb{R}$ and $1 \leq p \leq \infty$ the space $H^p_s(\Omega)$ is defined by

$$H^p_s = \left\{ u \in \mathcal{G}'(\Omega) \mid \|u\|_{H^p_s} = \left((2\pi)^{-n} \int_\Omega |\widehat{u}(\xi)|^p (1 + |\xi|^2)^{\frac{sp}{2}}\, d\xi \right)^{1/p} < \infty \right\},$$

where for $p = \infty$ we consider the norm $\|u\|_{H^\infty_s} = \text{ess.}\sup |(1 + |\xi|^2)^{s/2} \hat{u}(\xi)|$. The space $(H^p_s(\Omega), \|\cdot\|_{H^p_s})$ is the Banach space possessing the properties.

$(H1)$ One has $\mathcal{G} \subset H_s^p \subset \mathcal{G}'$ in the topological sense: the topology in $\mathcal{G}$ is stronger than that induced there by H_s^p and the topology in H_s^p is stronger than one induced from $\mathcal{G}'$.

$(H2)$ C_0^∞ is dense in H_s^p for $p < \infty$.

$(H3)$ If $-\infty < s_1 < s_2 < \infty$, then $H_{s_2}^p \subset H_{s_1}^p$. It follows from

$$(1 + |\xi|^2)^{s_1} \leq (1 + |\xi|^2)^{s_2}, \quad \xi \in \mathbb{R}^n.$$

If K is a compact in Ω, then the inclusion map $\iota: H_{s_2}^p \cap \mathcal{E}'(K) \to H_{s_1}^p$ is completely continuous. Conversely, if there is a non empty open set $V \subset \Omega$, such that $H_{s_2}^p \cap \mathcal{E}'(V) \subset H_{s_1}^p$ then $s_1 < s_2$.

$(H4)$ If $-\infty < s_1, s_2 < \infty$, then $H_{s_1}^p \cap H_{s_2}^p = H_{s_1+s_2}^p$ and

$$\max_{j=1,2} \|u\|_{H_{s_j}^p} \leq \|u\|_{H_{s_1+s_2}^p} \leq \|u\|_{H_{s_1}^p} + \|u\|_{H_{s_2}^p} \quad \text{for} \quad u \in H_{s_1}^p \cap H_{s_2}^p.$$

$(H5)$ Let $\phi \in C_0^\infty(\mathbb{R}^n)$ be such that $\int_{\mathbb{R}^n} \phi \, dx = 1$. Set $\phi_\delta(x) = \delta^{-n}\phi(\frac{x}{\delta})$. If $u \in H_s^p$, $p < \infty$, then the convolution $S_\delta u = u * \phi_\delta$ is a smooth function for any $\delta > 0$ and it converges to u in H_s^p as $\delta \to 0$.

$(H6)$ Let $\psi \in C_0^\infty(\mathbb{R}^n)$ be such that $\psi(0) = 1$. Set $\psi^\delta(x) = \psi(\delta x)$. If $u \in H_s^p$, $p < \infty$, then $\psi^\delta u$ converges to u in H_s^p as $\delta \to 0$.

We say that a function u belongs to $H_{s,loc}^p(\Omega)$ if $u \in H_s^p(V)$ for any open set $V \subset \Omega$ with $\overline{V} \subset \Omega$.

Space H_s

The space $H_s(\Omega)$ is a special case of $H_s^p(\Omega)$ when $p = 2$. The Banach structure is defined by the inner product and therefore it is a Hilbert space. Thus

$$H_s = \left\{ u \in \mathcal{G}'(\Omega) \mid \|u\|_{H_s} = \left((2\pi)^{-n} \int_\Omega |\widehat{u}(\xi)|^2 (1 + |\xi|^2)^s \, d\xi \right)^{1/2} < \infty \right\}.$$

When $s > 0$ is integer, we have

$$H_s(\Omega) = \{u \in L^2(\Omega) \mid \partial^\alpha u \in L^2(\Omega), \quad \text{for} \quad |\alpha| \leq s\}.$$

Also, since $\|u\|_{H_{s+1}}^2 = \|u\|_{H_s}^2 + \sum_{j=1}^n \|\partial_j u\|_{H_s}^2$, we conclude that

$$H_{s+1}(\Omega) = \{u \in H_s(\Omega) \mid \partial_j u \in H_s(\Omega), \quad \text{for} \quad j = 1, 2, \ldots, n\}.$$

If $u \in H_s$, $a \in \mathcal{G}(\Omega)$ and $\phi \in C_0^\infty(\Omega)$, then

$$aS_\delta u - S_\delta(au) = a(u * \phi_\delta) - (au) * \phi_\delta \to 0 \quad \text{in} \quad H_{s+1} \quad \text{as} \quad \delta \to 0.$$

Now we define the space $H_{s,loc}(M)$ on a C^∞ smooth manifold M. We say that a distribution u belongs to $H_{s,loc}(M)$ if for any chart (U,φ), the function $u\circ\varphi$ is in the space $H_s(\varphi(U))$.

Let Ω_1 and Ω_2 be two open sets in $\mathbb{R}^n$ and $f\colon\Omega_1\to\Omega_2$ be a diffeomorphism. Let $\Omega_2'\subset\Omega_2$ and $u\in\mathcal{E}'(\Omega_2')$. Then it follows that $u\circ f\in H_s(\Omega_1)$ if and only if $u\in H_s(\Omega_2')$ and we have

$$\|u\circ f\|_{H_s(\Omega_1)}\le C\|u\|_{H_s(\Omega_2')}\quad\text{for}\quad u\in H_s(\Omega_2')\cap\mathcal{E}'(\Omega_2')$$

and the constant C does not depend on u.

Let now M be a C^∞ smooth compact manifold with a boundary.

1. For integer positive s the natural inclusion map $H_{s+1}(M)\to H_s(M)$ is completely continuous.
2. Let P be a differential operator of order less than or equal to m with C^∞ coefficients defined on M. Then the mapping $u\mapsto Pu$ defines a continuous mapping of $H_{m,loc}(M)$ into $H_{0,loc}(M)$.

Differential Operators

Recall that $\partial_j=\frac{\partial}{\partial x_j}$ is the differentiation with respect to x_j in $\mathbb{R}^n$. Denote $D_j=-i\partial_j$, where i is the imaginary unit. Then we write $D^\alpha=D_1^{\alpha_1}\dots D_n^{\alpha_n}$ for a multi-index $\alpha=(\alpha_1,\dots,\alpha_n)$. Let $P(\xi)$ be a polynomial in n variables $\xi_1,\dots,\xi_n$, with complex coefficients. Denote by $P(D)$ the differential operator obtained if ξ_j is replaced by D_j. This gives a one-to-one correspondence between polynomials and differential operators with constant coefficients. Recall the classical *du Bois Reymond lemma*: if u and f are continuous functions in a domain $\Omega\subset\mathbb{R}^n$ and $D_ju=f$ in the distribution sense, then $D_ju=f$ in the classical sense also. Now we give a formal definition of the differential operator.

Definition 3.1 A linear map $u\mapsto P(D)u=\sum_\alpha a^\alpha D^\alpha u$, where a^α are complex valued functions defined in some domain $\Omega\subset\mathbb{R}^n$ is called a differential operator. Here it is also assumed that all but a finite number of coefficients a^α vanish identically on any compact subset of Ω, thus the sum $\sum_\alpha a^\alpha D^\alpha u$ is always finite.

To emphasise that the coefficients a^α depend on $x\in\Omega$, we write $P(x,D)$. If $a^\alpha\in C^k(\Omega)$, $0\le k\le\infty$ for every α, we say that P has C^k coefficients. Thus we interpret P as a linear map from $C^\infty(\Omega)$ to $C^k(\Omega)$.

We need to define a differential operator with C^k coefficients on a C^∞ smooth manifold M. We define the linear map $P\colon C^\infty(M)\to C^k(M)$ by choosing an arbitrary chart (U,φ) and setting

$$(Pu)\circ\varphi:=\sum_\alpha a^\alpha(\varphi)D^\alpha(u\circ\varphi)=P^\varphi(u\circ\varphi)=P^\varphi u_\varphi\quad\text{for}\quad u\in C^\infty(M).$$

Observe that a linear map $L: C^\infty(M) \to C^k(M)$ is a differential operator if and only if $\operatorname{supp}(Lv) \subset \operatorname{supp} v$ for every $v \in C^\infty(M)$, see [85].

The next steps are to define distributions on a manifold and then differential operators acting on the distributions. Let (U, φ) and (U', φ') be two arbitrary charts on M. If there is a distribution $u_\varphi = u \circ \varphi \in \mathcal{D}'(\varphi(U))$ such that

$$u_{\varphi'} = u \circ \varphi' = u_\varphi \circ (\varphi' \varphi^{-1}) \quad \text{is a distribution on} \quad \varphi'(U \cap U'), \tag{16}$$

then we call the system $\{u_\varphi\}$ *a distribution u on M*. The set of all distributions on M is denoted by $\mathcal{D}'(M)$. Let $u \in \mathcal{D}'(M)$ then u_φ is a distribution on $\varphi(U)$ and then $P^\varphi u_\varphi$ satisfies (16). Thus we call the system $\{P^\varphi u_\varphi\}$ the distribution $Pu \in \mathcal{D}'(M)$.

If P is a differential operator on a domain $\Omega \subset \mathbb{R}^n$, then the order m of P in Ω is defined by $m = \sup\{|\alpha| \mid a^\alpha \neq 0\}$. The order can be $+\infty$. If P is of finite order m in Ω, the *principal part* (or the *characteristic form*) of P is the following polynomial in ξ

$$P_m(x, \xi) = \sum_{|\alpha|=m} a^\alpha(x)\xi^\alpha, \quad x \subset \Omega,\ \xi \in \mathbb{C}^n.$$

The order of a differential operator on a manifold can also be defined.

Fundamental Solutions

Definition 3.2 A distribution $E \in \mathcal{D}'(\mathbb{R}^n)$ is called a fundamental solution for the differential operator $P(D)$ with constant coefficients if

$$P(D)E = \delta_0,$$

where δ_0 is the Dirac distribution at $0 \in \mathbb{R}^n$.

Notice that to every differential operator $P(D)$ with constant coefficients there exists a fundamental solution E in a certain functional class $\mathcal{F}$, which depends on P. The details can be found in [54, Theorem 3.1.1].

Let us assume that we are given a differential equation $P(D)u = f$ with $f \in \mathcal{E}'(\mathbb{R}^n)$ and let E be a fundamental solution: $P(D)E = \delta$. Then

$$P(D)(E * f) = f, \quad E * (P(D)u) = u, \qquad u, f \in \mathcal{E}'(\mathbb{R}^n).$$

The properties of the convolution lead to the following two facts

1. If $P(D)u = f$ with $f \in \mathcal{E}'(\mathbb{R}^n) \cap H_k^p(\mathbb{R}^n)$ and $E \in H_{s,loc}^\infty(\mathbb{R}^n)$, then the solution u is given by the formula $u = E * f$ and it belongs to $H_{ks,loc}^p(\mathbb{R}^n)$.

2. Let $u \in \mathcal{E}'(\mathbb{R}^n)$ and $-\infty < s < \infty$, $1 \leq p \leq \infty$ are given. Then $u \in H^p_{sK}(\mathbb{R}^n)$ if and only if $f = P(D)u \in H^p_s(\mathbb{R}^n)$, where the function K depends on the properties of the operator P.

We now will discuss the existence of solutions and the geometric properties of domain $\Omega \subset \mathbb{R}^n$. A solution of the differential equation $P(D)u = 0$ in $\mathbb{R}^n$ is called an *exponential solution* if it can be written in the form

$$u(x) = p(x)e^{i(x,\xi)} \quad \text{for some} \quad \zeta \in \mathbb{C}^n$$

and polynomial function p. It is known [54] that the closed linear hull in $C^\infty(\mathbb{R}^n)$ of exponential solutions consists of all solutions of $P(D)u = 0$ in $C^\infty(\mathbb{R}^n)$. If $\Omega \subset \mathbb{R}^n$ is an open convex set, then the exponential solutions are dense in the set of all solutions of $P(D)u = 0$ in $C^\infty(\Omega)$.

Consider the differential equation $P(D)u = f, f \in \mathcal{D}'(\Omega) \cap H^p_{s,\text{loc}}$. Then the solution exists if and only if the open set Ω satisfy the following definition of P-convexity: If for every compact set $K \subset \Omega$ there is another compact set $K' \subset \Omega$ such that for any chosen $\phi \in C_0^\infty(\Omega)$ the following is true:

$$\text{supp}(P(-D)\phi) \subset K \quad \text{implies} \quad \text{supp}(\phi) \subset K'.$$

As a consequence we obtain that if Ω is P-convex, then the equation $P(D)u = f$ has a solution $u \in C^\infty(\Omega)$ for every $f \in C^\infty$.

Hypoelliptic Operators with Constant and Variable Coefficients

We start from the historical observations. Hermann Weyl in [115] has shown that weak solutions of the Laplace equation are actually smooth functions. The fact that weak solutions are C^∞ solutions are very important for applications and particularly for variational methods to the solutions of differential equations. A generalisation of the Weyl result can be obtained by two ways. One of them is based on the properties of fundamental solutions, see [91]. Another way, using only a priori estimates, was developed in [43, 61, 81]. If we are given a partial differential equation with constant coefficients, then the classical solution is analytic if and only if the equation is elliptic, see [86]. In the same work it was shown that a solution of a non-linear elliptic partial differential equation with analytic coefficients is an analytic function, that extends the result in [17]. If we change the category of solutions from analytic to C^∞ smooth and ask the analogous questions, then the complete answer can be found in works of Hörmander [52], see also [54, 56, 57, 102]. We briefly explain the main idea to study the following problem. Consider the equation $Pu = f$, where u and f are given distributions. We ask how smooth is the solution u if f is C^∞ smooth? In the first step the analytic properties of the homogeneous solutions are transferred to the algebra-geometric properties of the polynomial $P(\xi)$ defining the differential operator. Let us consider the following subspace of the Fréchet space $H^p_{s,\mathrm{loc}}(\Omega)$, where Ω is an open subset of $\mathbb{R}^n$:

$$\mathcal{N} = \{u \in H^p_{s,\mathrm{loc}}(\Omega) \mid P(D)u = 0\}.$$

The space $\mathcal{N}$, that is the kernel of the linear operator $P(D)$, is a Fréchet subspace with respect to the topology induced from $H^p_{s,\mathrm{loc}}(\Omega)$.

Definition 3.3 A Fréchet topological space is called a Montel space if any bounded sequence contains a convergent subsequence.

With the functional space $\mathcal{N}$ we associate the algebraic hypersurface in $\mathbb{C}^n$ defined by the polynomial $P(\xi)$ associated with the differential operator $P(D)$:

$$\Lambda = \{\lambda \in \mathbb{C}^n \mid P(\lambda) = 0\}.$$

Theorem 3.4 *Let us assume that $\mathcal{N}$ is a Montel space, then the following conditions are equivalent.*

H1. *$Im(\lambda) \to \infty$ as $\lambda \to \infty$ on the surface Λ.*
H2. *We define the distance function from $\xi \in \mathbb{R}^n$ to the hypersurface Λ by*

$$d(\xi) = \mathrm{dist}(\xi, \Lambda) = \inf_{\lambda \in \Lambda}\{\mathrm{dist}(\xi, \lambda)\}.$$

Then $d(\xi) \to \infty$ as $\xi \to \infty$ in $\mathbb{R}^n$.
H3. *There are positive constants $c > 0$ and $C > 0$ such that $|\xi|^c \le Cd(\xi)$ for sufficiently large $|\xi|$, $\xi \in \mathbb{R}^n$.*

Proof We present a very rough draft of the proof that condition $H1$ is fulfilled. For the details we refer the reader to [54, Chap. IV]. We assume that there is a sequence $\lambda_j = \xi_j + i\eta_j \to \infty$ on Λ as $j \to \infty$, but $\eta_j = \mathrm{Im}(\lambda_j)$ is bounded. The rest of the proof is based on the two side estimate

$$\int_{\mathbb{R}^n} |\hat{\varphi}(\xi - i\eta_j)|^p |m(\xi)|^p \, d\xi \le \|\varphi u_j\|^p_{H^p_s(\Omega)} \le \int_{\mathbb{R}^n} |\hat{\varphi}(\xi - i\eta_j)|^p |M(\xi)|^p \, d\xi, \tag{17}$$

where $\varphi \in C_0^\infty(\Omega)$ is an arbitrary function, $u_j(x) = \frac{e^{i(x,\xi)}}{(1+|\xi_j|^2)^{s/2}} \in \mathcal{N}$, and

$$m(\xi) \le \left(\frac{1 + |\xi|^2}{1 + |\xi_j|^2}\right)^{s/2} \le M(\xi).$$

The boundedness of η_j implies the boundedness of the right hand side of (17) and, therefore, the boundedness of u_j in $H^p_{s,loc}(\Omega)$. Since the sequence $u_j \to 0$ in $\mathcal{D}'(\Omega)$

$$\left(\text{this follows from the fact that } \int \varphi u_j = \frac{\hat{\varphi}(-\lambda_j)}{(1 + |\xi_j|^2)^{s/2}} \to 0 \quad \text{as} \quad \lambda_j \to \infty\right),$$

the point 0 is the only limit point of the bounded sequence $\{u_j\}$ in $H^p_{s,loc}(\Omega)$. But if $\|\varphi u_j\|^p_{H^p_s(\Omega)}$ tends to zero, then the left hand side of (17) implies that $\hat{\varphi}(\xi - i\eta_0) \equiv 0$, where η_0 is a limit point of the sequence η_j. Thus $\varphi = 0$, that contradicts to the arbitrary choice of $\varphi \in C_0^\infty(\Omega)$.

The condition $H2$ is a reformulation of $H1$ and the $H3$ is equivalent to $H2$ from estimation on the distance function. □

L. Hörmander called the operators $P(D)$ satisfying to equivalent conditions $H1$ – $H3$ *hypoelliptic differential operators.*

Definition 3.5 A differential operator $P(D)$ of order m is called elliptic if its principal part $P_m(D)$ (that is the homogeneous part of $P(D)$ of order m exactly) satisfies the condition

$$P_m(\xi) \neq 0, \quad \text{for all} \quad 0 \neq \xi \in \mathbb{R}^n.$$

As a consequence, we immediately obtain that the elliptic operator is hypoelliptic.

We list consequences of Theorem 3.4, showing the advantages of the hypoelliptic operators. We assume that the equation $P(D)u = f$ is given for the distributions u and f, and that the operator $P(D)$ is hypoelliptic. Then

1. If $u \in \mathcal{D}'(\Omega)$ and $f \in H^p_{s,loc}(\Omega)$, then $u \in H^p_{\Phi(s,P),loc}(\Omega)$, where the function $\Phi(s,P)$ depends on the properties of the operator $P(D)$. Particularly, if $P(D)$ is an elliptic operator of order m, then $u \in H^p_{ms,loc}(\Omega)$.
2. If $u \in \mathcal{D}'(\Omega)$ and $f \in C^\infty(\Omega)$, then $u \in C^\infty(\Omega)$.
3. We have $\mathcal{N} = \{u \in \mathcal{D}'(\Omega) \mid P(D)u = 0\} \subset C^\infty(\Omega)$ and the topology induced in $\mathcal{N}$ by $C^\infty(\Omega)$ coincides with the topology induced by $H^p_{s,loc}(\Omega)$. Particularly, $\mathcal{N}$ is the Montel space.

In virtue of the above theorems we can give the now classical definition of the hypoellipticity.

Definition 3.6 Let $u,f \in \mathcal{D}'(\Omega)$ and $Pu = f$. If for any open set $V \subset \Omega$ the condition $f \in C^\infty(V)$ implies that $u \in C^\infty(V)$, then the differential operator P is called hypoelliptic.

Hypoelliptic Operators with Variable Coefficients

Now we turn to consider the differential operators with coefficients depending on the point $x \in \Omega \subset \mathbb{R}^n$ and study generalisations of the hypoellipticity property. We need to compare the differential operators with constant coefficients and the arbitrary differential operators. It can be achieved by freezing coefficients of an arbitrary differential operator at one point and considering it as a differential operator with constant coefficients. Let $P(\xi)$ be a polynomial of $\xi = (\xi_1, \dots, \xi_n) \in \mathbb{R}^n$. We need an auxiliary notation to proceed with the following definitions. We write

$$P^{(\alpha)}(\xi) = \frac{\partial^{|\alpha|} P(\xi)}{\partial^{\alpha_1}\xi_1 \dots \partial^{\alpha_n}\xi_n} = i^{|\alpha|} D^\alpha P(\xi), \qquad \widetilde{P}(x,\xi) = \Big(\sum_{|\alpha|\geq 0} |P^{(\alpha)}(\xi)|^2\Big)^{1/2}.$$

Definition 3.7 Let $P_1(D)$ and $P_2(D)$ be two differential operators (with constant coefficients). We say that $P_1(D)$ is weaker than $P_2(D)$ and write $P_1 \prec P_2$ if

$$\frac{\widetilde{P}_1(\xi)}{\widetilde{P}_2(\xi)} < C, \quad \xi \in \mathbb{R}^n.$$

Definition 3.8 A differential operator (with variable coefficients) $P(x,D)$ defined for $x \in \Omega$ is said to have constant strength in a domain Ω of $\mathbb{R}^n$ if for arbitrary fixed $y, y' \in \Omega$ the differential operators (with constant coefficients) $P(y,D)$ and $P(y',D)$ are equally strong, that is

$$\frac{\widetilde{P}(y,\xi)}{\widetilde{P}(y',\xi)} \leq C_{y,y'}, \quad y, y' \in \Omega, \;\; \xi \in \mathbb{R}^n.$$

We list the properties of operators of constant strength.

1. Let $P(x,D)$ have constant strength and $x_0 \in \Omega$ be a fixed point. Set $P_0(D) = P(x_0,D)$ and let $P_0,\dots,P_r$ be a basis for a finite dimensional vector space of operators with constant coefficients weaker than P_0. Then we have

$$P(x,D) = P_0(D) + \sum_{j=0}^{r} c_j(x) P_j(D),$$

where the coefficients c_j are uniquely determined, vanish at x_0, and have the same continuity and differentiability properties as the coefficients of $P(x,D)$.
2. Let $P(x,D)$ be a differential operator of order m with coefficients in $C^m(\Omega)$ and let it have constant strength in Ω. Then it follows that the adjoint operator ${}^tP(x,D)$ is also of constant strength and is as strong as $P(x,-D)$, for every $x \in \Omega$.
3. Let $P(x,D)$ and $Q(x,D)$ be of constant strength in Ω. Assume that $Q(x,D)$ is of order m and the coefficients of $P(x,D)$ is in $C^m(\Omega)$. Then the operator $R(x,D) = P(x,D)Q(x,D)$ is of constant strength. Moreover, $R(x,\xi)$ is as strong as $P(x,\xi)Q(x,\xi)$, for every $x \in \Omega$.
4. Let $P(x,D)$ be a differential operator of constant strength having continuous coefficients in a neighbourhood of $x_0 \in \mathbb{R}^n$. If Ω is a sufficiently small neighbourhood of x_0, then we can find a linear operator $E \in L^2(\Omega)$ such that

$$P(x,D)Ef = f, \quad f \in L^2(\Omega), \qquad EP(x,D)u = u, \quad u \in C_0^\infty(\Omega). \tag{18}$$

If we assume that the coefficients of $P(x,D)$ are C^∞ smooth, then there is a linear operator $E\colon \mathcal{E}'(\Omega) \to \mathcal{E}'(\Omega)$, such that (18) holds for $f, u \in \mathcal{E}'(\Omega)$.

The last property of the operators with constant strength is the hypoellipticity.

Theorem 3.9 ([54, 57]) *Let $P(x,D)$ be a differential operator of constant strength with C^∞ smooth coefficients and $P(x,D)u = f$. Assume that the operator $P_0(D) = P(x_0,D)$ is hypoelliptic for some $x_0 \in \Omega$. If $u \in \mathcal{D}'(\Omega)$ and $f \in H^p_{s,\mathrm{loc}}(\Omega)$, then it follows that $u \in H^p_{\Phi(P_0,s),\mathrm{loc}}(\Omega)$, where $\Phi(P_0,s)$ depends on the operator $P_0(D)$. Particularly, if $f \in C^\infty(\Omega)$, then $u \in C^\infty(\Omega)$ and therefore the operator $P(x,D)$ is hypoelliptic.*

Hörmander Theorem

We agree to denote all constants by the letter C and use the inequality for the equivalence. For example, the inequality $|u| \le C|Xu|$ simply means that $|u| = O(|Xu|)$.

Let Ω be a domain in $\mathbb{R}^n$ and consider a special type of the second order differential operator

$$Pu = \sum_{j=1}^{k} X_{ju}^2 + X_0 u + cu, \qquad X_j = \sum_{m=1}^{n} a_j^m(x)\frac{\partial}{\partial x^m}, \tag{19}$$

where the real valued functions a_j^m are of class $C^\infty(\Omega)$. Define the sub-bundle D of $T\Omega$, generated by the vector fields X_j, $j = 0,1,\ldots,k$, by setting $D_p = \mathrm{span}\{X_0,\ldots,X_k\}|_p$ for $p \in \Omega$. We aim to sketch the proof of Theorem 3.10, following [68].

Theorem 3.10 ([53]) *If the sub-bundle D generated by the vector fields X_j, $j = 0,1,\ldots,k$, is bracket generating at each point of an n-dimensional manifold M, then the operator* (19) *is hypoelliptic.*

Recall that the bracket generating property of vector fields $X_0,\ldots,X_k$ on a connected manifold M implies that any two points can be connected by a curve tangent to the sub-bundle D of TM, $D_p = \mathrm{span}\{X_0,\ldots,X_k\}|_p$, $p \in M$.

We restrict the consideration of Theorem 3.10 to a domain $\Omega \subset \mathbb{R}^n$ to avoid the additional technicalities. The main idea of the proof is to show that the bracket generating condition implies the *subelliptic estimate*:

$$\|u\|^2_{H_\epsilon(\Omega)} \le C\Big(\|Pu\|^2_{L^2(\Omega)} + \|u\|^2_{L^2(\Omega)}\Big) \quad \text{for any} \quad u \in C_0^\infty(\Omega), \tag{20}$$

for some $\epsilon > 0$, and $C > 0$. In general, subelliptic estimates imply the hypoellipticity property, see [70]. The main arguments in [70] are the following. An arbitrary distribution can be made into a smooth function by means of convolution. Then the nature of the differential operator P allows to localise an obtained smooth function for which the subelliptic estimate holds. In the last step the subelliptic estimate makes it possible to use the iteration process, showing that the solution actually belongs to H_s for any real s that implies C^∞ smoothness of the solution. Here we present the proof based on the theory of pseudo differential operators, which can be found in [68]. The original proof of Hörmander [53] is based on analytic tools and gives more precise estimates on how the value of ϵ depends on the number of steps of commutators in order to generate the tangent space at each point of the domain $\Omega \subset \mathbb{R}^n$. The reader can find different approaches to the proof of subelliptic estimates, for instance in [67, 69, 83, 88].

We first formulate auxiliary technical lemmas and then prove Theorem 3.10.

Auxiliary Material for Hörmander Theorem

Some Useful Inequalities

Cauchy-Schwartz inequality:

$$(x,y) \le |(x,y)| \le \|x\|\,\|y\| \le \frac{1}{2}(\|x\|^2 + \|y\|^2). \tag{21}$$

Generalised Schwartz inequality:

$$\begin{aligned}(x,y)_{L^2} \le |(x,y)_{L^2}| &\le \|x\|_{H_s}\|y\|_{H_{-s}} \\ &\le \text{small const}\|x\|^2_{H_s} + \text{large const}\|y\|^2_{H_s}.\end{aligned}$$

We have

$$x^2 + y^2 \le x^2 + y^2 + 2xy = (x+y)^2, \tag{22}$$

$$xy \le \frac{1}{2}(x^2 + y^2), \tag{23}$$

for positive real x and y.

Localization Lemma and Energy Estimate

Lemma 3.11 *Assume that estimate* (20) *is true. Then, for any* $\zeta, \zeta_1 \in C_0^\infty(\Omega)$ *with* $\zeta_1|_{\mathrm{supp}(\zeta)} = 1$, *there is a constant* $C > 0$ *such that*

$$\|\zeta u\|^2_{H_\epsilon(\Omega)} \le C\Big(\|\zeta_1 Pu\|^2_{L^2(\Omega)} + \|\zeta_1 u\|^2_{L^2(\Omega)}\Big) \quad \textit{for any} \quad u \in C^\infty(\Omega).$$

Proof We substitute u by ζu in (20) and obtain

$$\|\zeta u\|^2_{H_\epsilon(\Omega)} \le C\Big(\|\zeta Pu\|^2_{L^2(\Omega)} + \|[P,\zeta]u\|^2_{L^2(\Omega)} + \|\zeta u\|^2_{L^2(\Omega)}\Big). \tag{24}$$

Then

$$\|\zeta u\|^2_{L^2(\Omega)} \le \Big(\sup_{x\in \mathrm{supp}(\zeta)} |\zeta(x)|\Big) \int_{\mathrm{supp}(\zeta)} |u(x)|^2\,dx \le C\|\zeta_1 u\|^2_{L^2(\Omega)} \tag{25}$$

and, analogously, $\|\zeta Pu\|^2_{L^2(\Omega)} \le C\|\zeta_1 Pu\|^2_{L^2(\Omega)}$. To estimate $\|[P,\zeta]u\|^2_{L^2(\Omega)}$ we write

$$[P,\zeta]u = 2\sum_{j=1}^{k}[X_j,\zeta]X_j u + \sum_{j=1}^{k}[X_j,[X_j,\zeta]]u + [X_0,\zeta]u, \tag{26}$$

which follows from

$$\begin{aligned}
[X^2,\zeta] &= X(X\zeta) - (\zeta X)X = X(X\zeta - \zeta X + \zeta X) - (\zeta X - X\zeta + X\zeta)X \\
&= X([X,\zeta] + \zeta X) - (-[X,\zeta] + X\zeta)X = X[X,\zeta] - [X,\zeta]X + 2[X,\zeta]X \\
&= [X,[X,\zeta]] + 2[X,\zeta]X.
\end{aligned}$$

Since $[X_0,\zeta]u = X_0(\zeta u) - \zeta(X_0 u) = uX_0\zeta$, we estimate

$$\|[X_0,\zeta]u\|^2_{L^2(\Omega)} = \int_{\mathrm{supp}(\zeta)} |uX_0\zeta|^2 dx \le C\|\zeta_1 u\|^2_{L^2(\Omega)} \tag{27}$$

as in (25). Analogously, by making use of the observation

$$[X_j,[X_j,\zeta]]u = X_j([X_j,\zeta]u) - [X_j,\zeta]X_j u = X_j(uX_j\zeta) - X_j uX_j\zeta = uX_j^2\zeta$$

and the Schwartz inequality, we deduce

$$\|\sum_{j=1}^{k}[X_j,[X_j,\zeta]]u\|^2_{L^2(\Omega)} \le C\sum_{j=1}^{k} \||uX_j^2\zeta|^2\|^2_{L^2(\Omega)} \le C\|\zeta_1 u\|^2_{L^2(\Omega)}. \tag{28}$$

Now, for any $j = 1, \dots, k$, we estimate

$$\begin{aligned}
\|[X_j, \zeta]X_j u\|^2_{L^2(\Omega)} &= \int_{\mathrm{supp}(\zeta)} |X_j \zeta X_j u|^2\, dx \le C \int_{\mathrm{supp}(\zeta)} |\zeta_1^2 X_j u|^2\, dx \\
&= C \int_{\mathrm{supp}(\zeta)} |X_j(\zeta_1^2 u)|^2\, dx \\
&\le C\Big(\big|(P\zeta_1^2 u, \zeta_1^2 u)_{L^2(\Omega)}\big| + \|\zeta_1^2 u\|^2_{L^2(\Omega)} \Big) \qquad (29) \\
&= C\Big(\big|([P, \zeta_1^2]u, \zeta_1^2 u)_{L^2(\Omega)}\big| \\
&\quad + \big|(\zeta_1^2 Pu, \zeta_1^2 u)_{L^2(\Omega)}\big| + \|\zeta_1^2 u\|^2_{L^2(\Omega)} \Big),
\end{aligned}$$

where in the fourth step we used the "energy estimate", see Proposition 3.12. By the Cauchy-Schwartz inequality (21) the last two terms can be estimated as follows

$$\big|(\zeta_1^2 Pu, \zeta_1^2 u)_{L^2(\Omega)}\big| + \|\zeta_1^2 u\|^2_{L^2(\Omega)} \le C\Big(\|\zeta_1 Pu\|^2_{L^2(\Omega)} + \|\zeta_1 u\|^2_{L^2(\Omega)} \Big).$$

Now we turn to estimate $\big|([P, \zeta_1^2]u, \zeta_1^2 u)_{L^2(\Omega)}\big|$ by making use of (26). We obtain

$$\begin{aligned}
|([P, \zeta_1^2]u, \zeta_1^2 u)_{L^2(\Omega)}| &\le 2\Big| \sum_{j=1}^k (X_j u X_j \zeta_1^2, \zeta_1^2 u)_{L^2(\Omega)} \Big| + \Big| \sum_{j=1}^k (u X_j^2 \zeta_1^2, \zeta_1^2 u)_{L^2(\Omega)} \Big| \\
&\quad + \Big| (u X_0 \zeta_1^2, \zeta_1^2 u)_{L^2(\Omega)} \Big| = K_1 + K_2 + K_3.
\end{aligned}$$

We have

$$K_3 = \Big| (u X_0 \zeta_1^2, \zeta_1^2 u)_{L^2(\Omega)} \Big| \le \int_{\mathrm{supp}(\zeta_1)} \zeta_1^2 u^2 |X_0 \zeta_1^2|\, dx \le C\|\zeta_1 u\|^2_{L^2(\Omega)}.$$

Analogously,

$$K_2 = \Big| \sum_{j=1}^k (u X_j^2 \zeta_1^2, \zeta_1^2 u)_{L^2(\Omega)} \Big| \le \sum_{j=1}^k \int_{\mathrm{supp}(\zeta_1)} |\zeta_1^2 u^2| |X_j^2 \zeta_1^2|\, dx \le C\|\zeta_1 u\|^2_{L^2(\Omega)}.$$

Then we write

$$(X_j u X_j \zeta_1^2, \zeta_1^2 u)_{L^2(\Omega)} = \frac{1}{4} \int_{\mathrm{supp}(\zeta_1)} X_j u^2 X_j \zeta_1^4\, dx.$$

Using integration by parts, we finish the estimation as we did for K_2 and obtain

$$K_1 = \Big|\sum_{j=1}^{k}(X_j u X_j \zeta_1^2, \zeta_1^2 u)_{L^2}\Big| \leq C\|\zeta_1 u\|^2_{L^2(\Omega)}.$$

Now we sum up the estimations for K_1, K_2, K_3 and insert them into (29). In the last step we join the inequalities (27), (28), and (29) to obtain

$$\|[P,\zeta]u\|_{L^2(\Omega)} \leq C\Big(\|\zeta_1 Pu\|_{L^2(\Omega)} + \|\zeta_1 u\|^2_{L^2(\Omega)}\Big).$$

We estimated all three terms in the right hand side of (24). That finishes the proof.

□

We prove the "energy estimate".

Proposition 3.12 *There is a constant $C > 0$ such that for the operator* (19) *we have*

$$\sum_{j=1}^{k}\|X_j u\|^2_{L^2(\Omega)} \leq C\Big(|(Pu,u)_{L^2(\Omega)}| + \|u\|^2_{L^2(\Omega)}\Big) \text{ for any } u \in C_0^\infty(\Omega). \tag{30}$$

Observe that (30) implies

$$\sum_{j=1}^{k}\|X_j u\|^2_{L^2(\Omega)} \leq C\Big(\|Pu\|^2_{L^2(\Omega)} + \|u\|^2_{L^2(\Omega)}\Big) \tag{31}$$

for any $u \in C_0^\infty(\Omega)$.

Proof We start from finding the form of the adjoint X_j^* for the operator $X_j = \sum_{m=1}^{n} a_j^m \frac{\partial}{\partial x^m}$. Since for any $\phi \in C_0^\infty(\Omega)$ we have

$$\int_\Omega u X_j \phi\, dx = \sum_{m=1}^{n} a_j^m u\phi|_{b\Omega} - \int_\Omega X_j u \phi\, dx - \int_\Omega u\Big(\sum_{m=1}^{n}\frac{\partial a_j^m}{\partial x^m}\Big)\phi\, dx,$$

the adjoint operator has the form $X_j^* u = -X_j u + A_j u$, where $A_j = -\sum_{m=1}^{n}\frac{\partial a_j^m}{\partial x^m}$. Thus

$$\begin{aligned}(X_j^2 u, u)_{L^2(\Omega)} &= (X_j u, X_j^* u)_{L^2(\Omega)} = -\|X_j u\|^2_{L^2(\Omega)} + (X_j u, A_j u)_{L^2(\Omega)} \\ &\leq -\|X_j u\|^2_{L^2(\Omega)} + \text{small const}\|X_j u\|^2_{L^2(\Omega)} + \text{large const}\|u\|^2_{L^2(\Omega)},\end{aligned}$$

and we obtain

$$\sum_{j=1}^{k} \|X_j u\|^2_{L^2(\Omega)} \le C\Big(-(Pu,u)_{L^2(\Omega)} + (X_0 u, u)_{L^2(\Omega)} + \|u\|^2_{L^2(\Omega)}\Big) \tag{32}$$

from the definition of the operator P. Observe that because of $(X_0u,u)_{L^2(\Omega)} = -(X_0u,u)_{L^2(\Omega)} + (u, A_0u)_{L^2(\Omega)}$ we can estimate

$$|(X_0u,u)_{L^2(\Omega)}| \le C\|u\|^2_{L^2(\Omega)}.$$

We substitute the last inequality into (32) and finish the proof. □

Properties of Pseudo Differential Operators

Consider the following three linear operators from $C_0^\infty(\Omega)$ to $C^\infty(\Omega)$:

(Ψ_1) multiplication by $a \in C^\infty(\mathbb{R}^n)$,
(Ψ_2) differentiation $\partial^\alpha = \frac{\partial^{|\alpha|}}{\partial_1^{\alpha_1}\dots\partial_n^{\alpha_n}}$, $|\alpha| = \sum_{j=1}^n \alpha_j$,
(Ψ_3) $(\Lambda^s u)(\xi) = (1 + |\xi|^2)^{s/2}\hat{u}(\xi)$.

Let $\mathcal{P}$ be the algebra of operators generated by the operators of type ($\Psi_1 - \Psi_3$) by composition, summation, and taking formal adjoints. Notice that $\Lambda^s u$ is not necessarily in $C_0^\infty(\Omega)$ but this difficulty could be overcome carefully, see [55]. Elements of $\mathcal{P}$ are called pseudo differential operators.

Definition 3.13 A linear operator $P\colon C_0^\infty(\Omega) \to C^\infty(\Omega)$ has the order m, $m \in \mathbb{R}$, if for any $r \in \mathbb{R}$, there is a constant $C_r > 0$ such that

$$\|Pu\|_{H_r(\Omega)} \le C_r \|u\|_{H_{r+m}(\Omega)} \quad \text{for all} \quad u \in C_0^\infty(\Omega).$$

The following properties of pseudo differential operators are true. Let $P, P' \in \mathcal{P}$ be of order m and m', respectively, then

1. the product PP' is of order $m + m'$,
2. the commutator $[P, P'] = PP' - P'P$ is of order $m + m' - 1$,
3. the adjoint operator P^* is of order m,
4. the operators of type (Ψ_1) are of order $m = 0$,
5. the operators of type (Ψ_2) are of order $m = |\alpha|$,
6. the operators Λ^s of type (Ψ_3) are of order $m = s$.

Sketch of the Proof of Theorem 3.10

Step 1: Subelliptic Estimates Implies Iteration "Inequality"

We need some auxiliary facts.

I. Let $\varepsilon > 0$, $s \in \mathbb{R}$ and $N > 0$ be given. Then for any $\delta > 0$ there is a constant $C(s, N, \delta)$ such that

$$\|u\|_{H_s(\Omega)} \leq \delta\|u\|_{H_{s+\epsilon}(\Omega)} + C(s, N, \delta)\|u\|_{H_{-N}(\Omega)} \quad \text{for all} \quad u \in C_0^\infty(\Omega). \tag{33}$$

II. We will also use the subelliptic estimate in the form

$$\|\zeta u\|_{H_\epsilon(\Omega)} \leq C(\|\zeta_1 Pu\|_{L^2(\Omega)} + \|\zeta_1 u\|_{L^2(\Omega)}) \quad \text{for any} \quad u \in C^\infty(\Omega),$$

that is obtained from (20) and (22).

We aim to show an iteration "inequality". *Let $Pu = f$ and ϵ be chosen arbitrarily. Given $s \in \mathbb{R}$ and $N > 0$ there is a constant $C_{s,N}$ such that*

$$\|u\|_{H_{s+\epsilon}(\Omega)} \leq C_{s,N}\Big(\|Pu\|_{H_s(\Omega)} + \|u\|_{H_{-N}(\Omega)}\Big) \quad \textit{for all} \quad u \in C_0^\infty(\Omega). \tag{34}$$

Let $u \in C_0^\infty(\Omega)$ and $s \in \mathbb{R}$ be given. We apply (20) to $\Lambda^s u$ (we actually have to justify that $\Lambda^s u \in C_0^\infty(\Omega)$ but we omit it.) We obtain

$$\begin{aligned}\|u\|_{H_{s+\epsilon}(\Omega)} &= \|\Lambda^s u\|_{H_\epsilon(\Omega)} \leq C\Big(\|P\Lambda^s u\|_{L^2(\Omega)} + \|\Lambda^s u\|_{L^2(\Omega)}\Big) \\ &= C\Big(\|[P, \Lambda^s]u\|_{L^2(\Omega)} + \|\Lambda^s Pu\|_{L^2(\Omega)} + \|\Lambda^s u\|_{L^2(\Omega)}\Big).\end{aligned}$$

Now the second term in the last expression is exactly what we need in (34). The third term can be estimated by (33) with small enough δ in order to bring the term $\delta\|u\|_{H_{s+\epsilon}(\Omega)}$ into the left hand side of (34). Thus we need to estimate $\|[P, \Lambda^s]u\|_{L^2(\Omega)}$ by $C\Big(\|Pu\|_{H_s(\Omega)} + \|u\|_{H_s(\Omega)}\Big)$. The operator P includes the terms of order two, one and zero. The term of order 0 gives the vanishing part in the commutator, therefore we can write the operator $[P, \Lambda^s]$ in the form

$$[P, \Lambda^s] = \sum_{j=1}^{k} T_j^s X_j + T^s, \tag{35}$$

where T_j^s and T^s are some pseudo differential operators of order s. Then

$$\|[P, \Lambda^s]u\|_{L^2(\Omega)} \le \sum_{j=1}^{k} \|T_j^s X_j u\|_{L^2(\Omega)} + \|T^s u\|_{L^2(\Omega)}$$
$$\le C\Big(\sum_{j=1}^{k} \|X_j u\|_{H_s(\Omega)} + \|u\|_{H_s(\Omega)}\Big).$$

Thus we need to obtain $\sum_{j=1}^{k} \|X_j u\|_{H_s(\Omega)} \le C\Big(\|Pu\|_{H_s(\Omega)} + \|u\|_{H_s(\Omega)}\Big)$. Let us first estimate

$$\begin{aligned}\sum_{j=1}^{k} \|X_j u\|^2_{H_s(\Omega)} &\le \sum_{j=1}^{k} \|\Lambda^s X_j u\|^2_{L^2(\Omega)} \\ &\le \sum_{j=1}^{k} \|X_j \Lambda^s u\|^2_{L^2(\Omega)} + \sum_{j=1}^{k} \|[\Lambda^s, X_j]u\|^2_{L^2(\Omega)} \\ &\le C\Big(|(P\Lambda^s u, \Lambda^s u)_{L^2(\Omega)}| + \|u\|^2_{H_s(\Omega)}\Big),\end{aligned} \tag{36}$$

where we used Proposition 3.12 to obtain the first term and the fact that the operator $[\Lambda^s, X_j]$ is of order s to estimate the second term. We continue the estimation

$$\begin{aligned}|(P\Lambda^s u, \Lambda^s u)_{L^2(\Omega)}| &\le |([P, \Lambda^s]u, \Lambda^s u)_{L^2(\Omega)}| + |(\Lambda^s P u, \Lambda^s u)_{L^2(\Omega)}| \\ &\le C\Big(|([P, \Lambda^s]u, \Lambda^s u)_{L^2(\Omega)}| + \|Pu\|^2_{H_s(\Omega)} + \|u\|^2_{H_s(\Omega)}\Big),\end{aligned} \tag{37}$$

where in the last step we used the Cauchy-Schwartz inequality and (23). Now, to estimate the first term in the right hand side of (37), we involve (35) and (21). We calculate

$$\begin{aligned}|([P, \Lambda^s]u, \Lambda^s u)_{L^2(\Omega)}| &\le \sum_{j=1}^{k} |(T_j^s X_j u, \Lambda^s u)_{L^2(\Omega)}| + |(T^s u, \Lambda^s u)_{L^2(\Omega)}| \\ &\le \sum_{j=1}^{k} \|T_j^s X_j u\|_{L^2(\Omega)} \|\Lambda^s u\|_{L^2(\Omega)} + \|T^s u\|_{L^2(\Omega)} \|\Lambda^s u\|_{L^2(\Omega)} \\ &\le \text{small constant}\|X_j u\|^2_{H_s(\Omega)} + C\|u\|^2_{H_s(\Omega)}.\end{aligned} \tag{38}$$

Setting (38) into the estimate (37), then into (36) and moving the first term (small constant$\|X_j u\|^2_{H_s(\Omega)}$) into the left hand side, we deduce the inequality

$$\sum_{j=1}^{k} \|X_j u\|^2_{H_s(\Omega)} \leq C\Big(\|Pu\|^2_{H_s(\Omega)} + \|u\|^2_{H_s(\Omega)}\Big) \leq C\Big(\|Pu\|_{H_s(\Omega)} + \|u\|_{H_s(\Omega)}\Big)^2,$$

that leads to

$$\sum_{j=1}^{k} \|X_j u\|_{H_s(\Omega)} \cdot 1 \leq \Big(k \sum_{j=1}^{k} \|X_j u\|^2_{H_s(\Omega)}\Big)^{1/2} \leq C\Big(\|Pu\|_{H_s(\Omega)} + \|u\|_{H_s(\Omega)}\Big).$$

We use the auxiliary fact (33) and then we move the term $\delta\|u\|_{H_{s+\epsilon}(\Omega)}$ to the left hand side. With this we finish the proof of Step 1.

Step 2: Smoothing and Iteration Procedures

Let $Pu = f$, where u and f are distributions. We now show that if $f|_V \in C^\infty(V)$ for any open set $V \subset \Omega$, then also $u|_V \in C^\infty(V)$, or in other words the operator P is hypoelliptic.

First of all we apply the arguments of the localisation lemma to (34) to get

$$\|\zeta u\|_{H_{s+\epsilon}(\Omega)} \leq C\Big(\|\zeta_1 Pu\|_{H_s(\Omega)} + \|\zeta_1 u\|_{H_{-N}(\Omega)}\Big) \quad \text{for all} \quad u \in C^\infty(\Omega).$$

For the beginning we substitute ζu into (34) and then almost literally follow the proof of Lemma 3.11.

The second observation is about the smoothness process. Let $u \in \mathcal{D}'$ (or any other function in L^1_{loc}), then we write

$$S_\delta u(x) = \int_{\mathbb{R}^n} u(x+\delta y)\varphi(y)\,dy = \frac{1}{\delta}\int_{\mathbb{R}^n} u(y)\varphi(\frac{y-x}{\delta})\,dy$$

for $\varphi \in C_0^\infty(\mathbb{R}^n)$ with $\int_{\mathbb{R}^n}\varphi(y)dy = 1$. Then $S_\delta u \in C^\infty$ for any $\delta > 0$.

Let us present some well known facts that we will use.

I. Let $x_0 \in \Omega$ and W_{x_0} be a neighbourhood of x_0 in an open set $V \subset \Omega$. If $u \in \mathcal{D}'$, then $u \in H_s(W_{x_0})$ if and only if there is $\zeta_1 \in C_0^\infty(V)$, $\zeta_1|_{W_{x_0}} = 1$ such that $S_\delta\zeta_1 u \in H_s(V)$ and the bound of $\|S_\delta\zeta_1 u\|_{H_s(V)}$ does not depend on $\delta > 0$.

II. If $u \in H_{s,loc}(W_{x_0})$ and T^1 is a pseudo differential operator of order 1, then $\|[S_\delta, T^1]u\|_{H_s(V)} \leq C\|u\|_{H_s(V)}$, where C does not depend on δ.

For $u \in \mathcal{D}'$ we have $S_\delta u \in C^\infty(V)$. Now we take inequality (34) and apply Lemma 3.11 to the function $S_\delta u$. We write

$$\|\zeta S_\delta u\|_{H_{s+\epsilon}(\Omega)} \leq C\Big(\|\zeta_1 P S_\delta u\|_{H_s(\Omega)} + \|\zeta_1 S_\delta u\|_{H_{-N}(\Omega)}\Big) \quad \text{for all } \; S_\delta u \in C^\infty(\Omega).$$

Since $\|\zeta S_\delta u\|_{H_s} = O(\|S_\delta \zeta u\|_{H_s})$ for any $s \in \mathbb{R}$ and the operator $[\zeta_1, P]$ is the first order pseudo differential operator we apply fact II and obtain

$$\|S_\delta \zeta u\|_{H_{s+\epsilon}(V)} \leq C\Big(\|S_\delta \zeta_1 (Pu)\|_{H_s(V)} + \|S_\delta \zeta_1 u\|_{H_{-N}(V)} + \|\zeta_1 u\|_{H_s(V)}\Big), \tag{39}$$

if we know that $\zeta_1 u \in H_s(V)$. Here $\zeta, \zeta_1 \in C_0^\infty(V)$ and $\zeta_1|_{\mathrm{supp}(\zeta)} = 1$.

Now we start to apply the iteration arguments: If $u \in \mathcal{D}'$, then there is $N > 0$ such that $\zeta_1 u \in H_{-N}(V)$ (and we choose supports of ζ and ζ_1 small enough, that $\zeta_1 P u = \zeta_1 f \in C_0^\infty(V)$). Since $\zeta_1 u \in H_{-N}(V)$, then $S_\delta \zeta_1 u \in H_{-N}(V)$ by fact I. We set $s = -N$ in (39) and obtain

$$\|S_\delta \zeta u\|_{H_{-N+\epsilon}(V)} \leq C\Big(\|S_\delta \zeta_1 (Pu)\|_{H_{-N}(V)} + \|S_\delta \zeta_1 u\|_{H_{-N}(V)} + \|\zeta_1 u\|_{H_{-N}(V)}\Big).$$

Thus $S_\delta \zeta u \in H_{-N+\epsilon}(V) \implies S_\delta \zeta_1 u \in H_{-N+\epsilon}(V) \iff u, \zeta u \in H_{-N+\epsilon}(V)$ by fact I. Applying the iteration arguments, we conclude that there is a neighbourhood W_{x_0} of the point $x_0 \in V$ such that for any $\zeta \in C_0^\infty(W_{x_0})$ we have that $\zeta u \in H_s(W_{x_0})$. Therefore $u \in \bigcap_s H_s(W_{x_0}) = C^\infty(W_{x_0})$. Thus for any $x_0 \in V$ we obtain $u \in C^\infty(W_{x_0})$ which implies $u \in C^\infty(V)$ and the proof is finished.

Step 3: Bracket Generating Condition Implies the Subelliptic Estimates

We need to extract from the algebraic bracket generating condition of the Hörmander theorem the subelliptic estimate (20) for any $u \in C_0^\infty(\Omega)$. Denote by D_j, $j = 1, \dots, n$, linear independent vector fields in Ω. Then

$$\|u\|_{H_s(\Omega)} = \|\Lambda^{s-1}\Lambda^1 u\|_{L^2(\Omega)} \leq C \sum_{j=1}^{n} \|D_j u\|_{H_{s-1}(\Omega)}.$$

By hypothesis of the Hörmander theorem any D_j can be expressed as follows

$$D_j = \sum a^{i_1 \dots i_p} F_{i_1 \dots i_p}, \quad 0 \leq i_m \leq k,$$

where

$$F_{i_1 \dots i_p} = \begin{cases} X_{i_p}, & \text{if} \quad p = 1 \\ [X_{i_p}, F_{i_1 \dots i_{p-1}}], & \text{if} \quad p > 1. \end{cases}$$

Thus in order to prove the subelliptic estimate (20) we only need to bound the norm $\|F_{i_1\dots i_p}u\|_{H_{\epsilon-1}(\Omega)}$ by $C\Big(\|Pu\|_{L^2(\Omega)} + \|u\|_{L^2(\Omega)}\Big)$. For simplicity, we write $F^p = [X, F^{p-1}]$. Then

$$
\begin{aligned}
\|F^p u\|^2_{H_{\epsilon-1}} &= (\Lambda^{\epsilon-1}F^p u, \Lambda^{\epsilon-1}F^p u)_{L^2(\Omega)} = (F^p u, \underbrace{\Lambda^{2\epsilon-2}F^p}_{\sim T^{2\epsilon-1}} u)_{L^2(\Omega)} \\
&= (XF^{p-1}u, T^{2\epsilon-1}u)_{L^2(\Omega)} - (F^{p-1}Xu, T^{2\epsilon-1}u)_{L^2(\Omega)} = I_1 - I_2,
\end{aligned}
$$

where we used the notation $T^{2\epsilon-1} = \Lambda^{2\epsilon-2}F^p$ for a pseudo differential operator of order $2\epsilon - 1$. Thus we need to estimate I_1 and I_2 by $C\Big(\|Pu\|^2_{L^2} + \|u\|^2_{L^2}\Big)$. The vector field X can be one of the vector fields X_j, $j = 1, \dots, k$ that form the second order term of P and can be estimated by the "energy estimate". We also need to consider $X = X_0$, that is the first order part of the differential operator P. Thus we consider two cases.

CASE 1: $X = X_j$. Using the expression $X^* = -X + f$ for the dual operator we obtain

$$
\begin{aligned}
I_1 = (XF^{p-1}u, T^{2\epsilon-1}u)_{L^2} &= (F^{p-1}u, X^*T^{2\epsilon-1}u)_{L^2} \\
&= -(F^{p-1}u, XT^{2\epsilon-1}u)_{L^2} + (F^{p-1}u, f\,T^{2\epsilon-1}u)_{L^2} \\
&= -(F^{p-1}u, T^{2\epsilon-1}Xu)_{L^2} - (F^{p-1}u, \underbrace{[X, T^{2\epsilon-1}]}_{\sim T^{2\epsilon-1}} u)_{L^2} \\
&\quad + (F^{p-1}u, f\,T^{2\epsilon-1}u)_{L^2} \\
&\le C\Big(\|F^{p-1}u\|^2_{H_{2\epsilon-1}} + \|Pu\|^2_{L^2} + \|u\|^2_{L^2}\Big),
\end{aligned}
$$

where we used the equality $\|F^{p-1}u\|_{H_{2\epsilon-1}} = \|(T^{2\epsilon-1})^*F^{p-1}u\|_{L^2}$, the Cauchy-Schwartz inequality, (23), and (31).

Before we start to estimate I_2, we note that $(F^{p-1})^* = -F^{p-1} + f$ with $f \in C^\infty(\Omega)$. Therefore, we apply similar arguments as above and obtain

$$
\begin{aligned}
I_2 = (F^{p-1}Xu, T^{2\epsilon-1}u)_{L^2} &= -(Xu, F^{p-1}T^{2\epsilon-1}u)_{L^2} + (Xu, f\,T^{2\epsilon-1}u)_{L^2} \\
&= -(Xu, T^{2\epsilon-1}F^{p-1}u)_{L^2} + (Xu, \underbrace{[F^{p-1}, T^{2\epsilon-1}]}_{\sim T^{2\epsilon-1}} u)_{L^2} \\
&\quad + (Xu, f\,T^{2\epsilon-1}u)_{L^2} \\
&\le C\Big(\|F^{p-1}u\|^2_{H_{2\epsilon-1}} + \|Xu\|^2_{L^2} + \|u\|^2_{H_{2\epsilon-1}}\Big) \\
&\le C\Big(\|F^{p-1}u\|^2_{H_{2\epsilon-1}} + \|Pu\|^2_{L^2} + \|u\|^2_{L^2}\Big).
\end{aligned}
$$

In the last step we used (31) and the fact that $\|u\|^2_{H_{2\epsilon-1}(\Omega)} \leq C\|u\|^2_{L^2(\Omega)}$ for $u \in C_0^\infty$, since the function $(1+|\xi|^2)^s$ is bounded on the support of u and the Fourier transform is an L^2-isometry.

Summing up the estimations for I_1 and I_2 we deduce

$$\|F^p u\|^2_{H_{\epsilon-1}} \leq C\Big(\|F^{p-1}u\|^2_{H_{2\epsilon-1}} + \|Pu\|^2_{L^2} + \|u\|^2_{L^2}\Big). \tag{40}$$

Before turning to the Case 2, we observe that changing ϵ by 2ϵ in (40) we obtain

$$\|F^p u\|^2_{H_{2\epsilon-1}} \leq C\Big(\|F^{p-1}u\|^2_{H_{4\epsilon-1}} + \|Pu\|^2_{L^2} + \|u\|^2_{L^2}\Big). \tag{41}$$

CASE 2: $X = X_0$. The adjoint operator for P has the form

$$P^* = \sum_{j=1}^k X_j^2 + \sum_{j=1}^k b_j X_j - X_0 + d,$$

where $b_j, d \in C^\infty(\Omega)$. It is obtained by careful integration by parts. We estimate

$$\begin{aligned}(X_0 F^{p-1}u, T^{2\epsilon-1}u)_{L^2} &= -(P^* F^{p-1}u, T^{2\epsilon-1}u)_{L^2} \\ &\quad + \sum_{j=1}^k (X_j^2 F^{p-1}u, T^{2\epsilon-1}u)_{L^2} \\ &\quad + \sum_{j=1}^k (b_j X_j F^{p-1}u, T^{2\epsilon-1}u)_{L^2} \\ &\quad + (dF^{p-1}u, T^{2\epsilon-1}u)_{L^2} = J_1 + J_2 + J_3 + J_4.\end{aligned}$$

Thus we need to estimate each of the integrals J_k, $k = 1, 2, 3, 4$. We have

$$\begin{aligned}J_4 &= (dF^{p-1}u, T^{2\epsilon-1}u)_{L^2} \\ &\leq C\Big(-(u, T^{2\epsilon-1}F^{p-1}u)_{L^2} + (u, [F^{p-1}, T^{2\epsilon-1}]u)_{L^2} + (u, fT^{2\epsilon-1}u)_{L^2} \\ &\leq C\Big(\|F^{p-1}u\|^2_{H_{2\epsilon-1}} + \|u\|^2_{H_{2\epsilon-1}} + \|u\|^2_{L^2}\Big)\end{aligned}$$

by arguments similar to those in Case 1 above. Then

$$J_3 = \sum_{j=1}^k (b_j X_j F^{p-1}u, T^{2\epsilon-1}u)_{L^2} \leq C\Big(\|F^{p-1}u\|^2_{H_{2\epsilon-1}} + \|Pu\|^2_{L^2} + \|u\|^2_{L^2}\Big)$$

as we did for estimation of I_1. Before estimating J_1 we note that

$$[X_j^2, T^{2\epsilon-1}] = 2[X_j, T^{2\epsilon-1}]X_j + [X_j, [X_j, T^{2\epsilon-1}]],$$

where $[X_j, T^{2\epsilon-1}]$ and $[X_j, [X_j, T^{2\epsilon-1}]]$ are pseudo differential operators of order $2\epsilon - 1$. Thus we can write

$$[P, T^{2\epsilon-1}]u = \sum_{j=1}^{k} T_j^{2\epsilon-1} X_j u + T_{k+1}^{2\epsilon-1} u. \tag{42}$$

Now we estimate J_1:

$$\begin{aligned} J_1 &= (P^* F^{p-1}u, T^{2\epsilon-1}u)_{L^2} = (F^{p-1}u, T^{2\epsilon-1}Pu)_{L^2} + (F^{p-1}u, [P, T^{2\epsilon-1}]u)_{L^2} \\ &\leq C\Big(\|F^{p-1}u\|^2_{H_{2\epsilon-1}} + \|Pu\|^2_{L^2} + \sum_{j=1}^{k} \|X_j u\|^2_{L^2} + \|u\|^2_{L^2}\Big) \\ &\leq C\Big(\|F^{p-1}u\|^2_{H_{2\epsilon-1}} + \|Pu\|^2_{L^2} + \|u\|^2_{L^2}\Big). \end{aligned}$$

Now we turn to estimate J_2. We fix $j = 1, \ldots, k$ and write

$$\begin{aligned} (X_j^2 F^{p-1}u, T^{2\epsilon-1}u)_{L^2} &= -(X_j F^{p-1}u, X_j T^{2\epsilon-1}u)_{L^2} + \underbrace{(X_j F^{p-1}u, f T^{2\epsilon-1}u)_{L^2}}_{\text{as } I_1} \\ &\cong -(X_j F^{p-1}u, T^{2\epsilon-1}X_j u)_{L^2} \\ &\quad + (X_j F^{p-1}u, \underbrace{[X_j, T^{2\epsilon-1}]}_{\sim T^{2\epsilon-1} \text{and as } I_1} u)_{L^2} + \ldots \\ &\cong -(X_j \underbrace{(T^{2\epsilon-1})^*}_{\sim T^{2\epsilon-1}} F^{p-1}u, X_j u)_{L^2} \\ &\quad - (\underbrace{[(T^{2\epsilon-1})^*, X_j]}_{\sim T^{2\epsilon-1}} F^{p-1}u, X_j u)_{L^2} + \ldots, \end{aligned} \tag{43}$$

where we denoted by ... the terms which we already know how to estimate. So,

$$J_2 \leq C\Big(\|X_j T^{2\epsilon-1} F^{p-1}u\|^2_{L^2} + \|F^{p-1}u\|^2_{H_{2\epsilon-1}} + \|Pu\|^2_{L^2} + \|u\|^2_{L^2}\Big).$$

We need to estimate the first term in the right hand side:

$$\sum_{j=1}^{k} \|X_j T^{2\epsilon-1} F^{p-1} u\|_{L^2}^2 = -\sum_{j=1}^{k} (X_j^2 T^{2\epsilon-1} F^{p-1} u, T^{2\epsilon-1} F^{p-1} u)_{L^2}$$
$$+ \sum_{j=1}^{k} (f X_j T^{2\epsilon-1} F^{p-1} u, T^{2\epsilon-1} F^{p-1} u)_{L^2}. \qquad (44)$$

The second term in (44) is estimated by

$$\leq \text{small constant} \sum_{j=1}^{k} \|X_j T^{2\epsilon-1} F^{p-1} u\|_{L^2}^2 + \text{large constant } \|T^{2\epsilon-1} F^{p-1} u\|_{L^2}^2$$

and the part "small constant $\sum_{j=1}^{k} \|X_j T^{2\epsilon-1} F^{p-1} u\|_{L^2}^2$" can be moved to the left hand side of (44). The first term in the right hand side of (44) is estimated by Proposition 3.12. Finally, we obtain

$$\sum_{j=1}^{k} \|X_j T^{2\epsilon-1} F^{p-1} u\|_{L^2}^2 \leq -(P T^{2\epsilon-1} F^{p-1} u, T^{2\epsilon-1} F^{p-1} u)_{L^2} + \|F^{p-1} u\|_{H_{2\epsilon-1}}^2$$

and we only have to estimate the first term in the right hand side of the last expression:

$$(P \underbrace{T^{2\epsilon-1} F^{p-1}}_{\sim T^{2\epsilon}} u, T^{2\epsilon-1} F^{p-1} u)_{L^2} (P T^{2\epsilon} u, T^{2\epsilon-1} F^{p-1} u)_{L^2}$$
$$= (Pu, T^{4\epsilon-1} F^{p-1} u)_{L^2} + ([P, T^{2\epsilon}]u, T^{2\epsilon-1} F^{p-1} u)_{L^2}$$
$$\leq C\Big(\|F^{p-1}\|_{H_{4\epsilon-1}}^2 + \|Pu\|_{L^2}^2 + \|u\|_{L^2}^2\Big),$$

where in the last step we used

$$[P, T^{2\epsilon}] = \sum_{j=1}^{k} T_j^{2\epsilon} X_j + T_{k+1}^{2\epsilon}$$

analogous to (42). Thus

$$J_2 \leq C\Big(\|F^{p-1} u\|_{H_{4\epsilon-1}}^2 + \|F^{p-1} u\|_{H_{2\epsilon-1}}^2 + \|Pu\|_{L^2}^2 + \|u\|_{L^2}^2\Big).$$

Observe that

$$\|F^{p-1}u\|^2_{H_{2\epsilon-1}} \le C\|F^{p-1}u\|^2_{H_{4\epsilon-1}}$$

since the function $(1+|\xi|^2)^{-2\epsilon}$ is bounded on the support of $u \in C_0^\infty(\Omega)$. Summing up the estimations for J_j, $j = 1,2,3,4$, we arrive to the inequality for I_1 when $X = X_0$:

$$I_1 = (X_0F^{p-1}u, T^{2\epsilon-1}u)_{L^2} \le C\Big(\|F^{p-1}u\|^2_{H_{4\epsilon-1}} + \|Pu\|^2_{L^2} + \|u\|^2_{L^2}\Big).$$

It is left to estimate I_2 for $X = X_0$. We substitute $X_0u = Pu - \sum_{j=1}^k X_j^2u - cu$ and obtain

$$\begin{aligned} I_2 &= (F^{p-1}X_0u, T^{2\epsilon-1}u)_{L^2} \\ &= (F^{p-1}Pu, T^{2\epsilon-1}u)_{L^2} - \sum_{j=1}^k (F^{p-1}X_j^2u, T^{2\epsilon-1}u)_{L^2} - (F^{p-1}(cu), T^{2\epsilon-1}u)_{L^2}. \end{aligned}$$

The first term is estimated by applying $(F^{p-1})^*$ and then reasoning as in the proof of the estimation of J_4. We get

$$(F^{p-1}Pu, T^{2\epsilon-1}u)_{L^2} \le C\Big(\|F^{p-1}u\|^2_{H_{2\epsilon-1}} + \|Pu\|^2_{L^2} + \|u\|^2_{L^2}\Big).$$

We use the Cauchy-Schwartz inequality and (23) to bound the third term. We only have to estimate the second term in I_2. We write

$$(F^{p-1}X_j^2u, T^{2\epsilon-1}u)_{L^2} = (X_j^2F^{p-1}u, T^{2\epsilon-1}u)_{L^2} + ([F^{p-1}, X_j^2]u, T^{2\epsilon-1}u)_{L^2}.$$

The first term was estimated in (43) and to estimate the second one we use

$$[F^{p-1}, X_j^2] = [F^{p-1}, X_j]X_j + X_j[F^{p-1}, X_j] = -\widetilde{F^p}X_j - X_j\widetilde{F^p}, \tag{45}$$

where $\widetilde{F^p}$ is a commutator of the same type as we used before. Thus substituting (45) and using the same arguments as in Case 1, we obtain

$$([\widetilde{F}^{p-1}, X_j^2]u, T^{2\epsilon-1}u)_{L^2} \le C\Big(\|\widetilde{F^p}u\|^2_{H_{2\epsilon-1}} + \|Pu\|^2_{L^2} + \|u\|^2_{L^2}\Big).$$

Analogously to (41), we deduce

$$\|\widetilde{F^p}u\|_{H_{2\epsilon-1}} \le C\Big(\|F^{p-1}u\|^2_{H_{4\epsilon-1}} + \|Pu\|^2_{L^2} + \|u\|^2_{L^2}\Big).$$

Now, summing up the estimations for Cases 1 and 2, we conclude that

$$\|F^p u\|^2_{H_{\epsilon-1}} \le C\Big(\|F^{p-1}\|^2_{H_{4\epsilon-1}} + \|Pu\|^2_{L^2} + \|u\|^2_{L^2}\Big).$$

Applying the iteration process, we arrive to the conclusion

$$\|F^p u\|^2_{H_{\epsilon-1}} \le C\Big(\sum_{j=1}^{k} \|X_j u\|^2_{H_{4^{p-1}\epsilon-1}} + \|X_0 u\|^2_{H_{4^{p-1}\epsilon-1}} + \|Pu\|^2_{L^2} + \|u\|^2_{L^2}\Big).$$

Finally, if we could show that

$$\|X_j u\|^2_{H_{-1/2}} \le C\Big(\|Pu\|^2_{L^2} + \|u\|^2_{L^2}\Big), \quad j = 0, 1, \dots, \tag{46}$$

then we could finish the proof setting $\epsilon = \frac{2}{4^p}$ for any $p > 0$ (or even $\epsilon \le \frac{2}{4^p}$). Let us show (46). We have

$$\|X_j u\|^2_{H_{-1/2}} = (\Lambda^{-1/2}X_j u, \Lambda^{-1/2}X_j u)_{L^2} = (X_j u, \underbrace{\Lambda^{-1}X_j}_{\sim T^0} u)_{L^2}$$

for $j = 0, 1, \dots, k$. If $j = 1, \dots, k$, then we use the Cauchy-Schwartz inequality, (23), and (31) to finish the proof. For $j = 0$ we write $X_0 = P - \sum_{j=1}^{k} X_j^2 - cu$ and obtain

$$\begin{aligned}\|X_0 u\|^2_{H_{-1/2}} &= (X_0 u, T^0 u)_{L^2} \le (Pu, T^0 u)_{L^2} - \sum_{j=1}^{k}(X_j^2 u, T^0 u)_{L^2} - (cu, T^0 u)_{L^2}\\ &\le C\Big(\|Pu\|^2_{L^2} + \|u\|^2_{L^2}\Big),\end{aligned}$$

since

$$|(X_j^2 u, T^0 u)_{L^2}| \le |(X_j u, T^0 X_j u)_{L^2}| + |(X_j u, \underbrace{[X_j, T^0]}_{\sim T^0} u)_{L^2}| + |(X_j u, fT^0 u)_{L^2}|.$$

Thus $\|F^p u\|^2_{H_{\epsilon-1}} \le C\Big(\|Pu\|^2_{L^2} + \|u\|^2_{L^2}\Big)$. It proves the subelliptic estimate (20).

4 Heisenberg Group and Hypoellipticity

We have already defined an elliptic operator. A subelliptic operator, roughly speaking, is an operator that satisfies the subelliptic estimate.

Definition 4.1 A self-adjoint differential operator P (of order 2) with C^∞ smooth coefficients defined on a Riemannian manifold M is said to be subelliptic of order ε, $0 < \varepsilon < 1$, at $x \in M$, if there exist a neighbourhood $U \subset M$ of x and a constant $C > 0$ such that

$$\|u\|^2_{H_\varepsilon(U)} \leq C\Big((Pu,u)_{L^2(U)} + \|u\|^2_{L^2(U)}\Big) \leq C\Big(\|Pu\|^2_{L^2(U)} + \|u\|^2_{L^2(U)}\Big)$$

for all $u \in C_0^\infty(U)$.

The elliptic and subelliptic operators appear very naturally in the theory of complex variables. In the following subsections, after a necessary introduction, we explain how the geometry of the domain in $\mathbb{C}^n$ is related to subelliptic and therefore also to hypoelliptic operators.

Complexifications

Here we present definitions of complexifications of real vector spaces, real manifolds and Lie groups, see, for instance [21].

Complexification of a Real Vector Space

A *complexification* of a real vector space V is the tensor product $V \otimes \mathbb{C}$ over $\mathbb{R}$, where the generators are $v \otimes 1$ and $v \otimes i$, $v \in V$. So, $V \otimes \mathbb{C}$ are all possible linear combinations of $v \otimes 1$ and $v \otimes i$, $v \in V$ with real coefficients, modulo the equivalence relations

$$(v_1 + v_2) \otimes z \sim v_1 \otimes z + v_2 \otimes z,$$
$$v \otimes (z_1 + z_2) \sim v \otimes z_1 + v \otimes z_2,$$
$$av \otimes z \sim v \otimes az, \quad a \in \mathbb{R}.$$

The real dimension of $V \otimes \mathbb{C}$ is $2 \dim V$. The multiplication by complex numbers on $V \otimes \mathbb{C}$ is defined by

$$\alpha(v \otimes z) = v \otimes \alpha z, \quad \text{for} \quad \alpha, z \in \mathbb{C}, \quad \text{and} \quad v \in V.$$

It makes the space $V \otimes \mathbb{C}$ into a complex vector space of complex dimension $\dim V$. The generators for the complex vector space $V \otimes \mathbb{C}$ are $v \otimes 1$ and $v \otimes i$. The real space V is naturally imbedded into $V \otimes \mathbb{C}$ by identifying V with the space $V \otimes 1$. The *conjugation* for $V \otimes \mathbb{C}$ is defined by $\overline{v \otimes z} := v \otimes \bar{z}$.

As an application we consider a complexification of a smooth real manifold M of real dimension n. For any $q \in M$, the complex vector space $T_qM \otimes \mathbb{C}$ is called the *complexified tangent space* and $T_q^*M \otimes \mathbb{C}$ is called the *complexified co-tangent space*. The complex space $T_q^*M \otimes \mathbb{C}$ can also be regarded as the complex dual space of $T_qM \otimes \mathbb{C}$ by defining the pairing

$$\langle v \otimes z, \xi \otimes w \rangle := \langle v, \xi \rangle zw, \quad \text{for} \quad v \in T_qM,\ \xi \in T_q^*M,\ z, w \in \mathbb{C},$$

for any point $q \in M$. The *complexified tangent bundle* is $T^{\mathbb{C}}M = \cup_{q\in M}(T_qM \otimes \mathbb{C})$ and the *complexified co-tangent bundle* is $T^{*\mathbb{C}}M = \cup_{q\in M}(T_q^*M \otimes \mathbb{C})$. A *complexified vector field* L on M is a smooth section of $T^{\mathbb{C}}M$, which means that L assigns to each $q \in M$ a vector $L_q \in T_qM \otimes \mathbb{C}$. In any smooth coordinate system $(U, \varphi = (x^1, \dots, x^n))$ we can express L as $L_q = \sum_{j=1}^n L^j(q)\partial_{x^j}$, where L^j, $j = 1, \dots, n$ are smooth, complex valued functions defined on $U \subset M$.

If M is a complex manifold of complex dimension n, then it is important to distinguish between the real tangent bundle and the complexified tangent bundle. The real tangent bundle TM corresponds to a smooth manifold M of real dimension $2n$. Its fiber T_qM is a real vector space and has real dimension $2n$. The fiber $T_qM \otimes \mathbb{C}$ of the complexified tangent bundle is a complex space of complex dimension $2n$.

Complex Structures

If the real vector space V is of even dimension, then it is possible to define an *almost complex structure* J, that is, a map $J: V \to V$, such that $J^2 = -\mathrm{Id}_V$.

Example Let $V = T_q\mathbb{R}^{2n} \cong \mathbb{C}^n$. Take the coordinates $q = (x_1, y_1, \dots, x_n, y_n)$. The *standard almost complex* structure for $T_q\mathbb{R}^{2n}$ is defined by setting

$$J(\partial_{x^j}) = \partial_{y^j}, \qquad J(\partial_{y^j}) = -\partial_{x^j}, \qquad j = 1, \dots, n, \tag{47}$$

on the standard basis. Then J extends by linearity to all $T_q\mathbb{R}^{2n}$. This almost complex structure is designed to simulate the multiplication by $i = \sqrt{-1}$.

The standard almost complex structure J^* on the co-tangent space $T_q^*\mathbb{R}^{2n}$ is the following

$$J^*(dx^j) = dy^j, \qquad J^*(dy^j) = -dx^j, \qquad j = 1, \dots, n.$$

An almost complex structure can be defined on a real tangent space of a complex manifold M by pushing forward the complex structure from $\mathbb{C}^n$ to M via a coordinate chart. For $q \in M$ and a holomorphic chart (U, ζ), $\zeta: U \to \mathbb{C}^n$, we define $J_q: T_qM \to T_qM$ by

$$J_q(L_q) := d_{\zeta(q)}\zeta^{-1} J(d_q\zeta(L_q)), \quad L_q \in T_qM, \tag{48}$$

where J in the right hand side is the standard almost complex structure in $\mathbb{C}^n$. The definition implies that if $\zeta = (z_1, \dots, z_n)$, $z_j = x^j + iy^j$, then $J_q(\partial_{x^j}) = \partial_{y^j}$ and $J_q(\partial_{y^j}) = -\partial_{x^j}$.

If J is an almost complex structure on a real vector space V, then we can extend it to an almost complex structure $J_{\mathbb{C}}$ on the complexification $V \otimes \mathbb{C}$ by setting

$$J_{\mathbb{C}}(v \otimes z) := J(v) \otimes z, \quad v \in V, \ z \in \mathbb{C}.$$

Then

$$J_{\mathbb{C}}(\overline{w}) = \overline{J_{\mathbb{C}} w}, \quad \text{for} \quad w \in V \otimes \mathbb{C}. \tag{49}$$

The linear map $J_{\mathbb{C}}$ has two eigenvalues i and $-i$, since $J_{\mathbb{C}}^2 = -\operatorname{Id}$. The corresponding eigenspaces are denoted by $V^{(1,0)}$ and $V^{(0,1)}$. Thus we have

$$V \otimes \mathbb{C} = V^{(1,0)} \oplus V^{(0,1)}$$

from linear algebra. The property (49) implies $\overline{V^{(1,0)}} = V^{(0,1)}$. Let us construct bases for $V^{(1,0)}$ and $V^{(0,1)}$. First we observe that v and Jv are linearly independent over $\mathbb{R}$ in V, since J has no real eigenvalues. Then

$$\{v_1 - iJv_1, \dots, v_n - iJv_n\} \tag{50}$$

is a basis for the complex n-dimensional vector space $V^{(1,0)}$ and

$$\{v_1 + iJv_1, \dots, v_n + iJv_n\} \tag{51}$$

is a basis for the complex n-dimensional vector space $V^{(0,1)}$. Recall, that real $\dim V = 2n$.

We apply this for a complex n-dimensional manifold M. Let $(z_1, \dots, z_n)$ with $z_j = x^j + iy^j$ be a set of local holomorphic coordinates and the almost complex structure on T_qM, $q \in M$, is given by (48). Define the vector fields

$$\partial_{z_j} = \frac{1}{2}(\partial_{x^j} - i\partial_{y^j}) \qquad \partial_{\bar{z}_j} = \frac{1}{2}(\partial_{x^j} + i\partial_{y^j}), \ j = 1, \dots, n.$$

Then in view of the above discussions, a basis for $T_q^{(1,0)}M$ is given by $\{\partial_{z_1}, \dots, \partial_{z_n}\}$ and a basis for $T_q^{(0,1)}M$ is given by $\{\partial_{\bar{z}_1}, \dots, \partial_{\bar{z}_n}\}$. Due to the form of the bases, the spaces $T_q^{(1,0)}M$ and $T_q^{(0,1)}M$ received the names *holomorphic and antiholomorphic* tangent vector spaces. The Hermitian inner product on $T_qM \otimes \mathbb{C}$ is defined by declaring that $\{\partial_{z_1}, \dots, \partial_{z_n}, \partial_{\bar{z}_1}, \dots, \partial_{\bar{z}_n}\}$ is an orthonormal basis.

Let M now be a real manifold, such that at each $q \in M$ the tangent space T_qM admits an almost complex structure $J_q \colon T_qM \to T_qM$. Then it leads to the splitting $T^{\mathbb{C}}M = T^{(1,0)}M \oplus T^{(0,1)}M$ into the holomorphic and antiholomorphic bundles with respect to extension $(J_q)_{\mathbb{C}}$. Each of $T^{(1,0)}M$ and $T^{(0,1)}M$ considered as real vector spaces is isomorphic to the real tangent bundle TM. If $T^{(1,0)}M$ is integrable, that is, $[T^{(1,0)}M, T^{(1,0)}M] \subset T^{(1,0)}M$, then the pair $(M, T^{(1,0)}M)$ is called a complex manifold.

Complexification of Lie Algebras

Let us impose a Lie algebra structure on V and see how one can define a complexification $\mathfrak{g} \otimes \mathbb{C}$ of the Lie algebra $\mathfrak{g} = (V, [\cdot, \cdot])$. All that we need is to define the Lie bracket

$$[v \otimes \alpha, u \otimes \beta] := [v, u] \otimes \alpha\beta, \qquad v, u \in \mathfrak{g}, \ \alpha, \beta \in \mathbb{C}. \tag{52}$$

Next we consider the relation between the almost complex structure and the Lie algebra structure. Let $\mathbb{G}$ be a Lie group and $\mathfrak{g}$ be its Lie algebra. Let $J\colon T_e\mathbb{G} \to T_e\mathbb{G}$ be an almost complex structure. It determines the splitting $T_e\mathbb{G} \otimes \mathbb{C} = \mathfrak{g} \otimes \mathbb{C} = \mathfrak{g}^{(1,0)} \oplus \mathfrak{g}^{(0,1)}$. If the subspace $\mathfrak{g}^{(1,0)}$ is a Lie subalgebra of $\mathfrak{g} \otimes \mathbb{C}$, then the pair $(\mathbb{G}, \mathfrak{g}^{(1,0)})$ is called a left invariant complex structure.

Exercises

1. Show that the standard almost complex structure (47) in $T_q\mathbb{R}^{2n}$ is an isometry in $\mathbb{R}^{2n}$.
2. Show that the description of J_q given in (48) does not depend on the choice of coordinate chart. Conclude that the push forward of the standard almost complex structure J from $\mathbb{C}^n$ to a complex manifold is well defined.
3. Prove (49).
4. Show that $v - iJv \in V^{(1,0)}$ and $v + iJv \in V^{(0,1)}$ for any $v \in V$.
5. Prove that (50) and (51) are linearly independent systems.
6. Find the dual basis for $\{\partial_{z_1}, \dots, \partial_{z_n}, \partial_{\bar{z}_1}, \dots, \partial_{\bar{z}_n}\}$ with respect to the standard Hermitian product.
7. Verify that the Lie bracket defined by (52) is $\mathbb{C}$-linear, skew symmetric and satisfies the Jacobi identity.

Inhomogeneous Cauchy-Riemann Equation

Let $\Omega \subset \mathbb{C}^n$ be a domain, where we want to solve a (non-homogeneous) Cauchy-Riemann equation

$$\bar{\partial} u = f, \tag{53}$$

where $\bar{\partial} = \left(\frac{\partial}{\partial \bar{z}_1}, \dots, \frac{\partial}{\partial \bar{z}_n}\right)$ with $\frac{\partial}{\partial \bar{z}_k} = \frac{1}{2}\left(\frac{\partial}{\partial x^k} + i\frac{\partial}{\partial y^k}\right)$ for $z_k = x^k + iy^k$. If $f = 0$, the solution u is called a *holomorphic function*. Generally, given $f\colon \Omega \to \mathbb{C}^n$ satisfying some smoothness requirements, we want to find a function $u\colon \Omega \to \mathbb{C}$ also possessing some smoothness properties. Since $\frac{\partial^2 u}{\partial \bar{z}_k \partial \bar{z}_j} = \frac{\partial^2 u}{\partial \bar{z}_j \partial \bar{z}_k}$, the function f cannot be chosen arbitrarily, but it has to satisfy the necessary compatibility conditions

$$\frac{\partial f_j}{\partial \bar{z}_k} = \frac{\partial f_k}{\partial \bar{z}_j}, \quad k, j = 1, 2, \dots n. \tag{54}$$

Observe that if u is a solution to (53), then $u + h$, where h is a holomorphic function, is also a solution. To find a unique solution, we impose the "orthogonality condition", by making use of L^2-inner product in the domain Ω. Denoting the space of holomorphic functions by $\mathcal{H}(\Omega)$, we will look for a solution to (53), satisfying

$$(u, h)_{L^2(\Omega)} = \int_\Omega u(z)\bar{h}(z)\, dz = 0 \quad \text{for each} \quad h \in \mathcal{H}(\Omega). \tag{55}$$

Stated in this way, the problem can be reduced to the solution of the boundary-value problem for the Laplace equation in Ω. To explain this method, also called "the method of orthogonal projections", we start from the construction of the solution to (53) in the domain $\Omega \subset \mathbb{C}$ with smooth boundary $b\Omega$. Observe that $\Delta = \frac{\partial^2}{\partial x^2} + \frac{\partial^2}{\partial y^2} = 4\bar{\partial}\partial$. Let us assume that the Dirichlet boundary-value problem

$$\begin{cases} \Delta U(z) = f(z), & z \in \Omega, \\ U(z) = 0, & z \in b\Omega, \end{cases}$$

has a solution for f. Then, since $f = \Delta U = 4\bar{\partial}\partial U$, we can set $u = 4\partial U$ in order to find the solution to (53). Checking (55), we obtain

$$\int_\Omega u\bar{h}\, dz = 4\int_\Omega \partial U\bar{h}\, dz = 4U\bar{h}|_{b\Omega} - 4\int_\Omega U\overline{\bar{\partial} h}\, dz = 0,$$

since h is harmonic in Ω and $U|_{b\Omega} = 0$.

Observe that for a domain in $\mathbb{C}$ we did not use the condition (54), but for the several complex variables it becomes essential and we need to reformulate the problem in the language of differential forms.

$\bar{\partial}$-*Complex*

Let M be an n-dimensional complex manifold, that for the first reading can be thought of as a domain in $\mathbb{C}^n$. For $0 \le r \le 2n$, we denote by $\Lambda^r(T^{*\mathbb{C}}M)$ the vector bundle of the complexified r-forms on M and by $\mathcal{E}^r(M)$ the space of smooth sections of $\Lambda^r(T^{*\mathbb{C}}M)$. The vector space $\Lambda^{p,q}(T_z^{*\mathbb{C}}M) = \Lambda^p(T_z^{*(1,0)}M) \wedge \Lambda^q(T_z^{*(0,1)}M)$, $z \in M$, is the span over $\mathbb{C}$ of the set

$$\{dz^I \wedge d\bar{z}^J = dz_{i_1} \wedge \ldots \wedge dz_{i_p} \wedge d\bar{z}_{j_1} \wedge \ldots \wedge dz_{j_q}\},$$

where $I = \{i_1, \ldots, i_p\}$ and $J = \{j_1, \ldots, j_q\}$ run over the set of all increasing multi-indices of length p and q, respectively. The bundle $\Lambda^{p,q}(T^{*\mathbb{C}}M)$ is a sub-bundle of $\Lambda^r(T^{*\mathbb{C}}M)$ for $r = p + q$. The space of smooth sections of the vector bundle $\Lambda^{p,q}(T^{*\mathbb{C}}M)$ is denoted by $\mathcal{E}^{p,q}(M)$ and is called the space of differential (p, q)-forms on M. An element $\phi \in \mathcal{E}^{p,q}(M)$ can be expressed in local coordinates as

$$\phi = \sum_{|I|=p,|J|=q} \phi_{IJ}(z)\, dz^I \wedge dz^J,$$

where ϕ_{IJ} are smooth complex-valued functions on M. We let

$$\pi^{p,q} \colon \Lambda^r(T^{*\mathbb{C}}M) \to \Lambda^{p,q}(T^{*\mathbb{C}}M)$$

be the projection, for any $0 \leq p, q \leq n$, $p + q = r$.

Definition 4.2 The Cauchy-Riemann operator $\bar{\partial} \colon \mathcal{E}^{p,q}(M) \to \mathcal{E}^{p,q+1}(M)$ and the operator $\partial \colon \mathcal{E}^{p,q}(M) \to \mathcal{E}^{p+1,q}(M)$ are defined by

$$\bar{\partial} = \pi^{p,q+1} \circ d, \quad \partial = \pi^{p+1,q} \circ d.$$

For example, if $f \colon M \to \mathbb{C}$ is a smooth function, or, in other words, $f \in \mathcal{E}^{(0,0)}(M)$, then

$$df = \sum_{j=1}^{n} \frac{\partial f}{\partial x^j} dx^j + \frac{\partial f}{\partial y^j} dy^j = \sum_{j=1}^{n} \frac{\partial f}{\partial z_j} dz_j + \sum_{j=1}^{n} \frac{\partial f}{\partial \bar{z}_j} d\bar{z}_j,$$

where $\sum_{j=1}^{n} \frac{\partial f}{\partial z_j} dz_j \in \mathcal{E}^{1,0}(M)$ and $\sum_{j=1}^{n} \frac{\partial f}{\partial \bar{z}_j} d\bar{z}_j \in \mathcal{E}^{0,1}(M)$. Therefore

$$\partial f = \sum_{j=1}^{n} \frac{\partial f}{\partial z_j} dz_j, \quad \bar{\partial} f = \sum_{j=1}^{n} \frac{\partial f}{\partial \bar{z}_j} d\bar{z}_j, \quad df = \partial f + \bar{\partial} f.$$

Since for higher degree forms we have

$$\partial(\phi_{IJ}\, dz^I \wedge dz^J) = \partial\phi_{IJ} \wedge dz^I \wedge dz^J, \quad \bar{\partial}(\phi_{IJ}\, dz^I \wedge dz^J) = \bar{\partial}\phi_{IJ} \wedge dz^I \wedge dz^J,$$

for $\phi_{IJ} \in \mathcal{E}^{0,0}(M)$, we conclude that $d\phi = \partial\phi + \bar{\partial}\phi$ for $\phi \in \mathcal{E}^{p,q}(M)$.

Since $\bar{\partial}^2 = 0$, the operator $\bar{\partial}$ has the associated cohomology, related to the $\bar{\partial}$-complex, also called the Dolbeault complex:

$$0 \longrightarrow \mathcal{E}^{p,0}(M) \xrightarrow{\bar{\partial}} \mathcal{E}^{p,1}(M) \xrightarrow{\bar{\partial}} \dots \xrightarrow{\bar{\partial}} \mathcal{E}^{p,n}(M) \longrightarrow 0.$$

We define the cohomology

$$H^{p,q}(M,\mathbb{C}) = \frac{\{f \in \mathcal{E}^{p,q} \mid \bar{\partial} f = 0\}}{\{f \in \mathcal{E}^{p,q} \mid f = \bar{\partial}\psi,\ \psi \in \mathcal{E}^{p,q-1})\}}.$$

We define now the adjoint map with respect to L^2-metric. Assume that the complex manifold is endowed with an Hermitian inner product that will induce the inner product $(.,.)$ on $\Lambda^{p,q}(T_z^{*\mathbb{C}}M)$, $z \in M$. Let $\mathcal{E}_0^{p,q}(M)$ be the space of compactly supported elements of $\mathcal{E}^{p,q}(M)$. We endow $\mathcal{E}_0^{p,q}(M)$ with the metric

$$(\phi, \psi)_{L^2} = \int_M (\phi(z), \psi(z))\, d\mathrm{Vol}(z), \tag{56}$$

where $d\mathrm{Vol}$ is the volume form on M. The adjoint operator $\bar{\partial}^*\colon \mathcal{E}_0^{p,q}(M) \to \mathcal{E}_0^{p,q-1}(M)$ to $\bar{\partial}\colon \mathcal{E}^{p,q-1}(M) \to \mathcal{E}^{p,q}(M)$ is defined by

$$(\bar{\partial}^*\phi, \psi)_{L^2} = (\phi, \bar{\partial}\psi)_{L^2}. \tag{57}$$

We defined the adjoint operator $\bar{\partial}^*$ for compactly supported forms. Sometimes it is defined more formally, requiring that the formula (57) is true for all smooth $\psi \in \mathcal{E}^{p,q-1}(\Omega)$, that in its turn implies that the boundary terms arising from the integration by parts must vanish, which is equivalent to saying that a particular linear relation among the coefficients of $\phi \in \mathcal{E}^{p,q}(\Omega)$ must vanish on the boundary $b\Omega$. We abbreviate it by writing $\phi \in \mathrm{Dom}(\bar{\partial}^*_{p,q})$. Now we would like to reduce the problem of solving (53) for several complex variables to one that is analogous of the boundary value problem for the Laplace operator. The correct analogue of the Laplace operator is the box operator

$$\Box = \bar{\partial}^*\bar{\partial} + \bar{\partial}\bar{\partial}^*\colon \mathcal{E}^{p,q}(\Omega) \to \mathcal{E}^{p,q}(\Omega).$$

By making use of this formalism, equation (53) is understood for $u \in \mathcal{E}^{0,0}(\Omega)$ and $f \in \mathcal{E}^{0,1}(\Omega), f = \sum_{j=1}^n f_j d\bar{z}_j$. The compatibility condition (54) is written in the form $\bar{\partial} f = 0$, due to the skew symmetry $d\bar{z}_k \wedge d\bar{z}_j = -d\bar{z}_j \wedge d\bar{z}_k$. The Dirichlet boundary value problem is substituted by the $\bar{\partial}$-Neumann problem: for a given $f \in \mathcal{E}^{p,1}(\Omega)$

we want to find $U \in \mathcal{E}^{p,1}(\Omega)$ satisfying

$$\begin{cases} \Box U(z) = f(z), \quad z \in \Omega, \\ U \in \mathrm{Dom}(\bar\partial^*_{p,1}), \\ \bar\partial U \in \mathrm{Dom}(\bar\partial^*_{p,2}). \end{cases} \tag{58}$$

Let us show that if there exists a solution of (58) with $\bar\partial f = 0$, then the solution of (53), satisfying (54) and (55) exists and is equal to $u = \bar\partial^* U$. Indeed for any harmonic $(p,1)$-form h we obtain

$$(f,h)_{L^2} = (\Box U, h)_{L^2} = ((\bar\partial^*\bar\partial + \bar\partial\bar\partial^*)U, h)_{L^2} = (\bar\partial U, \bar\partial h)_{L^2} + (\bar\partial\bar\partial^* U, h)_{L^2}$$

since $\bar\partial U \in \mathrm{Dom}(\bar\partial^*_{p,2})$. Now $(\bar\partial U, \bar\partial h)_{L^2} = 0$ since h is a harmonic $(p,1)$-form and therefore if we denote $\bar\partial^* U = u$ we get $\bar\partial u = f$. Observe that we restricted ourselves to harmonic $(p,1)$-forms h, since f is also harmonic $(p,1)$-form. Therefore $(f,\psi)_{L^2} = 0$ for any ψ from the orthogonal complement to harmonic $(p,1)$-forms. We only need to check (55). We see that

$$(u,h)_{L^2} = (\bar\partial^* U, h)_{L^2} = (U, \bar\partial h)_{L^2(\Omega)} = 0$$

for any $(p,0)$-form h.

At the end we emphasise that (58) is decoupled into two boundary value problems for the Laplacian $\Box$. One is the usual Dirichlet problem, determined by $U \in \mathrm{Dom}(\bar\partial^*_{p,1})$. The other one $\bar\partial U \in \mathrm{Dom}(\bar\partial^*_{p,2})$ is "non-elliptic" boundary value problem for the Laplacian $\Box$, which shows a difference between the theory of one complex variable and the theory of several complex variables.

Exercises

1. By making use of $d\phi = \partial\phi + \bar\partial\phi$ for $\phi \in \mathcal{E}^{p,q}(M)$, show the following properties:

    1. $\partial^2 = 0$, $\bar\partial^2 = 0$, $\partial\bar\partial = -\bar\partial\partial$;
    2. and

    $$\bar\partial(\phi \wedge \psi) = \bar\partial\phi \wedge \psi + (-1)^{p+q}\phi \wedge \bar\partial\psi,$$
    $$\partial(\phi \wedge \psi) = \partial\phi \wedge \psi + (-1)^{p+q}\phi \wedge \partial\psi,$$

    for any $\phi \in \mathcal{E}^{p,q}(M)$ and $\psi \in \mathcal{E}^{r,s}(M)$.

2. Suppose M and N are complex manifolds and $F: M \to N$ is a holomorphic map, that is a smooth map such that $\bar\partial F: T^{(1,0)}(M) \to T^{(1,0)}(N)$. Show that if $\phi \in \mathcal{E}^{p,q}(N)$, then $F^*\phi \in \mathcal{E}^{p,q}(M)$. Moreover

$$F^* \circ \bar\partial = \bar\partial \circ F^*, \qquad F^* \circ \partial = \partial \circ F^*.$$

3. Let $f \in \mathcal{E}^{0,0}(M)$, and J be a complex structure on M. In this case, the adjoint map J^* is the complex structure map on the space of 1-forms. Show that

$$\partial f = \frac{1}{2}\Big(df - iJ^* df\Big), \qquad \bar{\partial} f = \frac{1}{2}\Big(df + iJ^* df\Big).$$

4. The adjoint operator $\bar{\partial}^*$ can be calculated by making use of the contraction operator $\rfloor$. Show that for $\phi = f\, dz^I \wedge d\bar{z}^J, f \in \mathcal{E}_0^{0,0}(M)$ we have

$$\bar{\partial}^* \phi = -\sum_{j=1}^{n} \frac{\partial f}{\partial z_j} \frac{\partial}{\partial \bar{z}_j} \rfloor (dz^I \wedge d\bar{z}^J),$$

where the contraction operator $\rfloor$ is defined by

$$\frac{\partial}{\partial \bar{z}_j} \rfloor (dz^I \wedge d\bar{z}^J) = \begin{cases} 0, & \text{if } \ j \notin J, \\ (-1)^{p+k-1} dz^I \wedge d\bar{z}^{J'} & \text{if } \ j \in J \end{cases}$$

for $|I| = p$, $|J| = q$. Here if $j \in J = (j_1, \dots, j_k, \dots, j_q)$, then $j = j_k$ and $J' = (j_1, \dots, \widehat{j_k}, \dots, j_q)$ is the multi-index of the length $q-1$, where j_k is omitted.

The Boundary $\bar{\partial}_b$-Complex

First we give the rough idea and then the precise definitions. The significant difference that occurs in several complex variables is the presence of tangential Cauchy-Riemann operators. The tangential Cauchy-Riemann operators are complex vector fields on the boundary $b\Omega$ of a domain $\Omega \subset \mathbb{C}^{n+1}$ that are characterised by the following two properties:

1. The vector field on $b\Omega$ is the restriction of a vector field defined in $\overline{\Omega}$, such that in coordinates $z = (z_1, \dots, z_{n+1})$ of $\mathbb{C}^{n+1}$ it can be written in the form

$$Z = \sum_{j=1}^{n+1} a_j(z) \frac{\partial}{\partial \bar{z}_j}, \qquad z \in \mathbb{C}^{n+1}, \tag{59}$$

with complex valued functions a_j. This is precisely the first order differential operators that annihilate holomorphic functions in Ω.

2. The vector field (59) is tangential in the following sense. If r is a defining function of the boundary, that is $r: \overline{\Omega} \to \mathbb{R}$, $dr \neq 0$, and $r(z) = 0$ for $z \in b\Omega$, then

$$Zr(z) = \sum_{j=1}^{n+1} a_j(z) \frac{\partial r(z)}{\partial \bar{z}_j} = 0, \qquad z \in b\Omega.$$

An easy counting argument shows that, at each point of $b\Omega$, the vector space of tangential Cauchy-Riemann operators has complex dimension n.

Bochner, Lewy and others discovered that a function u on the boundary $b\Omega$ of a suitable regular domain Ω is the restriction of some holomorphic function F on Ω if and only if u satisfies the tangential Cauchy-Riemann equations. It gives rise to an analogue of $\bar{\partial}$-complex, but on the boundary $b\Omega$, and therefore can be called the boundary $\bar{\partial}_b$-complex. The idea behind the definition of $\bar{\partial}_b$ is as follows. Suppose that u is a smooth function defined on the boundary $b\Omega$. To define $\bar{\partial}_b u$, we first extend u to a smooth function F on all of $\overline{\Omega}$ and calculate $\bar{\partial}F$. We then restrict $\bar{\partial}F$ to $b\Omega$ and focus on the part $\bar{\partial}F|_{b\Omega}$. Of course, the restriction $\bar{\partial}F|_{b\Omega}$ should not depend on the particular extension. Namely, let r be a defining function of the boundary. If F_1 and F_2 are two extensions of u, then $\bar{\partial}F_1|_{b\Omega}$ and $\bar{\partial}F_2|_{b\Omega}$ differ by a multiple of $\bar{\partial}r$. Thus the boundary $(0,1)$-forms can be defined as restrictions to $b\Omega$ of ordinary $(0,1)$-forms in Ω, modulo multiples of $\bar{\partial}r$.

Boundary (p,q)-forms can be defined similarly for $p > 0$, $q > 0$, as well as the operators $\bar{\partial}$, ∂. In the presence of a suitable inner product on $b\Omega$, we can also define the formal adjoint $\bar{\partial}_b^*$ and pass to the corresponding boundary Laplacian

$$\Box_b = \bar{\partial}_b^* \bar{\partial}_b + \bar{\partial}_b \bar{\partial}_b^*. \tag{60}$$

Inverting $\Box_b$ is closely related to solving the $\bar{\partial}$-Neumann problem, but is not identical to it, first of all, because there are no boundary conditions like $U \in \mathrm{Dom}(\bar{\partial}_{0,1}^*)$ or $\bar{\partial}U \in \mathrm{Dom}(\bar{\partial}_{0,2}^*)$. However the "nonellipticity" of $\bar{\partial}_b$-Neumann problem is reflected here in the fact that the operator $\Box_b$ itself is not elliptic: it involves second order differentiation only in the directions of the real and imaginary parts of the tangential Cauchy-Riemann operators, so there is always one "missing" direction.

Let us now present the abstract analogue of $\bar{\partial}_b$ operator. The boundary of a domain in $\mathbb{C}^{n+1}$ is replaced by a Cauchy-Riemann manifold (shortly CR-manifold). A CR-manifold M is a real oriented C^∞-manifold of real dimension $2n+1$, $n = 1, 2, \ldots$, together with a sub-bundle $H^{(1,0)}(M)$ of the complexified tangent bundle $T^{\mathbb{C}}M$, satisfying:

1. $\dim_{\mathbb{C}} H^{(1,0)}(M) = n$,
2. $H^{(1,0)}(M) \cap H^{(0,1)}(M) = \{0\}$, where $H^{(0,1)}(M) = \overline{H^{(1,0)}(M)}$,
3. the sub-bundle $H^{(1,0)}(M)$ is integrable in the sense that if Z_1, Z_2 are sections of $H^{(1,0)}(M)$, then the commutator $[Z_1, Z_2]$ is also the section of $H^{(1,0)}(M)$,
4. the complementary line bundle K such that $H^{(1,0)}(M) \oplus H^{(0,1)}(M) \oplus K = T^{\mathbb{C}}M$ has a global section (that is always satisfied in practise).

For instance, if M is realised as a real hypersurface in $\mathbb{C}^{n+1}$ or, more general, in a complex $(n+1)$-dimensional manifold N, as a zero set of defining function $r\colon N \to \mathbb{R}$, $dr \neq 0$, then the canonical CR-structure on M is given by taking $H^{(1,0)}(M) = T^{\mathbb{C}}M \cap T^{(1,0)}(N)$. Then, if we denote by $\iota\colon M \to N$ the inclusion map, then $\iota^*(i(\bar{\partial} -$

$\partial)r)$ is a global section for the sub-bundle K. Nirenberg [82] showed that an abstract CR-manifold cannot always be imbedded in a complex manifold, even locally.

We denote $\Lambda^{p,q}(M) = \Lambda^p(H^{*(1,0)}(M)) \wedge \Lambda^q(H^{*(0,1)}(M))$ for the CR-manifold M, and by $\mathcal{E}^{p,q}(M)$ the set of smooth sections of $\Lambda^{p,q}(M)$. Then the operator $\bar{\partial}_b : \mathcal{E}^{p,q}(M) \to \mathcal{E}^{p,q+1}(M)$ is defined by $\bar{\partial}_b \phi = \pi^{p,q+1} \circ d\phi$ for $\phi \in \mathcal{E}^{p,q}(M)$, where $\pi^{p,q+1}$ is the projection of the form $d\phi$ onto the subspace $\mathcal{E}^{p,q+1}(M)$. The operator $\bar{\partial}_b$ possess the analogous properties, listed in the previous exercises. We just mention one more: If $f \in \mathcal{E}^{0,0}(M)$ and Z is a smooth section of $H^{(0,1)}(M)$ which is analogue of tangential Cauchy-Riemann operators, then $\bar{\partial}_b f(Z) = Zf$.

We assume that on the CR-manifold M is given an Hermitian metric g, such that the decomposition $H^{(1,0)}(M) \oplus H^{(0,1)}(M) \oplus K$ is orthogonal and the Hermitian metric induces the Riemannian metric on M (we denote the Riemannian metric also by g). Define the L^2-metric as in (56). Now we are ready to define the formal adjoint as it was done in (57) and the "Laplace" operator (60). The operator $\Box_b$ is never elliptic, since it degenerates along the direction K orthogonal to $H^{(1,0)}(M) \oplus H^{(0,1)}(M)$. We say in this case that the direction K is characteristic. Nevertheless, the operator $\Box_b$ is subelliptic under certain geometric conditions on M.

Let ω be a non-vanishing real one form which annihilates the distribution $H^{(1,0)}(M) \oplus H^{(0,1)}(M)$. In the case when M is a hypersurface in the complex manifold N given by the equation $M = \{z \in N \mid r(z) = 0,\ dr \neq 0\}$, then we can take $\omega = \iota^* i\big(\bar{\partial} - \partial)r\big)$ as was mentioned above.

Definition 4.3 The Levi form $(.,.)_{Levi}$ is an Hermitian form defined for sections of $H^{(1,0)}(M)$ by

$$(Z_1, Z_2)_{Levi} = -id\omega(Z_1 \wedge \bar{Z}_2). \tag{61}$$

The Levi form is a quadratic form on the sub-bundle $H^{(1,0)}(M)$. Since the one form ω is a conformal class (it is a real one form defined up to a nonvanishing real factor), the Levi form is also. We say that the CR-manifold M is *nondegenerate* if the Levi form is non-degenerate for each section Z of $H^{(1,0)}(M)$. We call M *k-strongly pseudoconvex* if the Levi form has k, $k \geq \frac{n}{2}$ positive eigenvalues. In the case $k = n$ the CR-manifold is called *strongly pseudoconvex*.

What is the relation between the Levi form and the characteristic direction? The definition (61) implies that

$$(Z_1, Z_2)_{Levi} = -id\omega(Z_1 \wedge \bar{Z}_2) = -\frac{i}{2}\Big(Z_1 \underbrace{\omega(\bar{Z}_2)}_{=0} - \bar{Z}_2 \underbrace{\omega(Z_1)}_{=0} - \omega([Z_1, \bar{Z}_2])\Big)$$

$$= \frac{i}{2}\omega([Z_1, \bar{Z}_2]).$$

Thus when $(Z_1, Z_2)_{Levi} \neq 0$, we can obtain some control over the characteristic direction by using the commutator $[Z_1, \bar{Z}_2]$. This is an essential point in the result obtained by Kohn, see section "Hypoellipticity of $\Box_b$ on Pseudoconvex CR

Manifolds”. Before turning to this point we consider the particular example of a CR-manifold, the Heisenberg group or the boundary of the Siegel upper half space.

$\Box_b$ *Operator on Heisenberg Group*

We recall the identification of the Heisenberg group $\mathbb{H}^n$ with the boundary of the Siegel upper half space $\mathcal{U}^n \subset \mathbb{C}^{n+1}$ in section “Action of the Heisenberg Group on the Siegel Upper Half Space”. Let us present the connection between the tangential Cauchy-Riemann operators and the Lie algebra $\mathfrak{h}^n$ of the Heisenberg group. We write the basis for the Heisenberg Lie algebra as real left invariant vector fields:

$$X_j = \frac{\partial}{\partial x^j} + 2y^j \frac{\partial}{\partial t}, \quad Y_j = \frac{\partial}{\partial y^j} - 2x^j \frac{\partial}{\partial t}, \quad T = \frac{\partial}{\partial t}.$$

The basis satisfies the commutation relations $[Y_j, X_k] = 4\delta_{jk} T$ and others vanish. We define the complex vector fields in $\mathbb{C}^{n+1}$:

$$\bar{Z}_j = \frac{1}{2}\left(X_j + iY_j\right) = \frac{\partial}{\partial \bar{\zeta}_j} - i\zeta_j \frac{\partial}{\partial t}, \quad Z_j = \frac{1}{2}\left(X_j - iY_j\right) = \frac{\partial}{\partial \zeta_j} + i\bar{\zeta}_j \frac{\partial}{\partial t}, \tag{62}$$

$j = 1, \ldots, n$, where $\frac{\partial}{\partial \bar{\zeta}} = \frac{1}{2}\left(\frac{\partial}{\partial x^j} + i\frac{\partial}{\partial y^j}\right)$ and $\frac{\partial}{\partial \zeta} = \frac{1}{2}\left(\frac{\partial}{\partial x^j} - i\frac{\partial}{\partial y^j}\right)$. The commutation relations change to

$$[\bar{Z}_j, Z_k] = 2i\delta_{jk} T \tag{63}$$

and others remain vanishing. The vector fields defined in (62) are tangential Cauchy-Riemann operators on $b\mathcal{U}_n$ and form the basis for the space of such operators. It can be checked by definition or we can argue as follows: At the identity element of the Heisenberg group $\mathbb{H}^n$, which corresponds to the origin of $b\mathcal{U}_n$ we have $\bar{Z}_j = \frac{\partial}{\partial \bar{\zeta}_j}$, agreeing with the form (59). Next, the vector fields $\bar{Z}_j$ are left invariant, since the vector fields X_j, Y_j are left invariant. This left action on the group $\mathbb{H}^n$ is induced by translation on $\mathcal{U}_n$ given by (13), that is a holomorphic self mapping of $\mathbb{C}^{n+1}$. Since holomorphic mappings preserve the vector fields (59), we conclude that item 1 of the definition of tangential CR operators is true for $\bar{Z}_j$. Moreover, at each point of the boundary $b\mathcal{U}_n$, there are n linearly independent restrictions of $\bar{Z}_j$ showing that $\{\bar{Z}_j\}_{j=1}^n$ form a basis for the tangential Cauchy-Riemann operators on $b\mathcal{U}_n$. Now recall the change of coordinates: $\zeta_j = z_j, j = 1, \ldots, n$, $t = \operatorname{Re} z_{n+1}$, $r = \operatorname{Im} z_{n+1} - |z'|^2$. Then

$$\frac{\partial}{\partial \bar{z}_j} = \frac{\partial}{\partial \bar{\zeta}_j} - z_j \frac{\partial}{\partial r}, \; j = 1, \ldots, n, \qquad \frac{\partial}{\partial \bar{z}_{n+1}} = \frac{1}{2}\left(\frac{\partial}{\partial t} + i\frac{\partial}{\partial r}\right).$$

Observe that since $\frac{\partial r}{\partial \bar{z}_j} - 2iz_j \frac{\partial r}{\partial \bar{z}_{n+1}} = 0$, the operator $\frac{\partial}{\partial \bar{z}_j} - 2iz_j \frac{\partial}{\partial \bar{z}_{n+1}}$ is a tangential Cauchy-Riemann operator and

$$\frac{\partial}{\partial \bar{z}_j} - 2iz_j \frac{\partial}{\partial \bar{z}_{n+1}} = \frac{\partial}{\partial \bar{\zeta}_j} - i\zeta_j \frac{\partial}{\partial t} = \bar{Z}_j.$$

Using the coordinates (z, t) on $\mathbb{H}^n$, we write $z_j = x^j + iy^j$, $d\bar{z}_1, \dots, d\bar{z}_{n+1}$ for the basis of $(0, 1)$-forms, and $d\bar{z}^J = d\bar{z}_{j_1} \wedge \dots \wedge d\bar{z}_{j_q}$ for the basis of $(0, q)$-forms. For the moment, we restrict to $(0, q)$-forms and call them simply q-forms, nevertheless the considerations are valid for any $p > 0$. In general, any $f \in \mathcal{E}^{0,q}(\mathbb{H}^n)$ can be written as $f = \sum_{|J|=q} f_J d\bar{z}^J$, where f_J are smooth complex valued functions on $\mathbb{H}^n$. For the left invariant Cauchy-Riemann operators $\bar{Z}_j = \frac{\partial}{\partial \bar{z}_j} - iz_j \frac{\partial}{\partial t}$ on $\mathbb{H}^n$ we have

$$\bar{\partial}_b f = \sum_{j=1}^{n} \bar{Z}_j(f) dz_j, \qquad f \in \mathcal{E}^{0,0}(\mathbb{H}^n), \quad \text{and}$$

$$\bar{\partial}_b f = \bar{\partial}_b \Big(\sum_{|J|=q} f_J d\bar{z}^J \Big) = \sum_{j=1}^{n} \sum_{|J|=q} \bar{Z}_j(f_J) d\bar{z}_j \wedge d\bar{z}^J, \qquad f \in \mathcal{E}^{0,q}(\mathbb{H}^n). \tag{64}$$

By making use of definition of the dual $\bar{\partial}_b^*$ and Exercise 4, we obtain

$$\bar{\partial}_b^* \Big(\sum_{|J|=q} f_J d\bar{z}^J \Big) = - \sum_{j=1}^{n} \Big(\sum_{|J|=q} Z_j(f_J) \frac{\partial}{\partial \bar{z}_j} \rfloor d\bar{z}^J \Big), \tag{65}$$

and, particularly, for 1-form $f = \sum_{j=1}^{n} f_j d\bar{z}_j$, we obtain $\bar{\partial}_b^* f = - \sum_{j=1}^{n} Z_j(f_j)$. We now show that the operator $\Box_b$ is an analogous of the Laplace operator, i.e. it can be expressed as a sum of operators of the second order.

Theorem 4.4 ([94]) *We have the identity*

$$\Box_b \Big(\sum_{|J|=q} f_J d\bar{z}^J \Big) = \sum_{|J|=q} \mathcal{L}_\alpha(f_J) d\bar{z}^J,$$

where $\alpha = n - 2q$ *and*

$$\mathcal{L}_\alpha = -\frac{1}{2} \sum_{j=1}^{n} (Z_j \bar{Z}_j + \bar{Z}_j Z_j) + i\alpha T.$$

Proof Recall the formulas (64) and (65) that we apply to the q-form $f = f_J d\bar{z}^J$. We obtain

$$\bar{\partial}_b(f) = \sum_{l=1}^{n} \bar{Z}_l(f_J)\, d\bar{z}_l \wedge d\bar{z}^J, \tag{66}$$

$$\bar{\partial}_b^*(f) = -\sum_{k=1}^{n} Z_k(f_J)\, \frac{\partial}{\partial \bar{z}_k} \rfloor d\bar{z}^J. \tag{67}$$

Applying again formulas (64)–(67), we deduce

$$\begin{aligned}\Box_b f = (\bar{\partial}_b^* \bar{\partial}_b + \bar{\partial}_b \bar{\partial}_b^*) f &= -\sum_{k=1}^{n} \sum_{l=1}^{n} Z_k \bar{Z}_l(f_J)\, \frac{\partial}{\partial \bar{z}_k} \rfloor (d\bar{z}_l \wedge d\bar{z}^J) \\ &\quad - \sum_{l=1}^{n} \sum_{k=1}^{n} \bar{Z}_l Z_k(f_J)\, d\bar{z}_l \wedge (\frac{\partial}{\partial \bar{z}_k} \rfloor d\bar{z}^J).\end{aligned} \tag{68}$$

CASE $k \neq l$. We have $Z_k \bar{Z}_l = \bar{Z}_l Z_k$ due to (63). Moreover

$$\frac{\partial}{\partial \bar{z}_k} \rfloor (d\bar{z}_l \wedge d\bar{z}^J) = -d\bar{z}_l \wedge (\frac{\partial}{\partial \bar{z}_k} \rfloor d\bar{z}^J).$$

It is easy to remember the last formula, since the operation "$\wedge$" adds the element to the wedge product, while the operation "$\rfloor$" removes the element from the wedge product and then permuting $d\bar{z}_l$ and $d\bar{z}_k$ we gain a sign "$-$". Thus, if $k \neq l$, then the corresponding terms give zero impact into $\Box_b$ as it is easy to see from (68).

CASE $k = l$. We have

$$\frac{\partial}{\partial \bar{z}_k} \rfloor (d\bar{z}_k \wedge d\bar{z}^J) = \begin{cases} 0, & \text{if} \quad k \in J, \\ d\bar{z}^J, & \text{if} \quad k \notin J, \end{cases}$$

$$d\bar{z}_k \wedge (\frac{\partial}{\partial \bar{z}_k} \rfloor d\bar{z}^J) = \begin{cases} 0, & \text{if} \quad k \notin J, \\ d\bar{z}^J, & \text{if} \quad k \in J. \end{cases}$$

Thus formula (68) is written as

$$\Box_b f = -\Big(\sum_{k \notin J} Z_k \bar{Z}_k + \sum_{k \in J} \bar{Z}_k Z_k\Big)(f_J)\, d\bar{z}^J. \tag{69}$$

We use (63) and write

$$\bar{Z}_k Z_k = \frac{1}{2}(\bar{Z}_k Z_k + Z_k \bar{Z}_k) + iT, \qquad Z_k \bar{Z}_k = \frac{1}{2}\Big(\bar{Z}_k Z_k + Z_k \bar{Z}_k\Big) - iT.$$

Substituting last formulas to (69) we obtain

$$\begin{aligned}\Box_b f &= -\Big(\frac{1}{2}\sum_{k\notin J}(Z_k\bar{Z}_k + \bar{Z}_k Z_k) - (n-q)iT \\ &\quad + \frac{1}{2}\sum_{k\in J}(Z_k\bar{Z}_k + \bar{Z}_k Z_k) + qiT\Big) f_J\, d\bar{z}^J \\ &= \Big(-\frac{1}{2}\sum_{k=1}\Big(Z_k\bar{Z}_k + \bar{Z}_k Z_k\Big) + i\alpha T\Big) f_J\, d\bar{z}^J\end{aligned}$$

for $\alpha = n - 2q$. □

Observe that the operator $\mathcal{L}_\alpha$ possesses the symmetry properties analogous to the classical Laplacian $\Delta = \sum_{k=1}^n \frac{\partial^2}{\partial x_k^2}$.

1. The operator $\mathcal{L}_\alpha$ is left invariant, i.e. does not depend on the left translation on the Heisenberg group. It is a counterpart of the property of the Laplacian Δ to have constant coefficients or to be invariant in $\mathbb{R}^n$ with respect to the Euclidean translation.
2. Recall that we defined the dilation on the Heisenberg group by $\delta_\lambda(z,t) = (\lambda z, \lambda^2 t)$, that corresponds to the automorphism of the Heisenberg algebra:

$$d\delta_\lambda([Z_j, \bar{Z}_j]) = [d\delta_\lambda Z_j, d\delta_\lambda \bar{Z}_j]) = [\lambda Z_j, \lambda \bar{Z}_j] = \lambda^2 [Z_j, \bar{Z}_j] = 2i\lambda^2 T.$$

   Under this dilation the differential operator $\mathcal{L}_\alpha$ is a homogeneous of order 2, which is analogous to that the Laplace operator Δ is a homogeneous operator of order 2 with respect to the classical dilatation in $\mathbb{R}^n$.
3. Let $U \in \mathrm{U}(n)$ be a unitary transformation, then we define the rotation on the Heisenberg group $\mathbb{H}^n$ by $\mathrm{Rot}(z,t) = (Uz, t)$. The differential operator $\mathcal{L}_\alpha$ is invariant under this type of rotations. It corresponds to the rotation invariance of the Laplace operator Δ with respect to the orthogonal transformation in $\mathbb{R}^n$.

Analogously to the Laplace operator in $\mathbb{R}^n$, any operator on the Heisenberg group, possessing these three symmetry properties, equals the differential operator $\mathcal{L}_\alpha$ up to the constant multiple.

Hypoellipticity of $\Box_b$ *on Pseudoconvex CR Manifolds*

Let M be a CR-manifold. Then we have the tangential Cauchy-Riemann complex

$$0 \longrightarrow \mathcal{E}^{p,0}(M) \xrightarrow{\bar\partial_b} \mathcal{E}^{p,1}(M) \xrightarrow{\bar\partial_b} \dots \xrightarrow{\bar\partial_b} \mathcal{E}^{p,n}(M) \longrightarrow 0,$$

where $\mathcal{E}^{p,q}(M)$ is the space of smooth sections of (p,q)-forms $\Lambda^{p,q}(M)$ as was defined in section "The Boundary $\bar\partial_b$-Complex". More about $\bar\partial_b$-cohomology see in [71]. The operator $\Box_b$ is not elliptic, however Kohn [66] showed, that if M is strongly pseudoconvex, then the operator $\Box_b$ is subelliptic and satisfies a priori estimates

$$\|f\|^2_{H_{s+1}} \le C\Big(\|\Box_b f\|^2_{H_s} + \|f\|^2_{L^2}\Big), \quad f \in \mathcal{E}^{p,q}_0(M), \quad s = 0,1,2,\dots.$$

As it was shown in [70], this leads to the hypoellipticity of $\Box_b$. Notice that strong pseudoconvexity in this case is equivalent to bracket generating condition in the Hörmander sense. But there is more behind the properties of the differential operator $\Box_b$. Namely, if the pseudoconvex CR manifold M is compact, then there is an operator G defined on forms from $\mathcal{E}^{p,q}$ satisfying

$$\|Gf\|_{H_{s+1}} \le C\|f\|_{H_s} \quad \text{and} \quad G\Box_b = \Box_b G = I - \Pi,$$

where Π is the orthogonal projection onto the kernel of the operator $\Box_b$. The Kohn method did not give any hint on how to find a solution. We discuss the construction of the solution of it in section "Fundamental Solution for $\Box_b$" and now we concentrate on the Kohn proof of the hypoellipticity of $\Box_b$. In the rest of this section we give a rough scheme of the hypoellipticity of the operator $\Box_b$. For the complete proof we refer to [66].

The $\bar\partial_b$-Cohomology

Since the CR manifold M is assumed to be integrable, the following $\bar\partial_b$*-cohomology of* $\mathcal{E}^{p,q}(M)$, denoted by $H^{p,q}_b(M,\mathbb{C})$, can be defined:

$$H^{p,q}_b(M,\mathbb{C}) = \frac{\{f \in \mathcal{E}^{p,q}(M) \mid \bar\partial_b f = 0\}}{\{f \in \mathcal{E}^{p,q}(M) \mid f = \bar\partial_b \psi,\ \psi \in \mathcal{E}^{p,q-1}(M)\}}.$$

Note that, under certain circumstances, each cohomology class in $H^{p,q}_b(M,\mathbb{C})$ has a unique representative which minimises the L^2-norm. From the other side, if f minimises the L^2-norm in a cohomology class $H^{p,q}_b(M,\mathbb{C})$, then the kernel of $\bar\partial_b \colon \mathcal{E}^{p,q}(M) \to \mathcal{E}^{p,q+1}(M)$ is orthogonal to the image of $\bar\partial^{p,q-1}_b \colon \mathcal{E}^{p,q-1}(M) \to \mathcal{E}^{p,q}(M)$, that is a counterpart of (55). Indeed, let $f \in \mathcal{E}^{p,q}(M)$ with $\bar\partial_b f = 0$, or in other words $f \in \ker(\bar\partial_b)$, $\bar\partial_b\colon \mathcal{E}^{p,q}(M) \to \mathcal{E}^{p,q+1}(M)$. Assume also that f minimises the L^2-norm in a cohomology class $H_b^{p,q}(M,\mathbb{C})$, i.e.

$$\|f\|_{L^2} \le \|f + \bar\partial_b\psi\|_{L^2} \quad \text{for all} \quad \psi \in \mathcal{E}^{p,q-1}(M).$$

Then, replacing ψ by $a\psi$ with any $a \in \mathbb{C}$, we obtain

$$\|f\|_{L^2}^2 \le \|f + a\bar\partial_b\psi\|_{L^2}^2 = \|f\|_{L^2}^2 + 2\mathrm{Re}\bar a(f, \bar\partial_b\psi)_{L^2} + |a|^2\|\bar\partial_b\psi\|_{L^2}^2. \tag{70}$$

There are two options: Either $(f, \bar\partial_b\psi)_{L^2} = 0$, in which case we finish the proof, or $(f, \bar\partial_b\psi)_{L^2} \neq 0$. In the latter case we choose an arbitrary $\varepsilon > 0$ and set $a = -\varepsilon\frac{(f,\bar\partial_b\psi)_{L^2}}{|(f,\bar\partial_b\psi)_{L^2}|\|\bar\partial_b\psi\|_{L^2}^2}$. Substituting a in (70), we obtain

$$|(f, \bar\partial_b\psi)_{L^2}| \le \varepsilon,$$

and we conclude that $(f, \bar\partial_b\psi)_{L^2} = 0$ for all $\psi \in \mathcal{E}^{p,q-1}$. Now by making use of the definition of the formal dual to $\bar\partial_b$, we write

$$(f, \bar\partial_b\psi)_{L^2} = (\bar\partial_b^* f, \psi)_{L^2} \quad \text{for all } \psi \in \mathcal{E}^{p,q-1}(M)$$

and conclude that if f minimises the L^2-norm in a cohomology class $H_b^{p,q}(M,\mathbb{C})$, then $\bar\partial_b^* f = 0$.

Definition 4.5 A form $f \in \mathcal{E}^{p,q}(M)$ is called $\bar\partial_b$-harmonic if

$$\bar\partial_b f = 0 \quad \text{and} \quad \bar\partial_b^* f = 0.$$

The set of all $\bar\partial_b$-harmonic forms in $\mathcal{E}^{p,q}(M)$ is denoted by $\mathcal{H}_b^{p,q}(M)$.

Note that since $\Box_b = \bar\partial_b^*\bar\partial_b + \bar\partial_b\bar\partial_b^*$, then $\bar\partial_b$-harmonic forms, which can now be written as $\mathcal{H}_b^{p,q}(M) = \{f \in \mathcal{E}^{p,q} \mid \Box_b f = 0\}$, form the kernel of the operator $\Box_b$. This is an analogous definition of the harmonic function as a solution of the Laplace equation $\Delta f = 0$.

The Riemannian metric on M induces in a standard way an inner product $(.,.)$ on $\mathcal{E}^{p,q}(M)$. We define another inner product $(.,.)_b$ on $\mathcal{E}^{p,q}(M)$ by setting

$$(f,\phi)_b = (\bar\partial_b f, \bar\partial_b\phi) + (\bar\partial_b^* f, \bar\partial_b^*\phi) + (f,\phi).$$

We write $|f|_b^2 = (f,f)_b$ and say that the norm $|\cdot|_b$ in the space $(\mathcal{E}^{p,q}(M), |\cdot|_b)$ is *completely continuous* if any sequence $\{\phi_k\}$ from $\mathcal{E}^{p,q}(M)$ which is bounded in the norm $|\cdot|_b$ contains a subsequence convergent in L^2-norm $\|\cdot\|_{L^2}$. Recall also that a bounded linear operator $T\colon X \to Y$ between Banach spaces is called completely continuous if, for every weakly convergent sequence $\{x_n\}_{n=1}^\infty$ from X, the sequence

$\{Tx_n\}$ is norm-convergent in Y. If the norm $|\cdot|_b$ is completely continuous, then the space $\mathcal{H}_b^{p,q}(M)$ of $\bar{\partial}_b$-harmonic forms is finite dimensional.

Basic Estimates

Let $\zeta^1, \ldots, \zeta^n$ be a basis for $\mathcal{E}^{1,0}(U)$ and let η be a smooth one form on an open set $U \subset M$. Then $\bar{\zeta}^1, \ldots, \bar{\zeta}^n$ is a basis for $\mathcal{E}^{0,1}(U)$ and any one form on $U \subset M$ can be expressed as a linear combination of $\zeta^j, \bar{\zeta}^j, \eta$. If $x^1, \ldots, x^{2n-1}$ is a system of local coordinates on $U \subset M$ and $\zeta^j = \sum_k a_j^k dx^k$, $\eta = \sum_k b^k dx^k$, then for any function $u \in C^\infty(U)$ we write

$$u_{\zeta^j} = \sum_k a_j^k \frac{\partial u}{\partial x^k}, \quad u_{\bar{\zeta}^j} = \sum_k \bar{a}_j^k \frac{\partial u}{\partial x^k}, \quad u_\eta = \sum_k b^k \frac{\partial u}{\partial x^k}.$$

Let $f \in \mathcal{E}^{p,q}(U)$. Then in the basis $\zeta^j, \bar{\zeta}^j, \eta, j = 1, \ldots, n-1$ we have

$$f = \sum_{|I|=p,|J|=q} f_{IJ}\zeta^{IJ}, \qquad I = (i_1, \ldots, i_p), \quad J = (j_1, \ldots, j_q),$$

$1 \leq i_1 < \ldots < i_p \leq n-1$, $1 \leq j_1 < \ldots < j_q \leq n-1$, with $f_{IJ} \in C^\infty(U)$ and $\zeta^{IJ} = \zeta^1 \wedge \ldots \zeta^{i_p} \wedge \bar{\zeta}^{j_1} \wedge \ldots \wedge \bar{\zeta}^{j_q}$.

Theorem 4.6 ([66]) *If $x_0 \in M$ and if the Levi form at x_0 has $m = \max\{n-q, q+1\}$ non-zero eigenvalues of the same sign, then there is a neighbourhood U of x_0 and a constant C such that*

$$\sum_{|I|=p,|J|=q} \Big(\sum_{j=1}^{n-1} \int (|(f_{IJ})_{\zeta^j}|^2 + |(f_{IJ})_{\bar{\zeta}^j}|^2)\, d\mathrm{Vol} + Re \int (f_{IJ})_\eta (f_{IJ})\, d\mathrm{Vol}\Big) \leq C\|f\|_b^2$$

*for all $f \in \mathcal{E}^{p,q}(M)$ whose support lies in U. Here d*Vol *is the volume form on the CR-manifold M.*

Theorem 4.6 implies that $\mathrm{Re} \int (f_{IJ})_\eta (\psi_{IJ})\, d\mathrm{Vol} \leq C\Big(|f|_b^2 + |\psi|_b^2\Big)$ for all $f, \psi \in \mathcal{E}^{p,q}$ whose support lies in U. If we define the Sobolev norm by $\|f\|_{H_s}^2 = \sum_{|I|=p,|J|=q} \|f_{IJ}\|_{H_s}^2$, then Theorem 4.6 also implies that for any open set $V \subset U$ such that $\overline{V} \subset U$ there is a constant C_0 such that $\|f\|_{H_{1/2}}^2 \leq C_0|f|_b^2$ for all $f \in \mathcal{E}^{p,q}(M)$ whose support lies in V. All in all, summarising all of the above and applying the partition of the unity, we obtain the following result: *If M is m-strongly pseudoconvex with $m = \max\{n-q, q+1\}$, then there is a constant $C > 0$ such that*

$$\|f\|_{H_{1/2}} \leq C|f|_b \quad \textit{for all} \quad f \in \mathcal{E}^{p,q}(M). \tag{71}$$

Main Result

The estimate (71) implies that for any integer $s \geq 0$ there is a constant C such that

$$\|f\|_{H_{s+1/2}} \leq C_s \|\Box_b f + f\|_{H_{s-1/2}} \quad \text{for all} \quad f \in \mathcal{E}^{p,q}(M).$$

First of all we note that given $\beta \in \mathcal{E}^{p,q}(M)$, a form $f \in \mathcal{E}^{p,q}(M)$ satisfies

$$(\Box_b + I)f = \Box_b f + f = \beta \quad \text{if and only if} \quad (f, \psi)_b = (\beta, \psi) \quad \text{for any} \quad \psi \in \mathcal{E}^{p,q}(M).$$

Indeed

$$(f, \psi)_b = (\bar{\partial}_b f, \bar{\partial}_b \psi) + (\bar{\partial}_b^* f, \bar{\partial}_b^* \psi) + (f, \psi) = (\Box_b f + f, \psi) = (\beta, \psi).$$

In the next step, by some standard arguments we deduce the following: If (71) holds, then the norm $|\cdot|_b$ is completely continuous and so the operator $(\Box_b + I)^{-1}$ is completely continuous. The kernel $\mathcal{H}_b^{p,q}(M)$ of the operator $\Box_b$ is finite dimensional and we have the orthogonal decomposition into the image and the kernel of the operator $\Box_b$:

$$\mathcal{E}^{p,q}(M) = \Box_b \mathcal{E}^{p,q}(M) \oplus \mathcal{H}_b^{p,q}(M).$$

Thus the equation $\Box_b \phi = \beta$ has a solution $\phi \in \mathcal{E}^{p,q}(M)$, whenever β belongs to the image of the operator $\Box_b$, and $\beta \perp \mathcal{H}_b^{p,q}(M)$. Let us denote by $\Pi_b: \mathcal{E}^{p,q}(M) \to \mathcal{H}_b^{p,q}(M)$ the orthogonal projection to the kernel of the operator $\Box_b$. Then for any $\alpha \in \mathcal{E}^{p,q}(M)$ the form $\alpha - \Pi_b \alpha$ is in the image of $\Box_b$ and the solution of the equation $\Box_b \phi = \alpha - \Pi_b \alpha$ always exists by the discussions above. It allows us to define the operator

$$N_b: \mathcal{E}^{p,q}(M) \to \mathcal{E}^{p,q}(M)$$

by setting $N_b \alpha = \phi$, where $\Box_b \phi = \alpha - \Pi_b \alpha$, and $\Pi_b \phi = 0$. Thus to a form $\alpha \in \mathcal{E}^{p,q}(M)$ the operator N_b associates a unique solution of the equation $\Box_b \phi = \alpha - \Pi_b \alpha$ that is orthogonal to $\bar{\partial}_b$-harmonic forms. Now we formulate the final result.

Theorem 4.7 ([66]) *Let M be m-strongly pseudoconvex CR-manifold with $m = \max\{n - q, q + 1\}$. Then there is an operator $N_b: \mathcal{E}^{p,q}(M) \to \mathcal{E}^{p,q}(M)$ with the following properties:*

1. *If $f \in \mathcal{E}^{p,q}(M)$, then $f = \Box_b N_b f + \Pi_b f$,*
2. $\bar{\partial}_b N_b = N_b \bar{\partial}_b, \quad \bar{\partial}_b^* N_b = N_b \bar{\partial}_b^*, \quad \Box_b N_b = N_b \Box_b, \quad \Pi_b N_b = N_b \Pi_b.$
3. *The operator $N_b: \mathcal{E}^{p,q}(M) \to \mathcal{E}^{p,q}(M)$ is completely continuous and for each $s \geq 0$ we have $\|N_b f\|_{H_{s+1/2}} \leq C\|f\|_{H_{s-1/2}}$.*

We list some corollaries of Theorem 4.7.

- *Each $\bar\partial_b$-cohomology class $H_b^{p,q}(M,\mathbb{C})$ has a unique representative belonging* $\mathcal{H}_b^{p,q}(M)$. Indeed if $f \in H_b^{p,q}(M,\mathbb{C})$, then $\bar\partial_b f = 0$ by definition. Substituting $\bar\partial_b f$ into the decomposition for f from item 1 of Theorem 4.7 and using commutativity from item 2, we obtain that

$$f = \bar\partial_b \bar\partial_b^* N_b f + \Pi_b f.$$

In this case $\Pi_b f \in \mathcal{H}_b^{p,q}(M)$ is the unique representative from $H_b^{p,q}(M,\mathbb{C})$.

- In the previous corollary we found the solution of the equation $\bar\partial_b f = 0$. Now let $\alpha \in \mathcal{E}^{p,q}(M)$ and we want to find the solution $f \in \mathcal{E}^{p,q-1}(M)$ such that $\bar\partial_b f = \alpha$. Theorem 4.7 implies that

$$\bar\partial_b f = \alpha \quad \text{if and only if} \quad \Pi_b \alpha = \bar\partial_b \alpha = 0.$$

We may choose the solution $f = \bar\partial_b^* N_b \alpha$.

Fundamental Solution for $\Box_b$

We will not present all the details of calculations for the construction of the fundamental solution, but only give a hint regarding what one can expect. The full calculations can be found in [41, 94].

We are looking for the distribution E_α such that $\mathcal{L}_\alpha E_\alpha = \delta_0$. The first question that could be asked is: for which values of α is the operator $\mathcal{L}_\alpha$ invertible? Using the representation theory for the Heisenberg Lie algebra $\mathfrak{h}^n$ one associates to the left invariant vector fields X_j, Y_j, T linear operators acting on functions from $L^2(\mathbb{R}^n)$. Namely, for any $\lambda \in \mathbb{R}$ and X_j, Y_j, T we associate the operators

$$f \mapsto 2\pi i \lambda x_j f(x), \qquad f \mapsto \frac{\partial f}{\partial x_j}, \qquad f \mapsto \frac{\lambda \pi i}{2} f,$$

respectively. Then the operator $\mathcal{L}_\alpha$ can be written as

$$\mathcal{L}_\alpha = -\frac{1}{4}\sum_{j=1}^{n}(X_j^2 + Y_j^2) + i\alpha T = -\frac{1}{4}(\Delta f - 4\pi^2\lambda^2|x|^2 f) - \frac{\pi\alpha\lambda}{2} f.$$

Making the rescaling $x \mapsto |2\pi\lambda|^{1/2}x$, we obtain the n-dimensional analogue of the Hermite operator $x^2 - \frac{d^2}{dx^2}$:

$$-\frac{\pi|\lambda|}{2}\Big(\Delta - |x|^2 - \alpha\,\mathrm{sign}(\lambda)\Big)f. \tag{72}$$

It is known from the spectrum, that the operator (72) is invertible if $\alpha\,\mathrm{sign}(\lambda) \neq -n, -n-2, \ldots$. Thus, we conclude that the operator $\mathcal{L}_\alpha$ is invertible if $\alpha \neq \pm n, \pm(n+2), \ldots$. The values $\alpha = \pm n, \pm(n+2), \ldots$ are often called *forbidden values* for $\mathcal{L}_\alpha$.

We concentrate on the non forbidden values and observe that the fundamental solution has to respect the symmetries of the domain $b\mathcal{U}$, mentioned in section "Action of the Heisenberg Group on the Siegel Upper Half Space". Thus, we expect that the fundamental solution E_α has to be homogeneous of degree $-2n$, since the operator $\mathcal{L}_\alpha$ is homogeneous of order 2, it will give us the correct homogeneity of the delta function δ_0 which is $-2n-2$ with respect to the Heisenberg dilatation. Another observation is that E_α is invariant under the rotations on the Heisenberg group and therefore it has to be expressible in terms of the function $t \pm i|z|^2$. Finally, one comes to the theorem:

Theorem 4.8 ([94]) *The function* $E_\alpha(z,t) = \frac{1}{\gamma_\alpha}\phi_\alpha(z,t)$, *where*

$$\phi_\alpha(z,t) = (|z|^2 - it)^{-\frac{n+\alpha}{2}}(|z|^2 + it)^{-\frac{n-\alpha}{2}}$$

and

$$\gamma_\alpha = \frac{2^{2-n}\pi^{n+1}}{\Gamma(\frac{[n+\alpha]}{2})\Gamma(\frac{[n-\alpha]}{2})} \tag{73}$$

is the fundamental solution of $\mathcal{L}_\alpha$ *on the Heisenberg group* $\mathbb{H}^n$.

To prove the theorem we need to check that $\mathcal{L}_\alpha\phi_\alpha = \gamma_\alpha\delta_0$. For the beginning, we observe that $\phi_\alpha(z,t)$ is homogeneous of order $-2n$ with respect to the Heisenberg dilatation and therefore $\mathcal{L}_\alpha\phi_\alpha$ is a distribution homogeneous of order $-2n-2$. Moreover, since $\mathcal{L}_\alpha\phi_\alpha(z,t) = 0$ for all $(z,t) \neq (0,0)$ the distribution $\mathcal{L}_\alpha\phi_\alpha$ differs from δ_0 on a constant and one needs to calculate this constant that finally is given by (73). Notice that since $\Gamma(s) = \infty$ for $s = 0, -1, -2, \ldots$, the constant γ_α vanishes exactly for the forbidden values $\alpha = n \pm 0, 2, 4, \ldots$, that again justifies the form of the fundamental solution.

In the particular case of $\alpha = 0$ the operator $\mathcal{L}_0 = -\frac{1}{4}\sum_{j=1}^{n}(X_j^2 + Y_j^2)$ can be called nowadays the *classical sub-Laplacian*. It is the homogeneous operator of order 2 given by the sum of squares of vector fields, generating the tangent space of $\mathbb{H}^n$ at each point. The fundamental solution is given by

$$E_0 = \frac{\gamma_0}{(|z^4 + t^2|)^{\frac{n}{2}}} = \frac{\gamma_0}{\|(z,t)\|^{Q-2}}, \quad \gamma_0 = \frac{2^{2-n}\pi^{n+1}}{\left(\Gamma(\frac{n}{2})\right)^2}.$$

Here $Q = 2n + 2$ is the Hausdorff dimension, also called the homogeneous dimension, of the Heisenberg group. Recall that the fundamental solution of the Laplace equation $\Delta E = \delta_0$ in $\mathbb{R}^n$ is given by $E = \frac{c_0}{\|x\|^{n-2}}$.

Further Development

As we saw the Heisenberg group is identified with the boundary $\text{Im}(z_2) = |z_1|^2$ of the Siegel upper half space in $\mathbb{C}^2$ and the subelliptic operators $\mathcal{L}_\alpha$ are related to the boundary Laplacian. The question of finding fundamental solutions for more general domains was considered, for instance, in [10]. The authors found an explicit fundamental solution for sub-Laplacians associated to the hypersurface $\text{Im}(z_2) = |z_1|^{2k} \subset \mathbb{C}^2$ and developed new methods for constructing the fundamental solutions for subelliptic operators, written as a sum of the squares of vector fields satisfying the Hörmander condition plus the lower order terms, see also [11–13]. Fundamental solutions for some anisotropic domains in $\mathbb{C}^n$ and for H-type groups were studies in [27, 60]. Generalisations and fundamental solutions to powers of sub-Laplacians, twisted sub-Laplacians, translation and rotation invariant sub-Laplacians, and some other modifications can be found in [16, 26, 47, 74, 77, 78, 101].

Recall the Grušin sub-bundle from one of the examples of section "Distributions and Non-holonomic Constraints". The authors of [104–106] studied properties for operators of a specific form degenerating on a submanifold but still satisfying the hypoellipticity conditions. Nowadays, operators in this class are called Grušin type operators. Note that such kind of operators can be obtained as a certain projection of sub-Laplacians from higher step nilpotent Lie groups admitting so-called compact lattices. The fundamental solutions for Grušin type operators were constructed, for instance, in [8, 9, 18–20, 49, 58].

5 The Quaternion Cauchy-Szegö Kernel and Related PDE

In this section we will show how the ideas developed for the Euclidean space and the Heisenberg group, and, respectively, upper half space in $\mathbb{R}^n$, the Siegel upper half space in $\mathbb{C}^n$, see, for instance [96, Chap. 3] and [93, Chap. 2], can be extended to the quaternion Heisenberg group and analogous of the Siegel upper half space in the multidimensional quaternion space. Recently, the interest in developing a theory for the regular functions of several quaternion variables, as the counterpart of the theory of several complex variables for holomorphic functions increases, see [2–4, 22, 32, 33, 108, 109, 111], and references therein. The material of the present section is based on [29, 30].

Right Quaternion Vector Space

The space $\mathbb{Q}$ of quaternion numbers forms a division algebra with respect to the coordinate addition and the quaternion multiplication

$$\begin{aligned} q\sigma =&(x_1 + \mathbf{i}x_2 + \mathbf{j}x_3 + \mathbf{k}x_4)(\sigma_1 + \mathbf{i}\sigma_2 + \mathbf{j}\sigma_3 + \mathbf{k}\sigma_4) \\ = &\quad \sigma_1 x_1 - \sigma_2 x_2 - \sigma_3 x_3 - \sigma_4 x_4 + (\sigma_2 x_1 + \sigma_1 x_2 + \sigma_4 x_3 - \sigma_3 x_4)\mathbf{i} \\ + &\ (\sigma_3 x_1 - \sigma_4 x_2 + \sigma_1 x_3 + \sigma_2 x_4)\mathbf{j} + (\sigma_4 x_1 + \sigma_3 x_2 - \sigma_2 x_3 + \sigma_1 x_4)\mathbf{k}, \end{aligned}$$

for $q, \sigma \in \mathbb{Q}$. Denote by $\operatorname{Re} q = x_1$ the real part of q, and by $\operatorname{Im} q$ the imaginary part of q, which is the three dimensional vector $\overrightarrow{r} = (x_2, x_3, x_4)$.

The conjugate $\bar{q}$ of a quaternion $q = x_1 + x_2\mathbf{i} + x_3\mathbf{j} + x_4\mathbf{k}$ is defined by $\bar{q} = x_1 - x_2\mathbf{i} - x_3\mathbf{j} - x_4\mathbf{k}$ and the norm is $|q|^2 = \bar{q}q$. The conjugation inverses the product of quaternion numbers in the following sense $\overline{\sigma q} = \bar{q} \cdot \bar{\sigma}$ for any $\sigma, q \in \mathbb{Q}$.

Since the quaternion algebra $\mathbb{Q}$ is associative, although it is not commutative, there is a natural notion of a vector space over $\mathbb{Q}$, and many definitions and propositions for real or complex linear algebra also hold for quaternion linear spaces, see [5, 87, 110]. A *right quaternion vector space* is a set V with addition $+\colon V \times V \to V$ and *right scalar multiplication* $V \times \mathbb{Q} \to V : (v, \sigma) \mapsto v\sigma$. The space V is an abelian group with respect to addition, and satisfies the axioms of associativity and distributivity.

A *hyperhermitian semilinear form* on a right quaternion vector space V is a map $a\colon V \times V \longrightarrow \mathbb{Q}$ satisfying the following properties:

1. a is additive with respect to each argument,
2. $a(q, q'\sigma) = a(q, q')\sigma$ for any $q, q' \in V$ and any $\sigma \in \mathbb{Q}$,
3. $a(q, q') = \overline{a(q', q)}$.

Properties (2) and (3) imply $a(q\sigma, q') = \bar{\sigma}a(q, q')$.

A quaternion $(n{\times}n)$-matrix A is called *hyperhermitian* if $A^* = A$, where $(A^*)_{jk} = \overline{A}_{kj}$. For instance, for $q = (q_1, \dots, q_n)$, $p = (p_1, \dots, p_n) \in \mathbb{Q}^n$, set $a(q, p) = \sum_{i,j} \bar{q}_i A_{ij} p_j$. Then $a(\cdot, \cdot)$ defines a hyperhermitian semilinear form on $\mathbb{Q}^n$. A positive definite hyperhermitian semilinear form $a(\cdot, \cdot)$ on a right quaternion vector space is called an *inner product* and will be denoted from now on by $\langle v, w\rangle = a(v, w)$. We use the notations

$$\|v\| = \langle v, v\rangle^{\frac{1}{2}}, \qquad \text{and} \qquad \rho(v, w) = \|v - w\|, \qquad v, w \in V,$$

for the norm and distance on V in the rest of Sect. 5. If $\rho(\cdot\,, \cdot)$ is a complete distance, we call $(V, \langle .\,, . \rangle)$ a *right quaternion Hilbert space*. The quaternion version of Riesz's representation theorem is true. Namely, suppose that $(V, \langle .\,, . \rangle)$ is a right quaternion Hilbert space and $h\colon V \to \mathbb{Q}$ is a bounded right quaternion linear functional: h is additive and $h(v\sigma) = h(v)\sigma$ for any $v \in V$ and $\sigma \in \mathbb{Q}$. Then there exists a unique element $v_h \in V$ such that $h(v) = \langle v_h, v\rangle$ for any $v \in V$. The proof can be found in [29]. Particularly, the space $\mathbb{Q}^n$ is a right quaternion Hilbert space endowed with the inner product

$$\langle p,q\rangle = \sum_{l=1}^{n} \bar{p}_l q_l, \qquad p=(p_1,\ldots,p_n),\ \ q=(q_1,\ldots,q_n)\in\mathbb{Q}^n.$$

The Quaternion Siegel Upper Half-Space and the Quaternion Heisenberg Group

We present transformations acting on the quaternion Siegel upper half-space. A quaternion $(n\times n)$-matrix $\mathbf{a}=(a_{jk})$ acts on $\mathbb{Q}^n$ on the left as follows:

$$q\longmapsto \mathbf{a}q, \qquad (\mathbf{a}q)_j=\sum_{k=1}^{n} a_{jk}q_k \tag{74}$$

for $q=(q_1,\ldots q_n)^t$, where the upper index ^{t} denotes the transposition of the vector. Note that the transformation in (74) commutes with right multiplication by $\mathbf{i}_\beta$ ($\mathbf{i}_1=1, \mathbf{i}_2=\mathbf{i}, \mathbf{i}_3=\mathbf{j}, \mathbf{i}_4=\mathbf{k}$), i.e. $(\mathbf{a}q)\mathbf{i}_\beta=\mathbf{a}(q\mathbf{i}_\beta)$. The group $\mathrm{GL}(n,\mathbb{Q})$ is isomorphic to the group of all linear transformations of $\mathbb{R}^{4n}$ commuting with $\mathbf{i}_\beta$, while the compact Lie group $\mathrm{Sp}(n)$ consists of orthogonal transformations of $\mathbb{R}^{4n}$ commuting with $\mathbf{i}_\beta$. The quaternion Siegel upper half space is defined by

$$\mathcal{U}_n=\{q=(q_1,\ldots,q_n)\in\mathbb{Q}^n:\ \mathrm{Re}\,q_1\geq\sum_{k=2}^{n}|q_k|^2\}.$$

For shortness we write $q'=(q_2,\ldots,q_n)\in\mathbb{Q}^{n-1}$ and $|q'|^2=\sum_{k=2}^{n}|q_k|^2$. The boundary $b\mathcal{U}_n$ is a quadratic hypersurface given by the equation $\mathrm{Re}\,q_1=|q'|^2$.

Proposition 5.1 *The Siegel upper half space $\mathcal{U}_n$ is invariant under the following transformations.*

(1) *Translations:*

$$\tau_p\colon(q_1,q')\longmapsto\left(q_1+p_1+2\langle p',q'\rangle,q'+p'\right), \tag{75}$$

for $p=(p_1,p')=(p_1,\ldots,p_n)\in b\mathcal{U}_n$.

(2) *Rotations:*

$$R_{\mathbf{a}}\colon(q_1,q')\longmapsto(q_1,\mathbf{a}q'),\quad \textit{for}\quad \mathbf{a}\in\mathrm{Sp}(n-1), \tag{76}$$

$$R_\sigma:(q_1,q')\longmapsto(\overline{\sigma}q_1\sigma,q'\sigma),\quad \textit{for}\quad \sigma\in\mathbb{Q},\quad |\sigma|=1. \tag{77}$$

(3) *Dilations:*

$$\delta_r : (q_1, q') \longmapsto (r^2 q_1, rq'), \quad r > 0.$$

All the maps are extended to the boundary $b\mathcal{U}_n$ and transform the boundary $b\mathcal{U}_n$ to itself. Moreover, all the maps transform the hypersurface $b\mathcal{U}_n + \varepsilon\mathbf{e}$ to itself for each $\varepsilon > 0$.

Proof The formula (75) follows from

$$\begin{aligned} &\mathrm{Re}\,(q_1 + p_1 + 2\langle p', q'\rangle) - |q' + p'|^2 \\ &= \mathrm{Re}\, q_1 + \mathrm{Re}\, p_1 + 2\mathrm{Re}\,\langle p', q'\rangle - (|q'|^2 + |p'|^2 + 2\mathrm{Re}\,\langle p', q'\rangle) \\ &= \mathrm{Re}\, q_1 - |q'|^2 > 0 \end{aligned} \tag{78}$$

by $\mathrm{Re}\, p_1 = |p'|^2$. The rotations (76) obviously map $\mathcal{U}_n$ to itself. To show it for rotations (77), we note that

$$q_1^2 = -1 \quad \text{if and only if} \quad x_1 = 0 \text{ and } x_2^2 + x_3^2 + x_4^2 = 1 \tag{79}$$

for a quaternion number $q_1 = x_1 + \mathbf{i}x_2 + \mathbf{j}x_3 + \mathbf{k}x_4$. This is because of

$$q_1^2 = x_1^2 + 2x_1(\mathbf{i}x_2 + \mathbf{j}x_3 + \mathbf{k}x_4) + (\mathbf{i}x_2 + \mathbf{j}x_3 + \mathbf{k}x_4)^2$$

and

$$(\mathbf{i}x_2 + \mathbf{j}x_3 + \mathbf{k}x_4)^2 = -|\mathbf{i}x_2 + \mathbf{j}x_3 + \mathbf{k}x_4|^2 = -x_2^2 - x_3^2 - x_4^2. \tag{80}$$

Since

$$\overline{\sigma} q_1 \sigma = x_1 + \overline{\sigma}(\mathbf{i}x_2 + \mathbf{j}x_3 + \mathbf{k}x_4)\sigma, \tag{81}$$

and $\overline{\sigma}(\mathbf{i}x_2 + \mathbf{j}x_3 + \mathbf{k}x_4)\sigma\overline{\sigma}(\mathbf{i}x_2 + \mathbf{j}x_3 + \mathbf{k}x_4)\sigma = -x_2^2 - x_3^2 - x_4^2$, by (80), we see that the second term in the right hand side of (81) is pure imaginary by using (79). Consequently, $\mathrm{Re}\,(\overline{\sigma} q_1 \sigma) = \mathrm{Re}\, q_1$ and so

$$\mathrm{Re}\,(\overline{\sigma} q_1 \sigma) - |q'\sigma|^2 = \mathrm{Re}\, q_1 - |q'|^2. \tag{82}$$

The invariance of the hypersurface $b\mathcal{U}_n + \varepsilon\mathbf{e}$ under the maps τ_p and R_σ follows from (78) and (82). The other statements are obvious. □

The total group of rotations for $\mathcal{U}_n$ is $\mathrm{Sp}(n-1)\mathrm{Sp}(1)$ with $\mathrm{Sp}(1) \cong \{\sigma \in \mathbb{Q} \colon |\sigma| = 1\}$.

The *quaternion Heisenberg group* $q\mathbb{H}^{n-1}$ is the space $\mathbb{R}^{4n-1} = \mathbb{R}^3 \times \mathbb{R}^{4(n-1)}$, that is isomorphic to $\operatorname{Im}\mathbb{Q} \times \mathbb{Q}^{n-1}$, furnished with the non-commutative product

$$p \cdot q = (w, p') \cdot (v, q') = \left(w + v + 2\operatorname{Im}\langle p', q'\rangle, p' + q'\right),$$

where $p = (w, p'), q = (v, q') \in \operatorname{Im}\mathbb{Q} \times \mathbb{Q}^{n-1}$, and $\langle.,.\rangle$ is the inner product defined in Eq. (5) on $\mathbb{Q}^{n-1}$. Translation τ_p on $\mathcal{U}_n$ can be viewed as an action of the quaternion Heisenberg group $q\mathbb{H}^{n-1}$ on the quaternion Siegel upper half space $\mathcal{U}_n$. Let $p = (v, p') \in q\mathbb{H}^{n-1}$, then the translation (75) can be written as

$$\tau_p \colon (q_1, q') \longmapsto (q_1 + |p'|^2 + v + 2\langle p', q'\rangle, q' + p').$$

It is obviously extended to the boundary $b\mathcal{U}_n$. It is easy to see that the action on $b\mathcal{U}_n$ is transitive, for calculation see also [23]. Therefore, we can identify points in $q\mathbb{H}^{n-1}$ with points in $b\mathcal{U}_n$ by the result of the translation τ_p of the origin $(0, 0)$.

Regular Functions on the Quaternion Siegel Upper Half-Space

In the present section we show the invariance of the regularity under linear transformations in Proposition 5.1. For the lth coordinate of a point $q = (q_1, \cdots, q_n) \in \mathbb{Q}^n$ we write

$$q_l = x_{4l-3} + x_{4l-2}\mathbf{i} + x_{4l-1}\mathbf{j} + x_{4l}\mathbf{k}, \qquad l = 1, \ldots, n.$$

For a domain $\Omega \subset \mathbb{Q}^n$, a C^1-smooth function $f = f_1 + \mathbf{i}f_2 + \mathbf{j}f_3 + \mathbf{k}f_4 \colon \Omega \to \mathbb{Q}$ is called (*left-*) *regular on* Ω if it satisfies the Cauchy-Fueter equations

$$\overline{\partial}_{q_l} f(q) = \partial_{x_{4l-3}} f(q) + \mathbf{i}\partial_{x_{4l-2}} f(q) + \mathbf{j}\partial_{x_{4l-1}} f(q) + \mathbf{k}\partial_{x_{4l}} f(q) = 0,$$

where $l = 1, \ldots, n$, $q \in \Omega$.

Proposition 5.2 *Let $f \colon \Omega \to \mathbb{Q}$ be a C^1-smooth function, where Ω is a domain in $\mathbb{Q}^n$.*

(1) *Define the pull-back function $\widehat{f}$ of f under the mapping $q \to Q = \mathbf{a}q$ for $\mathbf{a} = (a_{jk}) \in \mathrm{GL}(n, \mathbb{Q})$ by $\widehat{f}(q) := f(\mathbf{a}q)$. Then we have*

$$\overline{\partial}_{q_j} \widehat{f}(q) = \sum_{k=1}^{n} \overline{a}_{kj} \overline{\partial}_{q_k} f(Q)\Big|_{Q = \mathbf{a}q}.$$

(2) *Define the pull-back function* $\widetilde{f}$ *of* f *under the mapping* $q \to Q = q\sigma$ *for* $\sigma \in \mathbb{Q}$ *by* $\widetilde{f}(q) := f(q_1\sigma, \dots, q_n\sigma)$. *Then*

$$\overline{\partial}_{q_l}\widetilde{f}(q) = \overline{\partial}_{q_l}[\overline{\sigma} f(Q)]\Big|_{Q=q\sigma}, \qquad l = 1, \dots, n.$$

The proof of the first statement can be found in [110, Proposition 3.1]. The second statement is analogous to the formula of one quaternion variable and the proof is left for the reader, or can be found in [29]. We obtain the following corollaries:

1. If f is regular, then $\hat{f} = f(\mathbf{a}q)$ for some $\mathbf{a} \in \mathrm{GL}(n, \mathbb{Q})$ and $\widetilde{f} = f(q\sigma)$ for some $\sigma \in \mathbb{Q}$ are both regular.
2. The space of all regular functions on $\mathcal{U}_n$ is invariant under the transformations defined in Proposition 5.1. Namely, if f is regular on the Siegel upper half-space $\mathcal{U}_n$, then the functions $f(\tau_p(q))$, $p \in b\mathcal{U}_n$; $f(R_{\mathbf{a}}(q))$, $\mathbf{a} \in \mathrm{Sp}(n-1)$; $\sigma f(R_\sigma(q))$ for some $\sigma \in \mathbb{Q}$ with $|\sigma| = 1$; and $f(\delta_r(q))$ are all regular on $\mathcal{U}_n$.

To prove the corollaries we argue as follows. The translation τ_p in (75) can be represented as a composition of the linear transformation given by the quaternion matrix

$$\begin{bmatrix} 1 & 2\overline{p}' \\ 0 & I_{n-1} \end{bmatrix},$$

and the Euclidean translation $(q_1, q') \mapsto (q_1 + p_1, q' + p')$. The first transformation preserves the regularity of a function by Proposition 5.2, while the later one obviously preserves the regularity of a function since the Cauchy-Fueter operators have constant coefficients.

The equation $\overline{\partial}_{q_l}[\sigma f(q\sigma)] = \overline{\partial}_{q_l}[\overline{\sigma}\sigma f(Q)]_{Q=q\sigma} = |\sigma|^2\overline{\partial}_{q_l} f(Q)|_{Q=q\sigma} = 0$ follows from Proposition 5.2 (2) and shows that $\sigma f(\overline{\sigma} q_1\sigma, q'\sigma)$ is regular.

Hardy Space $H^2(\mathcal{U}_n)$

The identification of the quaternion Heisenberg group and the boundary of the quaternion Siegel upper half-space allows us to define the Lebesgue measure $\beta(\cdot)$ on $b\mathcal{U}_n$ by pulling back the Haar measure from $q\mathbb{H}^{n-1}$. The latter measure, in its turn, is a pull back of the Lebesgue measure $d\mu(\cdot) = dx\,dq'$ from $\mathbb{R}^3 \times \mathbb{R}^{4(n-1)}$. Let $L^2(b\mathcal{U}_n)$ denote the space of all $\mathbb{Q}$-valued functions which are square integrable with respect to the measure β. It follows from the definition that $L^2(b\mathcal{U}_n)$ is a right quaternion Hilbert space with the following inner product:

$$\langle f, g\rangle_{L^2} = \int_{b\mathcal{U}_n} \overline{f(q)} g(q)\, d\beta(q).$$

For any function $F: \mathcal{U}_n \to \mathbb{Q}$, we write F_ε for its "vertical translate". We mean that the vertical direction is given by the positive direction of $\operatorname{Re} q_1$: $F_\varepsilon(q) = F(q + \varepsilon\mathbf{e})$, where $\mathbf{e} = (1, 0, 0, \ldots, 0)$. If $\varepsilon > 0$, then F_ε is defined in a neighbourhood of $\partial\mathcal{U}_n$. In particular, F_ε is defined on $\partial\mathcal{U}_n$. The *Hardy space* $H^2(\mathcal{U}_n)$ consists of all regular functions F on $\mathcal{U}_n$, for which

$$\sup_{\varepsilon>0} \int_{\partial\mathcal{U}_n} |F_\varepsilon(q)|^2 d\beta(q) < \infty. \tag{83}$$

The norm $\|F\|_{H^2(\mathcal{U}_n)}$ of F is the square root of the left-hand side of (83). A function $F \in H^2(\mathcal{U}_n)$ has boundary value F^b that belongs to $L^2(b\mathcal{U}_n)$ in the following sense.

Theorem 5.3 *Suppose that $F \in H^2(\mathcal{U}_n)$. Then*

1. *There exists a function $F^b \in L^2(b\mathcal{U}_n)$ such that $F(q + \varepsilon\mathbf{e})|_{b\mathcal{U}_n} \to F^b(q)$ as $\varepsilon \to 0$ in $L^2(b\mathcal{U}_n)$ norm.*
2. $\|F^b\|_{L^2(b\mathcal{U}_n)} = \|F\|_{H^2(\mathcal{U}_n)}$.
3. *The space of all boundary values is a closed subspace of the space $L^2(b\mathcal{U}_n)$.*

This theorem was proven in [25, Theorem 4.2] for $n = 2$. The arguments work for an arbitrary n if we consider the following slice functions. Let $H^2(\mathbb{R}^4_+)$ be the classical Hardy space, that is the set of all harmonic functions $u: \mathbb{R}^4_+ \to \mathbb{R}$ such that $\sup_{t>0} \|u(t, \cdot)\|_{L^2(\mathbb{R}^3)} < \infty$. Assume that $F = F_1 + \mathbf{i}F_2 + \mathbf{j}F_3 + \mathbf{k}F_4 \in H^2(\mathcal{U}_n)$. Then the slice function $f_j(q_1) := F_j(q_1 + |q'|^2, q')$ is harmonic and belongs to $H^2(\mathbb{R}^4_+)$ for each $j = 1, \ldots, 4$ and any fixed $q' \in \mathbb{Q}^{n-1}$. We omit further details.

Proposition 5.4 *The Hardy space $H^2(\mathcal{U}_n)$ is a right quaternion Hilbert space under the inner product $\langle F, G\rangle = \langle F^b, G^b\rangle_{L^2(b\mathcal{U}_n)}$.*

Proof Since the Cauchy-Fueter operator $\overline{\partial}_{q_l}$ is right quaternion linear, i.e., for a fixed σ one has $\overline{\partial}_{q_l}(f(q)\sigma) = (\overline{\partial}_{q_l} f(q))\sigma$, we see that $f(q)\sigma$ is regular if $f(q)$ is. Thus, the Hardy space $H^2(\mathcal{U}_n)$ is a right quaternion vector space.

Set

$$\partial_{q_{l+1}} f := \overline{\overline{\partial}_{q_{l+1}} \overline{f}} = \partial_{x_{4l+1}} f - \partial_{x_{4l+2}} f\mathbf{i} - \partial_{x_{4l+3}} f\mathbf{j} - \partial_{x_{4l+4}} f\mathbf{k}.$$

It is straightforward to see that

$$0 = \partial_{q_{l+1}} \overline{\partial}_{q_{l+1}} f = (\partial^2_{x_{4l+1}} + \partial^2_{x_{4l+2}} + \partial^2_{x_{4l+3}} + \partial^2_{x_{4l+4}}) f.$$

Consequently, the functions $f_1, \ldots, f_4$ are harmonic on the line $\{(0, \ldots q_l, \ldots, 0)\}$ and so they are harmonic on $\mathbb{Q}^n$. Thus

$$f_j(q) = \frac{1}{|B|} \int_B f_j(p) d\mathrm{Vol}(p), \qquad j = 1, 2, 3, 4,$$

for $q \in \mathcal{U}_n$, where B is a small ball centred at q and contained in $\mathcal{U}_n$. We see that

$$|f(q)| \leq \frac{1}{|B|}\int_B |f(p)| d\mathrm{Vol}(p) \leq \left(\frac{1}{|B|}\int_B |f(p)|^2 d\mathrm{Vol}(p)\right)^{\frac{1}{2}}.$$

There exist $a, b > 0$ such that $B \subset \mathcal{U}_{n;a,b} := \{q \in \mathcal{U}_n |\ a < \operatorname{Re} q_1 - |q'|^2 < b\}$, and so

$$\begin{aligned}
|f(q)|^2 &\leq \frac{1}{|B|}\int_{\mathcal{U}_{n;a,b}} |f(x_1, \cdots, x_{4n})|^2 dx_1 \cdots dx_{4n} \\
&\leq \frac{1}{|B|}\int_{(a,b)\times\mathbb{R}^{4n-1}} \left| f\left(x_1 + \sum_{j=5}^{4n} |x_j|^2, x_2 \cdots, x_{4n}\right)\right|^2 dx_1 dx_2 \cdots dx_{4n} \\
&\leq \frac{1}{|B|}\int_a^b dx_1 \int_{b\mathcal{U}_n} |f(p + x_1\mathbf{e})|^2 d\beta(p) \leq c\|f\|^2_{H^2(\mathcal{U}_n)},
\end{aligned} \tag{84}$$

where $c = (b - a)/|B|$ is a positive constant depending on q, and independent of the functions $f \in H^2(\mathcal{U}_n)$. Here we have used the coordinates transformation $(x_1, \cdots, x_{4n}) \to (x_1 + \sum_{j=5}^{4n} |x_j|^2, x_2 \cdots, x_{4n})$, whose Jacobian is the identity.

To prove the completeness, we suppose that a Cauchy sequence $\{f^{(k)}\}$ in the Hardy space $H^2(\mathcal{U}_n)$ is given. We need to show that some subsequence converges to an element in $H^2(\mathcal{U}_n)$. Apply the estimate (84) to regular functions $f^{(k)} - f^{(l)}$ to get that for any compact subset $K \subset \mathcal{U}_n$ and $q \in K$,

$$|f^{(k)}(q) - f^{(l)}(q)| \leq c_K \|f^{(k)} - f^{(l)}\|_{H^2(\mathcal{U}_n)},$$

where c_K is a positive constant only depending on K. It means that the sequence $\{f^{(k)}\}$ converges uniformly on any compact subset of $\mathcal{U}_n$. Denote by f the limit. Recall the well known estimate

$$\|u\|_{C^1(B(q,r))} \leq C_r \|u\|_{C^0(B(q,2r))} \tag{85}$$

for any harmonic function u defined on the ball $B(q, 2r)$, where C_r is a positive constant only depending on r and the dimension, and independent of the function u, see [95, pp. 307–312]. Now apply the estimate (85) to each component of regular function $f = f_1 + \mathbf{i}f_2 + \mathbf{j}f_3 + \mathbf{k}f_4$, which is harmonic. By the argument of finite covering and estimate (84), we easily see that

$$\|f\|_{C^1(K)} \leq C'_K \|f\|_{H^2(\mathcal{U}_n)}$$

for some constant C'_K only depending on the compact K. It follows that

$$|\partial_{x_j} f^{(k)}(q) - \partial_{x_j} f^{(l)}(q)| \leq C'_K \|f^{(k)} - f^{(l)}\|_{H^2(\mathcal{U}_n)}$$

for $q \in K$, $j = 1, \ldots 4n$. Consequently, the limit function f is also C^1 and $\lim_{k\to\infty} \partial_{x_j} f^{(k)}(q) = \partial_{x_j} f(q)$. Thus, $\overline{\partial}_{q_l} f(q) = \lim_{k\to\infty} \overline{\partial}_{q_l} f^{(k)}(q) = 0$ and the limit function f is regular.

Since on the compact subset $K_{R,\varepsilon} := b\mathcal{U}_n \cap \overline{B(0,R)} + \varepsilon\mathbf{e}$ for fixed $R, \varepsilon > 0$, the sequence $\{f^{(k)}\}$ is uniformly convergent, we find that

$$\int_{b\mathcal{U}_n \cap \overline{B(0,R)}} |f_\varepsilon(q)|^2 d\beta(q) = \int_{K_{R,\varepsilon}} |f(q)|^2 d\beta(q)$$

$$= \lim_{k\to\infty} \int_{K_{R,\varepsilon}} |f^{(k)}(q)|^2 d\beta(q) \leq \sup_k \|f^{(k)}\|_{H^2(\mathcal{U}_n)} < \infty.$$

Consequently, f_ε is square integrable on $b\mathcal{U}_n$ for any $\varepsilon > 0$, and $\int_{b\mathcal{U}_n} |f_\varepsilon(q)|^2 d\beta(q) \leq \sup \|f^{(k)}\|_{H^2(\mathcal{U}_n)}$. Thus $f \in H^2(\mathcal{U}_n)$. □

As a corollary we note that the Hardy space $H^2(\mathcal{U}_n)$ is invariant under the transformations of Proposition 5.1. The proof is obvious.

The Quaternion Cauchy-Szegö Kernel

In this section we introduce the notion of the Cauchy-Szegö kernel for the projection operator from the space $L^2(b\mathcal{U}_n)$ to the space of the boundary values of a function from the Hardy space $H^2(\mathcal{U}_n)$. The main result of the section is the following theorem.

Theorem 5.5 *The quaternion Cauchy-Szegö kernel is given by*

$$S(q,p) = s\left(q_1 + \overline{p}_1 - 2\sum_{k=2}^{n} \overline{p}'_k q'_k\right),$$

for $p = (p_1, p') = (p_1, \ldots, p_n) \in \mathcal{U}_n$, $q = (q_1, q') = (q_1, \ldots, q_n) \in \mathcal{U}_n$, *where*

$$s(\sigma) = c_{n-1} \frac{\partial^{2(n-1)}}{\partial x_1^{2(n-1)}} \frac{\overline{\sigma}}{|\sigma|^4}, \qquad \sigma = x_1 + x_2\mathbf{i} + x_3\mathbf{j} + x_4\mathbf{k} \in \mathbb{Q}. \tag{86}$$

Here

$$c_{n-1} = \frac{4n-5}{2^{2n+3}\pi^{2n-1}\left((2n-2)!\right)^2 (n+1)(2n+1)} \frac{1}{K_n},$$

where the constant

$$K_n = \sum_{k=0}^{2n-2} \frac{(2n-1-k)(2n-k)(4n-1+k)}{6} \sum_{l=0}^{k} C_k^{2l} \sum_{m=0}^{l} (-1)^{k+m} C_l^m$$
$$\sum_{s=0}^{k-2m} \frac{C_{k-2m}^s}{2^{k-2m-s+1}} \frac{(-1)^s\big(2(k-2m-s+1)\big)!}{(k-2m-s+1)!(4n+1+k-2m-s)!}$$

depends only on n.

The quaternion Cauchy-Szegö kernel satisfies the reproducing property in the following sense

$$F(q) = \int_{b\mathcal{U}_n} S(q,Q)F^b(Q)d\beta(Q), \qquad q \in \mathcal{U}_n,$$

whenever $F \in H^2(\mathcal{U}_n)$ *and* F^b *its boundary value on* $b\mathcal{U}_n$.

Existence and Characterization of the Cauchy-Szegö Kernel

Theorem 5.6 *The Cauchy-Szegö kernel* $S(q,p)$ *is a unique* $\mathbb{Q}$*-valued function, defined on* $\mathcal{U}_n \times \mathcal{U}_n$ *satisfying the following conditions.*

1. *For each* $p \in \mathcal{U}_n$*, the function* $q \mapsto S(q,p)$ *is regular for* $q \in \mathcal{U}_n$*, and belongs to the Hardy space* $H^2(\mathcal{U}_n)$*. This allows us to define the boundary value* $S^b(q,p)$ *for each* $p \in \mathcal{U}_n$ *and for almost all* $q \in b\mathcal{U}_n$.
2. *The kernel S is symmetric:* $S(q,p) = \overline{S(p,q)}$ *for each* $(q,p) \in \mathcal{U}_n \times \mathcal{U}_n$*. The symmetry permits us to extend the definition of* $S(q,p)$ *so that for each* $q \in \mathcal{U}_n$*, the function* $S_b(q,p)$ *is defined for almost every* $p \in b\mathcal{U}_n$ *(here we use the subscript b to indicate the boundary value with respect to the second argument).*
3. *The kernel S satisfies the reproducing property in the following sense*

$$F(q) = \int_{b\mathcal{U}_n} S_b(q,Q)F^b(Q)d\beta(Q), \qquad q \in \mathcal{U}_n, \tag{87}$$

whenever $F \in H^2(\mathcal{U}_n)$.

Proof First we need to show that the Hardy space $H^2(\mathcal{U}_n)$ is nontrivial, otherwise the Cauchy-Szegö kernel vanishes. We claim that $s\left(q_1 + \bar{p}_1 - 2\sum_{k=2}^n \bar{p}_k' q_k'\right)$ for fixed $(p_1,\dots,p_n) \in \mathcal{U}_n$, with $s(\cdot)$ given by (86), is in the Hardy space $H^2(\mathcal{U}_n)$.

We use the notation $q_1 = x_1 + x_2\mathbf{i} + x_3\mathbf{j} + x_4\mathbf{k}$ and apply the Laplace operator

$$\overline{\partial}_{q_1}\partial_{q_1} = \partial_{x_1}^2 + \partial_{x_2}^2 + \partial_{x_3}^2 + \partial_{x_4}^2$$

to the harmonic function $\frac{1}{|q_1|^2}$ on $\mathbb{Q} \setminus \{0\}$ to see that $\partial_{q_1} \frac{1}{|q_1|^2} = -\frac{2\overline{q}_1}{|q_1|^4} = h(q_1)$ is a regular function on $\mathbb{Q} \setminus \{0\}$, which is homogeneous of degree -3. Since $\frac{\partial^{2(n-1)}}{\partial x_1^{2(n-1)}}$ commutes with $\overline{\partial}_{q_1}$, the function $s(q_1) = c_{n-1} \frac{\partial^{2(n-1)}}{\partial x_1^{2(n-1)}} \frac{\overline{q_1}}{|q_1|^4}$ in (86) is regular on $\mathbb{Q} \setminus \{0\}$. Consequently, $s(q_1 + y_1)$ for fixed $y_1 > 0$ is also regular on $\mathbb{Q} \setminus \{-y_1\}$, and so $\widetilde{s}(q_1, \ldots, q_n) := s(q_1 + y_1)$ is regular on $\left(\mathbb{Q} \setminus \{-y_1\}\right) \times \mathbb{Q}^{n-1}$. In particular, $\widetilde{s}(\cdot)$ is regular on the quaternion Siegel upper half-space $\mathcal{U}_n$. Now by the invariance, we see that $\widetilde{s}(\tau_{p^{-1}}(q))$ is also regular for fixed $p = (y_2\mathbf{i} + y_3\mathbf{j} + y_4\mathbf{k}, p') \in b\mathcal{U}_n$. So $s\left(q_1 + \overline{p}_1 - 2\sum_{k=2}^{n} \overline{p}'_k q'_k\right) = \widetilde{s}(\tau_{p^{-1}}(q))$ with $p_1 = y_1 + y_2\mathbf{i} + y_3\mathbf{j} + y_4\mathbf{k}$ is regular.

Note that there exists a constant $C > 0$, only depending on the dimension n, such that

$$|\widetilde{s}_\varepsilon(q)|^2 \leq \frac{C}{|q_1 + y_1 + \varepsilon|^{4n+2}} \leq \frac{C}{((|q'|^2 + y_1)^2 + |\mathrm{Im} q_1|^2)^{2n+1}}$$

for $q \in b\mathcal{U}_n$, which is obviously integrable with respect to the measure $d\beta$. Namely, $\widetilde{s}(\cdot)$ is in the Hardy space $H^2(\mathcal{U}_n)$, and so is $\widetilde{s}(\tau_{p^{-1}}(q))$ by the invariance of the Hardy space under the translation. The claim is proved.

Define a quaternion-valued right linear functional

$$\begin{aligned} l_q \colon H^2(\mathcal{U}_n) &\longrightarrow \mathbb{Q}, \\ F &\longmapsto F(q) \end{aligned}$$

for each $q \in \mathcal{U}_n$. The linear functional is bounded by (84). We apply the quaternion version of Riesz's representation theorem to see that there exists an element, denoted by $K(\cdot, q) \in H^2(\mathcal{U}_n)$, such that

$$l_q(F) = \langle K(\cdot, q), F \rangle = \langle K^b(\cdot, q), F^b \rangle_{L^2(b\mathcal{U}_n)}.$$

Here $K(\cdot, \cdot)$ is nontrivial and the boundary value $K^b(p, q)$ exists for almost all $p \in b\mathcal{U}_n$. We have

$$F(q) = \int_{b\mathcal{U}_n} \overline{K^b(Q, q)} F^b(Q) d\beta(Q). \tag{88}$$

Applying (88) to $K(\cdot, p)$ and $K(\cdot, q)$, we see that

$$\begin{aligned} K(q, p) = (K^b(\cdot, q), K^b(\cdot, p)) &= \int_{b\mathcal{U}_n} \overline{K^b(Q, q)} K^b(Q, p) d\beta(Q) \\ &= \overline{\int_{b\mathcal{U}_n} \overline{K^b(Q, p)} K^b(Q, q) d\beta(Q)} = \overline{K(p, q)} \end{aligned}$$

for a fixed $p \in \mathcal{U}_n$. Denote $S(q,p) := \overline{K(p,q)}$ for $(q,p) \in \mathcal{U}_n \times \mathcal{U}_n$. Then $S(q,p)$ is regular in q, and $S(q,p) = \overline{K(p,q)} = \overline{S(p,q)}$. The function S has the boundary values as in Theorem 5.3. Moreover, we have

$$S_b(q,p) = \overline{S^b(p,q)} \tag{89}$$

for $q \in \mathcal{U}_n$, $p \in b\mathcal{U}_n$, which follows from the symmetry $S(q, p + \varepsilon\mathbf{e}) = \overline{S(p + \varepsilon\mathbf{e}, q)}$ by taking $\varepsilon \to 0+$.

To show the uniqueness, suppose that $\widetilde{S}(\cdot,\cdot)$ is another function satisfying Theorem 5.6. By definition $\widetilde{S}(\cdot, q) \in H^2(\mathcal{U}_n)$ for any fixed $q \in \mathcal{U}_n$. Choose an arbitrary $p \in \mathcal{U}_n$ and apply the reproducing formula (87) to obtain

$$\begin{aligned}
\widetilde{S}(p,q) &= \int_{b\mathcal{U}_n} S_b(p,Q)\widetilde{S}^b(Q,q)d\beta(Q) = \overline{\int_{b\mathcal{U}_n} \overline{\widetilde{S}^b(Q,q)}\,\overline{S_b(p,Q)}d\beta(Q)} \\
&= \overline{\int_{b\mathcal{U}_n} \widetilde{S}_b(q,Q)S^b(Q,p)d\beta(Q)} = \overline{S(q,p)} = S(p,q).
\end{aligned}$$

In the third identity, we used Eq. (89) for $S(\cdot,\cdot)$ and $\widetilde{S}(\cdot,\cdot)$. □

The function $S(q,p)$ is conjugate right regular in variables $p = (p_1, \dots, p_n)$:

$$\partial_{p_l} S(q,p) = \overline{\bar{\partial}_{p_l} K(p,q)} = 0.$$

Invariance of the Cauchy-Szegö Kernel

Since the Siegel upper half-space possesses some invariance properties, it is expected that the Cauchy-Szegö kernel also inherits them. Namely, the following proposition is true.

Proposition 5.7 *The Cauchy–Szegö kernel has the following invariance properties.*

$$\left.\begin{aligned}
&S(\tau_p(q), \tau_p(Q)) = S(q,Q), \\
&S(R_{\mathbf{a}}(q), R_{\mathbf{a}}(Q)) = S(q,Q), \\
&\sigma S(R_\sigma(q), R_\sigma(Q))\overline{\sigma} = S(q,Q), \\
&S(\delta_r(q), \delta_r(Q)) r^{4n+2} = S(q,Q),
\end{aligned}\right. \quad \textit{for} \quad q, Q \in \mathcal{U}_n, \tag{90}$$

where $p \in b\mathcal{U}_n$, $\mathbf{a} \in \mathrm{Sp}(n-1)$, $\sigma \in \mathbb{Q}$ *with* $|\sigma| = 1$ *and* $r > 0$.

Proof Note that the measure $\beta(Q)$ is invariant under the translation τ_p. If $F \in H^2(\mathcal{U}_n)$, then $F(\tau_{-p}(q)) \in H^2(\mathcal{U}_n)$. We get

$$F(\tau_{-p}(q)) = \int_{b\mathcal{U}_n} S_b(q,Q)F^b(\tau_{-p}(Q))d\beta(Q) = \int_{b\mathcal{U}_n} S_b(q,\tau_p(Q))F^b(Q)d\beta(Q),$$

and by substituting $\tau_{-p}(q) \mapsto q$, we obtain

$$F(q) = \int_{b\mathcal{U}_n} S_b(\tau_p(q),\tau_p(Q))F^b(Q)d\beta(Q).$$

We conclude that the function $S(\tau_p(q),\tau_p(Q))$ is also regular in q, belongs to $H^2(\mathcal{U}_n)$, and it is symmetric. The first identity in (90) follows by the uniqueness in Theorem 5.6.

It follows from $|\xi\sigma| = |\sigma\xi| = |\xi|$ for any quaternion numbers $\xi \in \mathbb{Q}$ and $\sigma \in \mathbb{Q}$, $|\sigma| = 1$, that $(q_1,q') \mapsto (q_1\sigma, q'\sigma)$ and $(q_1,q') \mapsto (\overline{\sigma}q_1,q')$ are both orthogonal maps, so is their composition $R_\sigma\colon (q_1,q') \mapsto (\overline{\sigma}q_1\sigma, q'\sigma)$. If $F \in H^2(\mathcal{U}_n)$, then $\sigma^{-1}F(R_{\sigma^{-1}}(q))$ is regular and is in $H^2(\mathcal{U}_n)$ by definition. Therefore,

$$\begin{aligned}\sigma^{-1}F(R_{\sigma^{-1}}(q)) &= \int_{b\mathcal{U}_n} S_b(q,Q)\sigma^{-1}F^b(R_{\sigma^{-1}}(Q))d\beta(Q)\\ &= \int_{b\mathcal{U}_n} S_b(q,R_\sigma(Q))\sigma^{-1}F^b(Q)d\beta(Q),\end{aligned}$$

since β is invariant under the orthogonal transformation R_σ. Substituting $R_{\sigma^{-1}}(q) \mapsto q$ and multiplying by σ on both sides, we get

$$F(q) = \int_{b\mathcal{U}_n} \sigma S_b(R_\sigma(q),R_\sigma(Q))\overline{\sigma}F^b(Q)d\beta(Q).$$

The function $\sigma S(R_\sigma(q),R_\sigma(Q))\overline{\sigma}$ is also regular in q, belongs to $H^2(\mathcal{U}_n)$, and it is symmetric. The third identity in (90) follows by the uniqueness in Theorem 5.6.

The second and the fourth identities are proved by similar arguments. □

Determination of the Cauchy-Szegö Kernel

It is sufficient to show that $S_b(q,0) = s(q_1)$ with s given by (86). This is because of

$$S_b(q,p) = S_b(\tau_{(\overline{p}_1,-p')}(q),0) = s\left(q_1 + \overline{p}_1 - 2\sum_{k=2}^{n}\overline{p}'_k q'_k\right) \tag{91}$$

for $p = (p_1, p') \in b\mathcal{U}_n$, $q \in \mathcal{U}_n$. Here $(\overline{p}_1, -p') \in b\mathcal{U}_n$ and $\tau_{(\overline{p}_1, -p')}(p) = 0$ by (75). Taking conjugation of both sides of (91), we see that

$$S^b(q,p) = s\left(q_1 + \overline{p}_1 - 2\sum_{k=2}^{n} \overline{p}'_k q'_k\right)$$

holds for $p \in \mathcal{U}_n$ and $q \in b\mathcal{U}_n$ by the symmetry of the Cauchy–Szegö kernel $S(q,p)$ in Theorem 5.6. Now we fix a point $(p_1, \ldots, p_n) \in \mathcal{U}_n$. We have seen that $s(q_1 + \overline{p}_1 - 2\sum_{k=2}^{n} \overline{p}'_k q'_k) \in H^2(\mathcal{U}_n)$ in the proof of Theorem 5.6. As elements of the Hardy space $H^2(\mathcal{U}_n)$, the functions $S(\cdot, p)$ and $s(q_1 + \overline{p}_1 - 2\sum_{k=2}^{n} \overline{p}'_k q'_k)$ coincide on the boundary $b\mathcal{U}_n$, therefore they must coincide on the whole $\mathcal{U}_n$ by the reproducing property (87).

Fix q_1 with $\operatorname{Re} q_1 > 0$. The equality

$$0 = \sum_{l=2}^{n} \partial_{q_l} \overline{\partial}_{q_l} u(q_1, q') = \sum_{l=2}^{n} (\partial^2_{x_{4l-3}} + \partial^2_{x_{4l-2}} + \partial^2_{x_{4l-1}} + \partial^2_{x_{4l}}) u(q_1, q'),$$

where $u(q) = S_b(q, 0)$, implies that each component of $u(q_1, \cdot)$ is a harmonic function on the ball $\{q' \in \mathbb{Q}^{n-1} \colon |q'| < \operatorname{Re} q_1\}$. On the other hand,

$$S_b((q_1, \mathbf{a}q'), 0) = S_b((q_1, q'), 0) \quad \text{for} \quad q \in \mathcal{U}_n,$$

by Proposition 5.7. Since $\mathrm{Sp}(n-1)$ acts on the sphere $\{q' \in \mathbb{Q}^{n-1} \colon |q'| = R,\ R < \operatorname{Re} q_1\}$ transitively, we see that $S_b((q_1, q'), 0)$ is constant on the sphere. Applying the maximum principle to each component of $S_b((q_1, q'), 0)$ as a harmonic function in q', we conclude that $S_b((q_1, q'), 0)$ is constant on the ball $\{q' \in \mathbb{Q}^{n-1} \colon |q'| < \operatorname{Re} q_1\}$, and so $S_b((q_1, q'), 0) \equiv S_b((q_1, 0), 0)$. Denote $s(q_1) := S_b((q_1, 0), 0)$ a $\mathbb{Q}$-valued function defined on the half-space $\mathbb{R}^4_+ = \{q_1 \in \mathbb{Q} \colon \operatorname{Re} q_1 > 0\}$.

We have $\sigma S_b((\overline{\sigma} q_1 \sigma, 0), 0)\overline{\sigma} = S_b((q_1, 0), 0)$ by the third identity in (90). More precisely,

$$s(\overline{\sigma} q_1 \sigma) = \overline{\sigma} s(q_1) \sigma, \tag{92}$$

for any $\sigma \in \mathbb{Q}$ with $|\sigma| = 1$, and similarly

$$s(r q_1) = r^{-2n-1} s(q_1), \tag{93}$$

by the fourth identity in (90) and $\delta_r \colon (q_1, 0) \mapsto (r^2 q_1, 0)$.

Take $q_1 = x_1 \in \mathbb{R}$ in (92) to get $s(x_1) = \overline{\sigma} s(x_1) \sigma$. Write $s(x_1) = \xi_1 + \xi_2 \mathbf{i} + \xi_3 \mathbf{j} + \xi_4 \mathbf{k}$ and choose $\sigma = \mathbf{i}$. Then $\xi_1 + \xi_2 \mathbf{i} + \xi_3 \mathbf{j} + \xi_4 \mathbf{k} = \overline{\mathbf{i}}(\xi_1 + \xi_2 \mathbf{i} + \xi_3 \mathbf{j} + \xi_4 \mathbf{k})\mathbf{i} = \xi_1 + \xi_2 \mathbf{i} - \xi_3 \mathbf{j} - \xi_4 \mathbf{k}$, and so $\xi_3 = \xi_4 = 0$. Similarly, $\xi_2 = 0$ by choosing $\sigma = \mathbf{j}$. Thus, (92) implies that $s(x_1)$ must be real.

Note that

$$\overline{\sigma}(x_1 + \mathbf{i}x_2)\sigma = x_1 + x_2[(2y_2^2 - 1)\mathbf{i} + 2y_2y_3\mathbf{j} + 2y_2y_4\mathbf{k}], \tag{94}$$

if $\sigma = y_2\mathbf{i} + y_3\mathbf{j} + y_4\mathbf{k}$ with $|\sigma| = 1$. It easily follows from (94) that the orbit of $x_1 + \mathbf{i}x_2$ under the adjoint action of unit quaternions is the 2-dimensional sphere

$$\{x_1 + \xi_2\mathbf{i} + \xi_3\mathbf{j} + \xi_4\mathbf{k} \mid \xi_2^2 + \xi_3^2 + \xi_4^2 = x_2^2\}.$$

Hence $s(q_1)$ is determined by its values on $\mathbb{R}_+^2 = \{(x_1, x_2) \in \mathbb{R}^2\colon x_1 > 0\}$ by (92). The homogeneous degree of s in (93) implies that the function s as a function of the variable q_1 is determined by its values in the semicircle $\{(x_1, x_2) \in \mathbb{R}^2\colon x_1 > 0\ x_1^2 + x_2^2 = 1\}$. At last, the Cauchy–Fueter equations for s gives four ordinary differential equations for four components of s along the semicircle. These ODEs together with the value $s(1)$ uniquely determine the function s.

Proposition 5.8 *On the half-space $\mathbb{R}_+^4 = \{q_1 \in \mathbb{Q}\colon Re\, q_1 > 0\}$, there exists a unique regular function up to a real constant satisfying* (92) *and* (93).

Proof Since the conjugation action of unit quaternions leaves the function s invariant, see (92), its infinitesimal action coincides. From one side, choose $\sigma_t = \cos t + \sin t\mathbf{j}$ for small t. Then

$$\begin{aligned}\overline{\sigma}_t q_1 \sigma_t &= q_1 - t\mathbf{j}(x_1 + x_2\mathbf{i} + x_3\mathbf{j} + x_4\mathbf{k}) + t(x_1 + x_2\mathbf{i} + x_3\mathbf{j} + x_4\mathbf{k})\mathbf{j} + O(t^2)\\ &= q_1 + 2t(-x_4\mathbf{i} + x_2\mathbf{k}) + O(t^2),\end{aligned}$$

where $q_1 = x_1 + x_2\mathbf{i} + x_3\mathbf{j} + x_4\mathbf{k}$, from which we get

$$\left.\frac{d}{dt}\right|_{t=0} s(\overline{\sigma}_t q_1 \sigma_t) = -2x_4\partial_{x_2}s(q_1) + 2x_2\partial_{x_4}s(q_1).$$

From the other side, taking derivatives of $\overline{\sigma}_t s(q_1)\sigma_t$ with respect to t at 0 we get

$$-2x_4\partial_{x_2}s(q_1) + 2x_2\partial_{x_4}s(q_1) = -\mathbf{j}s(q_1) + s(q_1)\mathbf{j}.$$

Similarly, choosing $\sigma_t = \cos t + \sin t\mathbf{k}$, we find that

$$2x_3\partial_{x_2}s(q_1) - 2x_2\partial_{x_3}s(q_1) = -\mathbf{k}s(q_1) + s(q_1)\mathbf{k}.$$

The homogeneity of degree $-2n - 1$ of the function s in (93) implies the Euler equation for s:

$$x_1\partial_{x_1}s(q_1) + x_2\partial_{x_2}s(q_1) + x_3\partial_{x_3}s(q_1) + x_4\partial_{x_4}s(q_1) = -(2n + 1)s(q_1). \tag{95}$$

Restricting to $q_1 = x_1 + x_2\mathbf{i} \in \mathbb{R}^2_+$, i.e. $x_3 = x_4 = 0$, we obtain

$$2x_2\partial_{x_4}s(q_1) = -\mathbf{j}s(q_1) + s(q_1)\mathbf{j}, \qquad -2x_2\partial_{x_3}s(q_1) = -\mathbf{k}s(q_1) + s(q_1)\mathbf{k}.$$

Substitute it into the Cauchy-Fueter equation

$$\partial_{x_1}s(q_1) + \mathbf{i}\partial_{x_2}s(q_1) + \mathbf{j}\partial_{x_3}s(q_1) + \mathbf{k}\partial_{x_4}s(q_1) = 0$$

to deduce

$$2x_2\partial_{x_1}s(q_1) + 2x_2\mathbf{i}\partial_{x_2}s(q_1) = -2\mathbf{i}s(q_1) + \mathbf{j}s(q_1)\mathbf{k} - \mathbf{k}s(q_1)\mathbf{j}. \tag{96}$$

Write $s(x_1 + \mathbf{i}x_2) = f_1 + f_2\mathbf{i} + f_3\mathbf{j} + f_4\mathbf{k}$ on $\mathbb{R}^2_+$. Then, Eq. (96) is equivalent to

$$\begin{aligned} x_2(\partial_{x_1}f_1 - \partial_{x_2}f_2) &= 2f_2, \qquad & x_2(\partial_{x_1}f_2 + \partial_{x_2}f_1) &= 0, \\ x_2(\partial_{x_1}f_3 - \partial_{x_2}f_4) &= f_4, \qquad & x_2(\partial_{x_1}f_4 + \partial_{x_2}f_3) &= -f_3. \end{aligned} \tag{97}$$

Euler's equation (95) implies

$$x_1\partial_{x_1}f_k + x_2\partial_{x_2}f_k = -(2n+1)f_k, \qquad k = 1,2,3,4, \tag{98}$$

on $\mathbb{R}^2_+$. Now we have four real functions f_1, f_2, f_3, f_4 on the upper half-plane $\mathbb{R}^2_+$ satisfying eight equations in (97)–(98) with conditions $f_2(x_1,0) = f_3(x_1,0) = f_4(x_1,0) = 0$ and $f_1(x_1,0)$ is real.

Taking the sum of the first identity in (97), multiplied by x_2, and the second one multiplied by $-x_1$, we obtain

$$x_2(x_2\partial_{x_1} - x_1\partial_{x_2})f_1 = x_2(2f_2 + x_1\partial_{x_1}f_2 + x_2\partial_{x_2}f_2) = -(2n-1)x_2f_2. \tag{99}$$

Set $x_1 = \cos\theta$, $x_2 = \sin\theta$, $\theta \in (-\pi,\pi)$, and $g_j(\theta) := f_j(\cos\theta, \sin\theta, 0, 0)$. Equality (99) implies

$$g_1'(\theta) = (2n-1)g_2. \tag{100}$$

Similarly, we have

$$\begin{aligned} x_2(x_2\partial_{x_1} - x_1\partial_{x_2})f_2 &= 2x_1f_2 + (2n+1)x_2f_1, \\ x_2(x_2\partial_{x_1} - x_1\partial_{x_2})f_3 &= x_1f_3 - 2nx_2f_4, \\ x_2(x_2\partial_{x_1} - x_1\partial_{x_2})f_4 &= x_1f_4 + 2nx_2f_3, \end{aligned}$$

and so

$$\begin{aligned}
\sin\theta g_2'(\theta) &= -2g_2\cos\theta - (2n+1)g_1\sin\theta,\\
\sin\theta g_3'(\theta) &= -g_3\cos\theta + 2ng_4\sin\theta, \qquad (101)\\
\sin\theta g_4'(\theta) &= -g_4\cos\theta - 2ng_3\sin\theta.
\end{aligned}$$

We obtain four real functions g_1, g_2, g_3, g_4 on $(-\pi,\pi)$ satisfying four ordinary differential equations (100)–(101) under the condition

$$g_1(0) \in \mathbb{R}^1, \qquad g_2(0) = g_3(0) = g_4(0) = 0.$$

To see that g_3 and g_4 vanish, note that s is real analytic since it is harmonic. So the functions $g_j, j = 3,4$, are real analytic in θ. Inductively, we can assume $g_j(\theta) = \sum_{m=N}^{\infty} a_m^{(j)}\theta^m, j = 3,4$. Compare the coefficients of term θ^N in the third and fourth equations in (101), we see that

$$Na_N^{(3)} = -a_N^{(3)}, \qquad Na_N^{(4)} = -a_N^{(4)},$$

and so $a_N^{(3)} = a_N^{(4)} = 0$. Therefore, $g_3 \equiv g_4 \equiv 0$. The uniqueness of g_1 and g_2 follows from the vanishing of the solutions g_1, g_2 to (100) and the first equation in (101) with the initial conditions $g_1(0) = g_2(0) = 0$ by the same arguments as above. The result follows. □

Corollary 5.9 *The function s is given by*

$$q_1 \mapsto c_{n-1}\frac{\partial^{2(n-1)}}{\partial x_1^{2(n-1)}}\frac{\overline{q}_1}{|q_1|^4} \qquad (102)$$

for some real constant c_{n-1}.

Indeed, in the proof of Theorem 5.6, we have seen that $h(q_1) = \partial_{q_1}\frac{1}{|q_1|^2} = -\frac{2\overline{q}_1}{|q_1|^4}$ is a regular function on $\mathbb{Q}\setminus\{0\}$, which obviously satisfies (92), and so is the function (102). The conjugation action $\overline{\sigma}q_1\sigma$, fixing x_1 for any $\sigma \in \mathbb{Q}$ with $|\sigma| = 1$, implies

$$\begin{aligned}
\left(\frac{\partial^{2(n-1)}}{\partial x_1^{2(n-1)}}h\right)(\overline{\sigma}q_1\sigma) &= \frac{\partial^{2(n-1)}}{\partial x_1^{2(n-1)}}[h(\overline{\sigma}q_1\sigma)] = \frac{\partial^{2(n-1)}}{\partial x_1^{2(n-1)}}[\overline{\sigma}h(q_1)\sigma]\\
&= \overline{\sigma}\frac{\partial^{2(n-1)}}{\partial x_1^{2(n-1)}}h(q_1)\sigma,
\end{aligned}$$

i.e., (102) satisfies the invariance (92). The function, defined by (102), is homogeneous of degree $-2n-1$. So, the function s is given by expression (102) due to the uniqueness in Proposition 5.8.

We verify now that $s(x_1+\mathbf{i}x_2)$ satisfies (97). Write $s(x_1+\mathbf{i}x_2)=f_1+f_2\mathbf{i}+f_3\mathbf{j}+f_4\mathbf{k}$ on $\mathbb{R}^2_+$. Then, $f_3\equiv 0, f_4\equiv 0$ and

$$f_1=\frac{\partial^{2(n-1)}}{\partial x_1^{2(n-1)}}\frac{x_1}{(x_1^2+x_2^2)^2},\qquad f_2=\frac{\partial^{2(n-1)}}{\partial x_1^{2(n-1)}}\frac{-x_2}{(x_1^2+x_2^2)^2},$$

up to a constant c_n. Functions f_j's satisfy (97)–(98). Note that

$$\partial_{x_1}\frac{-x_2}{(x_1^2+x_2^2)^2}+\partial_{x_2}\frac{x_1}{(x_1^2+x_2^2)^2}=0. \tag{103}$$

Taking derivatives $\frac{\partial^{2(n-1)}}{\partial x_1^{2(n-1)}}$ and multiplying by x_2 both sides of (103), one obtains the second equation in (97). Note that

$$\partial_{x_1}\frac{x_1}{(x_1^2+x_2^2)^2}-\partial_{x_2}\frac{-x_2}{(x_1^2+x_2^2)^2}=\frac{-2}{(x_1^2+x_2^2)^2}. \tag{104}$$

Taking derivatives $\frac{\partial^{2(n-1)}}{\partial x_1^{2(n-1)}}$ and multiplying by x_2 on both sides of (104), we get the first equation in (97).

In the last step we need to calculate the constant (5.5) in the quaternion Cauchy-Szegö kernel. The calculation of the constant is very tedious, using special functions and can be found in detail in [29].

k-Cauchy-Fueter Complex and Related Partial Differential Equations

On the one dimensional quaternionic space, the k-Cauchy-Fueter operators are the Euclidean version of helicity $\frac{k}{2}$ massless field operators [40, 114] on the Minkowski space (corresponding to the Dirac-Weyl equation for $k=1$, Maxwell's equation for $k=2$, the linearized Einstein's equation for $k=3$, etc.). They are the quaternionic counterpart of the Cauchy-Riemann operator in complex analysis. The k-Cauchy-Fueter complexes on multidimensional quaternionic space $\mathbb{Q}^n$, which play the role of Dolbeault complex in several complex variables, are now explicitly known [109], see also [7, 22, 32, 33]. It is quite interesting to develop a theory of several quaternionic variables by analyzing these complexes, as it was done for the $\bar{\partial}$-complex in the theory of several complex variables. We saw that a non-homogeneous $\bar{\partial}$-equation (cf., e.g., [48]), leads to the study of $\bar{\partial}$-Neumann problem. The non-homogeneous k-Cauchy-Fueter equation on the whole quaternionic space

$\mathbb{Q}^n$ was solved in [109]. See [62, 75, 109, 112] (also [4, 25, 29, 32, 110] for $k = 1$) and references therein for results about k-regular functions. Note that the non-homogeneous $\overline{\partial}$-equation on a smooth domain in the complex plane is always solvable. The non-homogeneous 1-Cauchy-Fueter equation on a smooth domain in $\mathbb{Q}$ is also always solvable since it is exactly the Dirac operator on $\mathbb{R}^4$. But even on one dimensional quaternionic space $\mathbb{Q}$, the k-Cauchy-Fueter operator for $k \geq 2$ is overdetermined. The non-homogeneous k-Cauchy-Fueter equation can only be solved under the compatibility condition given by the k-Cauchy-Fueter complex. The k-Cauchy-Fueter complex over a smooth domain $\Omega \in \mathbb{Q}$ is

$$0 \longrightarrow C^\infty(\Omega, \mathbb{C}^{k+1}) \xrightarrow{D_0^{(k)}} C^\infty(\Omega, \mathbb{C}^{2k}) \xrightarrow{D_1^{(k)}} C^\infty(\Omega, \mathbb{C}^{k-1}) \longrightarrow 0, \tag{105}$$

$k = 2, 3, \ldots$, where $D_j^{(k)}$, $j = 0, 1$ are the k-Cauchy-Fueter operators, which are analogous to the $\overline{\partial}$ operators in the several complex variables. In this section, we will investigate the non-homogeneous k-Cauchy-Fueter equation

$$D_0^{(k)} u = f, \tag{106}$$

on a smooth domain Ω in $\mathbb{Q}$ under the compatibility condition

$$D_1^{(k)} f = 0. \tag{107}$$

We define the first and the second cohomology groups of the k-Cauchy-Fueter complex as

$$H^1_{(k)}(\Omega, \mathbb{Q}) = \frac{\left\{f \in C^\infty(\Omega; \mathbb{C}^{2k}) \colon D_1^{(k)} f = 0\right\}}{\left\{D_0^{(k)} u \colon u \in C^\infty(\Omega; \mathbb{C}^{k+1})\right\}},$$

$$H^2_{(k)}(\Omega, \mathbb{Q}) = \frac{C^\infty(\Omega; \mathbb{C}^{k-1})}{\left\{D_1^{(k)} u \colon u \in C^\infty(\Omega; \mathbb{C}^{2k})\right\}}.$$

The 0th cohomology group is $H^0_{(k)}(\Omega, \mathbb{Q}) = \ker D_0^{(k)}$ and it is an infinite dimensional space of k-regular functions [62]. The first cohomology group can be also represented by Hodge-type elements:

$$\mathcal{H}^1_{(k)}(\Omega) = \left\{f \in C^\infty(\Omega, \mathbb{C}^{2k}) \colon D_1^{(k)} f = 0, D_0^{(k)*} f = 0\right\},$$

where $D_0^{(k)*}$ is the formal adjoint of $D_0^{(k)}$. The set $\mathcal{H}^1_{(k)}(\Omega)$ is an analogue of the space of harmonic functions. Denote by $H_s(\Omega, \mathbb{C}^n)$ the Sobolev space of all $\mathbb{C}^n$-valued functions, whose components are in $H_s(\Omega) = H_s(\Omega, \mathbb{C})$. We formulate two main results.

Theorem 5.10 *Suppose Ω is a domain in $\mathbb{Q}$ with a smooth boundary. Then*

(1) *the isomorphic spaces $H^1_{(k)}(\Omega, \mathbb{Q}) \cong \mathcal{H}^1_{(k)}(\Omega)$ are finite dimensional;*
(2) *if $f \in H_s(\Omega, \mathbb{C}^{2k})$, $s = 1, 2, \ldots$, then the non-homogeneous k-Cauchy-Fueter equation* (106) *is solvable by some $u \in H_{s+1}(\Omega, \mathbb{C}^{k+1})$ if and only if f is orthogonal to $\mathcal{H}^1_{(k)}(\Omega)$ in $L^2(\Omega, \mathbb{C}^{2k})$ and satisfies the compatibility condition* (107). *When it is solvable, it has a solution u satisfying the estimate*

$$\|u\|_{H_{s+1}(\Omega,\mathbb{C}^{k+1})} \leq C\|f\|_{H_s(\Omega,\mathbb{C}^{2k})} \tag{108}$$

for some constant C depending only on the domain Ω, k and s;
(3) *the equation $D_1^{(k)}\psi = \Psi$ is uniquely solved by a $\psi \in H_{s+1}(\Omega, \mathbb{C}^{2k})$ for any $\Psi \in H_s(\Omega, \mathbb{C}^{k-1})$ with estimate analogue* (108).

To prove Theorem 5.10, we consider the associated Laplacian of the k-Cauchy-Fuete complex (105)

$$\Box_1^{(k)} = D_0^{(k)}D_0^{(k)*} + D_1^{(k)*}D_1^{(k)},$$

where $D_0^{(k)*}$ and $D_1^{(k)*}$ are the formal adjoints of $D_0^{(k)}$ and $D_1^{(k)}$, respectively, and a natural boundary value problem

$$\begin{cases} \Box_1^{(k)}u = f, & \text{on } \Omega, \\ D_0^{(k)*}(\nu)u\Big|_{b\Omega} = 0, & \\ D_1^{(k)*}(\nu)D_1^{(k)}u\Big|_{b\Omega} = 0, & \end{cases} \tag{109}$$

where ν is the unit vector of outer normal to the boundary $b\Omega$, $u \in H_{s+2}(\Omega, \mathbb{C}^{2k})$ and $f \in H_s(\Omega, \mathbb{C}^{2k})$. We prove that this boundary value problem is regular and obtain the following result.

Theorem 5.11 *Suppose Ω is a domain in $\mathbb{Q}$ with a smooth boundary. If $f \in H_s(\Omega, \mathbb{C}^{2k})$, $s = 0, 1, 2, \ldots$, is orthogonal to $\mathcal{H}^1_{(k)}(\Omega)$ relative to the L^2 product, the boundary value problem* (109) *has a solution $u = N_1^{(k)}f$ such that*

$$\|u\|_{H_{s+2}(\Omega,\mathbb{C}^{2k})} \leq C\|f\|_{H_s(\Omega,\mathbb{C}^{2k})}$$

for some constant C only depending on the domain Ω, k and s.

Moreover, we have the Hodge-type orthogonal decomposition for any function $\psi \in H_s(\Omega, \mathbb{C}^{2k})$:

$$\psi = D_0^{(k)}D_0^{(k)*}N_1^{(k)}\psi + D_1^{(k)*}D_1^{(k)}N_1^{(k)}\psi + \Pi\psi = \Box_1^{(k)}N_1^{(k)}\psi + \Pi\psi, \tag{110}$$

where Π is the orthonormal projection to $\mathcal{H}^1_{(k)}(\Omega)$ under the $L^2(\Omega, \mathbb{C}^{2k})$ product.

Although for a smooth domain in the complex plane, its $\bar{\partial}$-cohomology always vanishes, its De Rham cohomology groups, which are isomorphic to its simplicial cohomology groups, may be nontrivial. We conjecture that the cohomology groups $H^1_{(k)}(\Omega, \mathbb{Q})$ may be nontrivial for some domains Ω with smooth boundaries in $\mathbb{Q}$. It is quite interesting to characterize the class of domains in $\mathbb{Q}$ on which the non-homogeneous k-Cauchy–Fueter equation is always solvable. On the higher dimensional quaternionic space $\mathbb{Q}^n$, there is no reason to expect the corresponding boundary value problem of the non-homogeneous k-Cauchy–Fueter equation to be regular, as in the case of several complex variables.

The k-Cauchy–Fueter Complexes on a Domain in $\mathbb{Q}$

We identify the one dimensional quaternion space $\mathbb{Q}$ with the Euclidean space $\mathbb{R}^4$, setting

$$\begin{pmatrix} \nabla_{00'} & \nabla_{01'} \\ \nabla_{10'} & \nabla_{11'} \end{pmatrix} = \begin{pmatrix} \partial_{x_0} + i\partial_{x_1} & -\partial_{x_2} - i\partial_{x_3} \\ \partial_{x_2} - i\partial_{x_3} & \partial_{x_0} - i\partial_{x_1} \end{pmatrix},$$

where $(x_0, x_1, x_2, x_3) \in \mathbb{R}^4$. The matrix

$$\epsilon = (\epsilon_{A'B'}) = \begin{pmatrix} 0 & 1 \\ -1 & 0 \end{pmatrix}$$

is used to raise or lower indices, e.g. $\nabla_A^{A'} \epsilon_{A'B'} = \nabla_{AB'}$. The *k-Cauchy-Fueter complex* [109] on a domain Ω in $\mathbb{R}^4$ for $k \geq 2$ is given by

$$0 \to C^\infty(\Omega, \odot^k \mathbb{C}^2) \xrightarrow{D_0^{(k)}} C^\infty(\Omega, \odot^{k-1} \mathbb{C}^2 \otimes \mathbb{C}^2) \xrightarrow{D_1^{(k)}} C^\infty(\Omega, \odot^{k-2} \mathbb{C}^2 \otimes \Lambda^2 \mathbb{C}^2) \to 0,$$

where

$$\begin{aligned} (D_0^{(k)} \phi)_{AB' \cdots C'} &:= \sum_{A'=0',1'} \nabla_A^{A'} \phi_{A'B' \cdots C'}, \\ (D_1^{(k)} \psi)_{ABB' \cdots C'} &:= \sum_{A'=0',1'} \left(\nabla_A^{A'} \psi_{BA'B' \cdots C'} - \nabla_B^{A'} \psi_{AA'B' \cdots C'} \right). \end{aligned} \tag{111}$$

Sections $\phi \in C^\infty(\Omega, \odot^k \mathbb{C}^2)$ have $(k+1)$ components $\phi_{0' \ldots 0'}, \phi_{1' \ldots 0'}, \ldots, \phi_{1' \ldots 1'}$, while $D_0^{(k)} \phi \in C^\infty(\Omega, \odot^{k-1} \mathbb{C}^2 \otimes \mathbb{C}^2)$ have $2k$ components

$$(D_0^{(k)} \phi)_{A0' \ldots 0'}, \ (D_0^{(k)} \phi)_{A1' \ldots 0'}, \ldots, (D_0^{(k)} \phi)_{A1' \ldots 1'},$$

where $A = 0, 1$. Note that $\phi_{A'B'\dots C'}$ is invariant under the permutation of subscripts, $A', B', \cdots, C' = 0', 1'$. Here by "$\odot^k$" we denote the symmetric tensor product.

There is a family of equations in physics, called the *helicity $\frac{k}{2}$ massless field* equations [40, 114]. The first one is the Dirac–Weyl equation for an electron of mass zero whose solutions correspond to neutrinos. The second one is the Maxwell's equation whose solutions correspond to photons. The third one is the linearized Einstein's equation whose solutions correspond to weak gravitational fields, and so on. The k-Cauchy–Fueter equations are the Euclidean version of these equations. The *affine Minkowski space* can be embedded into the space of (2×2)-matrices with complex entries $\mathbb{C}^{2\times 2}$ by the map

$$(x_0, x_1, x_2, x_3) \mapsto \begin{pmatrix} x_0 + x_1 & x_2 + ix_3 \\ x_2 - ix_3 & x_0 - x_1 \end{pmatrix}, \qquad i = \sqrt{-1},$$

while the quaternionic algebra $\mathbb{Q}$ can be embedded into $\mathbb{C}^{2\times 2}$ by

$$x_0 + x_1\mathbf{i} + x_2\mathbf{j} + x_3\mathbf{k} \mapsto \begin{pmatrix} x_0 + ix_1 & -x_2 - ix_3 \\ x_2 - ix_3 & x_0 - ix_1 \end{pmatrix}.$$

The helicity $\frac{k}{2}$ massless field equation [40, 109] is $D_0^{(k)}\phi = 0$, where the $D_0^{(k)}$ is also given by (111) with $\nabla_{AB'}$ replaced by

$$\begin{pmatrix} \nabla_{00'} & \nabla_{01'} \\ \nabla_{10'} & \nabla_{11'} \end{pmatrix} = \begin{pmatrix} \partial_{x_0} + \partial_{x_1} & \partial_{x_2} + i\partial_{x_3} \\ \partial_{x_2} - i\partial_{x_3} & \partial_{x_0} - \partial_{x_1} \end{pmatrix}.$$

In the sequel we concentrate on the 2-Cauchy–Fueter complex.

The 2-Cauchy–Fueter Complex

We write

$$\begin{pmatrix} \nabla_0^{0'} & \nabla_0^{1'} \\ \nabla_1^{0'} & \nabla_1^{1'} \end{pmatrix} = \begin{pmatrix} \nabla_{00'} & \nabla_{01'} \\ \nabla_{10'} & \nabla_{11'} \end{pmatrix} \begin{pmatrix} 0 & -1 \\ 1 & 0 \end{pmatrix} \begin{pmatrix} \nabla_{01'} & -\nabla_{00'} \\ \nabla_{11'} & -\nabla_{10'} \end{pmatrix}$$
$$= \begin{pmatrix} -\partial_{x_2} - i\partial_{x_3} & -\partial_{x_0} - i\partial_{x_1} \\ \partial_{x_0} - i\partial_{x_1} & -\partial_{x_2} + i\partial_{x_3} \end{pmatrix}.$$

In the case $k = 2$, we use the notation $D_0 = D_0^{(2)}$ and $D_1 = D_1^{(2)}$. The 2-Cauchy-Fueter complex on a domain Ω in $\mathbb{R}^4$ is

$$0 \longrightarrow C^\infty(\Omega, \odot^2\mathbb{C}^2) \xrightarrow{D_0} C^\infty(\Omega, \mathbb{C}^2 \otimes \mathbb{C}^2) \xrightarrow{D_1} C^\infty(\Omega, \Lambda^2\mathbb{C}^2) \longrightarrow 0, \qquad (112)$$

with

$$(D_0\phi)_{AB'} = \sum_{A'=0',1'} \nabla_A^{A'}\phi_{A'B'} = \nabla_A^{0'}\phi_{0'B'} + \nabla_A^{1'}\phi_{1'B'},$$

$$(D_1\psi)_{01} = \sum_{A'=0',1'} \nabla_0^{A'}\psi_{1A'} - \nabla_1^{A'}\psi_{0A'} = \nabla_0^{0'}\psi_{10'} + \nabla_0^{1'}\psi_{11'} - \nabla_1^{0'}\psi_{00'} - \nabla_1^{1'}\psi_{01'},$$

where $A = 0,1, B' = 0',1'$, $\phi \in C^\infty(\Omega, \odot^2\mathbb{C}^2)$ has three components $\phi_{0'0'}, \phi_{1'0'} = \phi_{0'1'}$ and $\phi_{1'1'}$, while $D_0\phi \in C^\infty(\Omega, \mathbb{C}^2 \otimes \mathbb{C}^2)$ has four components $(D_0\phi)_{00'}$,$(D_0\phi)_{10'}$, $(D_0\phi)_{01'}$, $(D_0\phi)_{11'}$, and $\Psi = \Psi_{01} \in C^\infty(\Omega, \Lambda^2\mathbb{C}^2)$ is a scalar function.

It is known that (112) is a complex: $D_1D_0 = 0$, see [109]. Indeed, we calculate, for any $\phi \in C^\infty(\Omega, \odot^2\mathbb{C}^2)$

$$\begin{aligned}(D_1D_0\phi)_{01} &= \sum_{A'=0',1'} \nabla_0^{A'}(D_0\phi)_{1A'} - \nabla_1^{A'}(D_0\phi)_{0A'}\\ &- \sum_{A',C'=0',1'} \nabla_0^{A'}\nabla_1^{C'}\phi_{C'A'} - \nabla_1^{A'}\nabla_0^{C'}\phi_{C'A'} = 0\end{aligned}$$

by $\phi_{C'A'} = \phi_{A'C'}$ and the commutativity $\nabla_1^{A'}\nabla_0^{C'} = \nabla_0^{C'}\nabla_1^{A'}$ of scalar differential operators with constant coefficients.

The operator D_0 in (112) can be written as a (4×3)-matrix operator

$$D_0\phi = \begin{pmatrix}(D_0\phi)_{00'}\\(D_0\phi)_{10'}\\(D_0\phi)_{01'}\\(D_0\phi)_{11'}\end{pmatrix} = \begin{pmatrix}\nabla_0^{0'} & \nabla_0^{1'} & 0\\ \nabla_1^{0'} & \nabla_1^{1'} & 0\\ 0 & \nabla_0^{0'} & \nabla_0^{1'}\\ 0 & \nabla_1^{0'} & \nabla_1^{1'}\end{pmatrix}\begin{pmatrix}\phi_{0'0'}\\ \phi_{1'0'}\\ \phi_{1'1'}\end{pmatrix},$$

and the operator D_1 takes the form

$$D_1\psi = (-\nabla_1^{0'}, \nabla_0^{0'}, -\nabla_1^{1'}, \nabla_0^{1'})\begin{pmatrix}\psi_{00'}\\ \psi_{10'}\\ \psi_{01'}\\ \psi_{11'}\end{pmatrix}.$$

We define

$$\partial_{z_0} = \partial_{x_0} - i\partial_{x_1}, \quad \partial_{\bar z_0} = \partial_{x_0} + i\partial_{x_1}, \quad \partial_{z_1} = \partial_{x_2} - i\partial_{x_3}, \quad \partial_{\bar z_1} = \partial_{x_2} + i\partial_{x_3}$$

for $z_0 = x_0 + ix_1$, and $z_1 = x_2 + ix_3$. Using these notations, and the isomorphisms

$$\odot^2\mathbb{C}^2 \cong \mathbb{C}^3, \qquad \mathbb{C}^2 \otimes \mathbb{C}^2 \cong \mathbb{C}^4, \qquad \Lambda^2\mathbb{C}^2 \cong \mathbb{C}^1,$$

we rewrite $D_0: C^\infty(\Omega, \mathbb{C}^3) \to C^\infty(\Omega, \mathbb{C}^4)$ as

$$D_0\phi = \begin{pmatrix} -\partial_{\bar{z}_1} & -\partial_{\bar{z}_0} & 0 \\ \partial_{z_0} & -\partial_{z_1} & 0 \\ 0 & -\partial_{\bar{z}_1} & -\partial_{\bar{z}_0} \\ 0 & \partial_{z_0} & -\partial_{z_1} \end{pmatrix} \begin{pmatrix} \phi_0 \\ \phi_1 \\ \phi_2 \end{pmatrix},$$

and $D_1: C^\infty(\Omega, \mathbb{C}^4) \to C^\infty(\Omega, \mathbb{C})$ with $D_1\psi = \begin{pmatrix} -\partial_{z_0} & -\partial_{\bar{z}_1} & \partial_{z_1} & -\partial_{\bar{z}_0} \end{pmatrix} \begin{pmatrix} \psi_0 \\ \psi_1 \\ \psi_2 \\ \psi_3 \end{pmatrix}$.

The Laplacian Associated to 2-Cauchy-Fueter Complex

It is easy to see that

$$\overline{\begin{pmatrix} -\partial_{\bar{z}_1} & -\partial_{\bar{z}_0} \\ \partial_{z_0} & -\partial_{z_1} \end{pmatrix}}^t \begin{pmatrix} -\partial_{\bar{z}_1} & -\partial_{\bar{z}_0} \\ \partial_{z_0} & -\partial_{z_1} \end{pmatrix} = \begin{pmatrix} -\partial_{z_1} & \partial_{\bar{z}_0} \\ -\partial_{z_0} & -\partial_{\bar{z}_1} \end{pmatrix} \begin{pmatrix} -\partial_{\bar{z}_1} & -\partial_{\bar{z}_0} \\ \partial_{z_0} & -\partial_{z_1} \end{pmatrix} = \begin{pmatrix} \Delta & 0 \\ 0 & \Delta \end{pmatrix},$$

where ^{t} is the transpose, and $\Delta = \partial_{z_0}\partial_{\bar{z}_0} + \partial_{z_1}\partial_{\bar{z}_1} = \partial^2_{x_0} + \partial^2_{x_1} + \partial^2_{x_2} + \partial^2_{x_4}$ is the Laplacian on $\mathbb{R}^4$.

Let $\mathfrak{D}: C^1(\overline{\Omega}, \mathbb{C}^{n_1}) \to C^0(\overline{\Omega}, \mathbb{C}^{n_2})$ be a differential operator of the first order with constant coefficients. Recall that an operator $\mathfrak{D}^*$ is a formal adjoint of $\mathfrak{D}$ if

$$\int_\Omega \langle \mathfrak{D}u, v\rangle d\mathrm{Vol} = \int_\Omega \langle u, \mathfrak{D}^* v\rangle d\mathrm{Vol}, \quad \text{for any} \quad u \in C^1_0(\Omega, \mathbb{C}^{n_1}),\ v \in C^1_0(\Omega, \mathbb{C}^{n_2}),$$

where $\langle .,. \rangle$ is the Hermitian inner product in $\mathbb{C}^{n_j}, j = 1, 2$. It is easy to see that the formal adjoints of D_0 and D_1 are $D_0^* = -\overline{D_0}^t$ and $D_1^* = -\overline{D_1}^t$, respectively. Then,

$$\begin{aligned} D_0 D_0^* &= -\begin{pmatrix} -\partial_{\bar{z}_1} & -\partial_{\bar{z}_0} & 0 \\ \partial_{z_0} & -\partial_{z_1} & 0 \\ 0 & -\partial_{\bar{z}_1} & -\partial_{\bar{z}_0} \\ 0 & \partial_{z_0} & -\partial_{z_1} \end{pmatrix} \begin{pmatrix} -\partial_{z_1} & \partial_{\bar{z}_0} & 0 & 0 \\ -\partial_{z_0} & -\partial_{\bar{z}_1} & -\partial_{z_1} & \partial_{\bar{z}_0} \\ 0 & 0 & -\partial_{z_0} & -\partial_{\bar{z}_1} \end{pmatrix} \\ &= -\begin{pmatrix} \Delta & 0 & \partial_{\bar{z}_0}\partial_{z_1} & -\partial^2_{\bar{z}_0} \\ * & \Delta & \partial^2_{z_1} & -\partial_{\bar{z}_0}\partial_{z_1} \\ * & * & \Delta & 0 \\ * & * & * & \Delta \end{pmatrix}, \end{aligned} \tag{113}$$

where $*$-entries in the matrix are known by Hermitian symmetry of $D_0^*D_0$, and

$$\begin{aligned} D_1^*D_1 &= -\begin{pmatrix} -\partial_{\bar z_0} \\ -\partial_{z_1} \\ \partial_{\bar z_1} \\ -\partial_{z_0} \end{pmatrix} \begin{pmatrix} -\partial_{z_0} & -\partial_{\bar z_1} & \partial_{z_1} & -\partial_{\bar z_0} \end{pmatrix} \\ &= -\begin{pmatrix} \partial_{z_0}\partial_{\bar z_0} & \partial_{\bar z_0}\partial_{\bar z_1} & -\partial_{\bar z_0}\partial_{z_1} & \partial^2_{\bar z_0} \\ * & \partial_{z_1}\partial_{\bar z_1} & -\partial^2_{z_1} & \partial_{\bar z_0}\partial_{z_1} \\ * & * & \partial_{z_1}\partial_{\bar z_1} & -\partial_{\bar z_0}\partial_{\bar z_1} \\ * & * & * & \partial_{z_0}\partial_{\bar z_0} \end{pmatrix}. \end{aligned} \tag{114}$$

The sum of (113) and (114) gives the operator

$$\begin{aligned} \Box_1 = D_0D_0^* + D_1^*D_1 &= -\begin{pmatrix} \Delta + \partial_{z_0}\partial_{\bar z_0} & \partial_{\bar z_0}\partial_{\bar z_1} & 0 & 0 \\ \partial_{z_0}\partial_{z_1} & \Delta + \partial_{z_1}\partial_{\bar z_1} & 0 & 0 \\ 0 & 0 & \Delta + \partial_{z_1}\partial_{\bar z_1} & -\partial_{\bar z_0}\partial_{\bar z_1} \\ 0 & 0 & -\partial_{z_0}\partial_{z_1} & \Delta + \partial_{z_0}\partial_{\bar z_0} \end{pmatrix} \\ &= -\begin{pmatrix} \Delta + \Delta_1 & L & 0 & 0 \\ \overline{L} & \Delta + \Delta_2 & 0 & 0 \\ 0 & 0 & \Delta + \Delta_2 & -L \\ 0 & 0 & -\overline{L} & \Delta + \Delta_1 \end{pmatrix}, \end{aligned}$$

where

$$\Delta_1 := \partial_{z_0}\partial_{\bar z_0} = \partial^2_{x_0} + \partial^2_{x_1}, \qquad \Delta_2 := \partial_{z_1}\partial_{\bar z_1} = \partial^2_{x_2} + \partial^2_{x_3},$$
$$L := \partial_{\bar z_0}\partial_{\bar z_1} = (\partial_{x_0} + i\partial_{x_1})(\partial_{x_2} + i\partial_{x_3}).$$

The operator $\Box_1$ is obviously elliptic, i.e., its symbol for any $\xi \neq 0$ is positive definite.

Domains of the Adjoint Operators D_0^* and D_1^*

For the first order operator with constant coefficients $\mathfrak{D}\colon C^1(\overline{\Omega}, \mathbb{C}^{n_1}) \to C^0(\overline{\Omega}, \mathbb{C}^{n_2})$, and functions $u \in C^1(\overline{\Omega}, \mathbb{C}^{n_1})$ and $v \in C^1(\overline{\Omega}, \mathbb{C}^{n_2})$, we have

$$\int_\Omega \langle \mathfrak{D}u, v\rangle d\mathrm{Vol} = \int_\Omega \langle u, \mathfrak{D}^*v\rangle d\mathrm{Vol} + \int_{b\Omega} \langle u, \mathfrak{D}^*(\nu)v\rangle dS \tag{115}$$

by Green's formula, where $\nu = (\nu_0, \ldots, \nu_4)$ is the unit vector of outer normal to the boundary, and $\mathfrak{D}^*(\nu)$ is obtained by replacing ∂_{x_j} in $\mathfrak{D}^*$ by ν_j.

By abuse of notations, we also denote by $\mathfrak{D}^*$ the adjoint operator of the extended operator $\mathfrak{D}: L^2(\Omega, \mathbb{C}^{n_1}) \to L^2(\Omega, \mathbb{C}^{n_2})$. Now let Ω be $\mathbb{R}^4_+ = \{x = (x_0, \dots, x_3) \in \mathbb{R}^4\colon x_0 > 0\}$. Then the unit inner normal vector is $\nu = (1, 0, 0, 0)$. By definition of the adjoint operator, a function $\psi = (\psi_0, \psi_1, \psi_2, \psi_3)^t \in \mathrm{Dom}(D_0^*) \cap C^1(\Omega, \mathbb{C}^4)$ if and only if the integral over the boundary in (115) vanishes for any u, i.e., $D_0^*(\nu)\psi = 0$ on the boundary. Then,

$$0 = \begin{pmatrix} -\partial_{z_1} & \partial_{\bar{z}_0} & 0 & 0 \\ -\partial_{z_0} & -\partial_{\bar{z}_1} & -\partial_{z_1} & \partial_{\bar{z}_0} \\ 0 & 0 & -\partial_{z_0} & -\partial_{\bar{z}_1} \end{pmatrix} (\nu)\psi|_{b\Omega} = \begin{pmatrix} 0 & 1 & 0 & 0 \\ -1 & 0 & 0 & 1 \\ 0 & 0 & -1 & 0 \end{pmatrix} \psi|_{b\Omega},$$

from which we get

$$\psi_1 = \psi_2 = 0, \qquad \psi_0 - \psi_3 = 0 \qquad \text{on} \quad b\Omega. \tag{116}$$

Similarly, $\Psi \in \mathrm{Dom}(D_1^*) \cap C^1(\Omega, \mathbb{C})$ if and only if $D_1^*(\nu)\Psi = 0$ on the boundary, i.e.,

$$0 = \begin{pmatrix} -\partial_{\bar{z}_0} \\ -\partial_{z_1} \\ \partial_{\bar{z}_1} \\ -\partial_{z_0} \end{pmatrix} (\nu)\Psi|_{b\Omega} = \begin{pmatrix} -1 \\ 0 \\ 0 \\ -1 \end{pmatrix} \Psi|_{\partial\Omega},$$

which leads to $\Psi|_{b\Omega} = 0$. Now $D_1\psi \in \mathrm{Dom}(D_1^*) \cap C^1(\Omega, \mathbb{C})$ implies that

$$-\partial_{z_0}\psi_0 - \partial_{\bar{z}_1}\psi_1 + \partial_{z_1}\psi_2 - \partial_{\bar{z}_0}\psi_3 = 0 \quad \text{on} \quad b\Omega.$$

Note that $\partial_{\bar{z}_1}\psi_1 = \partial_{z_1}\psi_2 = 0$ since $\partial_{\bar{z}_1}$ and ∂_{z_1} are tangential derivatives, and ψ_1, ψ_2 both vanish on the boundary by (116). Therefore,

$$\partial_{z_0}\psi_0 + \partial_{\bar{z}_0}\psi_3 = \partial_{x_0}(\psi_0 + \psi_3) = 0, \qquad \text{on} \quad b\Omega \tag{117}$$

by using (116) again. So we need to solve the system $\Box_1^{(2)}\psi = f$ in Ω under the boundary conditions (116) and (117).

We need to define more operators. We obtain that

$$\begin{aligned} \Box_0 = D_0^* D_0 &= - \begin{pmatrix} -\partial_{z_1} & \partial_{\bar{z}_0} & 0 & 0 \\ -\partial_{z_0} & -\partial_{\bar{z}_1} & -\partial_{z_1} & \partial_{\bar{z}_0} \\ 0 & 0 & -\partial_{z_0} & -\partial_{\bar{z}_1} \end{pmatrix} \begin{pmatrix} -\partial_{\bar{z}_1} & -\partial_{\bar{z}_0} & 0 \\ \partial_{z_0} & -\partial_{z_1} & 0 \\ 0 & -\partial_{\bar{z}_1} & -\partial_{\bar{z}_0} \\ 0 & \partial_{z_0} & -\partial_{z_1} \end{pmatrix} \\ &= - \begin{pmatrix} \Delta & 0 & 0 \\ 0 & 2\Delta & 0 \\ 0 & 0 & \Delta \end{pmatrix}, \end{aligned} \tag{118}$$

and

$$\square_2 = D_1 D_1^* = -(-\partial_{z_0}, -\partial_{\bar{z}_1}, \partial_{z_1}, -\partial_{\bar{z}_0}) \begin{pmatrix} -\partial_{\bar{z}_0} \\ -\partial_{z_1} \\ \partial_{\bar{z}_1} \\ -\partial_{z_0} \end{pmatrix} = -2\Delta,$$

with the boundary condition $\Psi \in \mathrm{Dom}(D_1^*) \cap C^1(\Omega, \mathbb{C}^1)$, i.e., $\Psi|_{b\Omega} = 0$.

Corollary 5.12 *Suppose that* $u \in H_1(\Omega, \mathbb{C}^{n_1})$, $v \in H_1(\Omega, \mathbb{C}^{n_2})$, *and* $\mathfrak{D}(v)u|_{b\Omega} = 0$ *or* $\mathfrak{D}^*(v)v|_{b\Omega} = 0$. *Then*

$$(\mathfrak{D}u, v) = (u, \mathfrak{D}^* v), \qquad (v, \mathfrak{D}u) = (\mathfrak{D}^* v, u). \tag{119}$$

Proof The operator of restriction to the boundary $H_s(\Omega, \mathbb{C}^n) \to H_{s-\frac{1}{2}}(b\Omega, \mathbb{C}^n)$ for $s > \frac{1}{2}$ is a bounded operator by the trace theorem [100, Proposition 4.5, Chap. 4]. Moreover, $C^\infty(\overline{\Omega}, \mathbb{C}^n)$ is dense in $H_s(\Omega, \mathbb{C}^n)$ for $s \geq 0$. Approximating $u \in H_1(\Omega, \mathbb{C}^{n_1})$, $v \in H_1(\Omega, \mathbb{C}^{n_2})$, by functions from $C^\infty(\overline{\Omega}, \mathbb{C}^{n_j})$, we see that integration by parts (115) holds for $u \in H_1(\Omega, \mathbb{C}^{n_1})$, $v \in H_1(\Omega, \mathbb{C}^{n_2})$, [100, (7.2) in Chap. 5]). The boundary term vanishes by the assumption. □

Shapiro-Lopatinskii Condition

Assume that the differential operator

$$P(x, D) \colon C^\infty(\overline{\Omega}, E_0) \to C^\infty(\overline{\Omega}, E_1)$$

is elliptic of order m, and that the operators $B_j(x, D) \colon C^\infty(\overline{\Omega}, E_0) \to C^\infty(b\Omega, G_j)$, $j = 1, \ldots, l$, are differential operators of order $m_j \leq m - 1$, where $E_0, E_1, G_j, j = 1, \ldots, l$, are finite dimensional complex vector spaces. Let Ω be a domain in $\mathbb{R}^n$ with smooth boundary $b\Omega$. Consider the boundary value problem

$$\begin{cases} P(x, D)u = f, & \text{on} \quad \Omega, \\ B_j(x, D)u = g_j, & \text{on} \quad b\Omega, \quad j = 1, \ldots, l. \end{cases} \tag{120}$$

For fixed $x \in b\Omega$, define the half space $V_x := \{y \in \mathbb{R}^n; \langle y, \nu_x \rangle > 0\}$, where ν_x is the unit vector of inner normal to $b\Omega$ at point x. By a rotation if necessary, we can assume $\nu_x = (1, 0, \ldots, 0)$ and $P(x, D)$ can be written as

$$P(x, D) = \frac{\partial^m}{\partial x_1^m} + \sum_{\alpha=0}^{m-1} A_\alpha(x, D_{x'}) \frac{\partial^\alpha}{\partial x_1^\alpha},$$

up to the multiple of an invertible matrix, where the order of $A_\alpha(x, D_{x'})$ is equal to $m-\alpha$, $x' = (x_2, \ldots, x_n)$. For the elliptic operator $P(x, D)$, the boundary value problem (109) is called *regular* if for any $\xi \in \mathbb{R}^{n-1}$ and $\eta_j \in G_j$, there is a unique bounded solution on $\mathbb{R}_+ = [0, \infty)$ to the Cauchy problem

$$\frac{d^m\Phi}{dt^m} + \sum_{\alpha=0}^{m-1} \widetilde{A}_\alpha(\xi)\frac{d^\alpha\Phi}{dt^\alpha} = 0, \quad \widetilde{B}_j\left(\xi, \frac{d}{dt}\right)\Phi(0) = \eta_j, \quad j = 1, \ldots, l.$$

Here Φ is a E_0-valued function over $\mathbb{R}_+$, $\widetilde{A}_\alpha(\xi)$ is the homogeneous part of $A_\alpha(x, \xi)$ of degree $m-\alpha$, and $A_\alpha(x, \xi)$ is obtained by replacing $\partial_{y'}$ in $A_\alpha(x, \partial_{y'})$ by $i\xi$ (this condition is the same if it is replaced by $\frac{1}{i}\xi$). The operator $\widetilde{B}_j(\xi, d/dt)$ is defined similarly. The regularity property is equivalent to the fact that there is no nonzero bounded solution on $\mathbb{R}_+$ to the Cauchy problem

$$\frac{d^m\Phi}{dt^m} + \sum_{\alpha=0}^{m-1} \widetilde{A}_\alpha(\xi)\frac{d^\alpha\Phi}{dt^\alpha} = 0, \quad \widetilde{B}_j\left(\xi, \frac{d}{dt}\right)\Phi(0) = 0, \quad j = 1, \ldots, l. \tag{121}$$

Furthermore, it is equivalent to the fact that there is no nonzero rapidly decreasing solution on $\mathbb{R}_+$ to the Cauchy problem (121), see also [100, (ii') in p. 454]. This condition is usually called the *Lopatinski-Shapiro condition*.

The latter condition can also be stated without using rotations, see, for instance [55, §20.1.1]. The map

$$M_{x,\xi} \ni u \longrightarrow (B_1(x, i\xi + \nu_x\partial_t)u(0), \ldots, B_l(x, i\xi + \nu_x\partial_t)u(0))$$

for $x \in b\Omega$, and $\xi \perp \nu_x$, is bijective. Here $M_{x,\xi}$ is the set of all solutions $u \in C^\infty(\mathbb{R}_+, E_0)$ satisfying

$$P(x, i\xi + \nu_x\partial_t)u(t) = 0 \tag{122}$$

and which are bounded on $\mathbb{R}_+$. Here for a differential operator $P(x, D)$, the notation $P(\xi + \nu\partial_t)$ means that ∂_{x_j} is replaced by $i\xi_j + \nu_j\partial_t, j = 1, \ldots, n$. Equivalently, there is no nonzero rapidly decreasing solution on $\mathbb{R}_+$ to the ODE (122) under the initial condition

$$B_j(x, i\xi + \nu_x\partial_t)u(0) = 0, \qquad j = 1, \ldots, l.$$

Let us check the Shapiro-Lopatinskii condition for $k = 2$.

Proposition 5.13 *Suppose Ω is a smooth domain in $\mathbb{R}^4$. The boundary value problem*

$$\begin{cases} (D_0D_0^* + D_1^*D_1)\psi = 0, \quad on \quad \Omega, \\ D_0^*(\nu)\psi|_{b\Omega} = 0, \\ D_1^*(\nu)D_1\psi|_{b\Omega} = 0, \end{cases}$$

is regular.

Proof Here we check the Lopatinski-Shapiro condition by generalizing the method proposed by Dain in [35], which was also used in [107]. Originally, this method works for operator of type K^*K for some differential operator K of first order, while here our operator has the form $D_0D_0^* + D_1^*D_1$.

Fix a point in the boundary $b\Omega$. Without loss of generality, we assume this point to be the origin. Denote by $\nu \in \mathbb{R}^4$ the unit vector of inner normal to the boundary at the origin. Let $\mathfrak{V}_\nu = \{x \in \mathbb{R}^4 \colon x \cdot \nu > 0\}$ be a half space. For any fixed vector $\xi \perp \nu$, suppose that $u(t)$ is a rapidly decreasing solution on $[0, \infty)$ to the following ODE under the initial condition:

$$\begin{cases} (D_0D_0^* + D_1^*D_1)(i\xi + \nu\partial_t)u(t) = 0, \\ D_0^*(\nu)u(0) = 0, \\ D_1^*(\nu)D_1(i\xi + \nu\partial_t)u(0) = 0. \end{cases} \tag{123}$$

Let us prove that u vanishes. Define a function $U\colon \mathfrak{V}_\nu \to \mathbb{C}^4$ by $U(x) = e^{ix\cdot\xi}u(x\cdot\nu)$ for $x \in \mathfrak{V}_\nu$. Note that for a differential operator $Q = \sum_{j=0}^{3} Q_j\partial_{x_j}$, where the Q_j's are (4×3)-matrices, we have

$$QU(x) = \sum_{j=0}^{3} Q_j \left(i\xi_j u(x\cdot\nu) + \nu_j u'(x\cdot\nu)\right) e^{ix\cdot\xi}.$$

Then it is easy to see that (123) implies

$$\begin{cases} (D_0D_0^* + D_1^*D_1)U(x) = 0, \quad \text{on} \quad \mathfrak{V}_\nu, \\ D_0^*(\nu)U(x)|_{b\mathfrak{V}_\nu} = 0, \\ D_1^*(\nu)D_1U(x)|_{b\mathfrak{V}_\nu} = 0. \end{cases} \tag{124}$$

It is sufficient to show that U vanishes. Consider the interval $I_\xi = \{s\xi \in b\mathfrak{V}_\nu; |s| \leq \frac{\pi}{|\xi|}\}$, the ball $B_\xi = \{y' \in b\mathfrak{V}_\nu; y' \perp \xi, |y'| \leq r\}$ for any fixed $r > 0$, and the domain $\mathfrak{D}_\xi = I_\xi \times B_\xi \times \mathbb{R}_+\nu$, where $\mathbb{R}_+\nu = \{t\nu; t \in \mathbb{R}_+\}$.

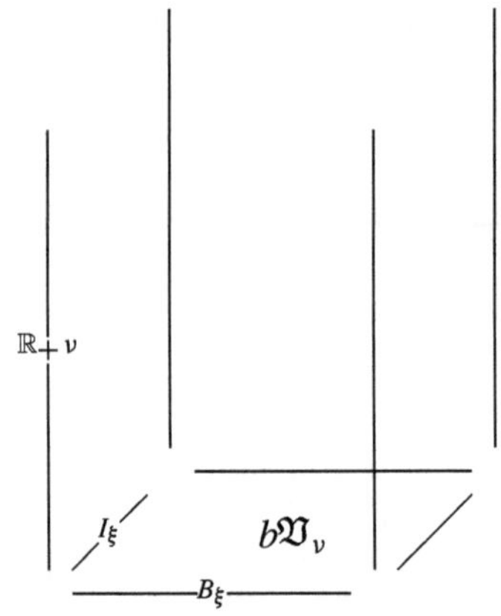

Since $U = e^{ix\cdot\xi}u(x\cdot\nu)$ rapidly decays in direction ν, we have

$$\int_{\mathfrak{D}_\xi}\langle(D_0D_0^*+D_1^*D_1)U,U\rangle = \int_{\mathfrak{D}_\xi}\langle D_0^*U,D_0^*U\rangle + \int_{\mathfrak{D}_\xi}\langle D_1U,D_1U\rangle$$
$$-\int_{I_\xi\times B_\xi\times\{0\}\cup bI_\xi\times B_\xi\times\mathbb{R}_+\nu\cup I_\xi\times bB_\xi\times\mathbb{R}_+\nu}(\langle D_0^*U,D_0^*(\nu)U\rangle \tag{125}$$
$$-\langle D_1^*(\nu)D_1U,U\rangle)dS,$$

by Green's formula (115), where $\langle.,.\rangle$ is the standard Hermitian inner product in $\mathbb{C}^4$. We observe the following:

(1) The integral $\int_{I_\xi\times B_\xi\times\{0\}}$ in Eq. (125) vanishes by the boundary condition $D_0^*(\nu)U = 0$ and $D_1^*(\nu)D_1U = 0$ on $b\mathfrak{V}_\nu$ in (124);
(2) The integral $\int_{bI_\xi\times B_\xi\times\mathbb{R}_+\nu}$ vanishes since U, D_0^*U and D_1U are periodic in direction ξ, and on the opposite surface, the identity $D_j^*(\nu)|_{\{\xi\}\times B_\xi\times\mathbb{R}_+\nu} = -D_j^*(\nu)|_{\{-\xi\}\times B_\xi\times\mathbb{R}_+\nu}$ holds for $j = 0, 1$;
(3) Similarly, the integral $\int_{I_\xi\times bB_\xi\times\mathbb{R}_+\nu}$ vanishes since U, D_0^*U and D_1U are constant in any direction in B_ξ, and as well on the opposite direction. We have the identity $D_j^*(\nu)|_{I_\xi\times\{v\}\times\mathbb{R}_+\nu} = -D_j^*(\nu)|_{I_\xi\times\{-v\}\times\mathbb{R}_+\nu}$ for any $v\in B_\xi$.

Obviously, the integral in the left hand side of (125) vanishes by the first equation in (124). Consequently,

$$\int_{\mathfrak{D}_\xi}\langle D_0^*U,D_0^*U\rangle + \langle D_1U,D_1U\rangle = 0,$$

i.e.,

$$D_0^*U = 0, \qquad D_1U = 0, \qquad \text{on} \quad \mathfrak{V}_\nu. \tag{126}$$

By applying Proposition 5.14 to the convex domain $\mathfrak{V}_\nu$, we see that there exists a function $\widetilde{U} \in C^\infty(\mathfrak{V}_\nu, \mathbb{C}^3)$ such that $D_0\widetilde{U} = U$ on $\mathfrak{V}_\nu$, and so $D_0^* D_0 \widetilde{U} = 0$ by the first identity in (126). By the explicit form of $D_0^* D_0$ in (118), we see that each component of $\widetilde{U}$ is harmonic on $\mathfrak{V}_\nu$. Consequently, each component of $U = D_0\widetilde{U}$ is also harmonic on $\mathfrak{V}_\nu$, since $\Delta U = \Delta D_0 \widetilde{U} = D_0 \Delta \widetilde{U} = 0$ by D_0 being a differential operator with constant coefficients and Δ being a scalar differential operator with constant coefficients. This implies that

$$\begin{cases} \Delta U_0 = \Delta U_1 = \Delta U_2 = \Delta U_3 = 0, \quad \text{on} \quad \mathfrak{V}_\nu, \\ D_0^*(\nu) U|_{b\mathfrak{V}_\nu} = 0, \\ D_1^*(\nu) D_1 U|_{b\mathfrak{V}_\nu} = 0. \end{cases} \tag{127}$$

In particular, when $\nu = (1, 0, 0, 0)$, we have

$$\begin{cases} \Delta U_0 = \Delta U_1 = \Delta U_2 = \Delta U_3 = 0, \quad \text{on} \quad \mathbb{R}^4_+, \\ U_1|_{\mathbb{R}^3} = U_2|_{\mathbb{R}^3} = 0, \\ (U_0 - U_3)|_{\mathbb{R}^3} = 0, \\ \partial_{x_0}(U_0 + U_3)|_{\mathbb{R}^3} = 0, \end{cases}$$

by the boundary conditions (116)–(117) for the upper half space. Note that a harmonic function on $\mathbb{R}^4_+$ with vanishing boundary value must vanish. We see that $U_1 \equiv U_2 \equiv U_0 - U_3 \equiv 0$ and $\partial_{x_0}(U_0 + U_3) \equiv 0$. Consequently, $U_0 + U_3$ is independent of x_0, and so vanishes since it is rapidly decreasing in x_0. Therefore, $U \equiv 0$.

For the general case of ν, we set $\zeta_0 = \nu_0 - i\nu_1$, $\zeta_1 = \nu_2 - i\nu_3$. Then

$$D_0(\nu) = \begin{pmatrix} -\overline{\zeta_1} & -\overline{\zeta_0} & 0 \\ \zeta_0 & -\zeta_1 & 0 \\ 0 & -\overline{\zeta_1} & -\overline{\zeta_0} \\ 0 & \zeta_0 & -\zeta_1 \end{pmatrix}, \quad \text{and} \quad D_1(\nu) = (-\zeta_0, -\overline{\zeta_1}, \zeta_1, -\overline{\zeta_0}). \tag{128}$$

It is direct to check that $D_1(\nu) D_0(\nu) = 0$, that follows from $D_1 D_0 = 0$. Note that

$$\det \begin{pmatrix} -\overline{\zeta_1} & -\overline{\zeta_0} \\ \zeta_0 & -\zeta_1 \end{pmatrix} = |\zeta_0|^2 + |\zeta_1|^2, \tag{129}$$

and therefore $D_0(\nu)$ in (128) has rank 3. The vector $D_1(\nu)$ in (128) does not vanish for non zero ν, i.e., $D_1(\nu)$ has rank 1. Hence, $\operatorname{Image} D_0(\nu) = \ker D_1(\nu)$ and $\operatorname{Image} D_1(\nu)^*$ is a 1-dimensional space orthogonal to $\ker D_1(\nu)$. Namely we have the orthogonal decomposition $\mathbb{C}^4 = \operatorname{Image} D_0(\nu) \oplus \operatorname{Image} D_1(\nu)^*$. We rewrite U as $U = D_0(\nu)U' + D_1(\nu)^* U''$, for some $\mathbb{C}^3$-valued function U' and scalar function

U''. Then,

$$D_0^*(\nu)U = D_0^*(\nu)(D_0(\nu)U' + D_1(\nu)^*U'') = D_0^*(\nu)D_0(\nu)U'. \tag{130}$$

Here $D_0^*(\nu)D_0(\nu)$ is an invertible (3×3)-matrix because $D_0(\nu)$ has rank 3. It follows from $D_0^*(\nu)D_0(\nu)\Delta U' = D_0^*(\nu)\Delta U = 0$ that U' is harmonic. So is U'' if we choose $U''(x)$ orthogonal to $\ker D_1(\nu)^*$ at each point x. The second equation in (127) together with (130) implies that $U' = 0$ on the boundary $b\mathfrak{V}_\nu$, and so it vanishes as a harmonic function on the whole half space $\mathfrak{V}_\nu$. Now we have $U = D_1(\nu)^*U''$.

The third equation in (127) implies that $D_1U|_{b\mathfrak{V}_\nu} = 0$. Then,

$$D_1U = D_1D_1(\nu)^*U'' = (-\partial_{z_0}, -\partial_{\bar{z}_1}, \partial_{z_1}, -\partial_{\bar{z}_0})\, D_1(\nu)^*U''$$

$$= \big(-(\partial_{x_0} - i\partial_{x_1}), -(\partial_{x_2} + i\partial_{x_3}), \partial_{x_2} - i\partial_{x_3}, -(\partial_{x_0} + i\partial_{x_1})\big) \begin{pmatrix} -(\nu_0 + i\nu_1) \\ -(\nu_2 - i\nu_3) \\ \nu_2 + i\nu_3 \\ -(\nu_0 - i\nu_1) \end{pmatrix} U''$$

$$= 2(\nu_0\partial_{x_0} + \nu_1\partial_{x_1} + \nu_2\partial_{x_2} + \nu_3\partial_{x_3})U'' = 2\partial_\nu U'' = 0$$

on the boundary $b\mathfrak{V}_\nu$. As U'' is a harmonic function, we must have $\partial_\nu U'' \equiv 0$ on the whole half space $\mathfrak{V}_\nu$. So U'' is constant in the direction ν. But it is also rapidly decreasing along this direction. Hence $U'' \equiv 0$ on $\mathfrak{V}_\nu$. Thus U vanishes on $\mathfrak{V}_\nu$. □

The Solvability of the Non-homogeneous *k*-Cauchy-Fueter Equations on Convex Domains Without Estimate

The following proposition is proved in [109] for any dimension by using twistor transformations. Here we give an elementary proof.

Proposition 5.14 *The sequence*

$$C^\infty(\Omega, \mathbb{C}^3) \xrightarrow{D_0} C^\infty(\Omega, \mathbb{C}^4) \xrightarrow{D_1} C^\infty(\Omega, \mathbb{C}^1),$$

is exact for any convex domain Ω. Namely, for any $\psi \in C^\infty(\Omega, \mathbb{C}^4)$ satisfying $D_1\psi = 0$, there exists $\phi \in C^\infty(\Omega, \mathbb{C}^3)$ such that $D_0\phi = \psi$ on Ω.

Let $\mathfrak{E}(\Omega)$ be the set of C^∞ functions on Ω and let $\mathfrak{R}$ be the ring of polynomials $\mathbb{C}[\xi_0, \xi_1, \ldots, \xi_n]$. For a positive integer p, $\mathfrak{R}^p$ denotes the space of all vectors $(f_1, \ldots, f_p)^t$ with $f_1, \ldots, f_p \in \mathfrak{R}$, and $\mathfrak{E}^p(\Omega)$ is defined similarly. The following result is essentially due to Ehrenpreis-Malgrange-Palamodov.

Theorem 5.15 ([79, Theorem A]) *Let $A(\xi)$, $B(\xi)$ be respectively $(q \times p)$ and $(r \times q)$ matrices of polynomials, and let $A(D)$ and $B(D)$ be differential operators* *obtained by substituting* ∂_{x_j} *to* $\frac{1}{i}\xi_j$ *to* $A(\xi)$ *and* $B(\xi)$*, respectively. Then the following statements are equivalent:*

(1) *the sequence* $\mathfrak{R}^p \xleftarrow{A(\xi)^t} \mathfrak{R}^q \xleftarrow{B(\xi)^t} \mathfrak{R}^r$ *is exact,*

(2) *the sequence* $\mathfrak{E}^p(\Omega) \xrightarrow{A(D)} \mathfrak{E}^q(\Omega) \xrightarrow{B(D)} \mathfrak{E}^r(\Omega)$ *is exact for any convex and non empty domain* $\Omega \subset \mathbb{R}^{n+1}$.

We also claim that the sequence

$$0 \leftarrow \mathbb{C}^3 \xleftarrow{D_0(\xi)^t} \mathbb{C}^4 \xleftarrow{D_1(\xi)^t} \mathbb{C}^1 \leftarrow 0 \tag{131}$$

is exact for any nonzero $\xi \in \mathbb{C}^4$. To show it we set $\eta_0 = \xi_0 - i\xi_1$, $\eta_1 = \xi_2 - i\xi_3$. Then

$$D_0(\xi)^t = \frac{1}{i}\begin{pmatrix} -\overline{\eta_1} & \eta_0 & 0 & 0 \\ -\overline{\eta_0} & -\eta_1 & -\overline{\eta_1} & \eta_0 \\ 0 & 0 & -\overline{\eta_0} & -\eta_1 \end{pmatrix}, \qquad D_1(\xi)^t = \frac{1}{i}\begin{pmatrix} -\eta_0 \\ -\overline{\eta_1} \\ \eta_1 \\ -\overline{\eta_0} \end{pmatrix}. \tag{132}$$

The proof of Image $D_1(\xi)^t = \ker D_0(\xi)^t$ is similar to the paragraph below (129).

Proposition 5.16 *The sequence* $\mathfrak{R}^3 \xleftarrow{D_0(\xi)^t} \mathfrak{R}^4 \xleftarrow{D_1(\xi)^t} \mathfrak{R}^1$ *is exact.*

Proof It is obvious that $D_0(\xi)^t D_1(\xi)^t = 0$ by (132). Suppose $D_0(\xi)^t \begin{pmatrix} p_1(\xi) \\ \vdots \\ p_4(\xi) \end{pmatrix} = 0$, where p_j are polynomials. Since the sequence (131) is exact, for each $\xi \neq 0$, there exists an element of $\mathbb{C}^1$, say f_ξ, such that

$$\begin{pmatrix} p_1(\xi) \\ \vdots \\ p_4(\xi) \end{pmatrix} = D_1(\xi)^t f_\xi = \frac{1}{i}\begin{pmatrix} -\xi_0 + i\xi_1 \\ -\xi_2 - i\xi_3 \\ \xi_2 - i\xi_3 \\ -\xi_0 - i\xi_1 \end{pmatrix} f_\xi.$$

It follows from the first two equations that $(\xi_0 + i\xi_1)p_1(\xi) + (\xi_2 - i\xi_3)p_2(\xi) = i(\xi_1^2 + \ldots + \xi_4^2)f_\xi$ on $\mathbb{R}^4 \setminus \{0\}$. Then f_ξ is a rational function $Q(\xi)/(\xi_1^2 + \ldots + \xi_4^2)$ for some polynomial $Q(\xi)$. The first equation above implies the following identity of polynomials:

$$ip_1(\xi)(\xi_1^2 + \ldots + \xi_4^2) = (-\xi_0 + i\xi_1)Q(\xi).$$

This equation also holds on $\mathbb{C}^4$ by natural extension of polynomials. By comparison of zero loci, we see that $-\xi_0 + i\xi_1$ must be a factor of $p_1(\xi)$. Namely,

$p_1(\xi) = (-\xi_0 + i\xi_1)q(\xi)$ for some polynomial $q(\xi)$. Consequently, $f_\xi = iq(\xi)$ is a polynomial on $\mathbb{R}^4$. □

Applying Theorem 5.15 to the exact sequence in Proposition 5.16, we get the Proposition 5.14. Analogous considerations can be done for the case $k > 2$, see details in [30].

The Fredholm Property and the Proof of the Main Theorem

Theorem 5.17 ([100, Propositions 11.14 and 11.16], [55, Theorem 20.1.8]) *Suppose that the boundary value problem* (120) *is regular. Then the operator*

$$T: H_{m+s}(\Omega, E_0) \longrightarrow H_m(\Omega, E_1) \oplus \bigoplus_{j=1}^{l} H_{m+s-m_j-\frac{1}{2}}(b\Omega, G_j),$$

$s = 0, 1, \ldots$, *defined by* $Tu = (P(x, D)u, B_1(x, D)u, \ldots, B_l(x, D)u)$ *is Fredholm, and satisfies the estimate*

$$\|u\|^2_{H_{m+s}(\Omega)} \leq C \left(\|Pu\|^2_{H_s(\Omega)} + \sum_{j=1}^{l} \|B_j u\|^2_{H_{m+s-m_j-\frac{1}{2}}(b\Omega)} + \|u\|^2_{H_{m-1}(\Omega)} \right) \tag{133}$$

for some positive constant C. Moreover, the kernel and the space orthogonal to the range consist of smooth functions.

By adding the boundary value condition (109), we consider the closed subspace $H_{s,b}(\Omega, \mathbb{C}^{2k})$ of Sobolev spaces $H_s(\Omega, C^{2k})$ defined by

$$H_{s,b}(\Omega, \mathbb{C}^{2k}) = \left\{ u \in H_s(\Omega, \mathbb{C}^{2k}) \colon D_0^{(k)*}(\nu)u = 0, D_1^{(k)*}(\nu)D_1^{(k)}u = 0 \text{ on } b\Omega \right\},$$

for $s > \frac{3}{2}$. The boundary value conditions above are well defined for $s > \frac{3}{2}$ by the trace theorem.

We know that the associated Laplacian $\Box_1^{(k)}$ is an elliptic operator. We already showed that boundary value problem (109) is regular. So we can apply Theorem 5.17 to obtain the Fredholm operator

$$T: H_{2+s}(\Omega, \mathbb{C}^{2k}) \longrightarrow H_s\left(\Omega, \mathbb{C}^{2k}\right) \oplus H_{s+\frac{3}{2}}\left(b\Omega, \mathbb{C}^{k+1}\right) \oplus H_{s+\frac{1}{2}}\left(b\Omega, \mathbb{C}^{k-1}\right)$$

defined by

$$Tu = \left(\Box_1^{(k)}u, D_0^{(k)*}(\nu)u\Big|_{b\Omega}, D_1^{(k)*}(\nu)D_1^{(k)}u\Big|_{b\Omega} \right). \tag{134}$$

Restricted to the closed subspace $H_{2+s,b}(\Omega, \mathbb{C}^{2k}) \subset H_{2+s}(\Omega, \mathbb{C}^{2k})$, the operator T gets the form $Tu = \left(\Box_1^{(k)} u, 0, 0\right)$ for $u \in H_{2+s,b}(\Omega, \mathbb{C}^{2k})$. Let us prove that the operator

$$\Box_1^{(k)} : H_{2+s,b}(\Omega, \mathbb{C}^{2k}) \longrightarrow H_s(\Omega, \mathbb{C}^{2k}) \tag{135}$$

is Fredholm. Suppose that $\Box_1^{(k)}$ in (135) is not Fredholm. Identifying $H_s(\Omega, \mathbb{C}^{2k})$ with the subspace $\{(f, 0, 0); f \in H_s(\Omega, \mathbb{C}^{2k})\}$ of

$$\mathfrak{W}_s = H_s\left(\Omega, \mathbb{C}^{2k}\right) \oplus H_{s+\frac{3}{2}}\left(b\Omega, \mathbb{C}^{k+1}\right) \oplus H_{s+\frac{1}{2}}\left(b\Omega, \mathbb{C}^{k-1}\right),$$

we see that the kernel of $\Box_1^{(k)}$ is contained in the kernel of the operator T in (134), and so its dimension must be finite. Thus the cokernel of $\Box_1^{(k)}$ should be infinite dimensional. To show that it is not true, we denote by M_0 the subspace of the Hilbert space $H_s(\Omega, \mathbb{C}^{2k})$ orthogonal to the range of $\Box_1^{(k)}$, and denote by M the subspace of the Hilbert space $\mathfrak{W}_s$ orthogonal to the range of T. Note that $H_s(\Omega, \mathbb{C}^{2k})$ is a closed subspace of the Hilbert space $\mathfrak{W}_s$ by the above identification, and the range of T in $\mathfrak{W}_s$ is closed because it is Fredholm. So as the intersection of $H_s(\Omega, \mathbb{C}^{2k})$ and the range of T, the range of $\Box_1^{(k)}$ is also closed. The space M is of finite dimension. Let $\{v_1, \dots v_m\}$ be a basis for M. Vectors $v_1, \dots v_m$ define linear functionals on $\mathfrak{W}_s$, in particular on M_0, by the inner product of $\mathfrak{W}_s$. Because M_0 is infinite dimensional, there must be some nonzero vector $v \in M_0$ in the kernel of these functionals, i.e., orthogonal to M. Consequently, $(v, 0, 0)$ belongs to the range of T. Namely, there exists $u \in H_{2+s}(\Omega, \mathbb{C}^{2k})$ such that $Tu = (v, 0, 0)$. This also implies that $u \in H_{2+s,b}(\Omega, \mathbb{C}^{2k})$ and $\Box_1^{(k)} u = v$, i.e., v is in the range of $\Box_1^{(k)}$. This contradicts to $v \in M_0$. Thus $\Box_1^{(k)}$ has finite dimensional cokernel. The result follows.

We are ready now to prove the main theorems.

Proof of Theorem 5.11 It is sufficient to prove the theorem for $s = 0$. We showed that the map $\Box_1^{(k)}: H_{2,b}(\Omega, \mathbb{C}^{2k}) \to L^2(\Omega, \mathbb{C}^{2k})$ is Fredholm. So its kernel, denoted by $\mathfrak{K}$, is finite dimensional. Denote by $\mathfrak{K}^\perp$ the orthogonal complement to $\mathfrak{K}$ in $H_{2,b}(\Omega, \mathbb{C}^{2k})$ under the inner product of $H_{2,b}(\Omega, \mathbb{C}^{2k})$. Denote by $\mathfrak{R}$ the range of $\Box_1^{(k)}$ in $L^2(\Omega, \mathbb{C}^{2k})$. It is a closed subspace since the cokernel of $\Box_1^{(k)}$ is also finite dimensional. Then $\Box_1^{(k)}: \mathfrak{K}^\perp \to \mathfrak{R}$ is bijective, and so there exists an inverse linear operator $\widetilde{N}_1^{(k)}: \mathfrak{R} \to \mathfrak{K}^\perp$. As the Fredholm operator, $\Box_1^{(k)}: H_{2,b}(\Omega, \mathbb{C}^{2k}) \to \mathfrak{R}$ is bounded, so is its inverse $\widetilde{N}_1^{(k)}$ by the inverse operator theorem. Moreover, $\widetilde{N}_1^{(k)}$ can be extended to a bounded operator $N_1^{(k)}: L^2(\Omega, \mathbb{C}^{2k}) \to \mathfrak{K}^\perp \subset H_{2,b}(\Omega, \mathbb{C}^{2k})$ by setting

$$N_1^{(k)} f = \begin{cases} \widetilde{N}_1^{(k)} f, & \text{if } f \in \mathfrak{R}, \\ 0, & \text{if } f \in \mathfrak{R}^\perp. \end{cases} \tag{136}$$

Moreover, there exists a positive constant C such that

$$\|N_1^{(k)} f\|_{H_2(\Omega,\mathbb{C}^{2k})} \leq C\|f\|_{L^2(\Omega,\mathbb{C}^{2k})} \quad \text{for any} \quad f \in L^2(\Omega,\mathbb{C}^{2k}). \tag{137}$$

Now we can establish the Hodge-type orthogonal decomposition following the ideas in [100, Chap. 5 §9]. By using the identity (119) twice, we see that if $\varphi, \varphi' \in H_{2,b}(\Omega,\mathbb{C}^{2k})$, then

$$\begin{aligned}\left(\Box_1^{(k)}\varphi,\varphi'\right) &= \left((D_0^{(k)}D_0^{(k)*} + D_1^{(k)*}D_1^{(k)})\varphi,\varphi'\right)\\ &= \left(D_0^{(k)*}\varphi, D_0^{(k)*}\varphi'\right) + \left(D_1^{(k)}\varphi, D_1^{(k)}\varphi'\right)\\ &= \left(\varphi,\left(D_0^{(k)}D_0^{(k)*} + D_1^{(k)*}D_1^{(k)}\right)\varphi'\right) = \left(\varphi,\Box_1^{(k)}\varphi'\right),\end{aligned} \tag{138}$$

by $D_0^{(k)*}(\nu)\varphi'|_{b\Omega} = D_1^{(k)*}(\nu)D_1^{(k)}\varphi|_{b\Omega} = 0$, $D_0^{(k)*}(\nu)\varphi|_{b\Omega} = D_1^{(k)*}(\nu)D_1^{(k)}\varphi'|_{b\Omega} = 0$.

We show that $\widetilde{N}_1^{(k)}$ is a self adjoint operator on $\mathfrak{R}$. For any $u, v \in \mathfrak{R}$, we can write $u = \Box_1^{(k)}\varphi$, $v = \Box_1^{(k)}\varphi' \in \mathfrak{R}$ for some $\varphi, \varphi' \in H_{2,b}(\Omega,\mathbb{C}^{2k})$. Then,

$$\left(\widetilde{N}_1^{(k)}u, v\right) = \left(\widetilde{N}_1^{(k)}\Box_1^{(k)}\varphi, \Box_1^{(k)}\varphi'\right) = \left(\varphi, \Box_1^{(k)}\varphi'\right) = \left(\Box_1^{(k)}\varphi, \varphi'\right) = \left(u, \widetilde{N}_1^{(k)}v\right)$$

by using (138). Consequently, $N_1^{(k)}$, as a trivial extension of $\widetilde{N}_1^{(k)}$, is also a self adjoint operator on $L^2(\Omega,\mathbb{C}^{2k})$. Because of the estimate (137), $N_1^{(k)}$ is compact on $L^2(\Omega,\mathbb{C}^{2k})$ by Rellich's theorem. Hence there is an orthonormal basis $\{u_j\}_{j=1}^{\infty}$ for $\mathfrak{R} \subset L^2(\Omega,\mathbb{C}^{2k})$ consisting of eigenfunctions of $N_1^{(k)}$ such that $N_1^{(k)}u_j = \lambda_j u_j$, $\lambda_j \searrow 0$. Here $\lambda_j \neq 0$ since $N_1^{(k)}$ is the inverse of $\Box_1^{(k)}: \mathfrak{K}^{\perp} \to \mathfrak{R}$. We have $u_j \in H_{2,b}(\Omega,\mathbb{C}^{2k})$ for each j by (136). Obviously, $\Box_1^{(k)}u_j = \frac{1}{\lambda_j}u_j$. Then any element of $\mathfrak{K}^{\perp}$ can be written as $\sum_{j=1}^{\infty}\lambda_j a_j u_j$ for some a_j's with $\sum_{j=1}^{\infty}|a_j|^2 < \infty$. Denote by $u_l^0 \in H_{2,b}(\Omega,\mathbb{C}^{2k})$, $l = 1,\ldots,\dim\mathfrak{K}$, a basis for $\mathfrak{K}$. Then $\{u_j\} \cup \{u_l^0\}$ is a basis for $H_{2,b}(\Omega,\mathbb{C}^{2k})$. Because $C_0^{\infty}(\Omega,\mathbb{C}^{2k}) \subset H_{2,b}(\Omega,\mathbb{C}^{2k})$ and $C_0^{\infty}(\Omega,\mathbb{C}^{2k})$ is dense in $L^2(\Omega,\mathbb{C}^{2k})$, we see that $H_{2,b}(\Omega,\mathbb{C}^{2k})$ is dense in $L^2(\Omega,\mathbb{C}^{2k})$. So $\{u_j\} \cup \{u_l^0\}$ is also a basis for $L^2(\Omega,\mathbb{C}^{2k})$. Consequently,

$$L^2(\Omega,\mathbb{C}^{2k}) = \mathfrak{K} \oplus \mathfrak{R}. \tag{139}$$

If $\psi \in \mathfrak{K}$, then

$$0 = \left(\left(D_0^{(k)}D_0^{(k)*} + D_1^{(k)*}D_1^{(k)}\right)\psi,\psi\right) = \left(D_0^{(k)*}\psi, D_0^{(k)*}\psi\right) + \left(D_1^{(k)}\psi, D_1^{(k)}\psi\right)$$

by using the identity (119) since $\psi \in H_{2,b}(\Omega, \mathbb{C}^{2k})$. Thus $D_0^{(k)*}\psi = 0, D_1^{(k)}\psi = 0$. Note that since functions are $\mathfrak{K}$ is C^∞-smooth functions on Ω, we conclude that $\mathfrak{K} = \mathcal{H}^1_{(k)}(\Omega)$ by applying the elliptic estimate (133).

By the construction of the operator $N_1^{(k)}$ above and the decomposition (139), any $\psi \in H_s(\Omega, \mathbb{C}^{2k})$ admits the Hodge-type decomposition:

$$\psi = \Box_1^{(k)} N_1^{(k)}\psi + \Pi\psi = D_0^{(k)} D_0^{(k)*} N_1^{(k)}\psi + D_1^{(k)*} D_1^{(k)} N_1^{(k)}\psi + \Pi\psi, \tag{140}$$

where Π is the orthonormal projection to $\mathfrak{K} = \mathcal{H}^1_{(k)}(\Omega)$ with respect to the L^2 inner product.

It is sufficient to prove the orthogonality of the first two terms in (140) for smooth functions, since $C^\infty(\overline{\Omega}, \mathbb{C}^{2k})$ is dense in $L^2(\Omega, \mathbb{C}^{2k})$ and operators $D_0^{(k)} D_0^{(k)*} N_1^{(k)}$ and $D_1^{(k)*} D_1^{(k)} N_1^{(k)}$ are both bounded in $L^2(\Omega, \mathbb{C}^{2k})$. The orthogonality follows from

$$\left(D_0^{(k)} D_0^{(k)*} N_1^{(k)}\psi, D_1^{(k)*} D_1^{(k)} N_1^{(k)}\psi\right) = \left(D_1^{(k)} D_0^{(k)} D_0^{(k)*} N_1^{(k)}\psi, D_1^{(k)} N_1^{(k)}\psi\right) = 0,$$

where we used the definition of duality applied for the functions

$$u = D_0^{(k)} D_0^{(k)*} N_1^{(k)}\psi \in H_1(\Omega, \mathbb{C}^{2k}) \quad \text{and} \quad v = D_1^{(k)} N_1^{(k)}\psi \in H_2(\Omega, \mathbb{C}^{2k})$$

when $\psi \in H_1(\Omega, \mathbb{C}^{2k})$, and we also used the equality $D_1^{(k)*}(\nu) D_1^{(k)} N_1^{(k)}\psi|_{b\Omega} = 0$ and $D_1^{(k)} D_0^{(k)} = 0$. The theorem is proved.

Proof of Theorem 5.10 We claim that if $D_1^{(k)}\psi = 0$ and ψ is orthogonal to $\mathcal{H}^1_{(k)}(\Omega)$, then $\phi = D_0^{(k)*} N_1^{(k)}\psi$ satisfies $D_0^{(k)}\phi = \psi$. Under the condition $D_1^{(k)}\psi = 0$, the second term in the decomposition (110) vanishes because

$$\begin{aligned} \left\| D_1^{(k)*} D_1^{(k)} N_1^{(k)}\psi \right\|_{L^2}^2 &= \left(D_1^{(k)*} D_1^{(k)} N_1^{(k)}\psi, D_1^{(k)*} D_1^{(k)} N_1^{(k)}\psi\right) \\ &= \left(\psi, D_1^{(k)*} D_1^{(k)} N_1^{(k)}\psi\right) = \left(D_1^{(k)}\psi, D_1^{(k)} N_1^{(k)}\psi\right) = 0, \end{aligned}$$

by using identity (119). Here ψ, $D_1^{(k)} N_1^{(k)}\psi \in H_s(\Omega, \mathbb{C}^{2k})$, $s \geq 1$, and $N_1^{(k)}\psi \in H_{2+s,b}(\Omega, \mathbb{C}^{2k})$ implies that $D_1^{(k)*}(\nu) D_1^{(k)} N_1^{(k)}\psi|_{b\Omega} = 0$. The second identity comes from the orthogonality in the Hodge-type decomposition (140). The claim follows by $\Pi\psi = 0$.

The estimate (108) follows from the estimate for the solution operator $N_1^{(k)}$ in Theorem 5.11.

Conversely, if $\psi = D_0^{(k)}\phi$ for some $\phi \in H_{s+1}(\Omega, \mathbb{C}^{k+1})$. Then $\psi \perp \mathcal{H}^1_{(k)}(\Omega)$ because of

$$(\psi, u) = \left(D_0^{(k)}\phi, u\right) = \left(\phi, D_0^{(k)*} u\right) = 0 \quad \text{for any} \quad u \in \mathcal{H}^1_{(k)}(\Omega),$$

by using the identity (119), since $D_0^{(k)*}(\nu)u = 0$ on the boundary and u and ϕ are both from $H_1(\Omega, \mathbb{C}^{k+1})$. The proof is finished.

Acknowledgements It is a pleasure to thank the organisers of the school "Analytic, Algebraic and Geometric Aspects of Differential Equations" in Będlewo, Poland, for the invitation, warm hospitality, and extraordinarily interesting school. I am grateful to the Analysis and PDE Group at the University of Bergen for the creative friendly atmosphere and the infinite source of ideas, questions and emotions. I appreciate Anja Eidsheim for helping me to correct the language. Special sincere thanks to my husband Alexander Vasiliev for his infinite patience and support of all my initiatives.

The author has been partially supported by the grant of the Norwegian Research Council #239033/F20, NFR-DAAD project 267630/F10 and EU FP7 IRSES program STREVCOMS, grant no. PIRSES-GA-2013-612669.

References

1. R.A. Adams, J.J. Fournier, *Sobolev Spaces*, 2nd edn. Pure and Applied Mathematics, vol. 140 (Elsevier/Academic, Amsterdam, 2003), 305pp.
2. W. Adams, P. Loustaunau, Analysis of the module determining the properties of regular functions of several quaternionic variables. Pac. J. Math. **196**, 1–15 (2001)
3. W. Adams, P. Loustaunau, V. Palamodov, D. Struppa, Hartogs' phenomenon for polyregular functions and projective dimension of related modules over a polynomial ring. Ann. Inst. Fourier **47**, 623–640 (1997)
4. W. Adams, C. Berenstein, P. Loustaunau, I. Sabadini, D. Struppa, Regular functions of several quaternionic variables and the Cauchy–Fueter complex. J. Geom. Anal. **9**, 1–15 (1999)
5. S. Alesker, Non-commutative linear algebra and plurisubharmonic functions of quaternionic variables. Bull. Sci. Math. **127**(1), 1–35 (2003)
6. N. Arcozzi, A. Baldi, From Grushin to Heisenberg via an isoperimetric problem. J. Math. Anal. Appl. **340**(1), 165–174 (2008)
7. R. Baston, Quaternionic complexes. J. Geom. Phys. **8**, 29–52 (1992)
8. W. Bauer, K. Furutani, C. Iwasaki, Fundamental solution of a higher step Grushin type operator. Adv. Math. **271**, 188–234 (2015)
9. W. Bauer, K. Furutani, C. Iwasaki, The inverse of a parameter family of degenerate operators and applications to the Kohn-Laplacian. Adv. Math. **277**, 283–337 (2015)
10. R. Beals, B. Gaveau, P. Greiner, The Green function of model step two hypoelliptic operators and the analysis of certain tangential Cauchy Riemann complexes. Adv. Math. **121**(2), 288–345 (1996)
11. R. Beals, B. Gaveau, P. Greiner, Complex Hamiltonian mechanics and parametrices for subelliptic Laplacians. I. Bull. Sci. Math. **121**(1), 1–36 (1997)
12. R. Beals, B. Gaveau, P. Greiner, Complex Hamiltonian mechanics and parametrices for subelliptic Laplacians. II. Bull. Sci. Math. **121**(2), 97–149 (1997)
13. R. Beals, B. Gaveau, P. Greiner, Complex Hamiltonian mechanics and parametrices for subelliptic Laplacians. III. Bull. Sci. Math. **121**(3), 195–259 (1997)
14. A. Bellaïche, The tangent space in Sub-Riemannian geometry, in *Sub-Riemannian Geometry*, ed. by A. Bellaïche, J.-J. Risler. Progress in Mathematics, vol. 144 (Birkhäuser, Basel, 1996), 393pp.
15. E. Beltrami, Delle variabili complesse sopra una superficie qualunque. Ann. Mat. Pura Appl. **2**(1), 329–366 (1867)
16. C. Benson, A.H. Dooley, G. Ratcliff, Fundamental solutions for powers of the Heisenberg sub-Laplacian. Ill. J. Math. **37**(3), 455–476 (1993)

17. S. Bernstein, Sur la nature analytique des solutions des équations aux dérivées partielles du second ordre. Math. Ann. (French) **59**(1–2), 20–76 (1904)
18. T. Bieske, Fundamental solutions to the p-Laplace equation in a class of Grushin vector fields. Electron. J. Diff. Equat. **2011**(84), 10pp. (2011)
19. T. Bieske, K. Childers, Generalizations of a Laplacian-type equation in the Heisenberg group and a class of Grushin-type spaces. Proc. Am. Math. Soc. **142**(3), 989–1003 (2014)
20. T. Bieske, J. Gong, The P-Laplace equation on a class of Grushin-type spaces. Proc. Am. Math. Soc. **134**(12), 3585–3594 (2006). (electronic)
21. A. Boggess, *CR Manifolds and the Tangential Cauchy-Riemann Complex.* Studies in Advanced Mathematics (CRC, Boca Raton, FL, 1991), 364pp.
22. J. Bureš, A. Damiano, I. Sabadini, Explicit resolutions for several Fueter operators. J. Geom. Phys. **57**, 765–775 (2007)
23. O. Calin, D.C. Chang, I. Markina, Geometric analysis on H-type groups related to division algebras. Math. Nachr. **282**(1), 44–68 (2009)
24. L. Capogna, D. Danielli, S.D. Pauls, J.T. Tyson, *An Introduction to the Heisenberg Group and the Sub-Riemannian Isoperimetric Problem.* Progress in Mathematics, vol. 259 (Birkhäuser, Basel, 2007), 223pp.
25. D.C. Chang, I. Markina, Quaternion H-type group and differential operator Δ_λ. Sci. China Ser. A **51**(4), 523–540 (2008)
26. D.C. Chang, J.Z. Tie, Estimates for powers of sub-Laplacian on the non-isotropic Heisenberg group. J. Geom. Anal. **10**(4), 653–678 (2000)
27. D.C. Chang, J.Z. Tie, A note on Hermite and subelliptic operators. Acta Math. Sin. (Engl. Ser.) **21**(4), 803–818 (2005)
28. C.H. Chang, D.C. Chang, B. Gaveau, P. Greiner, H.P. Lee, Geometric analysis on a step 2 Grusin operator. Bull. Inst. Math. Acad. Sin. (N.S.) **4**(2), 119–188 (2009)
29. D.C. Chang, I. Markina, W. Wang, On the Cauchy-Szegö kernel for quaternion Siegel upper half-space. Complex Anal. Oper. Theory **7**(5), 1623–1654 (2013)
30. D.C. Chang, I. Markina, W. Wang, On the Hodge-type decomposition and cohomology groups of k-Cauchy-Fueter complexes over domains in the quaternionic space. J. Geom. Phys. **107**, 15–34 (2016)
31. W.L. Chow, Über Systeme von linearen partiellen Differentialgleichungen erster Ordnung. Math. Ann. **117**, 98–105 (1939)
32. F. Colombo, I. Sabadini, F. Sommen, D. Struppa, *Analysis of Dirac Systems and Computational Algebra.* Progress in Mathematical Physics, vol. 39 (Birkhäuser, Boston, 2004), 332pp.
33. F. Colombo, V. Souček, D. Struppa, Invariant resolutions for several Fueter operators. J. Geom. Phys. **56**(7), 1175–1191 (2006)
34. M. Cowling, A.H. Dooley, A. Korányi, F. Ricci, H-type groups and Iwasawa decompositions. Adv. Math. **87**(1), 1–41 (1991)
35. S. Dain, Elliptic systems, in analytical and numerical approaches to mathematical relativity. Lect. Notes Phys. **692**, 117–139 (2006)
36. D. Danielli, N. Garofalo, D.M. Nhieu, Sub-Riemannian calculus on hypersurfaces in Carnot groups. Adv. Math. **215**(1), 292–378 (2007)
37. M.P. do Carmo, *Riemannian Geometry.* Mathematics: Theory & Applications (Birkhäuser, Boston, MA, 1992), 300pp.
38. J. Dou, P. Niu, J. Han, Polar coordinates for the generalized Baouendi-Grushin operator and applications. J. Partial Differ. Equat. **20**(4), 322–336 (2007)
39. J.J. Duistermaat, J.A.C. Kolk, *Lie Groups.* Universitext (Springer, Berlin, 2000), 344pp.
40. M. Eastwood, R. Penrose, R. Wells, Cohomology and massless fields. Commun. Math. Phys. **78**(3), 305–351 (1980)
41. G.B. Folland, A fundamental solution for a subelliptic operator. Bull. Am. Math. Soc. **79**, 373–376 (1973)
42. G.B. Folland, E.M. Stein, *Hardy Spaces on Homogeneous Groups.* Mathematical Notes, vol. 28 (Princeton University Press, University of Tokyo Press, Tokyo, Princeton, NJ, 1982), 285pp.

43. K.O. Friedrichs, On the differentiability of the solutions of linear elliptic differential equations. Commun. Pure Appl. Math. **6**, 299–326 (1953)
44. G. Frobenius, Über die Pfaffsche probleme. J. Reine Angew. Math. **82**, 230–315 (1877)
45. N. Garofalo, D. Vassilev, Strong unique continuation properties of generalized Baouendi-Grushin operators. Commun. Partial Differ. Equat. **32**(4–6), 643–663 (2007)
46. V. Gershkovich, A. Vershik, Nonholonomic manifolds and nilpotent analysis. J. Geom. Phys. **5**(3), 407–452 (1988)
47. P. Gorelli, Fundamental solutions for translation and rotation invariant differential operators on the Heisenberg group. Colloq. Math. **83**(2), 183–200 (2000)
48. P.C. Greiner, E.M. Stein, *Estimates for the* $\bar{\partial}$*-Neumann Problem.* Mathematical Notes, vol. 19 (Princeton University Press, Princeton, NJ, 1977), 195pp.
49. P. Greiner, D. Holcman, Y. Kannai, Wave kernels related to second-order operators. Duke Math. J. **114**(2), 329–386 (2002)
50. M. Gromov, Carnot-Carathéodory spaces seen from within, in *Sub-Riemannian Geometry*, ed. by A. Bellaïche, J.-J. Risler. Progress in Mathematics, vol. 144 (Birkhäuser, Basel, 1996), 393pp.
51. V.V. Grušin, A certain class of hypoelliptic operators. Mat. Sb. (N.S.) **83**(125), 456–473 (1970) (Russian)
52. L. Hörmander, On the theory of general partial differential operators. Acta Math. **94**, 161–248 (1955)
53. L. Hörmander, Hypoelliptic second order differential equations. Acta Math. **119**, 147–171 (1967)
54. L. Hörmander, *Linear Partial Differential Operators.* Classics in Mathematics (Springer, Berlin-New York, 1976), 285pp.
55. L. Hörmander, *The Analysis of Linear Partial Differential Operators III.* Grundlehren der Mathematischen Wissenschaften, vol. 274 (Springer, Berlin, 1994), 525pp.
56. L. Hörmander, *The Analysis of Linear Partial Differential Operators. I. Distribution Theory and Fourier Analysis.* Classics in Mathematics (Springer, Berlin, 2003), 440pp.
57. L. Hörmander, *The Analysis of Linear Partial Differential Operators. II. Differential Operators with Constant Coefficients.* Classics in Mathematics (Springer, Berlin, 2005), 392pp.
58. H. Hueber, A note on the computation of Green functions for sub-Laplacians on Lie algebras of type H, in *Potential Theory (Prague, 1987)* (Plenum, New York, 1988), pp. 121–126
59. D. Husemoller, *Fibre Bundles*, 3rd edn. Graduate Texts in Mathematics, vol. 20 (Springer, New York, 1994), 353pp.
60. Y. Jin, G. Zhang, Fundamental solutions of Kohn sub-Laplacians on anisotropic Heisenberg groups and H-type groups. Can. Math. Bull. **54**(1), 126–140 (2011)
61. F. John, *Plane Waves and Spherical Means Applied to Partial Differential Equations* (Interscience, New York, London, 1955), 172pp.
62. Q.Q. Kang, W. Wang, On Penrose integral formula and series expansion of k-regular functions on the quaternionic space $\mathbb{H}^n$. J. Geom. Phys. **64**, 192–208 (2013)
63. A. Kaplan, Fundamental solutions for a class of hypoelliptic PDE generated by composition of quadratics forms. Trans. Am. Math. Soc. **258**(1), 147–153 (1980)
64. A. Kaplan, On the geometry of groups of Heisenberg type. Bull. Lond. Math. Soc. **15**(1), 35–42 (1983)
65. A.W. Knapp, *Lie Groups Beyond an Introduction*, 2nd edn. Progress in Mathematics, vol. 140 (Birkhäuser, Boston, MA, 2002), 812pp.
66. J.J. Kohn, Boundaries of Complex Manifolds, in *1965 Proceedings of Conference Complex Analysis*, Minneapolis (Springer, Berlin, 1964), pp. 81–94
67. J.J. Kohn, Pseudo-differential operators and non-elliptic problems, in *1969 Pseudo-Differential Operators (C.I.M.E., Stresa)* (Edizioni Cremonese, Rome, 1968), pp. 157–165
68. J.J. Kohn, Pseudo-differential operators and hypoellipticity, in *Partial Differential Equations.* Proceedings of Symposium in Pure Mathematics, vol. XXIII, University of California, Berkeley, CA (1971), pp. 61–69 (American Mathematical Society, Providence, RI, 1973)

69. J.J. Kohn, Lectures on degenerate elliptic problems, in *Pseudo-Differential Operator with Applications*, Bressanone, 1977 (Liguori, Naples, 1978), pp. 89–151
70. J.J. Kohn, L. Nirenberg, Non-coercive boundary value problems. Commun. Pure Appl. Math. **18**, 443–492 (1965)
71. J.J. Kohn, H. Rossi, On the extension of holomorphic functions from the boundary of a complex manifold. Ann. Math. **81**(2), 451–472 (1965)
72. A. Korányi, Geometric properties of Heisenberg-type groups. Adv. Math. **56**(1), 28–38 (1985)
73. A. Korányi, H.M. Reimann, Quasiconformal mappings on the Heisenberg group. Invent. Math. **80**(2), 309–338 (1985)
74. A. Kumar, M.M. Mishra, Powers of sub-Laplacian on step two nilpotent Lie groups. J. Geom. Anal. **23**(3), 1559–1570 (2013)
75. H. Liu, G. Zhang, Realization of quaternionic discrete series on the unit ball in $\mathbb{H}^d$. J. Funct. Anal. **262**, 2979–3005 (2012)
76. R. Montgomery, *A Tour of Subriemannian Geometries, Their Geodesics and Applications*. Mathematical Surveys and Monographs, vol. 91 (American Mathematical Society, Providence, RI, 2002), 259pp.
77. D. Müller, F. Ricci, Analysis of second order differential operators on Heisenberg groups. I. Invent. Math. **101**(3), 545–582 (1990)
78. D. Müller, F. Ricci, Analysis of second order differential operators on Heisenberg groups. II. J. Funct. Anal. **108**(2), 296–346 (1992)
79. M. Nacinovich, Complex analysis and complexes of differential operators. Springer Lect. Notes Math. **950**, 105–195 (1982)
80. A. Nagel, E.M. Stein, S. Wainger, Balls and metrics defined by vector fields. I. Basic properties. Acta Math. **155**(1–2), 103–147 (1985)
81. L.A. Nirenberg, Remarks on strongly elliptic partial differential equations. Commun. Pure Appl. Math. **8**, 649–675 (1955)
82. L.A. Nirenberg, A certain problem of Hans Lewy. I. Uspehi Mat. Nauk **29**(2(176)), 241–251 (1974). (Russian) Translated from the English by Ju.V. Egorov. Collection of articles dedicated to the memory of Ivan Georgievich Petrovskii Georgievich Petrovskii (1901–1973)
83. O.A. Oleĭnik, E.V. Radkevič, *Second Order Equations with Nonnegative Characteristic Form.* (Russian). Mathematical Analysis, 1969 (Russian), pp. 7–252. (errata insert) Akad. Nauk SSSR Vsesojuzn. Inst. Nauchn. i Tehn. Informacii, Moscow, 1971
84. B. O'Neill, *Semi-Riemannian Geometry. With Applications to Relativity*. Pure and Applied Mathematics, vol. 103 (Academic, [Harcourt Brace Jovanovich], New York, 1983), 468pp.
85. J. Peetre, Réctification á l'article "Une caractérisation abstraite des opérateurs différentiels". (French) Math. Scand. **8**, 116–120 (1960)
86. I. Petrowsky, Sur l'analyticité des solutions des systémes d'équations différentielles. (French) Rec. Math. N. S. [Mat. Sbornik] **5**(47), 3–70 (1939)
87. I.R. Porteous, *Clifford Algebras and the Classical Groups*. Cambridge Studies in Advanced Mathematics, vol. 50 (Cambridge University Press, Cambridge, 1995), 295pp.
88. E.V. Radkevič, Hypoelliptic operators with multiple characteristics. (Russian) Mat. Sb. (N.S.) **79**(121), 193–216 (1969)
89. P.K. Rashevskiĭ, About connecting two points of complete nonholonomic space by admissible curve. Uch. Zapiski Ped. Inst. K. Liebknecht **2**, 83–94 (1938)
90. A.A. Sagle, R.E. Walde, *Introduction to Lie Groups and Lie Algebras*. Pure and Applied Mathematics, vol. 51 (Academic, New York, London, 1973), 361pp.
91. L. Schwartz, *Théorie des distributions*. (French) Publications de l'Institut de Mathématique de l'Université de Strasbourg, vol. IX-X (Hermann, Paris, 1966), 420pp.
92. I.M. Singer, J.A. Thorpe, *Lecture Notes on Elementary Topology and Geometry*. Undergraduate Texts in Mathematics (Springer, New York, Heidelberg, 1976), 232pp.

93. E.M. Stein, *Boundary Behaviour of Holomorphic Functions of Several Complex Variables* (Princeton University Press, University of Tokyo Press, Princeton, NJ, Tokyo, 1972), 72pp.
94. E.M. Stein, *Harmonic Analysis: Real-Variable Methods, Orthogonality, and Oscillatory Integrals*, With the assistance of Timothy S. Murphy. Princeton Mathematical Series, vol. 43. Monographs in Harmonic Analysis, III (Princeton University Press, Princeton, NJ, 1993), 695pp.
95. E.M. Stein, R. Shakarchi, *Functional Analysis: Introduction to Further Topics in Analysis.* Princeton Lectures in Analysis, vol. 4 (Princeton University Press, Princeton and Oxford, 2011), 423pp.
96. E.M. Stein, G. Weiss, Introduction to Fourier Analysis on Euclidean Spaces. Princeton Mathematical Series, vol. 32 (Princeton University Press, Princeton, NJ, 1971), 297pp.
97. R.S. Strichartz, Sub-Riemannian geometry. J. Differential Geom. **24**, 221–263 (1986); Correction, J. Differential Geom. **30**, 595–596 (1989)
98. R.S. Strichartz, *A Guide to Distribution Theory and Fourier Transforms* (World Scientific, River Edge, NJ, 2003), 226pp.
99. H.J. Sussmann, Orbits of families of vector fields and integrability of distributions. Trans. Am. Math. Soc. **180**, 171–188 (1973)
100. M. Taylor, *Partial Differential Equations I.* Applied Mathematical Sciences, vol. 115 (Springer, New York, 1996), 563pp.
101. J. Tie, The twisted Laplacian on $\mathbb{C}^n$ and the sub-Laplacian on $\mathbb{H}_n$. Commun. Partial Differ. Equat. **31**(7–9), 1047–1069 (2006)
102. F. Trèves, *Linear Partial Differential Equations with Constant Coefficients: Existence, Approximation and Regularity of Solutions*. Mathematics and its Applications, vol. 6 (Gordon and Breach, New York, London, Paris, 1966), 534pp.
103. J.T. Tyson, Sharp weighted Young's inequalities and Moser-Trudinger inequalities on Heisenberg type groups and Grushin spaces. Potential Anal. **24**(4), 357–384 (2006)
104. M.I. Višik, V.V. Grušin, A certain class of degenerate elliptic equations of higher orders. (Russian) Mat. Sb. (N.S.) **79**(121), 3–36 (1969)
105. M.I. Višik, V.V. Grušin, Boundary value problems for elliptic equations which are degenerate on the boundary of the domain. (Russian) Mat. Sb. (N.S.) **80**(122), 455–491 (1969)
106. M.I. Višik, V.V. Grušin, Elliptic pseudodifferential operators on a closed manifold which are degenerate on a submanifold. (Russian) Dokl. Akad. Nauk SSSR **189**, 16–19 (1969)
107. W. Wang, Korn's inequality and Donati's theorem for the conformal Killing operator on pseudo-Euclidean space. J. Math. Anal. Appl. **345**, 777–782 (2008)
108. W. Wang, On non-homogeneous Cauchy-Fueter equations and Hartogs' phenomenon in several quaternionic variables. J. Geom. Phys. **58**, 1203–1210 (2008)
109. W. Wang, The k-Cauchy-Fueter complex, Penrose transformation and Hartogs phenomenon for quaternionic k-regular functions. J. Geom. Phys. **60**, 513–530 (2010)
110. W. Wang, On the optimal control method in quaternionic analysis. Bull. Sci. Math. **135**, 988–1010 (2011)
111. W. Wang, The tangential Cauchy-Fueter complex on the quaternionic Heisenberg group. J. Geom. Phys. **61**, 363–380 (2011)
112. H.Y. Wang, G.B. Ren, Bochner-Martinelli formula for k-Cauchy-Fueter operator. J. Geom. Phys. **84**, 43–56 (2014)
113. F.W. Warner, *Foundations of Differentiable Manifolds and Lie Groups.* Graduate Texts in Mathematics, vol. 94 (Springer, New York, Berlin, 1983), 272pp.
114. R. Wells, J. Wolf, Complex manifolds and mathematical physics. Bull. Am. Math. Soc. **1**(2), 296–336 (1979)
115. H. Weyl, The method of orthogonal projection in potential theory. Duke Math. J. **7**, 411–444 (1940)

Asymptotic Analysis and Summability of Formal Power Series

Javier Sanz

Abstract For many problems (ODEs, PDEs, difference equations, etc.) it makes sense to look for formal power series solutions which, if found, could well be divergent. However, these formal solutions will frequently have an asymptotic meaning, being representations, in a precise sense, of actual, analytic solutions of the corresponding problem. Summability techniques aim at reconstructing such proper solutions from the formal ones. We will present a slight extension of the successful and well-known technique of k-summability in a direction of the complex plane, put forward by J.-P. Ramis and which was the building block for multisummability, a procedure able to sum any formal solution to a system of meromorphic ordinary differential equations at an irregular singular point. The extension concerns the consideration of Carleman ultraholomorphic classes in sectors, more general than the Gevrey classes appearing in Ramis' theory, and which consist of holomorphic functions whose derivatives' growth is governed in terms of a sequence of real numbers, say $\mathbb{M}$. Whenever $\mathbb{M}$ is subject to standard conditions, flat functions in the class are constructed on sectors of optimal opening and, resting on the work of W. Balser on moment summability methods, suitable kernels and Laplace and Borel-type transforms are introduced which lead to a tractable concept of $\mathbb{M}$-summability. We will comment on some applications of this tool to the study of the summability properties of formal solutions to some classes of so-called moment differential equations.

Keywords Asymptotic expansions • Carleman ultraholomorphic classes • Integral transforms • Summability of formal power series

Mathematics Subject Classification (2000). Primary 40C10; Secondary 35C15, 35C20, 34K25

J. Sanz (✉)
Departamento de Álgebra, Análisis Matemático, Geometría y Topología, Instituto de Investigación en Matemáticas de la Universidad de Valladolid (IMUVA), Facultad de Ciencias, Universidad de Valladolid, Paseo de Belén 7, 47011 Valladolid, Spain
e-mail: jsanzg@am.uva.es

G. Filipuk et al. (eds.), *Analytic, Algebraic and Geometric Aspects of Differential Equations*, Trends in Mathematics, DOI 10.1007/978-3-319-52842-7_4

1 Introduction

The study of the existence and meaning of formal power series solutions to differential or difference equations has a long history, going back at least to the works of L. Euler in the eighteenth century. Although these solutions are frequently divergent, under fairly general conditions the rate of growth of their coefficients is not arbitrary. Indeed, a remarkable result of Maillet [35] in 1903 states the following:

Let $G(x, Y, Y_1, \dots, Y_n)$ be an analytic function at the origin in $n+2$ variables and $\hat{f} = \sum_{p\geq 0} a_p z^p$ be a formal power series solution of the differential equation

$$G(x, y, y', \dots, y^{(n)}) = 0.$$

Then, there exist $C, A, k > 0$ such that $|a_p| \leq CA^p(p!)^{1/k}$ for every $p \geq 0$.

The series subject to these kind of estimates are named of Gevrey type of order $1/k$. In order to give an analytical meaning to these formal solutions, appearing in this and many different contexts, a whole theory of asymptotic expansions has been developed during the past century, both in the general and in the Gevrey sense, and these series turn out to be asymptotic representations (in a sense to be made precise) of actual solutions defined in suitable domains. Moreover, generically there is the possibility of reconstructing such analytic solutions from the formal ones by a process known as multisummability, developed in the 1980s by J.-P. Ramis, J. Écalle, W. Balser, et al. The following result by Braaksma [9] in 1992 is certainly one of the major achievements in this body of knowledge:

Every formal power series solution to a nonlinear meromorphic system of ordinary differential equations at an irregular singular point is multisummable.

A fundamental tool in this respect is that of k-summability, and the main aim of this survey is to provide an introduction to a slight generalization of this concept, whose applicability is currently under study. The extension concerns the consideration of series, respectively asymptotic expansions, whose coefficients, resp. remainders, are governed by a sequence more general than, but sharing its fundamental properties with, the powers of the factorial.

The Carleman ultraholomorphic classes $\mathcal{A}_{\mathbb{M}}(S)$ in a sector S of the Riemann surface of the logarithm consist of those holomorphic functions f in S whose derivatives of order $p \geq 0$ are uniformly bounded there by, essentially, the values $p!M_p$, where $\mathbb{M} = (M_p)_{p\in\mathbb{N}_0}$ is a sequence of positive real numbers. In case bounds are not uniform on S but are valid and depend on every proper subsector of S to which the function is restricted, we obtain the class $\tilde{\mathcal{A}}_{\mathbb{M}}(S)$ of functions with a (non-uniform) $\mathbb{M}$-asymptotic expansion at 0 in S, given by a formal power series $\hat{f} = \sum_{p\geq 0} a_p z^p/p!$ whose coefficients are again suitably bounded in terms of $\mathbb{M}$ (we write $f \sim_{\mathbb{M}} \hat{f}$ and $(a_p)_{p\in\mathbb{N}_0} \in \Lambda_{\mathbb{M}}$). The map sending f to $(a_p)_{p\in\mathbb{N}_0}$ is the asymptotic Borel map $\tilde{\mathcal{B}}$, and f is said to be flat if $\tilde{\mathcal{B}}(f)$ is the null sequence. See Sect. 3 for the precise definitions of all these classes and concepts.

In order to obtain good properties for these classes, the sequence $\mathbb{M}$ is usually subject to some standard conditions; in particular, we will mainly consider strongly regular sequences as defined by Thilliez [57], see Sect. 2. The best known example is that of Gevrey classes, appearing when the sequence is chosen to be $\mathbb{M}_\alpha = (p!^\alpha)_{p\in\mathbb{N}_0}$, $\alpha > 0$, and for which we use the notations $\mathcal{A}_\alpha(S)$, $\tilde{\mathcal{A}}_\alpha(S)$, Λ_α, $f \sim_\alpha \hat{f}$ and so on, for simplicity. Let us denote by S_γ the sector bisected by the direction $d = 0$ and with opening $\pi\gamma$. It is well known that $\tilde{\mathcal{B}} : \tilde{\mathcal{A}}_\alpha(S_\gamma) \to \Lambda_\alpha$ is surjective if, and only if, $\gamma \leq \alpha$ (Borel–Ritt–Gevrey theorem, see [41, 50, 51], [1, Theorem 2.2.1]). It is natural to call this an extension result, and to think then about the possibility of obtaining linear and continuous right inverses for $\tilde{\mathcal{B}}$ in suitably topologized classes. On the other hand, $\tilde{\mathcal{B}}$ is injective (i.e., $\tilde{\mathcal{A}}_\alpha(S_\gamma)$ does not contain nontrivial flat functions, and then the class $\tilde{\mathcal{A}}_\alpha(S_\gamma)$ is said to be quasianalytic) if, and only if, $\gamma > \alpha$ (Watson's lemma, see for example [2, Proposition 11]). Our main aim in the first part of this exposition is to provide generalizations of this kind of results in the framework of Carleman ultraholomorphic classes associated with strongly regular sequences inducing a proximate order. We will mainly follow the work [54]. Let us start with an overview of the existing literature in this respect.

In 1995 Thilliez [56] obtained right inverses in the Gevrey case when $\gamma < \alpha$ by applying techniques from the ultradifferentiable setting (i.e. regarding extension results for classes of smooth functions on open subsets of $\mathbb{R}^n$, determined by imposing a suitable growth of the derivatives), and the same was done by the author in [53] by adapting the truncated Laplace transform procedure already used by J.P. Ramis in Borel–Ritt–Gevrey theorem [50]. Regarding general classes, Schmets and Valdivia [55] extended some results of Petzsche [49] for ultradifferentiable classes, and applied them in order to provide the first powerful results in the present framework. Subsequently, Thilliez [57] improved the results in [55] in several respects (see Sect. 3.1 in his paper for the details) by relying on a double application of suitable Whitney's extension results for Whitney ultradifferentiable jets on compact sets with Lipschitz boundary appearing in [8, 14, 49]. In particular, he introduced a growth index $\gamma(\mathbb{M}) \in (0, \infty)$ for every strongly regular sequence $\mathbb{M}$ (which for $\mathbb{M}_\alpha$ equals α), and proved the following facts: if $\gamma < \gamma(\mathbb{M})$, then $\mathcal{A}_{\mathbb{M}}(S_\gamma)$ is not quasianalytic, and there exist right inverses for $\tilde{\mathcal{B}}$, which are obtained due to the explicit construction of nontrivial flat functions in the class $\mathcal{A}_{\mathbb{M}}(S_\gamma)$. Indeed, these flat functions allowed Lastra et al. [29] to define suitable kernels and moment sequences by means of which to obtain again right inverses by the classical truncated Laplace transform technique. Because of the integral form of the solution, this procedure admits an easy generalization to the several variable case, and does not rest on any result from the ultradifferentiable setting.

However, the preceding results for general classes are not fully satisfactory. Firstly, the equivalences stated in Borel–Ritt–Gevrey theorem and Watson's lemma for the Gevrey case are now only one-side implications. Secondly, and strongly related to the previous remark, the need to restrict the opening of the sector S_γ to $\gamma < \gamma(\mathbb{M})$ in order to obtain flat functions in $\mathcal{A}_{\mathbb{M}}(S_\gamma)$ does not allow one to treat the apparently limit situation in which $\gamma = \gamma(\mathbb{M})$. Note that, in the Gevrey case, the function $e^{-z^{-1/\alpha}}$ is flat in the class $\tilde{\mathcal{A}}_\alpha(S_\alpha)$, and of course in every $\mathcal{A}_\alpha(S_\gamma)$ for $\gamma < \alpha$. So, our main objective will be to obtain flat functions in sectors of optimal opening.

In this sense, we first introduce for every strongly regular sequence $\mathbb{M}$ a new constant $\omega(\mathbb{M})$, measuring the rate of growth of the sequence $\mathbb{M}$, in terms of which quasianalyticity in the classes $\mathcal{A}_{\mathbb{M}}(S_\gamma)$ may be properly characterized due to a classical result of Korenbljum ([25]; see Theorem 4.3). This constant is easily computed in concrete situations [see (14)], and indeed it is the inverse of the order of growth of the classical function $M(t)$ associated with $\mathbb{M}$, namely $M(t) = \sup_{p\in\mathbb{N}_0}\log(t^p/M_p)$, $t > 0$ [see (15)]. Regarding the construction of flat functions, Thilliez [58] had characterized flatness in $\tilde{\mathcal{A}}_{\mathbb{M}}(S_\gamma)$ in terms of the existence of non-uniform estimates governed by the function $e^{-M(1/|z|)}$, much in the same way as the function $e^{-z^{-1/\alpha}}$ expresses flatness in the Gevrey case. So, it became clear to us the need to construct functions in sectors whose growth is accurately specified by the function $M(t)$. The classical theory of growth for holomorphic functions defined in sectorial regions, based on the notion of (constant) exponential order, showed itself not profound enough to deal with the general case. Luckily, the theory of proximate orders, allowing to change the constant order $\rho > 0$ into a function $\rho(r)$ more closely specifying the desired rate of growth, is available since the 1920s, with results by E. Lindelof, G. Valiron, V. Bernstein, M.M. Dzhrbashyan, M.A. Evgrafov, A.A. Gol'dberg, I.V. Ostrovskii (see [17, 33]) and, in our regards, mainly Maergoiz [34], which have been the key for our success. The problem of characterizing those sequences $\mathbb{M}$ associated with a proximate order has been solved (Proposition 4.28), and it turns out that all the interesting examples of strongly regular sequences appearing in the literature belong to this class. Whenever this is the case, the results of L.S. Maergoiz allow us to obtain the desired flat functions in $\tilde{\mathcal{A}}_{\mathbb{M}}(S_{\omega(\mathbb{M})})$ (see Theorem 4.26) and, immediately, we may generalize Watson's lemma, see Corollary 4.30. Subsequently, in Sect. 5 suitable kernels and moment sequences are introduced, by means of which we may prove that $\tilde{\mathcal{B}}$ is surjective in $\tilde{\mathcal{A}}_{\mathbb{M}}(S_\gamma)$ if, and only if, $\gamma \le \omega(\mathbb{M})$, so generalizing Borel–Ritt–Gevrey theorem (see Theorem 5.11).

It should be mentioned that for the standard strongly regular sequences appearing in the literature, the value of the constants $\gamma(\mathbb{M})$ and $\omega(\mathbb{M})$ agree, and the equality remains true as long as $\mathbb{M}$ induces a proximate order, as it has been recently obtained by J. Jiménez-Garrido and the author (see Proposition 4.28, and [22] for its proof). We have $\gamma(\mathbb{M}) \le \omega(\mathbb{M})$ in general, and $\gamma(\mathbb{M}) < \omega(\mathbb{M})$ can actually occur, as it has been proved in [23], so for some sequences our results definitely improve those of V. Thilliez by enlarging the sectors for which non-quasianalyticity holds. In any case, the equivalences stated in Theorem 5.11 and Corollary 4.30 are new. To end this topic, in Theorem 5.13 we gather the information concerning the existence of right inverses for $\tilde{\mathcal{B}}$ in $\mathcal{A}_{\mathbb{M}}(S_\gamma)$ as long as $\mathbb{M}$ induces a proximate order: they exist whenever $\gamma < \omega(\mathbb{M})$, and their existence, under some specific condition (satisfied, for instance, in the Gevrey case), implies that $\gamma < \omega(\mathbb{M})$.

The aim of the second part of this survey is to put forward a concept of summability of formal (i.e. divergent in general) power series in the framework of general Carleman ultraholomorphic classes in sectors, so generalizing the by-now classical and powerful tool of k-summability of formal Gevrey power series, introduced by Ramis [50, 51]. We will follow a paper of Lastra et al. [31].

Observe that the injectivity of the asymptotic Borel map $\tilde{\mathcal{B}}$ in $\tilde{\mathcal{A}}_{\mathbb{M}}(G)$, for a sectorial region G, means that every function f in such a class is determined by its asymptotic expansion $\hat{f}$, and it makes sense to call f the sum of $\hat{f}$; this is the idea behind summability methods in this context.

The classical Watson's Lemma is the departure point for the definition of $1/\alpha$-summability in a direction. The paradigmatic example of flat function in $\tilde{\mathcal{A}}_\alpha(S_\alpha)$ is $f_\alpha(z) = \exp(-z^{-1/\alpha})$, and it gives rise to kernels in terms of which one may define formal and analytic Laplace and Borel transforms permitting the reconstruction of the sum of a given Gevrey formal power series belonging to $\tilde{\mathcal{B}}(\tilde{\mathcal{A}}_\alpha(S_\gamma))$ for some $\gamma > \alpha$. The technique of multisummability (in a sense, an iteration of a finite number of $1/\alpha$-summability procedures) has been proven to apply successfully to a plethora of situations concerning the study of formal power series solutions at a singular point for linear and nonlinear (systems of) meromorphic ordinary differential equations in the complex domain (see, to cite but a few, the works [1, 2, 7, 9, 42, 52]), for partial differential equations (for example, [3, 4, 18, 38, 48]), as well as for singular perturbation problems (see [5, 12, 30], among others).

However, it is known that non-Gevrey formal power series solutions may appear for different kinds of equations. For example, Thilliez has proven some results on solutions within these general classes for algebraic equations in [58]. Also, Immink in [20, 21] has obtained some results on summability for solutions of difference equations whose coefficients grow at an intermediate rate between Gevrey classes, called of 1^+ level, that is governed by a strongly regular sequence. Very recently, Malek [39] has studied some singularly perturbed small step size difference-differential nonlinear equations whose formal solutions with respect to the perturbation parameter can be decomposed as sums of two formal series, one with Gevrey order 1, the other of 1^+ level, a phenomenon already observed for difference equations [10].

All these results invite one to try to extend summability tools so that they are able to deal with formal power series whose coefficients' growth is controlled by a general strongly regular sequence, so including Gevrey, 1^+ level and other interesting examples. Our approach will be inspired by the study of moment summability methods, equivalent in a sense to $1/\alpha$-summability, developed by W. Balser in [2, Sect. 5.5] and which relies on the determination of a pair of kernel functions with suitable asymptotic and growth properties, in terms of which to define formal and analytic Laplace- and Borel-like transforms. These summability methods have already found its application to the analysis of formal power series solutions of different classes of partial differential equations (for example, by Malek [36, 37] and by Michalik [43, 44]), and also for so-called moment-partial differential equations, introduced by Balser and Yoshino [6] and subsequently studied by Michalik [45–47].

Thanks to the results in the first sections, a definition of $\mathbb{M}$-summability in a direction d may be easily put forward, see Definition 6.1.

Since the existence of flat functions in $S_{\omega(\mathbb{M})}$ is known whenever $\mathbb{M}$ induces a proximate order, we devote Sect. 6 to the introduction of kernels of $\mathbb{M}$-summability (see Definition 6.2), and the associated formal and analytic transforms, in terms of which to reconstruct the sums of $\mathbb{M}$-summable formal power series in a direction, as stated in Theorem 6.18. Once our tool has been designed, it is necessary to test its applicability to the study of formal solutions of different types of algebraic and differential equations in the complex domain. Our first attempt is contained in the last section. The notion of formal moment-differential operator was firstly introduced by Balser and Yoshino in [6]. Generally speaking, given the sequence of moments $\mathfrak{m}_e = (m_e(p))_{p\in\mathbb{N}_0}$ of a kernel function e of order $k > 0$ (in other words and according to Remark 6.19(i), a kernel for $\mathbb{M}_{1/k}$-summability), one can define $\partial_{\mathfrak{m}_e,z}$ as an operator from $\mathbb{C}[[z]]$ into itself given by

$$\partial_{\mathfrak{m}_e,z}\left(\sum_{p\geq 0}\frac{f_p}{m_e(p)}z^p\right) = \sum_{p\geq 0}\frac{f_{p+1}}{m_e(p)}z^p,$$

in much the same way as, for the usual derivative ∂, one has $\partial\left(\sum_{p\geq 0}\frac{f_p}{p!}z^p\right) = \sum_{p\geq 0}\frac{f_{p+1}}{p!}z^p$. For two sequences of moments $\mathfrak{m}_1 = (m_1(p))_{p\in\mathbb{N}_0}$, of Gevrey order k_1, and $\mathfrak{m}_2 = (m_2(p))_{p\in\mathbb{N}_0}$, of Gevrey order k_2, they study the Gevrey order of the formal power series solutions of an inhomogeneous moment-partial differential equation with constant coefficients in two variables,

$$p(\partial_{\mathfrak{m}_1,t},\partial_{\mathfrak{m}_2,z})\hat{u}(t,z) = \hat{f}(t,z),$$

where $p(\lambda,\xi)$ is a given polynomial. Subsequently, Michalik [45] considers the corresponding initial value problem

$$P(\partial_{\mathfrak{m}_1,t},\partial_{\mathfrak{m}_2,z})u(t,z) = 0, \qquad \partial^j_{\mathfrak{m}_1,t}u(0,z) = \varphi_j(z) \quad \text{for } j = 0,\dots,n-1,$$

where $P(\lambda,\xi)$ is a polynomial of degree n with respect to λ and the Cauchy data are analytic in a neighborhood of $0 \in \mathbb{C}$. A formal solution $\hat{u}$ is constructed, and a detailed study is made of the relationship between the summability properties of $\hat{u}$ and the analytic continuation properties and growth estimates for the Cauchy data. We will generalize his results for strongly regular moment sequences of a general kernel of summability. A significant part of the statements are given without proof, since the arguments do not greatly differ from those in [45]. On the other hand, complete details are provided when the differences between both situations are worth stressing.

The interested reader may find in [32] a recent application of the tools described here to the asymptotic study of the solutions of a class of singularly perturbed partial differential equations in whose coefficients there appear sums of formal power series in this generalized sense.

Notation

We set $\mathbb{N} := \{1, 2, \ldots\}$, $\mathbb{N}_0 := \mathbb{N} \cup \{0\}$. $\mathcal{R}$ stands for the Riemann surface of the logarithm, and $\mathbb{C}[[z]]$ is the space of formal power series in z with complex coefficients.

For $\gamma > 0$, we consider unbounded sectors

$$S_\gamma := \{z \in \mathcal{R} : |\arg(z)| < \frac{\gamma\pi}{2}\}$$

or, in general, bounded or unbounded sectors

$$S(d, \alpha, r) := \{z \in \mathcal{R} : |\arg(z) - d| < \frac{\alpha\pi}{2},\ |z| < r\},$$

$$S(d, \alpha) := \{z \in \mathcal{R} : |\arg(z) - d| < \frac{\alpha\pi}{2}\}$$

with bisecting direction $d \in \mathbb{R}$, opening $\alpha\pi$ and (in the first case) radius $r \in (0, \infty)$.

A sectorial region $G(d, \alpha)$ with bisecting direction $d \in \mathbb{R}$ and opening $\alpha\pi$ will be a domain in $\mathcal{R}$ such that $G(d, \alpha) \subset S(d, \alpha)$, and for every $\beta \in (0, \alpha)$ there exists $\rho = \rho(\beta) > 0$ with $S(d, \beta, \rho) \subset G(d, \alpha)$. In particular, sectors are sectorial regions.

A sector T is a bounded proper subsector of a sectorial region G (denoted by $T \ll G$) whenever the radius of T is finite and $\overline{T} \setminus \{0\} \subset G$. Given two unbounded sectors T and S, we say T is an unbounded proper subsector of S, and we write $T \prec S$, if $\overline{T} \setminus \{0\} \subset S$.

$D(z_0, r)$ stands for the disk centered at z_0 with radius $r > 0$.

For an open set $U \subset \mathcal{R}$, $\mathcal{O}(U)$ denotes the set of holomorphic functions defined in U.

$\Re(z)$ stands for the real part of a complex number z, and we write $\lfloor x \rfloor$ for the integer part of $x \in \mathbb{R}$, i.e. the greatest integer not exceeding x.

2 Strongly Regular Sequences: Associated Functions and Growth Index

Most of the information in this subsection is taken from the works of Chaumat and Chollet [14], Goldberg and Ostrovskii [17], Komatsu [24] and Thilliez [57], which we refer to for further details. In what follows, $\mathbb{M} = (M_p)_{p \in \mathbb{N}_0}$ will stand for a sequence of positive real numbers, and we will always assume that $M_0 = 1$.

Definition 2.1 We say:

(i) $\mathbb{M}$ is *logarithmically convex* (for short, (lc)) if

$$M_p^2 \le M_{p-1} M_{p+1}, \qquad p \in \mathbb{N}.$$

(ii) $\mathbb{M}$ is *derivation closed* (for short, (dc)) if there exists $A > 0$ such that

$$M_{p+1} \le A^{p+1} M_p, \qquad p \in \mathbb{N}_0.$$

(iii) $\mathbb{M}$ is of *moderate growth* (briefly, (mg)) if there exists $B > 0$ such that

$$M_{p+q} \le B^{p+q} M_p M_q, \qquad p, q \in \mathbb{N}_0.$$

(iv) $\mathbb{M}$ satisfies the *non-quasianalyticity condition* (briefly, (nq)) if

$$\sum_{q \ge 0} \frac{M_q}{(q+1)M_{q+1}} < \infty.$$

(v) $\mathbb{M}$ satisfies the *strong non-quasianalyticity condition* [for short, (snq)] if there exists $C > 0$ such that

$$\sum_{q \ge p} \frac{M_q}{(q+1)M_{q+1}} \le C \frac{M_p}{M_{p+1}}, \qquad p \in \mathbb{N}_0.$$

The terminology will be explained by some of the following comments or results. Obviously, (mg) implies (dc), and (snq) implies (nq).

Definition 2.2 (Thilliez [57]) We say $\mathbb{M}$ is *strongly regular* if it verifies (lc), (mg) and (snq).

Remark 2.3 In the literature a different set of conditions appears frequently when dealing with ultraholomorphic or ultradifferentiable classes of functions. Let us clarify the relationship between these two approaches. H. Komatsu (see [24, 40]) uses the following terminology: for him, (lc) is (M.1); (dc) is (M.2)'; (mg) is (M.2); and he introduces:

(i) $\mathbb{M}$ verifies (M.3)' if

$$\sum_{q \ge 0} \frac{M_q}{M_{q+1}} < \infty.$$

(ii) $\mathbb{M}$ satisfies (M.3) if there exists $C > 0$ such that

$$\sum_{q \ge p} \frac{M_q}{M_{q+1}} \le C \frac{pM_p}{M_{p+1}}, \qquad p \in \mathbb{N}_0.$$

Again it is clear that (M.3) implies (M.3)'. Moreover, if $\mathbb{M}$ is strongly regular, then $\mathbb{M}^* := (p!M_p)_{p \in \mathbb{N}_0}$ verifies the conditions (M.1), (M.2) and (M.3) of H. Komatsu. On the other hand, if a sequence of positive real numbers $\mathbb{M}^* = (M_p^*)_{p \in \mathbb{N}_0}$, with

$M_0^* = 1$, verifies (M.2) and (M.3), and moreover $\mathbb{M} := (M_p^*/p!)_{p\in\mathbb{N}_0}$ is (lc), then $\mathbb{M}$ is strongly regular.

Definition 2.4 For every sequence $\mathbb{M} = (M_p)_{p\in\mathbb{N}_0}$ the associated *sequence of quotients* $\boldsymbol{m} = (m_p)_{p\in\mathbb{N}_0}$ is defined by

$$m_p := \frac{M_{p+1}}{M_p}, \qquad p \in \mathbb{N}_0.$$

Remark 2.5 The properties (lc), (dc), (nq) and (snq) can be easily stated in terms of the sequence of quotients, and we will see in Proposition 2.6(ii.3) that the same holds for (mg) as long as the given sequence $\mathbb{M}$ is (lc). Moreover, observe that for every $p \in \mathbb{N}$ one has

$$M_p = \frac{M_p}{M_{p-1}}\frac{M_{p-1}}{M_{p-2}}\cdots\frac{M_2}{M_1}\frac{M_1}{M_0} = m_{p-1}m_{p-2}\ldots m_1 m_0. \qquad (1)$$

So, one may recover the sequence $\mathbb{M}$ (with $M_0 = 1$) once $\boldsymbol{m}$ is known, and so the knowledge of one of the sequences amounts to that of the other. Sequences of quotients of sequences $\mathbb{M}$, $\mathbb{L}$, etc. will be denoted by lowercase letters $\boldsymbol{m}$, $\boldsymbol{\ell}$ and so on. Whenever some statement refers to a sequence denoted by a lowercase letter such as $\boldsymbol{m}$, it will be understood that we are dealing with a sequence of quotients [of the sequence $\mathbb{M}$ given by (1)].

Proposition 2.6 *Let* $\mathbb{M} = (M_p)_{p\in\mathbb{N}_0}$ *be a sequence. Then, we have:*

(i) $\mathbb{M}$ *is (lc) if, and only if,* $\boldsymbol{m}$ *is nondecreasing. If, moreover,* $\mathbb{M}$ *satisfies (nq) then* $\boldsymbol{m}$ *tends to infinity.*

(ii) *Suppose from now on that* $\mathbb{M}$ *is (lc). Then*

(ii.1) $(M_p^{1/p})_{p\in\mathbb{N}}$ *is nondecreasing, and* $M_p^{1/p} \le m_{p-1}$ *for every* $p \in \mathbb{N}$ *(hence,* $M_p \le m_p^p$ *for every* $p \in \mathbb{N}_0$*). Moreover,* $\lim_{p\to\infty} m_p = \infty$ *if, and only if,* $\lim_{p\to\infty} M_p^{1/p} = \infty$.

(ii.2) $M_pM_q \le M_{p+q}$ *for every* $p, q \in \mathbb{N}_0$.

(ii.3) *The following statements are equivalent*

(ii.3.a) $\mathbb{M}$ *is (mg),*

(ii.3.b) $\sup_{p\in\mathbb{N}} \frac{m_p}{M_p^{1/p}} < \infty$,

(ii.3.c) $\sup_{p\in\mathbb{N}_0} \frac{m_{2p}}{m_p} < \infty$,

(ii.3.d) $\sup_{p\in\mathbb{N}} \left(\frac{M_{2p}}{M_p^2}\right)^{1/p} < \infty$.

Indeed, if $B > 0$ *is the constant involved in (mg), then*

$$m_p^p \le B^{2p}M_p, \quad p \in \mathbb{N}_0. \qquad (2)$$

Proof We only prove the last statement in (ii.3) (and, in particular, that (ii.3.a) implies (ii.3.b)). From (lc), for $p \in \mathbb{N}$ we have that

$$m_p^p \leq m_p m_{p+1} \ldots m_{2p-1} = \frac{M_{2p}}{M_p} = \frac{M_{p+p}}{M_p},$$

and applying (mg) we deduce that $m_p^p \leq B^{2p} M_p$. □

In the next definitions and results we take into account the conventions adopted in Remark 2.5(ii).

Definition 2.7 Let $\mathbb{M} = (M_p)_{p\in\mathbb{N}_0}$ and $\mathbb{L} = (L_p)_{p\in\mathbb{N}_0}$ be sequences, we say that $\mathbb{M}$ *is bounded from above by* $\mathbb{L}$, and we write $\mathbb{M} \ll \mathbb{L}$, if there exists $C > 0$ such that

$$M_p \leq C^p L_p, \qquad p \in \mathbb{N}_0.$$

Equivalently,

$$\sup_{p\in\mathbb{N}_0} \left(\frac{M_p}{L_p}\right)^{1/p} < \infty.$$

Definition 2.8 Let $\boldsymbol{m} = (m_p)_{p\in\mathbb{N}_0}$ and $\boldsymbol{\ell} = (\ell_p)_{p\in\mathbb{N}_0}$ be sequences of positive real numbers, we say that $\boldsymbol{m}$ *is bounded from above by* $\boldsymbol{\ell}$, and we write $\boldsymbol{m} \preccurlyeq \boldsymbol{\ell}$, if there exists $c > 0$ such that

$$m_p \leq c\ell_p, \qquad p \in \mathbb{N}_0$$

or, equivalently, if

$$\sup_{p\in\mathbb{N}_0} \frac{m_p}{\ell_p} < \infty.$$

Proposition 2.9 *Let* $\mathbb{M}$ *and* $\mathbb{L}$ *be sequences.*

(i) *If* $\boldsymbol{m} \preccurlyeq \boldsymbol{\ell}$ *then* $\mathbb{M} \ll \mathbb{L}$.
(ii) *If* $\mathbb{M}$ *and* $\mathbb{L}$ *are (lc) and* $\mathbb{M}$ *is (mg), then* $\mathbb{M} \ll \mathbb{L}$ *implies* $\boldsymbol{m} \preccurlyeq \boldsymbol{\ell}$.

Proof The proof of (i) follows immediately from (1). By Proposition 2.6 we see that $L_p^{1/p} \leq \ell_p$ and there exists $A > 0$ such that $m_p \leq A^2 M_p^{1/p}$, for every $p \in \mathbb{N}$. If one has $\mathbb{M} \ll \mathbb{L}$, then there exists $C > 0$ such that $M_p \leq C^p L_p$ for every $p \in \mathbb{N}_0$, and so

$$m_p \leq A^2 M_p^{1/p} \leq A^2 C L_p^{1/p} \leq A^2 C \ell_p$$

for every $p \in \mathbb{N}$, as desired. □

Definition 2.10 ([14, 49]) Let $\mathbb{M}$ and $\mathbb{L}$ be sequences, we say that $\mathbb{M}$ *is equivalent to* $\mathbb{L}$, and we write $\mathbb{M} \approx \mathbb{L}$, if $\mathbb{M} \ll \mathbb{L}$ and $\mathbb{L} \ll \mathbb{M}$.

Definition 2.11 Let $\boldsymbol{m}$ and $\boldsymbol{\ell}$ be sequences, we say that $\boldsymbol{m}$ *is equivalent to* $\boldsymbol{\ell}$, and we write $\boldsymbol{m} \simeq \boldsymbol{\ell}$, if $\boldsymbol{m} \preccurlyeq \boldsymbol{\ell}$ and $\boldsymbol{\ell} \preccurlyeq \boldsymbol{m}$.

Using the previous result we easily obtain

Proposition 2.12 *Let* $\mathbb{M}$ *and* $\mathbb{L}$ *be sequences.*

(i) *If* $\boldsymbol{m} \simeq \boldsymbol{\ell}$ *then* $\mathbb{M} \approx \mathbb{L}$.
(ii) *If* $\mathbb{M}$ *and* $\mathbb{L}$ *are (lc) and (mg), then* $\mathbb{M} \approx \mathbb{L}$ *amounts to* $\boldsymbol{m} \simeq \boldsymbol{\ell}$.

Remark 2.13 In particular, if $\mathbb{M}$ and $\mathbb{L}$ are strongly regular sequences we have that $\boldsymbol{m} \preccurlyeq \boldsymbol{\ell}$ is equivalent to $\mathbb{M} \ll \mathbb{L}$, and $\boldsymbol{m} \simeq \boldsymbol{\ell}$ is equivalent to $\mathbb{M} \approx \mathbb{L}$.

The next proposition is easily proved from the very definitions of (mg) and (snq).

Proposition 2.14 *Property (mg) is preserved by the relation* $\approx$*, and property (snq) is preserved by the relation* $\simeq$.

Remark 2.15 From the previous proposition we see that if a sequence $\mathbb{M}$ is (mg) and (snq), and another sequence $\mathbb{L}$ is (lc) and such that $\boldsymbol{\ell} \simeq \boldsymbol{m}$, then $\mathbb{L}$ is strongly regular. In particular, whenever $\mathbb{M}$ is (mg) and (snq), and $\boldsymbol{m}$ is eventually nondecreasing, it is easy to construct a strongly regular sequence $\mathbb{L}$ such that $\boldsymbol{\ell} \simeq \boldsymbol{m}$ and, in fact, $\ell_p = m_p$ for every p greater than or equal to some suitable p_0.

Example 2.16

(i) Given $\alpha > 0$, the best known example of strongly regular sequence is $\mathbb{M}_\alpha = (p!^\alpha)_{p\in\mathbb{N}_0}$, called the *Gevrey sequence of order* α.
(ii) The sequences $\mathbb{M}_{\alpha,\beta} = \big(p!^\alpha \prod_{m=0}^{p} \log^\beta(e+m)\big)_{p\in\mathbb{N}_0}$, where $\alpha > 0$ and $\beta \in \mathbb{R}$, are strongly regular (in case $\beta < 0$, the sequence has to be suitably modified according to Remark 2.15).
(iii) For $q > 1$, $\mathbb{M} = (q^{p^2})_{p\in\mathbb{N}_0}$ is (lc) and (snq), but not (mg).

The following result of H.-J. Petzsche will be useful in the sequel.

Corollary 2.17 (Petzsche [49], Corollary 1.3(a)) *Let* $\mathbb{M}$ *be (lc) and (snq). Then there exists* $\varepsilon > 0$ *and a sequence* $\mathbb{L}$ *such that* $\boldsymbol{\ell} \simeq \boldsymbol{m}$ *and the sequence* $(L_p p!^{-\varepsilon})_{p\in\mathbb{N}_0}$ *is (lc) and (snq).*

For the sake of completeness we include the following result.

Proposition 2.18 *Given two strongly regular sequences* $\mathbb{M} = (M_p)_{p\in\mathbb{N}_0}$ *and* $\mathbb{L} = (L_p)_{p\in\mathbb{N}_0}$*, the product sequence* $\mathbb{M}\cdot\mathbb{L} := (M_pL_p)_{p\in\mathbb{N}_0}$ *is also strongly regular.*

Proof Properties (lc) and (mg) are easily checked for $\mathbb{M}\cdot\mathbb{L}$. Regarding (snq), we will use that, according to Lemma 1.3.4 in Thilliez [57], $\mathbb{M}^s := (M_p^s)_{p\in\mathbb{N}_0}$ is strongly regular for every $s > 0$ (properties (lc) and (mg) are obvious, and (snq) is obtained thanks to Corollary 2.17). Hence, $\mathbb{M}^2$ and $\mathbb{L}^2$ are (snq). Now, Cauchy-Schwarz inequality gives that, for every $p \geq 0$,

$$\sum_{q\geq p} \frac{M_q L_q}{(q+1)M_{q+1}L_{q+1}} \leq \Big(\sum_{q\geq p} \frac{M_q^2}{(q+1)M_{q+1}^2}\Big)^{1/2}\Big(\sum_{q\geq p} \frac{L_q^2}{(q+1)L_{q+1}^2}\Big)^{1/2}$$
$$\leq \sqrt{BB'}\frac{M_p}{M_{p+1}}\frac{L_p}{L_{p+1}},$$

where the positive constants B and B' are the ones appearing in (snq) for $\mathbb{M}^2$ and $\mathbb{L}^2$, respectively. □

Many of the properties of the sequences $\mathbb{M}$ (or $\boldsymbol{m}$) may be better studied, or expressed, by means of any of the two following auxiliary functions associated with $\mathbb{M}$. We assume that $\mathbb{M}$ is (lc) and that $\lim_{p\to\infty} m_p = \infty$ [this happens, for example, if $\mathbb{M}$ satisfies (nq)].

The map $h_{\mathbb{M}} : [0,\infty) \to \mathbb{R}$ is defined by

$$h_{\mathbb{M}}(t) := \inf_{p\in\mathbb{N}_0} M_p t^p, \quad t > 0; \qquad h_{\mathbb{M}}(0) = 0, \tag{3}$$

and it turns out to be a non-decreasing continuous map in $[0,\infty)$ onto $[0,1]$. In fact

$$h_{\mathbb{M}}(t) = \begin{cases} t^p M_p \text{ if } t \in \big[\frac{1}{m_p}, \frac{1}{m_{p-1}}\big),\ p = 1,2,\ldots, \\ 1 \qquad\qquad \text{if } t \geq 1/m_0. \end{cases}$$

One may also consider the function

$$M(t) := \sup_{p\in\mathbb{N}_0} \log\big(\frac{t^p}{M_p}\big) = -\log\big(h_{\mathbb{M}}(1/t)\big), \quad t > 0; \qquad M(0) = 0, \tag{4}$$

which is a non-decreasing continuous map in $[0,\infty)$ with $\lim_{t\to\infty} M(t) = \infty$. Indeed,

$$M(t) = \begin{cases} p\log t - \log(M_p) \text{ if } t \in [m_{p-1}, m_p),\ p = 1,2,\ldots, \\ 0 \qquad\qquad\qquad\qquad \text{if } t \in [0, m_0), \end{cases}$$

and one can easily check that M is convex in $\log t$, i.e., the map $t \mapsto M(e^t)$ is convex in $\mathbb{R}$. For every $p \in \mathbb{N}_0$, the continuity of M at m_p amounts to the trivial equality $m_p^p/M_p = m_p^{p+1}/M_{p+1}$. Moreover, since the sequence $\boldsymbol{m} = (m_p)_{p\in\mathbb{N}_0}$ (respectively, the function $M(t)$) increases to infinity as p (resp. t) tends to infinity, the sequence $(M(m_p))_{p\in\mathbb{N}_0} = \big(\log(m_p^p/M_p)\big)_{p\in\mathbb{N}_0}$, and consequently also $(m_p^p/M_p)_{p\in\mathbb{N}_0}$, increase to infinity, starting at the value 0 and 1, respectively. Note also that the p-th and $(p+1)$-th terms of any of these two sequences are equal if, and only if, $m_p = m_{p+1}$.

The condition of moderate growth plays a fundamental role in the proof of (5), which will in turn be crucial in many of our arguments.

Lemma 2.19 ([14, 57]) *Let $\mathbb{M} = (M_p)_{p\in\mathbb{N}_0}$ be (lc), (mg) and such that* $\boldsymbol{m}$ *tends to infinity. Let s be a real number with $s \geq 1$. There exists $\rho(s) \geq 1$ (only depending on s and $\mathbb{M}$) such that*

$$h_{\mathbb{M}}(t) \leq (h_{\mathbb{M}}(\rho(s)t))^s \qquad \textit{for } t \geq 0. \tag{5}$$

Proof If $B > 0$ is the constant appearing in (mg), it is clear that $M_{2k} \leq B^{2k}M_k^2$ for every $k \in \mathbb{N}_0$, and so for every $t > 0$ one has

$$2M(t) = \sup_{k\geq 0} \log \frac{t^{2k}}{M_k^2} \leq \sup_{k\geq 0} \log \frac{B^{2k}t^{2k}}{M_{2k}} \leq M(Bt).$$

Inductively, one obtains that $2^q M(t) \leq M(B^q t)$ for every $q \in \mathbb{N}_0$ and $t > 0$, or equivalently, $M(t/B^q) \leq M(t)/2^q$. Now, given $s \geq 1$ there exists a unique $q \in \mathbb{N}_0$ such that $s \in [2^q, 2^{q+1})$ (indeed, $q = \lfloor \log_2 s \rfloor$), and if we put $\rho(s) := B^{q+1}$ we have that

$$M\Big(\frac{t}{\rho(s)}\Big) \leq \frac{1}{2^{q+1}} M(t) < \frac{1}{s} M(t), \quad t > 0.$$

This is precisely the inequality (5). □

Remark 2.20 If the sequences $\mathbb{M} = (M_p)_{p\in\mathbb{N}_0}$ and $\mathbb{L} = (L_p)_{p\in\mathbb{N}_0}$ are equivalent, there exist $C, D > 0$ such that

$$C^p M_p \leq L_p \leq D^p M_p, \qquad p \in \mathbb{N}_0.$$

It is straightforward to check that

$$h_{\mathbb{M}}(Ct) \leq h_{\mathbb{L}}(t) \leq h_{\mathbb{M}}(Dt), \qquad t \geq 0, \tag{6}$$

or, equivalently,

$$M(t/C) \geq L(t) \geq M(t/D), \qquad t \geq 0. \tag{7}$$

We now recall the following definitions and facts, mainly taken from the book of Goldberg and Ostrovskii [17], which will allow us to determine the rate of growth (resp. decrease) of the function $M(t)$ (resp. $h_{\mathbb{M}}(t)$) as t tends to infinity (resp. to 0).

Definition 2.21 ([17], p. 43) Let $\alpha(t)$ be a nonnegative and nondecreasing function in (c, ∞) for some $c \geq 0$ (we write $\alpha \in \Lambda$). The *order* of α is defined as

$$\rho = \rho[\alpha] := \limsup_{t\to\infty} \frac{\log^+ \alpha(t)}{\log t} \in [0, \infty],$$

where $\log^+ = \max(\log, 0)$. $\alpha(t)$ is said to have finite order if $\rho < \infty$.

Remark 2.22 We note that this concept already appears in the definition of exponential order of a non constant entire function f. If $M_f(r) := \max_{|z|=r} |f(z)|$, $r > 0$, then we know that the function M_f is nondecreasing and tends to infinity as $r \to \infty$, and so $\log(M_f) \in \Lambda$. The exponential order of f is defined precisely as $\rho[\log(M_f)]$.

We are firstly interested in determining the order of the function $M(t) \in \Lambda$ defined in (4) and associated with a sequence $\mathbb{M}$ which is (lc) and such that $\lim_{p\to\infty} m_p = \infty$. To this end, we need to recall now the definition of exponent of convergence of a sequence and how it may be computed.

Proposition 2.23 ([19], p. 65) *Let $(c_p)_{p\in\mathbb{N}_0}$ be a nondecreasing sequence of positive real numbers tending to infinity. The exponent of convergence of $(c_p)_p$ is defined as*

$$\lambda_{(c_p)} := \inf\{\mu > 0 : \sum_{p=0}^{\infty} \frac{1}{c_p^{\mu}} \text{ converges}\}$$

(if the previous set is empty, we put $\lambda_{(c_p)} = \infty$). Then, one has

$$\lambda_{(c_p)} = \limsup_{p\to\infty} \frac{\log(p)}{\log(c_p)}. \tag{8}$$

We will also need the following fact, which can be found in [40, p. 21]: if we consider the *counting function* for the sequence of quotients $\boldsymbol{m}$, $\nu : (0,\infty) \to \mathbb{N}_0$ given by

$$\nu(t) := \#\{p : m_p \le t\}, \tag{9}$$

then one has that

$$M(t) = \int_0^t \frac{\nu(s)}{s}\, ds, \qquad t \ge 0. \tag{10}$$

We may now compute the order of $M(t)$ in terms of the sequence $\boldsymbol{m}$.

Theorem 2.24 ([54]) *Let $\mathbb{M}$ be (lc) and such that $\lim_{p\to\infty} m_p = \infty$. Then, the order of $M(t)$ is given by*

$$\rho[M] = \limsup_{p\to\infty} \frac{\log(p)}{\log(m_p)}. \tag{11}$$

Proof We take into account the link given in (10) between $M(t)$ and the counting function $\nu(t)$ for $\boldsymbol{m}$ [as defined in (9)], which also belongs to Λ. The following inequalities are evident for every $t > m_0$:

$$M(et) \geq \int_t^{et} \frac{\nu(s)}{s}\,ds \geq \nu(t)\int_t^{et} \frac{1}{s}\,ds = \nu(t),$$
$$M(t) = \int_{m_0}^{t} \frac{\nu(s)}{s}\,ds \leq \nu(t)\int_{m_0}^{t} \frac{1}{s}\,ds = \nu(t)\log(t/m_0).$$

Then, from the very definition we may deduce that the order of $M(t)$ equals that of $\nu(t)$. Now, it is a general fact for functions α in Λ that

$$\rho[\alpha] = \inf\{\mu > 0 : \int^{\infty} \frac{\alpha(t)}{t^{1+\mu}}dt \text{ converges}\}$$

(the lower endpoint in the integral is inessential; when the set in the right hand side is empty it is understood that $\rho[\alpha] = \infty$). On the other hand, for every $t > 0$ it is straightforward to prove that

$$\sum_{m_p \leq t} \frac{1}{m_p^{\mu}} = \frac{\nu(t)}{t^{\mu}} + \mu \int_{m_0}^{t} \frac{\nu(s)}{s^{\mu+1}}\,ds,$$

from where one may deduce that the series

$$\sum_{p=0}^{\infty} \frac{1}{m_p^{\mu}}$$

converges if, and only if, the integral

$$\int_{m_0}^{\infty} \frac{\nu(t)}{t^{1+\mu}}dt$$

does. These two last facts together lead to the conclusion that $\rho[M] = \rho[\nu] = \lambda_{(m_p)}$, which is given by the formula in (8). $\square$

Remark 2.25

(i) Let $\mathbb{M}_{\alpha}$ be the Gevrey sequence of order $\alpha > 0$, and $M_{\alpha}(t)$ its associated function. By means of (11), it is obvious that $\rho[M_{\alpha}] = 1/\alpha$.
(ii) Let $\mathbb{M}$ and $\mathbb{L}$ satisfy the assumptions in the previous result, and suppose first that $\mathbb{M} \ll \mathbb{L}$, i.e. $M_p \leq C^p L_p$ for some $C > 0$ and every $p \in \mathbb{N}_0$. By the very definition, one has that $M(t) \geq L(t/C)$ for every $t \geq 0$, and consequently $\rho[M] \geq \rho[L]$. Observe that this fact is not clear from formula (11), unless we admit that $\mathbb{M}$ is (mg) (and so $\boldsymbol{m} \preccurlyeq \boldsymbol{\ell}$ by Proposition 2.9).

If we suppose that $\mathbb{M} \approx \mathbb{L}$, then we conclude that $\rho[M] = \rho[L]$. Again, only if $\mathbb{M}$ and $\mathbb{L}$ are (mg) we may be sure that also $\boldsymbol{m} \simeq \boldsymbol{\ell}$, and then formula (11) makes the equality of the respective orders evident.

(iii) Given a strongly regular sequence $\mathbb{M}$, from Corollary 2.17 it is plain to see that there exist $\varepsilon, a_1 > 0$ such that

$$a_1^p p!^\varepsilon \leq M_p, \qquad p \in \mathbb{N}_0.$$

On the other hand, choose $\delta > 0$ such that $e^\delta \geq B^2$, where B is the constant in (mg). Using (2) and the estimate

$$\Big(\frac{p}{p+1}\Big)^{p+1} \leq \frac{1}{e},$$

one can prove by induction that $M_p \leq M_1^p p^{\delta p}$ for every $p \in \mathbb{N}$. Stirling's formula implies that there exists $a_2 > 0$ such that

$$M_p \leq a_2^p p!^\delta, \qquad p \in \mathbb{N}_0.$$

From the two previous remarks we deduce that $1/\delta \leq \rho[M] \leq 1/\varepsilon$, and, in particular, $\rho[M] \in (0, \infty)$.

(iv) If we put $\mathbb{M} = (q^{p^2})_{p\in\mathbb{N}_0}$, with $q > 1$, then $\mathbb{M}$ is (lc) and (nq) but (mg) is not satisfied, and we easily obtain from (11) that $\rho[M] = 0$, which is not possible for strongly regular sequences.

In case $\mathbb{M} = (\prod_{k=0}^{p} \log(e+k))_{p\in\mathbb{N}_0}$, $\mathbb{M}$ is (lc), (mg) and $\boldsymbol{m}$ tends to infinity, but (nq) is not satisfied. From (11) we see that $\rho[M] = \infty$.

3 Asymptotic Expansions and Ultraholomorphic Classes

Given a sequence of positive real numbers $\mathbb{M} = (M_p)_{p\in\mathbb{N}_0}$, a constant $A > 0$ and a sector S in the Riemann surface of the logarithm, we define

$$\mathcal{A}_{\mathbb{M},A}(S) = \{f \in \mathcal{O}(S) : \|f\|_{\mathbb{M},A} := \sup_{z\in S, p\in\mathbb{N}_0} \frac{|f^{(p)}(z)|}{A^p p! M_p} < \infty\}.$$

$(\mathcal{A}_{\mathbb{M},A}(S), \|\ \|_{\mathbb{M},A})$ is a Banach space, and $\mathcal{A}_{\mathbb{M}}(S) := \cup_{A>0}\mathcal{A}_{\mathbb{M},A}(S)$ is called a *Carleman ultraholomorphic class* in the sector S.

One may accordingly define classes of sequences

$$\Lambda_{\mathbb{M},A} = \Big\{\boldsymbol{\mu} = (\mu_p)_{p\in\mathbb{N}_0} \in \mathbb{C}^{\mathbb{N}_0} : |\boldsymbol{\mu}|_{\mathbb{M},A} := \sup_{p\in\mathbb{N}_0} \frac{|\mu_p|}{A^p p! M_p} < \infty\Big\}.$$

$(\Lambda_{\mathbb{M},A}, |\ |_{\mathbb{M},A})$ is again a Banach space, and we put $\Lambda_{\mathbb{M}} := \cup_{A>0}\Lambda_{\mathbb{M},A}$, which we call a *Carleman class* of sequences.

Since the derivatives of $f \in \mathcal{A}_{\mathbb{M},A}(S)$ are Lipschitzian, for every $p \in \mathbb{N}_0$ one may define

$$f^{(p)}(0) := \lim_{z\in S, z\to 0} f^{(p)}(z) \in \mathbb{C},$$

and it is clear that the sequence

$$\tilde{\mathcal{B}}(f) := (f^{(p)}(0))_{p\in\mathbb{N}_0} \in \Lambda_{\mathbb{M},A}, \qquad f \in \mathcal{A}_{\mathbb{M},A}(S).$$

The map $\tilde{\mathcal{B}} : \mathcal{A}_{\mathbb{M}}(S) \longrightarrow \Lambda_{\mathbb{M}}$ so defined is called the *asymptotic Borel map*. It has nice properties under suitable assumptions on the sequence $\mathbb{M}$.

Proposition 3.1 *Let $\mathbb{M}$ be a sequence.*

(i) *If $\mathbb{M}$ is (lc) (respectively, (dc)), then $\mathcal{A}_{\mathbb{M}}(S)$ is closed under products (resp. under derivatives). So, if $\mathbb{M}$ is (lc) and (dc), $\mathcal{A}_{\mathbb{M}}(S)$ is a differential algebra.*
(ii) *Let us endow $\Lambda_{\mathbb{M}}$ with the operations induced by derivation and product in $\mathcal{A}_{\mathbb{M}}(S)$ through the map $\tilde{\mathcal{B}}$ (i.e. those which make the operations to commute with $\tilde{\mathcal{B}}$). If $\mathbb{M}$ is (lc) and (dc), $\Lambda_{\mathbb{M}}$ is a differential algebra, and $\tilde{\mathcal{B}}$ is a homomorphism of differential algebras.*

Proof Regarding products, (i) is a consequence of Leibniz's formula and the property in Proposition 2.6.(ii.1). On the other hand, if we have (dc) with a constant $B > 0$ and take $f \in \mathcal{A}_{\mathbb{M},A}(S)$, it is immediate to prove that $f' \in \mathcal{A}_{\mathbb{M},AB}(S)$, and so $\mathcal{A}_{\mathbb{M}}(S)$ is derivation closed.

The operations in $\Lambda_{\mathbb{M}}$ are the left shift operator (corresponding to derivation) and the product induced by Leibniz's rule: if $\boldsymbol{\lambda} = (\lambda_p)_{p\in\mathbb{N}_0}$ and $\boldsymbol{\mu} = (\mu_p)_{p\in\mathbb{N}_0}$ belong to $\Lambda_{\mathbb{M}}$, we put

$$\boldsymbol{\lambda}\cdot\boldsymbol{\mu} = (\nu_p)_{p\in\mathbb{N}_0}, \qquad \text{where } \nu_p := \sum_{k=0}^{p}\binom{p}{k}\lambda_k\mu_{p-k}, \quad p \in \mathbb{N}_0.$$

The rest of the statements is straightforward. □

Although it will not be used in our approach, from the topological viewpoint the spaces $\mathcal{A}_{\mathbb{M}}(S)$ and $\Lambda_{\mathbb{M}}$ have a natural structure of (LB)-spaces, that is, they are inductive limits of Banach spaces.

Next, we will recall the relationship between these classes and the concept of asymptotic expansion.

Definition 3.2 We say a holomorphic function f in a sectorial region G admits the formal power series $\hat{f} = \sum_{p=0}^{\infty} a_p z^p \in \mathbb{C}[[z]]$ as its $\mathbb{M}$-*asymptotic expansion* in G (when the variable tends to 0) if for every $T \ll G$ there exist $C_T, A_T > 0$ such that for every $p \in \mathbb{N}_0$, one has

$$\left| f(z) - \sum_{k=0}^{p-1} a_k z^k \right| \le C_T A_T^p M_p |z|^p, \qquad z \in T. \tag{12}$$

We will write $f \sim_{\mathbb{M}} \sum_{p=0}^{\infty} a_p z^p$ in G. $\tilde{\mathcal{A}}_{\mathbb{M}}(G)$ stands for the space of functions admitting $\mathbb{M}$-asymptotic expansion in G.

Definition 3.3 Given a sector S, we say $f \in \mathcal{O}(S)$ admits $\hat{f}$ as its *uniform* $\mathbb{M}$-*asymptotic expansion in* S *of type* $A > 0$ if there exists $C > 0$ such that for every $p \in \mathbb{N}_0$, one has

$$\left| f(z) - \sum_{k=0}^{p-1} a_k z^k \right| \le C A^p M_p |z|^p, \qquad z \in S. \tag{13}$$

We have the following result relating the spaces with uniform, or not, bounds for the remainders or the derivatives (see [2, 15]).

Proposition 3.4 *Let S be a sector and G a sectorial region.*

(i) *If* $f \in \mathcal{A}_{\mathbb{M},A}(S)$, *then* f *admits the series* $\hat{f} = \sum_{p\in\mathbb{N}_0} \frac{1}{p!} f^{(p)}(0) z^p$ *as its uniform* $\mathbb{M}$-*asymptotic expansion in S of type A.*
(ii) $f \in \tilde{\mathcal{A}}_{\mathbb{M}}(G)$ *if, and only if, for every* $T \ll G$ *there exists* $A_T > 0$ *such that* $f|_T \in \mathcal{A}_{\mathbb{M},A_T}(T)$. *Hence, the map* $\tilde{\mathcal{B}} : \tilde{\mathcal{A}}_{\mathbb{M}}(G) \longrightarrow \Lambda_{\mathbb{M}}$ *is also well defined.*

Proof (i) is a consequence of Taylor's formula. (ii) Emanates from Cauchy's integral formula for the derivatives, and the fact that, given $T \ll G$, there exist a sector $U \ll G$ and $r > 0$ such that for every $z \in T$ one has $D(z, r|z|) \subset U$. □

The map $\tilde{\mathcal{B}} : \tilde{\mathcal{A}}_{\mathbb{M}}(G) \longrightarrow \Lambda_{\mathbb{M}}$ is again a homomorphism of differential algebras, as long as $\mathbb{M}$ is (lc) and (dc).

Remark 3.5 Consider a pair of equivalent sequences $\mathbb{M}$ and $\mathbb{L}$. It is obvious that $\Lambda_{\mathbb{M}} = \Lambda_{\mathbb{L}}$, the spaces $\mathcal{A}_{\mathbb{M}}(S)$ and $\mathcal{A}_{\mathbb{L}}(S)$ coincide for a sector S, and also $\tilde{\mathcal{A}}_{\mathbb{M}}(G)$ and $\tilde{\mathcal{A}}_{\mathbb{L}}(G)$ agree for a sectorial region G.

Remark 3.6 As a consequence of Cauchy's integral formula for the derivatives, given a sector S one can prove that whenever $T \ll S$, there exists a constant $c = c(T, S) > 0$ such that the restriction to T, f_T, of functions f defined on S and admitting uniform $\mathbb{M}$-asymptotic expansion in S of type $A > 0$, belongs to $\mathcal{A}_{\mathbb{M},cA}(T)$, and moreover, if one has (13) then $\|f_T\|_{\mathbb{M},cA} \le C$.

4 Injectivity of the Borel Map: Korenbljum's Result and Watson's Lemma

In the definition of general asymptotic expansions in a sectorial region G, the form of the bounds for the remainders given in (12) is just $C_{T,p}|z|^p$, i.e., no precision is made about the way in which the constants $C_{T,p}$ depend on $T \ll G$ or $p \in \mathbb{N}$. In this case, the classical Borel-Ritt theorem states the surjectivity of the asymptotic Borel map for whatever region G, while it is well known that this map is never injective (observe that the function $\exp(-1/z)$ is flat in the right half plane, and one may reason by ramification in any other situation). So, given a sectorial region it is not possible to assign to a given formal power series a natural, unambiguous sum (i.e. an analytic function in the region asymptotically represented by the series). However, the situation changes drastically when $\mathbb{M}$-asymptotic expansions are considered, as we will see in this section.

Quasianalyticity and Korenbljum's Result

We are interested in characterizing those classes in which the asymptotic Borel map is injective; these will be called quasianalytic Carleman classes.

Definition 4.1 A function f in any of the classes $\tilde{\mathcal{A}}_{\mathbb{M}}(G)$ or $\mathcal{A}_{\mathbb{M},A}(S)$ is said to be *flat* if $\tilde{\mathcal{B}}(f)$ is the null sequence or, in other words, $f \sim_{\mathbb{M}} \hat{0}$ (uniformly or not, depending on the class), where $\hat{0}$ denotes the null power series.

Definition 4.2 Let S be a sector and $\mathbb{M} = (M_p)_{p\in\mathbb{N}_0}$ be a sequence of positive numbers. We say that $\mathcal{A}_{\mathbb{M}}(S)$ (or $\tilde{\mathcal{A}}_{\mathbb{M}}(G)$) is *quasianalytic* if it does not contain nontrivial flat functions.

Characterizations of quasianalyticity for general sequences $\mathbb{M}$, in one and several variables, and for the case $\mathcal{A}_{\mathbb{M}}(S)$, are available in [27], generalizing the work of Korenbljum [25]. In this paper, we restrict our attention to the one-variable case.

Note that a suitable rotation allows one to consider only sectors (or sectorial regions) bisected by the direction $d = 0$. As shown in the next result, concerning classes of functions with uniform bounds for the derivatives throughout the sector S_γ, quasianalyticity is governed by the opening of the sector, i.e. by γ. Moreover, one easily observes that if $\mathcal{A}_{\mathbb{M}}(S_\gamma)$ is quasianalytic (respectively, non quasianalytic), then $\mathcal{A}_{\mathbb{M}}(S_\beta)$ will also be for every $\beta > \gamma$ (resp. for every $\beta < \gamma$). So, one expects to find a value of the opening that separates both situations, and this will indeed happen.

Although we mainly focus on strongly regular sequences, in many of the results in this section weaker assumptions on $\mathbb{M}$ suffice, as it will be indicated. Moreover, we remark that the condition (lc) is not necessary in the next statement, but its inclusion eludes the consideration of technically cumbersome regularizations of the sequence involved (the minorant of Faber or the log-convex regularization).

Theorem 4.3 ([25]) *Let $\mathbb{M}$ be (lc) and let $\gamma > 0$ be given. The following statements are equivalent:*

(i) *The class $\mathcal{A}_{\mathbb{M}}(S_\gamma)$ is quasianalytic.*

(ii) $\displaystyle\sum_{p=0}^{\infty}\Big(\frac{M_p}{(p+1)M_{p+1}}\Big)^{1/(\gamma+1)} = \infty.$

Remark 4.4 The previous result is a consequence of a version of Watson's Lemma due to Mandelbrojt [40, Sect. 2.4.III]. Note that the result is easy if the sequence $\boldsymbol{m}$ does not tend to infinity. In this case, there would exist $C > 0$ such that $M_p^{1/p} \le m_p \le C$ for every $p \in \mathbb{N}$. So, on one hand the series in Theorem 4.3.(ii) would be divergent for every $\gamma > 0$. On the other hand, by Proposition 3.4.(i) and the estimates in (13), for a flat function $f \in \mathcal{A}_{\mathbb{M}}(S_\gamma)$ we would have

$$|f(z)| \le C_0 A^p M_p |z|^p \le C_0 (AC)^p |z|^p$$

for every $z \in S_\gamma$, every $p \in \mathbb{N}$ and some $C_0, A > 0$. We deduce that $f(z)$ vanishes for every $z \in S_\gamma \cap D(0, 1/(AC))$, and so the class is quasianalytic for every $\gamma > 0$.

In view of this last remark, we will consider in the following sequences $\mathbb{M}$ which are (lc) and with $\boldsymbol{m}$ tending to infinity.

Accordingly, we introduce a new quantity regarding quasianalyticity.

Definition 4.5 For a sequence $\mathbb{M}$, we put

$$Q_{\mathbb{M}} = \{\gamma > 0 : \mathcal{A}_{\mathbb{M}}(S_\gamma) \text{ is quasianalytic}\}.$$

The *order of quasianalyticity* of $\mathbb{M}$ is defined as $\omega(\mathbb{M}) := \inf Q_{\mathbb{M}}$.

We can obtain its value due to the following result.

Theorem 4.6 ([54]) *Let $\mathbb{M}$ be an (lc) sequence such that $\boldsymbol{m}$ tends to infinity. Then,*

$$\omega(\mathbb{M}) = \liminf_{p\to\infty} \frac{\log(m_p)}{\log(p)} = \frac{1}{\lambda_{(m_p)}}. \tag{14}$$

Consequently, we have that

$$\omega(\mathbb{M}) = \frac{1}{\rho[M]}, \tag{15}$$

and if $\mathbb{M}$ is strongly regular, then $\omega(\mathbb{M}) \in (0, \infty)$.

Proof Since $\mathbb{M}$ is (lc), the sequence $(p!M_p)_{p\in\mathbb{N}_0}$ also is. So, the sequence of its quotients, $((p+1)m_p)_{p\in\mathbb{N}_0}$, is nondecreasing and, moreover, tends to infinity because $\boldsymbol{m}$ does. In view of (8), the exponent of convergence of the sequence $((p+1)m_p)_{p\in\mathbb{N}_0}$ and that of the sequence $(m_p)_{p\in\mathbb{N}_0}$ are related as follows:

$$\lambda_{((p+1)m_p)} = \limsup_{p\to\infty} \frac{\log(p)}{\log((p+1)m_p)} = \frac{1}{1+\liminf_{p\to\infty}\frac{\log(m_p)}{\log(p)}} = \frac{1}{1+1/\lambda_{(m_p)}}$$

(with the usual conventions regarding division by 0 or ∞). On the other hand, from Theorem 4.3 and the definition of $\omega(\mathbb{M})$ it is clear that

$$\frac{1}{\omega(\mathbb{M})+1} = \lambda_{((p+1)m_p)},$$

hence

$$\omega(\mathbb{M}) = \liminf_{p\to\infty}\frac{\log(m_p)}{\log(p)} = \frac{1}{\lambda_{(m_p)}}.$$

Comparing this to (11), and by Remark 2.25.(iii), we conclude. □

Remark 4.7 Observe that $\pi\omega(\mathbb{M})$ is the optimal opening for quasianalyticity, in the sense that the class $\mathcal{A}_{\mathbb{M}}(S)$ is (respectively, is not) quasianalytic whenever the opening of S exceeds (resp. is less than) this quantity. When the opening of the sector equals $\pi\omega(\mathbb{M})$ both cases are possible, as shown in the forthcoming Example 4.9.

Remark 4.8 If the sequences $\mathbb{M}$ and $\mathbb{L}$ are such that $\mathbb{M} \ll \mathbb{L}$, then $\mathcal{A}_{\mathbb{M}}(S) \subset \mathcal{A}_{\mathbb{L}}(S)$ for any sector S, and so $Q_{\mathbb{L}} \subset Q_{\mathbb{M}}$ and $\omega(\mathbb{M}) \leq \omega(\mathbb{L})$. Note that, if $\mathbb{M}$ and $\mathbb{L}$ are (lc) and $\boldsymbol{m}$ and $\boldsymbol{\ell}$ tend to infinity, this last inequality is not at all clear from the formula (14), unless conditions are verified by $\mathbb{M}$ and $\mathbb{L}$ which imply $\boldsymbol{m} \preccurlyeq \boldsymbol{\ell}$ (for example, if $\mathbb{M}$ and $\mathbb{L}$ are strongly regular).

If $\mathbb{M}$ and $\mathbb{L}$ are equivalent we obviously have $\omega(\mathbb{M}) = \omega(\mathbb{L})$, what may be deduced from (14), whenever applicable, if we also know that $\boldsymbol{m} \simeq \boldsymbol{\ell}$.

Example 4.9 Consider the sequences $\mathbb{M}_{\alpha,\beta}$, $\alpha > 0$, $\beta \in \mathbb{R}$, introduced in Examples 2.16.(ii). Applying Theorem 4.3, it is easy to check that

$$Q_{\mathbb{M}_{\alpha,\beta}} = \begin{cases} [\alpha,\infty) & \text{if } \alpha \geq \beta-1, \\ (\alpha,\infty) & \text{if } \alpha < \beta-1, \end{cases}$$

so that $\omega(\mathbb{M}_{\alpha,\beta}) = \alpha$ (in particular, for Gevrey sequences, appearing when $\beta = 0$).

We next recall the notion of growth index defined and studied by Thilliez in [57, Sect. 1.3].

Definition 4.10 Let $\mathbb{M} = (M_p)_{p\in\mathbb{N}_0}$ be a strongly regular sequence and $\gamma > 0$. We say $\mathbb{M}$ satisfies property (P_γ) if there exists a sequence of real numbers $\mathbb{M}' = (M'_p)_{p\in\mathbb{N}_0}$ such that $\boldsymbol{m}' \simeq \boldsymbol{m}$ and $(M'_p p!^{-\gamma})_{p\in\mathbb{N}_0}$ is (lc).

The *growth index* of $\mathbb{M}$ is

$$\gamma(\mathbb{M}) := \sup\{\gamma \in \mathbb{R} : (P_\gamma) \text{ is fulfilled}\}.$$

Remark 4.11 According to Corollary 2.17, for every strongly regular $\mathbb{M}$ there exists $\varepsilon > 0$ such that (P_ε) holds. On the other hand, as indicated in Remark 2.25(iii) there exist $\delta, a > 0$ such that $M_p \le a^p p!^\delta$ for every $p \in \mathbb{N}_0$. It is easy to conclude that $\gamma(\mathbb{M})$ is always a positive real number.

Example

(i) For the Gevrey sequence of order $\alpha > 0$, one has $\gamma(\mathbb{M}_\alpha) = \alpha$.
(ii) For the sequences $\mathbb{M}_{\alpha,\beta}$ in Example 2.16.(ii) one can check that $\gamma(\mathbb{M}_{\alpha,\beta}) = \alpha$.

Regarding the relationship between $\omega(\mathbb{M})$ and $\gamma(\mathbb{M})$, we have the following result.

Proposition 4.12 *For every strongly regular sequence $\mathbb{M}$ one has $\omega(\mathbb{M}) \ge \gamma(\mathbb{M})$.*

Proof Suppose $\mathbb{M}$ verifies (P_γ) for some $\gamma > 0$. This easily implies the existence of a constant $a > 0$ such that $a^p p!^\gamma \le M_p$ for every $p \in \mathbb{N}_0$. Hence, by Remark 4.8 we have $\gamma = \omega((a^p p!^\gamma)_{p\in\mathbb{N}_0}) \le \omega(\mathbb{M})$, and the definition of $\gamma(\mathbb{M})$ is enough to conclude. □

According to the very definition of $\omega(\mathbb{M})$, the previous result is indeed equivalent to the following one, proved by Thilliez [57] and, subsequently, by A. Lastra and the author [27]. However, the present argument seems to be simpler than the ones involved in the previous proofs of this theorem.

Theorem 4.13 ([27, 57]) *Let $0 < \gamma < \gamma(\mathbb{M})$. Then, the class $\mathcal{A}_{\mathbb{M}}(S_\gamma)$ is not quasianalytic.*

Flat Functions via Proximate Orders and Watson's Lemma

In the previous subsection quasianalyticity for the classes $\mathcal{A}_{\mathbb{M}}(S)$ (those defined by imposing uniform bounds for the derivatives) has been completely characterized, thanks to Korenbljum's result. Regarding the classes $\tilde{\mathcal{A}}_{\mathbb{M}}(G)$ of functions with $\mathbb{M}$-asymptotic expansion in a sectorial region G (with non-uniform bounds), the aforementioned result by Mandelbrojt [40, Sect. 2.4.III] makes it easy to obtain the following statement (for more details, the reader may consult [22]).

Theorem 4.14 (Generalized Watson's Lemma, Partial Version) *If the opening of G is larger (respectively, smaller) than $\pi\omega(\mathbb{M})$, then $\tilde{\mathcal{B}} : \tilde{\mathcal{A}}_{\mathbb{M}}(G) \longrightarrow \Lambda_{\mathbb{M}}$ is (resp. is not) injective.*

So, the case when the opening equals $\pi\omega(\mathbb{M})$ remains unsolved. In order to treat this situation, we will make use of the following result of V. Thilliez which characterizes flatness in the non-uniform case.

Theorem 4.15 ([58], Proposition 4) *Let $\mathbb{M}$ be (lc) and such that $\boldsymbol{m}$ tends to infinity, and let S be a sector. For $f \in \mathcal{O}(S)$, the following are equivalent:*

(i) $f \in \tilde{\mathcal{A}}_{\mathbb{M}}(S)$ *and* $f \sim_{\mathbb{M}} \hat{0}$.
(ii) *For every bounded proper subsector T of S there exist $c_1, c_2 > 0$ with*

$$|f(z)| \leq c_1 h_{\mathbb{M}}(c_2|z|) = c_1 e^{-M(1/(c_2|z|))}, \qquad z \in T.$$

Looking at the condition in (ii), if we find a holomorphic function $V(z)$ in S whose real part's growth is suitably governed by the auxiliary function $M(t)$, then $\exp(-V(1/z))$ is expected to be flat. Indeed, this may be easily achieved in the Gevrey case, as we now show. Consider the Gevrey sequence of order $1/k > 0$, i.e. $\mathbb{M}_{1/k} = (p!^{1/k})_{p\in\mathbb{N}_0}$; we write $\tilde{\mathcal{A}}_{1/k}(S)$, $\Lambda_{1/k}$, $f \sim_{1/k} \hat{f}$ and so on for simplicity. It turns out (see, for example, [16]) that there exist $c_1, c_2 > 0$ such that for small t,

$$e^{-c_1 t^{-k}} \leq h_{1/k}(t) \leq e^{-c_2 t^{-k}},$$

and so, for large t, $c_2 t^k \leq M_{1/k}(t) \leq c_1 t^k$. Accordingly, flat functions are those exponentially decreasing of order k.

Theorem 4.16 (Gevrey Null Asymptotics) *Let S be a sector and $f \in \mathcal{O}(S)$. The following are equivalent:*

(i) $f \in \tilde{\mathcal{A}}_{1/k}(S)$ *and* $f \sim_{1/k} \hat{0}$.
(ii) *For every bounded proper subsector T of S there exist $c_1, c_2 > 0$ with*

$$|f(z)| \leq c_1 e^{-c_2|z|^{-k}}, \qquad z \in T.$$

If we choose $V(z) = z^k$, $z \in S_{1/k}$, then for every subsector T of $S_{1/k}$ there exists $b_1, b_2 > 0$ such that

$$\Re(V(z)) \geq b_1|z|^k \geq b_2 M_{1/k}(|z|), \qquad z \in T,$$

where $\Re$ stands for the real part. So, $\exp(-V(1/z)) = \exp(-z^{-k})$ is a nontrivial flat function in $\tilde{\mathcal{A}}_{1/k}(S_{1/k})$, and we can state the classical Watson's Lemma.

Theorem 4.17 (Watson's Lemma) *The class $\tilde{\mathcal{A}}_{1/k}(S_\gamma)$ is quasianalytic if, and only if, $\gamma > 1/k$.*

In this subsection we show how one can construct, in a similar way, nontrivial flat functions in the classes $\tilde{\mathcal{A}}_{\mathbb{M}}(S_{\omega(\mathbb{M})})$, for a strongly regular sequence $\mathbb{M}$ and for optimal unbounded sectors (and consequently for any sectorial region with the same opening), by relying on the notion of analytic proximate orders, appearing in the theory of growth of entire functions and developed, among others, by E. Lindelöf, G. Valiron, B. Ja. Levin, A.A. Goldberg, I.V. Ostrovskii and L.S. Maergoiz (see the references [17, 33, 34, 60]).

The main idea is to study the growth of a holomorphic function f in a sector by comparing $\log|f(z)|$, for $|z| = r$, to an expression of the form $r^{\rho(r)}$ for some suitably chosen function $\rho(r)$ in $(0,\infty)$ (instead of comparing to r^k, $k > 0$, what corresponds to the functions of finite exponential order). The admissible functions $\rho(r)$ are called proximate orders.

Definition 4.18 ([60]) We say a real function $\rho(r)$, defined on (c,∞) for some $c \geq 0$, is a *proximate order* if the following hold:

(i) $\rho(r)$ is continuous and piecewise continuously differentiable in (c,∞) (meaning that it is differentiable except at a sequence of points, tending to infinity, at any of which it is continuous and has distinct finite lateral derivatives),
(ii) $\rho(r) \geq 0$ for every $r > c$,
(iii) $\lim_{r\to\infty} \rho(r) = \rho < \infty$,
(iv) $\lim_{r\to\infty} r\rho'(r)\log(r) = 0$.

Observe that positive constant functions are proximate orders.

Definition 4.19 Two proximate orders $\rho_1(r)$ and $\rho_2(r)$ are said to be *equivalent* if

$$\lim_{r\to\infty} \big(\rho_1(r) - \rho_2(r)\big)\log(r) = 0.$$

Remark 4.20 If $\rho_1(r)$ and $\rho_2(r)$ are equivalent and $\lim_{r\to\infty} \rho_1(r) = \rho$, then $\lim_{r\to\infty} \rho_2(r) = \rho$ and $\lim_{r\to\infty} r^{\rho_1(r)}/r^{\rho_2(r)} = 1$.

Definition 4.21 Let $\rho(r)$ be a proximate order and f be an entire function. The *type of f associated with* $\rho(r)$ is

$$\sigma_f(\rho(r)) = \sigma_f := \limsup_{r\to\infty} \frac{\log \max_{|z|=r} |f(z)|}{r^{\rho(r)}}.$$

We say $\rho(r)$ is a *proximate order of f* if $0 < \sigma_f < \infty$.

Remark 4.22

(i) If $\rho(r) \to \rho > 0$ is a proximate order of f, then f is of exponential order ρ and there exists $K > 0$ such that for every $z \in \mathbb{C}$ with $|z|$ large enough one has

$$|f(z)| \leq \exp(K|z|^{\rho(|z|)}).$$

Moreover, and according to Remark 4.20, the type of f does not change if we substitute a proximate order of f by an equivalent one.

(ii) Every entire function of finite exponential order admits a proximate order, and so this notion allows one to forget about functions of minimal or maximal type. As an example, we note that the function $h(z) = 1/\Gamma(z)$, where $\Gamma(z)$

is Euler's Gamma function, is entire of order 1 and maximal type, and $\rho(r) = 1 + \log_2(r)/\log(r)$ is a proximate order for h (where $\log_2(r) = \log(\log(r))$).

The following result of Maergoiz [34] will be the key for our construction. For an arbitrary sector bisected by the positive real axis, it provides holomorphic functions whose restriction to $(0, \infty)$ is real and has a growth at infinity specified by a prescribed proximate order.

Theorem 4.23 ([34], Theorem 2.4) *Let $\rho(r)$ be a proximate order with $\rho(r) \to \rho > 0$ as $r \to \infty$. For every $\gamma > 0$ there exists an analytic function $V(z)$ in S_γ such that:*

(i) For every $z \in S_\gamma$,

$$\lim_{r\to\infty} \frac{V(zr)}{V(r)} = z^\rho,$$

uniformly in the compact sets of S_γ (in other words, V is of regular variation).

(ii) $\overline{V(z)} = V(\overline{z})$ for every $z \in S_\gamma$ (where, for $z = (|z|, \arg(z))$, we put $\overline{z} = (|z|, -\arg(z))$).

(iii) $V(r)$ is positive in $(0, \infty)$, monotone increasing and $\lim_{r\to 0} V(r) = 0$.

(iv) The function $t \in \mathbb{R} \to V(e^t)$ is strictly convex (i.e. V is strictly convex relative to $\log(r)$).

(v) The function $\log(V(r))$ is strictly concave in $(0, \infty)$.

(vi) The function $\rho_0(r) := \log(V(r))/\log(r)$, $r > 0$, is a proximate order equivalent to $\rho(r)$.

We denote by $\mathfrak{B}(\gamma, \rho(r))$ the class of such functions V. They share a property that will be crucial in the construction of flat functions.

Proposition 4.24 ([34], Property 2.9) *Let $\rho > 0$, $\rho(r)$ be a proximate order with $\rho(r) \to \rho$, $\gamma \geq 2/\rho$ and $V \in \mathfrak{B}(\gamma, \rho(r))$. Then, for every $\alpha \in (0, 1/\rho)$ there exist constants $b > 0$ and $R_0 > 0$ such that*

$$\Re(V(z)) \geq bV(|z|), \quad z \in S_\alpha,\ |z| \geq R_0.$$

Note that, given $V \in \mathfrak{B}(\gamma, \rho(r))$ and according to Theorem 4.23(vi), the rate of growth of $V(r)$ at infinity is the same as that of $r^{\rho(r)}$. So, coming back to the comments following Theorem 4.15, we are very close to our objective of obtaining flat functions in case the auxiliary function $M(r)$ may be written as $r^{d(r)}$, $d(r)$ being a proximate order. It is then natural to introduce the following function.

Definition 4.25 Given an (lc) sequence $\mathbb{M}$ with $\lim_{p\to\infty} m_p = \infty$ and with associated function $M(r)$, we define the function $d(r)$, for $r > \max\{1, m_0\}$, by

$$d(r) = \frac{\log(M(r))}{\log(r)}.$$

In the next result we obtain the desired flat functions in case $\omega(\mathbb{M}) < 2$ and $d(r)$ is a proximate order. Subsequently, we will indicate how to deal with the case $\omega(\mathbb{M}) \geq 2$. Finally, we will determine conditions on $\mathbb{M}$ amounting to $d(r)$ being a proximate order.

Theorem 4.26 ([54]) *Suppose $\mathbb{M}$ is a strongly regular sequence with $\omega(\mathbb{M}) < 2$ and such that $d(r)$ is a proximate order. Then, for every $V \in \mathfrak{B}(2\omega(\mathbb{M}), d(r))$ the function G defined in $S_{\omega(\mathbb{M})}$ by*

$$G(z) = \exp(-V(1/z))$$

belongs to $\tilde{\mathcal{A}}_{\mathbb{M}}(S_{\omega(\mathbb{M})})$ and it is a (nontrivial) flat function.

Proof It is enough to reason with sectors $S(0, \omega, r_0) \ll S_{\omega(\mathbb{M})}$, where $0 < \omega < \omega(\mathbb{M})$ and $r_0 > 0$. If $z \in S(0, \omega, r_0)$, we have $1/z \in S_\omega$. By our assumptions, $d(r)$ is a proximate order, and by (11) and (15), we have that

$$\lim_{r\to\infty} d(r) = \rho[M] = \frac{1}{\omega(\mathbb{M})}.$$

We are in a position to apply Proposition 4.24 with $\rho = 1/\omega(\mathbb{M})$, $\rho(r) = d(r)$, $\gamma = 2\omega(\mathbb{M})$ and $\alpha = \omega$, and deduce the existence of constants $R_0 > 0$ and $b > 0$ such that $\Re(V(\zeta)) \geq bV(|\zeta|)$ whenever $\zeta \in S_\omega$ with $|\zeta| \geq R_0$. Then, for $z \in S(0, \omega, 1/R_0)$ we obtain

$$|G(z)| = e^{-\Re(V(1/z))} \leq e^{-bV(1/|z|)},$$

and for a suitable $C > 0$ we will have $|G(z)| \leq Ce^{-bV(1/|z|)}$ for $z \in S(0, \omega, r_0)$. Now observe that, by the definition of $\mathfrak{B}(2\omega(\mathbb{M}), d(r))$, we know that the function $\log(V(r))/\log(r)$ is a proximate order equivalent to $d(r) = \log(M(r))/\log(r)$, so that, as a consequence of Remark 4.20, we have that there exists $c > 0$ such that for $r > 1/r_0$ one has $V(r) > cM(r)$, and

$$|G(z)| \leq Ce^{-bcM(1/|z|)} = C(h_{\mathbb{M}}(|z|))^{bc} \leq Ch_{\mathbb{M}}(D|z|), \qquad z \in S(0, \omega, r_0),$$

where $D > 0$ is a positive constant, suitably chosen according to whether $bc > 1$ or not [see property (5)]. It suffices to take into account Theorem 4.15 in order to conclude. □

Remark 4.27 In case $\omega(\mathbb{M}) \geq 2$, we may also construct nontrivial flat functions by taking into account the following facts:

(i) Given a strongly regular sequence $\mathbb{M} = (M_p)_{p\in\mathbb{N}_0}$ and a positive real number $s > 0$, the sequence of s-powers $\mathbb{M}^{(s)} := (M_p^s)_{p\in\mathbb{N}_0}$ is strongly regular (see Lemma 1.3.4 in [57]) and one easily checks that, with self-explaining notation, $\boldsymbol{m}^{(s)} = (m_p^s)_{p\in\mathbb{N}_0}$, $M^{(s)}(t) = sM(t^{1/s})$ for every $t \geq 0$, $\omega(\mathbb{M}^{(s)}) = s\omega(\mathbb{M})$,

$d^{(s)}(r) = d(r^{1/s})/s + \log(s)/\log(r)$ for r large enough, and

$$r(d^{(s)})'(r)\log(r) = \frac{1}{s}r^{1/s}d'(r^{1/s})\log(r^{1/s}) - \frac{\log(s)}{\log(r)}$$

whenever both sides are defined. So, it is clear that $d(r)$ is a proximate order if, and only if, $d^{(s)}(r)$ is.

(ii) If $\mathbb{M}$ is strongly regular, $\omega(\mathbb{M}) \geq 2$ and $d(r)$ is a proximate order, choose $s > 0$ such that $s\omega(\mathbb{M}) < 2$. By (i), we may apply Theorem 4.26 to $\mathbb{M}^{(s)}$ and obtain $G_0 \in \tilde{\mathcal{A}}_{\mathbb{M}^{(s)}}(S_{\omega(\mathbb{M}^{(s)})})$ which is flat. Now, the function G, given in $S_{\omega(\mathbb{M})}$ by $G(z) = G_0(z^s)$, is well-defined and it is plain to see that it is a nontrivial flat element in $\tilde{\mathcal{A}}_{\mathbb{M}}(S_{\omega(\mathbb{M})})$.

Our next objective is to characterize those $\mathbb{M}$ such that $d(r)$ is a proximate order. The function $d(r)$ is clearly continuous and piecewise continuously differentiable in its domain. By Theorem 2.24, it is clear that we need to show that

$$\rho[M] = \lim_{r\to\infty} d(r) \qquad \text{(instead of } \limsup_{r\to\infty}\text{)}$$

and that $rd'(r)\log(r) \to 0$ as $r \to \infty$. The following result provides us with statements equivalent to these facts.

Proposition 4.28 *Let $\mathbb{M}$ be a strongly regular sequence, and $d(r)$ its associated function. Consider the following conditions:*

(i) *$d(r)$ is a proximate order,*
(ii) $\lim_{r\to\infty} d(r) = \rho[M]$,
(iii) $\lim_{p\to\infty} \dfrac{\log(m_p)}{\log(p)} = \omega(\mathbb{M})$,
(iv) $\lim_{p\to\infty} \log\left(\dfrac{m_p}{M_p^{1/p}}\right) = \omega(\mathbb{M})$.

Then, (i) and (iv) are equivalent, and any of them implies (ii) and (iii), which are themselves equivalent. Moreover, in case (i) holds we have that $\omega(\mathbb{M}) = \gamma(\mathbb{M})$.

The proof being lengthy, we omit it and refer the reader to a work by J. Jiménez-Garrido and the author [22] and to its correction and continuation in [23].

Remark 4.29

(i) The previous condition (iv) holds for every sequence $\mathbb{M}_{\alpha,\beta}$, so that in any of these cases $d(r)$ is a proximate order and it is possible to construct flat functions in the corresponding classes. Nevertheless, in [23] we have been able to provide examples of strongly regular sequences for which $d(r)$ is not a proximate order.
(ii) In the Gevrey case, $\mathbb{M}_{1/k} = (p!^{1/k})_{\in\mathbb{N}_0}$, let us put $M_{1/k}(r), d_{1/k}(r)$, and so on, to denote the corresponding associated functions. Since for large r we have $c_2r^k \leq M_{1/k}(r) \leq c_1r^k$ for suitable constants $c_1, c_2 > 0$, then $\log(c_2) \leq (d_{1/k}(r) -$

$k)\log(r) \le \log(c_1)$ eventually. This shows one can work with the constant proximate order $\rho(r) \equiv k$, and any $V \in \mathfrak{B}(2/k, \rho(r))$ will provide us (due to Theorem 4.26, and since $V(r)$ will be bounded above and below by r^k times some suitable constants) with a flat function in the class $\tilde{\mathcal{A}}_{1/k}(S_{1/k})$. It is easy to see that $V(z) = z^k$ belongs to $\mathfrak{B}(2/k, \rho(r))$, and we obtain in this way the classical flat function in this situation, namely $G(z) = \exp(-z^{-k})$.

We are in a position to characterize quasianalyticity in the classes $\tilde{\mathcal{A}}_{\mathbb{M}}(S_\gamma)$.

Corollary 4.30 (Watson's Lemma) *Suppose* $\mathbb{M}$ *is strongly regular and such that* $d(r)$ *is a proximate order, and let* $\gamma > 0$ *be given. Then,* $\tilde{\mathcal{A}}_{\mathbb{M}}(S_\gamma)$ *is quasianalytic if, and only if,* $\gamma > \omega(\mathbb{M})$.

Remark 4.31 One may observe the difference with respect to the classes $\mathcal{A}_{\mathbb{M}}(S_\gamma)$, which could be quasianalytic for $\gamma = \omega(\mathbb{M})$ (see Example 4.9).

5 Surjectivity and Right Inverses for the Borel Map

As a next step in our study, we now devote ourselves to extend to general Carleman classes $\tilde{\mathcal{A}}_{\mathbb{M}}(S_\gamma)$ the following well-known result, characterizing the surjectivity of the Borel map in Gevrey classes.

Theorem 5.1 (Borel–Ritt–Gevrey, Ramis [50]) *For* $\alpha > 0$ *and* $\gamma > 0$, $\tilde{\mathcal{B}} : \tilde{\mathcal{A}}_\alpha(S_\gamma) \longrightarrow \Lambda_\alpha$ *is surjective if, and only if,* $\gamma \le \alpha$.

Our proof will be constructive, and will rest on the use of truncated Laplace-like transforms whose kernels are intimately related to the nontrivial flat functions obtained in Theorem 4.26. With any such kernel we will associate a sequence of moments which, in turn, will be equivalent to the sequence $\mathbb{M}$ we departed from. The results in this section which have not been endorsed to any author come from the paper [54].

Definition 5.2 Let $\mathbb{M} = (M_p)_{p\in\mathbb{N}_0}$ be a strongly regular sequence such that $d(r)$ is a proximate order, and consider $V \in \mathfrak{B}(2\omega(\mathbb{M}), d(r))$. We define the *kernel associated with* V as $e_V : S_{\omega(\mathbb{M})} \to \mathbb{C}$ given by

$$e_V(z) := ze^{-V(z)}, \qquad z \in S_{\omega(\mathbb{M})}.$$

Remark 5.3

(i) In a paper by Lastra et al. [29], similar kernels were obtained from flat functions constructed by Thilliez in [57]. The main difference with respect to the present one, which will be extremely important in forthcoming applications of these ideas to summability theory of formal power series, is that Thilliez needed to slightly restrict the opening of the optimal sector in order to construct such flat functions, while here we have been able to do it in the whole of $S_{\omega(\mathbb{M})}$.

(ii) The factor z appearing in e_V takes care of the integrability of $z^{-1}e_V(z)$ at the origin (see *(i)* in the next lemma). Indeed, it could be changed into any power z^α for positive α, where the principal branch of the power is to be considered. Our choice aims at simplicity.

Lemma 5.4 *The function e_V enjoys the following properties:*

(i) *$z^{-1}e_V(z)$ is integrable at the origin, it is to say, for any $t_0 > 0$ and $\tau \in \mathbb{R}$ with $|\tau| < \frac{\pi\omega(\mathbb{M})}{2}$ the integral $\int_0^{t_0} t^{-1}|e_V(te^{i\tau})|dt$ is finite.*

(ii) *For every $T \prec S_{\omega(\mathbb{M})}$ there exist $C, K > 0$ such that*

$$|e_V(z)| \le Ch_{\mathbb{M}}\left(\frac{K}{|z|}\right), \qquad z \in T. \tag{16}$$

(iii) *For every $x \in \mathbb{R}$, $x > 0$, the value $e_V(x)$ is positive real.*

Proof

(i) Let $t_0 > 0$ and $\tau \in \mathbb{R}$ with $|\tau| < \frac{\pi\omega(\mathbb{M})}{2}$. Since the function $\exp(-V(1/z))$ is flat, from Theorem 4.15 we obtain $c_1, c_2 > 0$ (depending on τ and t_0) such that

$$\int_0^{t_0} \frac{|e_V(te^{i\tau})|}{t}dt \le \int_0^{t_0} c_1 h_{\mathbb{M}}(c_2/t)dt.$$

As $h_{\mathbb{M}}$ is continuous and $h_{\mathbb{M}}(s) \equiv 1$ when $s \ge \frac{1}{m_1}$, this integral converges.

(ii) As before, given $T \prec S_{\omega(\mathbb{M})}$ and $R > 0$ there exist $c_1, c_2 > 0$ (depending on T and R) such that

$$|e_V(z)| \le c_1|z|h_{\mathbb{M}}(c_2/|z|), \qquad z \in T,\ |z| \ge R.$$

If $|z| \ge R$, we may apply (5) for $s = 2$ and the definition of $h_{\mathbb{M}}$ to deduce that

$$\begin{aligned}|e_V(z)| &\le c_1|z|\Big(h_{\mathbb{M}}\big(\frac{\rho(2)c_2}{|z|}\big)\Big)^2\\ &\le c_1|z|h_{\mathbb{M}}\big(\frac{\rho(2)c_2}{|z|}\big)M_2\big(\frac{\rho(2)c_2}{|z|}\big)^2 \le \frac{\rho(2)^2c_1c_2^2M_2}{R}h_{\mathbb{M}}\big(\frac{\rho(2)c_2}{|z|}\big).\end{aligned}$$

On the other hand, since V is bounded at the origin (because of property (iii) in Theorem 4.23), for $z \in T$ with $|z| < R$ we deduce that $e_V(z) = ze^{-V(z)}$ is bounded, and, in order to conclude, it suffices to observe that $h_{\mathbb{M}}(\rho(2)c_2/|z|)$ is bounded below by some positive constant for $|z| < R$.

(iii) $V(x)$ is real if $x > 0$, so $e_V(x) = xe^{-V(x)} > 0$.

□

Remark 5.5 As suggested in Remarks 4.29.(ii) and 5.3.(ii), in the Gevrey case $\mathbb{M}_{1/k}$, $k > 0$, it is natural and standard to consider the kernel

$$e_k(z) = kz^k \exp(-z^k), \qquad z \in S_{1/k}.$$

Definition 5.6 Let $V \in \mathfrak{B}(2\omega(\mathbb{M}), d(r))$. We define the *moment function associated with V* (or to e_V) as

$$m_V(\lambda) := \int_0^\infty t^{\lambda-1} e_V(t)dt = \int_0^\infty t^\lambda e^{-V(t)}dt.$$

From Lemma 5.4 we see that m_V, well defined in $\{z \in \mathbb{C} : \Re(\lambda) \geq 0\}$, is continuous in its domain, and holomorphic in $\{z \in \mathbb{C} : \Re(\lambda) > 0\}$. Moreover, $m_V(x) > 0$ for every $x \geq 0$. So, the following definition makes sense.

Definition 5.7 The sequence of positive real numbers $\mathfrak{m}_V = (m_V(p))_{p\in\mathbb{N}_0}$ is the *sequence of moments associated with V* (or to e_V).

Proposition 5.8 *Let e_V be a kernel associated with the strongly regular sequence $\mathbb{M}$, and $\mathfrak{m}_V = (m_V(p))_{p\in\mathbb{N}_0}$ the sequence of moments associated with V. Then $\mathbb{M}$ and $\mathfrak{m}_V$ are equivalent.*

Proof It suffices to work with $p \geq 1$. From (16) we have $C, K > 0$ such that

$$m_V(p) \leq C\int_0^\infty t^{p-1} h_{\mathbb{M}}(K/t)dt = C\int_0^{m_p} t^{p-1} h_{\mathbb{M}}(K/t)dt + C\int_{m_p}^\infty t^{p-1} h_{\mathbb{M}}(K/t)dt.$$

In the first integral of the right-hand side we take into account that $h_{\mathbb{M}}$ is bounded by 1, while in the second one we use the definition of $h_{\mathbb{M}}$ to obtain that $h_{\mathbb{M}}(K/t) \leq K^{p+1}M_{p+1}/t^{p+1}$, $t > m_p$. This yields

$$m_V(p) \leq C\frac{t^p}{p}\Big|_0^{m_p} - CK^{p+1}M_{p+1}\frac{1}{t}\Big|_{m_p}^\infty = \frac{Cm_p^p}{p} + CK^{p+1}\frac{M_{p+1}}{m_p}.$$

We have $M_{p+1} = m_p M_p$, and we may apply (ii.3) in Proposition 2.6 to obtain that

$$m_V(p) \leq C(A^{2p} + K^{p+1})M_p \leq 2C\max\{1, K\}(\max\{A^2, K\})^p M_p,$$

what concludes the first part of the proof.

On the other hand, Maergoiz [34, Theorem 3.3] has shown that the function $\log(V(r))/\log(r)$ is a proximate order of the entire function

$$F_V(z) = \sum_{p=0}^\infty \frac{z^p}{m_V(p)}, \qquad z \in \mathbb{C}. \tag{17}$$

By Remark 4.22(i) there exist constants $C_1, K_1 > 0$ such that for every $z \in \mathbb{C}$ one has

$$|F_V(z)| \le C_1 \exp(K_1 V(|z|)).$$

Now, recall from Theorem 4.23(vi) that $\log(V(r))/\log(r)$ is a proximate order equivalent to $d(r) = \log(M(r))/\log(r)$. Consequently, by Remark 4.20 there exists $K_2 > 0$ such that $V(r) \le K_2 M(r)$ for large r, and so we have

$$|F_V(z)| \le \tilde{C} \exp(\tilde{K} M(|z|)) \tag{18}$$

for every $z \in \mathbb{C}$ and suitably large constants $\tilde{C}, \tilde{K} > 0$. Finally, we take into account the following result by H. Komatsu.

Proposition 5.9 ([24], Proposition 4.5) *Let $M(r)$ be the function associated with a sequence $\mathbb{M}$ verifying (lc) and $\lim_{p\to\infty} m_p = \infty$. Given an entire function $F(z) = \sum_{p=0}^{\infty} a_p z^p$, $z \in \mathbb{C}$, the following statements are equivalent:*

(i) *There exist $C, K > 0$ such that $|F(z)| \le Ce^{M(K|z|)}$, $z \in \mathbb{C}$.*
(ii) *There exist $c, k > 0$ such that for every $p \in \mathbb{N}_0$, $|a_p| \le ck^p/M_p$.*

It suffices to apply this equivalence to the function F_V, by virtue of (18) and of Lemma 2.19, and we end the second part of the proof. □

Remark 5.10

(i) We record for the future that, as a consequence of the first part of the previous proof, given $K > 0$ there exist $C, D > 0$ such that for every $p \in \mathbb{N}$ one has

$$\int_0^\infty t^{p-1} h_{\mathbb{M}}(K/t)dt \le CD^p M_p. \tag{19}$$

(ii) In the Gevrey case $\mathbb{M}_{1/k}$ and with the kernel e_k introduced in Remark 5.5, we obtain the moment function $m_{1/k}(\lambda) = \Gamma(1 + \lambda/k)$ for $\Re(\lambda) \ge 0$, and we immediately check that $\mathbb{M}_{1/k}$ and $\mathfrak{m}_{1/k} = (m_{1/k}(p))_{p\in\mathbb{N}_0}$ are equivalent.

The proof of the next result, a generalization of the classical Borel–Ritt–Gevrey theorem, will only be sketched, since it is similar to the original one in the Gevrey case (see [2, 13, 50, 59]; in the several variables case, see [53]). Indeed, in a previous work by Lastra et al. [29, Theorem 4.1], this same technique was applied by using kernels derived from the flat functions of Thilliez [57], what obliged us to work in sectors of non-optimal opening. This drawback is now overcome under the additional assumption that the sequence $\mathbb{M}$ defines a proximate order $d(r)$, which is the case for all the examples appearing in applications.

Theorem 5.11 (Generalized Borel–Ritt–Gevrey Theorem) *Let $\mathbb{M}$ be a strongly regular sequence such that $d(r)$ is a proximate order, and let $\gamma > 0$ be given. The following statements are equivalent:*

(i) $\gamma \leq \omega(\mathbb{M})$,
(ii) *For every $\boldsymbol{a} = (a_p)_{p\in\mathbb{N}_0} \in \Lambda_{\mathbb{M}}$ there exists a function $f \in \tilde{\mathcal{A}}_{\mathbb{M}}(S_\gamma)$ such that*

$$f \sim_{\mathbb{M}} \hat{f} = \sum_{p\in\mathbb{N}_0} \frac{a_p}{p!} z^p,$$

i.e., $\tilde{\mathcal{B}}(f) = \boldsymbol{a}$. In other words, the Borel map $\tilde{\mathcal{B}} : \tilde{\mathcal{A}}_{\mathbb{M}}(S_\gamma) \longrightarrow \Lambda_{\mathbb{M}}$ is surjective.

Proof (i) $\Longrightarrow$ (ii) It is enough to treat the case $\gamma = \omega(\mathbb{M})$. Choose a function $V \in \mathfrak{B}(2\omega(\mathbb{M}), d(r))$, and consider the associated kernel e_V (see Definition 5.2) and sequence of moments $\mathfrak{m}_V = (m_V(p))_{p\in\mathbb{N}_0}$ (see Definition 5.7). Given $(a_p)_{p\in\mathbb{N}_0} \in \Lambda_{\mathbb{M}}$, there exist $C_1, D_1 > 0$ such that

$$|a_p| \leq C_1 D_1^p p! M_p, \quad p \in \mathbb{N}_0,$$

so that, by Proposition 5.8, the series

$$\hat{g} = \sum_{p\in\mathbb{N}_0} \frac{a_p}{p! m_V(p)} z^p \tag{20}$$

converges in a disc $D(0, R)$ for some $R > 0$, to a holomorphic function g. Choose $0 < R_0 < R$, and define

$$f(z) := \int_0^{R_0} e_V\left(\frac{u}{z}\right) g(u) \frac{du}{u}, \qquad z \in S_{\omega(\mathbb{M})}, \tag{21}$$

which turns out to be a holomorphic function in $S_{\omega(\mathbb{M})}$. Given $T \prec S_{\omega(\mathbb{M})}$, $N \in \mathbb{N}$ and $z \in T$, by standard arguments we have

$$\begin{aligned}
f(z) - \sum_{p=0}^{N-1} a_p \frac{z^p}{p!} &= f(z) - \sum_{p=0}^{N-1} \frac{a_p}{m_V(p)} m_V(p) \frac{z^p}{p!} \\
&= \int_0^{R_0} e_V\left(\frac{u}{z}\right) \sum_{k=0}^{\infty} \frac{a_k}{m_V(k)} \frac{u^k}{k!} \frac{du}{u} - \sum_{p=0}^{N-1} \frac{a_p}{m_V(p)} \int_0^{\infty} u^{p-1} e_V(u) du \frac{z^p}{p!} \\
&= \int_0^{R_0} e_V\left(\frac{u}{z}\right) \sum_{k=N}^{\infty} \frac{a_k}{m_V(k)} \frac{u^k}{k!} \frac{du}{u} - \int_{R_0}^{\infty} e_V\left(\frac{u}{z}\right) \sum_{p=0}^{N-1} \frac{a_p}{m_V(p)} \frac{u^p}{p!} \frac{du}{u} \\
&= f_1(z) + f_2(z).
\end{aligned}$$

By Proposition 5.8 there exist $C_2, D_2 > 0$ such that

$$\frac{|a_k|}{m_V(k)k!} \leq \frac{C_1 D_1^k k! M_k}{m_V(k)k!} \leq C_2 D_2^k \tag{22}$$

for all $k \in \mathbb{N}_0$, and so, taking $R_0 \leq (1-\epsilon)/D_2$ for some $\epsilon > 0$ if necessary, we get

$$|f_1(z)| \leq C_2 \int_0^{R_0} \left| e_V\left(\frac{u}{z}\right) \right| \sum_{k=N}^{\infty} (D_2 u)^k \frac{du}{u} \leq \frac{1}{\epsilon} C_2 D_2^N \int_0^{R_0} \left| e_V\left(\frac{u}{z}\right) \right| u^{N-1} du. \tag{23}$$

On the other hand, we have $u^p \leq R_0^p u^N / R_0^N$ for $u \geq R_0$ and $0 \leq p \leq N-1$. So, according to (22), we may write

$$\begin{aligned} \sum_{p=0}^{N-1} \frac{|a_p| u^p}{m_V(p)p!} &\leq \sum_{p=0}^{N-1} \frac{C_1 D_1^p p! M_p u^p}{m_V(p)p!} \leq \sum_{p=0}^{N-1} C_1 D_1^p C_2 D_2^p u^p \\ &\leq \frac{u^N}{R_0^N} \sum_{p=0}^{N-1} C_1 D_1^p C_2 D_2^p R_0^p \leq C_3 D_3^N u^N \end{aligned}$$

for some positive constants C_3, D_3, and deduce that

$$|f_2(z)| \leq C_3 D_3^N \int_{R_0}^{\infty} \left| e_V\left(\frac{u}{z}\right) \right| u^{N-1} du. \tag{24}$$

In view of (23) and (24), we are done if we prove that

$$\int_0^{\infty} \left| e_V\left(\frac{u}{z}\right) \right| u^{N-1} du \leq C_4 D_4^N m_V(N) |z|^N$$

for every $z \in T$ and for suitable $C_4, D_4 > 0$. But this is a straightforward consequence of Lemma 5.4.(ii) and the estimates in (19).

(ii) $\Longrightarrow$ (i) We will not provide all the details, but the argument could be completed easily with some of the results in the next sections. Anyway, the idea is similar to the one in the Gevrey case, see [2, p. 99]. For $\gamma > \omega(\mathbb{M})$, consider a path $\delta_{\omega(\mathbb{M})}$ in S_γ like the ones used in the classical Borel transform (see Sect. 6, just before Proposition 6.13), consisting of a segment from the origin to a point z_0 with $\arg(z_0) = \omega(\mathbb{M})(\pi + \varepsilon)/2$ (for some $\varepsilon \in (0, \pi)$), then the circular arc $|z| = |z_0|$ from z_0 to the point z_1 on the ray $\arg(z) = -\omega(\mathbb{M})(\pi + \varepsilon)/2$, and finally the segment from z_1 to the origin. Choose any lacunary series $\hat{g} = \sum_{p=0}^{\infty} b_p z^p / p!$ convergent in the unit disc to a function g that has no analytic continuation beyond that disc (for example, $\hat{g} = \sum_{p=0}^{\infty} z^{2^p}$). Then, the equivalence of $\mathbb{M}$ and $\mathbb{m}_V$ implies that $\boldsymbol{a} = (m_V(p) b_p)_{p \in \mathbb{N}_0}$ belongs to $\Lambda_{\mathbb{M}}$. If there would exist a function

$f \in \tilde{\mathcal{A}}_{\mathbb{M}}(S_\gamma)$ such that $f \sim_{\mathbb{M}} \hat{f} := \sum_{p\in\mathbb{N}_0} m_V(p)b_p z^p/p!$, one may check that the function

$$G(u) := \frac{-1}{2\pi i}\int_{\delta_{\omega(\mathbb{M})}} F_V(u/z)f(z)\frac{dz}{z}, \quad u \in S_\varepsilon,$$

where F_V is the function introduced in (17), is an analytic continuation of g into the unbounded sector S_ε. Since this is not possible, we deduce $\tilde{\mathcal{B}}$ is not surjective in this case. □

Finally, we will state a result concerning the surjectivity of the asymptotic Borel map $\tilde{\mathcal{B}}$ in the classes $\mathcal{A}_{\mathbb{M}}(S_\gamma)$, and the existence of suitably defined linear continuous right inverses for $\tilde{\mathcal{B}}$.

We mention that V. Thilliez obtained the following result.

Theorem 5.12 (Thilliez [57]) *Let $\mathbb{M}$ be a strongly regular sequence and γ a real number with $0 < \gamma < \gamma(\mathbb{M})$. Then there exist $c \geq 1$, depending only on $\mathbb{M}$ and γ, such that for every $A > 0$ there exists a linear continuous map $U_{\mathbb{M},A,\gamma} : \Lambda_{\mathbb{M},A} \to \mathcal{A}_{\mathbb{M},cA}(S_\gamma)$ with $\tilde{\mathcal{B}} \circ U_{\mathbb{M},A,\gamma} = Id_{\Lambda_{\mathbb{M},A}}$.*

In particular, $\tilde{\mathcal{B}} : \mathcal{A}_{\mathbb{M}}(S_\gamma) \to \Lambda_{\mathbb{M}}$ is surjective.

His proof is based on the works of Bruna [11], Petzsche [49], Bonet et al. [8], and Chaumat and Chollet [14], on the existence of linear continuous versions of Whitney's extension theorem for Whitney ultradifferentiable jets. We will reprove this result (see the implication (i) $\Longrightarrow$ (ii) in Theorem 5.13) with a completely different technique, resting on the truncated Laplace transform with kernel generated by a flat function, as before. The integral expression for the operators obtained is well suited for their extension to the several variable case. The interested reader may compare this and other approaches in [28, 29, 53].

Although $\gamma(\mathbb{M}) \leq \omega(\mathbb{M})$ in general, our result does not mean an improvement in this respect, since under our assumption that $d(r)$ is a proximate order one always has $\gamma(\mathbb{M}) = \omega(\mathbb{M})$ (see Proposition 4.28).

Theorem 5.13 *Let $\mathbb{M}$ be strongly regular and such that $d(r)$ is a proximate order, and let $\gamma > 0$ be given.*

(a) Each of the following assertions implies the next one:

- (i) $\gamma < \omega(\mathbb{M})$.
- (ii) *There exists $d \geq 1$ such that for every $A > 0$ there is a linear continuous operator*

$$T_{\mathbb{M},A,\gamma} : \Lambda_{\mathbb{M},A} \to \mathcal{A}_{\mathbb{M},dA}(S_\gamma)$$

such that $\tilde{\mathcal{B}} \circ T_{\mathbb{M},A,\gamma} = \mathrm{Id}_{\Lambda_{\mathbb{M},A}}$, the identity map in $\Lambda_{\mathbb{M},A}$.

- (iii) *The Borel map $\tilde{\mathcal{B}} : \mathcal{A}_{\mathbb{M}}(S_\gamma) \to \Lambda_{\mathbb{M}}$ is surjective.*

(iv) *There exists a function* $f \in \mathcal{A}_{\mathbb{M}}(S_\gamma)$ *such that for every* $m \in \mathbb{N}_0$ *we have* $f^{(m)}(0) = \delta_{1,m}$ *(where* $\delta_{1,m}$ *stands for Kronecker's delta).*

(b) If one has

$$\sum_{p=0}^{\infty}\Big(\frac{1}{(p+1)m_p}\Big)^{1/(\omega(\mathbb{M})+1)} = \infty, \tag{25}$$

then (i) is equivalent to:

(v) *The Borel map* $\tilde{\mathcal{B}} : \mathcal{A}_{\mathbb{M}}(S_\gamma) \to \Lambda_{\mathbb{M}}$ *is not injective, i.e.,* $\mathcal{A}_{\mathbb{M}}(S_\gamma)$ *is not quasianalytic.*

(c) If one has

$$\sum_{p=0}^{\infty}\Big(\frac{1}{m_p}\Big)^{1/\omega(\mathbb{M})} = \infty, \tag{26}$$

then all the conditions (i)–(v) are equivalent to each other.

Proof

(a) (i) $\Longrightarrow$ (ii) Fix $A > 0$. For every $\boldsymbol{a} = (a_p)_{p\in\mathbb{N}_0} \in \Lambda_{\mathbb{M},A}$, the series $\hat{g}$ given in (20) converges in a disc $D(0,R)$ not depending on $\boldsymbol{a}$. We define $T_{\mathbb{M},A,\gamma}(\boldsymbol{a})$ as the restriction to S_γ of the function defined in (21), which was shown to belong to $\tilde{\mathcal{A}}_{\mathbb{M}}(S_{\omega(\mathbb{M})})$ with uniform estimates in every $T \prec S_{\omega(\mathbb{M})}$. By combining the information in Proposition 3.4 with that in Remark 3.6, we conclude that there exists $d \geq 1$ such that $T_{\mathbb{M},A,\gamma}$ sends $\Lambda_{\mathbb{M},A}$ into $\mathcal{A}_{\mathbb{M},dA}(S_\gamma)$ and solves the problem.
(ii) $\Longrightarrow$ (iii) and (iii) $\Longrightarrow$ (iv) are immediate.

(b) By the definition of $\omega(\mathbb{M})$, we always have that (i) implies (v), and that (v) implies $\gamma \leq \omega(\mathbb{M})$. But condition (25) excludes equality by Theorem 4.3.

(c) Under condition (26), the fact that (iv) implies (i) may be obtained in the same way as Proposition 3.3 in [26]. So, (i)–(iv) are all equivalent to each other. According to (b), in order to conclude it suffices to prove that condition (26) implies condition (25), but this was obtained in Proposition 4.8.(i) in [27].

□

Remark 5.14 For Gevrey sequences, condition (26) holds, since it amounts to the divergence of the harmonic series. In general, condition (25) does not imply (26). For instance, as stated in Example 4.9, the sequence $\mathbb{M}_{\alpha,\beta}$ satisfies (25) if, and only if, $\alpha \geq \beta - 1$. One easily checks that it satisfies (26) if, and only if, $\alpha \geq \beta$. So, if $\beta - 1 \leq \alpha < \beta$ we have that $\mathbb{M}_{\alpha,\beta}$ satisfies (25) and not (26). Whenever this is the case, it is an open problem to decide whether (iv) in the previous theorem implies (i).

6 Summability in a Direction

If the opening of a sectorial region G is greater than $\pi\omega(\mathbb{M})$, then $\tilde{\mathcal{B}} : \tilde{\mathcal{A}}_{\mathbb{M}}(G) \longrightarrow \Lambda_{\mathbb{M}}$ is injective. So, we are ready for the introduction of a new concept of summability of formal power series in a direction.

Definition 6.1 Let $d \in \mathbb{R}$. We say $\hat{f} = \sum_{p\geq 0} \frac{f_p}{p!} z^p$ is $\mathbb{M}$-*summable in direction* d if there exist a sectorial region $G = G(d, \gamma)$, with $\gamma > \omega(\mathbb{M})$, and a function $f \in \tilde{\mathcal{A}}_{\mathbb{M}}(G)$ such that $f \sim_{\mathbb{M}} \hat{f}$.

In this case, by virtue of Proposition 3.4(ii) we have $(f_p)_{p\in\mathbb{N}_0} \in \Lambda_{\mathbb{M}}$. According to Theorem 4.6, f is unique with the property stated, and will be denoted

$$f = \mathcal{S}_{\mathbb{M},d}\hat{f}, \text{ the } \mathbb{M} - \text{sum of } \hat{f} \text{ in direction } d.$$

Our next aim in this section will be to develop suitable tools in order to recover f from $\hat{f}$ by means of formal and analytic transforms, in the same vein as in the classical theory for the Gevrey case and the so-called k-summability. We will follow the ideas in the theory of general moment summability methods put forward by W. Balser [2]. The case $\omega(\mathbb{M}) < 2$ is mainly treated, and indications will be given on how to work in the opposite situation.

Definition 6.2 Let $\mathbb{M}$ be strongly regular with $\omega(\mathbb{M}) < 2$. A pair of complex functions e, E are said to be *kernel functions for* $\mathbb{M}$-*summability* if:

(I) e is holomorphic in $S_{\omega(\mathbb{M})}$.
(II) $z^{-1}e(z)$ is locally uniformly integrable at the origin, i.e., there exists $t_0 > 0$, and for every $z_0 \in S_{\omega(\mathbb{M})}$ there exists a neighborhood U of z_0, $U \subset S_{\omega(\mathbb{M})}$, such that the integral $\int_0^{t_0} t^{-1} \sup_{z\in U} |e(t/z)| dt$ is finite.
(III) For every $\varepsilon > 0$ there exist $c, k > 0$ such that

$$|e(z)| \leq c h_{\mathbb{M}}\left(\frac{k}{|z|}\right) = c\, e^{-M(|z|/k)}, \qquad z \in S_{\omega(\mathbb{M})-\varepsilon}, \tag{27}$$

where $h_{\mathbb{M}}$ and M are the functions associated with $\mathbb{M}$, defined in (3) and (4), respectively.
(IV) For $x \in \mathbb{R}$, $x > 0$, the values of $e(x)$ are positive real.
(V) If we define the *moment function* associated with e,

$$m_e(\lambda) := \int_0^{\infty} t^{\lambda-1} e(t) dt, \qquad \Re(\lambda) \geq 0,$$

then the function E given by

$$E(z) := \sum_{p=0}^{\infty} \frac{z^p}{m_e(p)}, \qquad z \in \mathbb{C},$$

is entire, and there exist $C, K > 0$ such that

$$|E(z)| \leq \frac{C}{h_{\mathbb{M}}(K/|z|)} = Ce^{M(|z|/K)}, \quad z \in \mathbb{C}. \tag{28}$$

(VI) $z^{-1}E(1/z)$ is locally uniformly integrable at the origin in the sector $S(\pi, 2-\omega(\mathbb{M}))$, in the sense that there exists $t_0 > 0$, and for every $z_0 \in S(\pi, 2-\omega(\mathbb{M}))$ there exists a neighborhood U of z_0, $U \subset S(\pi, 2-\omega(\mathbb{M}))$, such that the integral $\int_0^{t_0} t^{-1} \sup_{z\in U} |E(z/t)| dt$ is finite.

From (I)–(IV) we see that the function m_e is continuous in $\{\lambda \in \mathbb{C} : \Re(\lambda) \geq 0\}$, holomorphic in $\{\lambda \in \mathbb{C} : \Re(\lambda) > 0\}$, and $m_e(x) > 0$ for every $x \geq 0$.

Remark 6.3

(i) The existence of such kernels may be deduced, as we will show, by taking into account the construction of flat functions in $\tilde{\mathcal{A}}_{\mathbb{M}}(S_{\omega(\mathbb{M})})$ whenever $d(r)$ is a proximate order.

(ii) According to Definition 6.2(V), the knowledge of e is enough to determine the pair of kernel functions. So, in the sequel we will frequently omit the function E in our statements.

(iii) In case $\omega(\mathbb{M}) \geq 2$, condition (VI) in Definition 6.2 does not make sense. However, we note that for a positive real number $s > 0$ the sequence of $1/s$-powers $\mathbb{M}^{(1/s)} := (M_p^{1/s})_{n\in\mathbb{N}_0}$ is also strongly regular and, as it is easy to check,

$$h_{\mathbb{M}^{(1/s)}}(t) = \left(h_{\mathbb{M}}(t^s)\right)^{1/s}, \quad t \geq 0, \tag{29}$$

and $\omega(\mathbb{M}^{(1/s)}) = \omega(\mathbb{M})/s$. So, following the ideas of Sect. 5.6 in [2], we will say that a complex function e is a kernel for $\mathbb{M}$-summability if there exist $s > 0$ with $\omega(\mathbb{M})/s < 2$, and a kernel $\tilde{e} : S_{\omega(\mathbb{M})/s} \to \mathbb{C}$ for $\mathbb{M}^{(1/s)}$-summability such that

$$e(z) = \tilde{e}(z^{1/s})/s, \qquad z \in S_{\omega(\mathbb{M})}.$$

If one defines the moment function m_e as before, it is plain to see that $m_e(\lambda) = m_{\tilde{e}}(s\lambda)$, $\Re(\lambda) \geq 0$. The properties verified by $\tilde{e}$ and $m_{\tilde{e}}$ are easily translated into similar ones for e, but in this case the function

$$E(z) = \sum_{p=0}^{\infty} \frac{z^p}{m_e(p)} = \sum_{p=0}^{\infty} \frac{z^p}{m_{\tilde{e}}(sp)}$$

does not have the same properties as before, and one rather pays attention to the kernel associated with $\tilde{e}$,

$$\tilde{E}(z) = \sum_{p=0}^{\infty} \frac{z^p}{m_{\tilde{e}}(p)} = \sum_{p=0}^{\infty} \frac{z^p}{m_e(p/s)}, \tag{30}$$

which will behave as indicated in (v) and (vi) of Definition 6.2 for such a kernel for $\mathbb{M}^{(1/s)}$-summability.

It is worth remarking that, once such an s as in the definition exists, one easily checks that for any real number $t > \omega(\mathbb{M})/2$ a kernel $\hat{e}$ for $\mathbb{M}^{(1/t)}$-summability exists with $e(z) = \hat{e}(z^{1/t})/t$.

Definition 6.4 Let e be a kernel for $\mathbb{M}$-summability and m_e its associated moment function. The sequence of positive real numbers $\mathfrak{m}_e = (m_e(p))_{p\in\mathbb{N}_0}$ is known as the *sequence of moments* associated with e.

The following result is a consequence of the estimates, for the kernels e and E, appearing in (27) and (28) respectively. We omit its proof, since it may be easily adapted from the proof of Proposition 5.8.

Proposition 6.5 *Let $\mathbb{M} = (M_p)_{p\in\mathbb{N}_0}$ be a strongly regular sequence, e be a kernel function for $\mathbb{M}$-summability, and $\mathfrak{m}_e = (m_e(p))_{p\in\mathbb{N}_0}$ the sequence of moments associated with e. Then $\mathbb{M}$ and $\mathfrak{m}_e$ are equivalent.*

Remark 6.6 For any kernel e for $\mathbb{M}$-summability one may prove that the sequence of moments $\mathfrak{m}_e = (m_e(p))_{p\in\mathbb{N}_0}$ is also strongly regular: Firstly, up to multiplication by a constant scaling factor, one may always suppose that $m_e(0) = 1$. Property (lc) is a consequence of Hölder's inequality, since for every $p \in \mathbb{N}$ one has

$$m_e(p)^2 = \|t^{p-1}e(t)\|_1^2 \le \|(t^{p-2}e(t))^{1/2}\|_2^2\|(t^pe(t))^{1/2}\|_2^2 = m_e(p-1)m_e(p+1)$$

(where $\|\cdot\|_1$ and $\|\cdot\|_2$ are the standard L^1 and L^2 norms). Regarding conditions (mg) and (snq), they are preserved as we know that $\mathfrak{m}_e \approx \mathbb{M}$. Note that the statement for (snq) has not been proved, but may be found in the work of Petzsche [49, Corollary 3.2].

Bearing this fact in mind, in Definition 6.2 one could depart not from a strongly regular sequence $\mathbb{M}$, but from a kernel e, initially defined and positive in direction $d = 0$, whose moment function $m_e(\lambda)$ is supposed to be well-defined for $\lambda \ge 0$, and such that the sequence $\mathfrak{m}_e$ is strongly regular. This allows one to consider the constant $\omega(\mathfrak{m}_e)$, which would equal $\omega(\mathbb{M})$ according to Proposition 5.8 and Remark 4.8, and also the function $h_{\mathfrak{m}_e}$, in terms of which one may rephrase all the items in Definition 6.2, specially the estimates in (27) and (28), with exactly the same meaning, according to the relationship between $h_{\mathfrak{m}_e}$ and $h_{\mathbb{M}}$ indicated in (6). This insight will be fruitful in Sect. 7, when dealing with so called moment-partial differential equations.

We are ready for the construction of kernels.

Theorem 6.7 *Suppose $\mathbb{M}$ is a strongly regular sequence with $\omega(\mathbb{M}) < 2$ and such that the function $d(r)$ is a proximate order. Then, for every $V \in \mathfrak{B}(2\omega(\mathbb{M}), d(r))$, the function e_V defined in $S_{\omega(\mathbb{M})}$ by*

$$e_V(z) = z \exp(-V(z))$$

is a kernel of $\mathbb{M}$-summability.

Proof The properties (I) to (IV) in Definition 6.2 are clear or have been obtained, with some slight modification, in Lemma 5.4. Now, reasoning as in the proof of Proposition 5.8 we see that, if $(m_V(p))_{p\in\mathbb{N}_0}$ is the sequence of moments of e_V, the function

$$E_V(z) = \sum_{p=0}^{\infty} \frac{z^p}{m_V(p)}, \qquad z \in \mathbb{C},$$

is entire and we have

$$|E_V(z)| \le \tilde{C} \exp(\tilde{K} M(|z|))$$

for every $z \in \mathbb{C}$ and suitably large constants $\tilde{C}, \tilde{K} > 0$, and so condition (V) in Definition 6.2 is satisfied. Finally, we take into account the following result.

Proposition 6.8 ([34], (3.25)) *Let $\rho(r)$ be a proximate order with $\rho > 1/2$, $\gamma \ge 2/\rho$ and $V \in \mathfrak{B}(\gamma, \rho(r))$. Then, for every $\varepsilon > 0$ such that $\varepsilon < \pi(1 - 1/(2\rho))$ we have, uniformly as $|z| \to \infty$, that (in Landau's notation)*

$$E_V(z) = O\left(\frac{1}{|z|}\right), \qquad \frac{\pi}{2\rho} + \varepsilon \le |\arg z| \le \pi.$$

This information easily implies that also condition (VI) in Definition 6.2 is fulfilled, what concludes the proof. □

Remark 6.9 In case $\omega(\mathbb{M}) \ge 2$, we consider $s > 0$ and $\mathbb{M}^{(1/s)} := (M_n^{1/s})_{n\in\mathbb{N}_0}$ as in Remark 6.3(iii), in such a way that $\omega(\mathbb{M}^{(1/s)}) = \omega(\mathbb{M})/s < 2$. By Remark 4.27(i), $d(r)$ is a proximate order if, and only if, $d^{(1/s)}(r)$ is. Were this the case, by the previous result we would have kernels $\tilde{e}$ for $\mathbb{M}^{(1/s)}$-summability, and the function $e(z) = \tilde{e}(z^{1/s})/s$ will be a kernel for $\mathbb{M}$-summability.

The next definition resembles that of functions of exponential growth, playing a fundamental role when dealing with Laplace and Borel transforms in the theory of k-summability for Gevrey classes. For convenience, we will say a holomorphic function f in a sector S is *continuous at the origin* if $\lim_{z\to 0,\, z\in T} f(z)$ exists for every $T \ll S$.

Definition 6.10 Let $\mathbb{M} = (M_p)_{p\in\mathbb{N}_0}$ be a sequence of positive real numbers which is (lc) and such that $\boldsymbol{m}$ tends to infinity, and consider an unbounded sector S in $\mathcal{R}$.

The set $\mathcal{O}^{\mathbb{M}}(S)$ consists of the holomorphic functions f in S, continuous at 0 and having $\mathbb{M}$-growth in S, i.e. such that for every unbounded proper subsector T of S there exist $r, c, k > 0$ such that for every $z \in T$ with $|z| \geq r$ one has

$$|f(z)| \leq \frac{c}{h_{\mathbb{M}}(k/|z|)} = c\exp(M(|z|/k)). \tag{31}$$

Remark 6.11 Since continuity at 0 has been asked for, $f \in \mathcal{O}^{\mathbb{M}}(S)$ implies that for every $T \prec S$ there exist $c, k > 0$ such that for every $z \in T$ one has (31).

We are ready for the introduction of the $\mathbb{M}$-Laplace transform.

Given a kernel e for $\mathbb{M}$-summability, a sector $S = S(d, \alpha)$ and $f \in \mathcal{O}^{\mathbb{M}}(S)$, for any direction τ in S we define the operator $T_{e,\tau}$ sending f to its *e-Laplace transform in direction* τ, defined as

$$(T_{e,\tau}f)(z) := \int_0^{\infty(\tau)} e(u/z)f(u)\frac{du}{u}, \quad |\arg(z) - \tau| < \omega(\mathbb{M})\pi/2, \ |z| \text{ small enough}, \tag{32}$$

where the integral is taken along the half-line parameterized by $t \in (0, \infty) \mapsto te^{i\tau}$. We have the following result.

Proposition 6.12 *For a sector $S = S(d, \alpha)$ and $f \in \mathcal{O}^{\mathbb{M}}(S)$, the family $\{T_{e,\tau}f\}_{\tau \text{ in } S}$ defines a holomorphic function $T_e f$ in a sectorial region $G(d, \alpha + \omega(\mathbb{M}))$.*

Proof Let $\tau \in \mathbb{R}$ be a direction in S, i.e., such that $|\tau - d| < \alpha\pi/2$. We will show that for every β with $0 < \beta < \omega(\mathbb{M})$, there exists $r = r(f, \tau, \beta) > 0$ such that $T_{e,\tau}f$ is holomorphic in $S(\tau, \beta, r)$. Hence, $T_{e,\tau}f$ will be holomorphic in $G_\tau := \cup_{0<\beta<\omega(\mathbb{M})} S(\tau, \beta, r)$, which is a sectorial region $G_\tau = G(\tau, \omega(\mathbb{M}))$.

For every $u, z \in \mathcal{R}$ with $\arg(u) = \tau$ and $|\arg(z) - \tau| < \omega(\mathbb{M})\pi/2$ we have that $u/z \in S_{\omega(\mathbb{M})}$, so that the expression under the integral sign in (32) makes sense. We fix $a > 0$, and write

$$\int_0^{\infty(\tau)} e(u/z)f(u)\frac{du}{u} = \int_0^{ae^{i\tau}} e(u/z)f(u)\frac{du}{u} + \int_{ae^{i\tau}}^{\infty(\tau)} e(u/z)f(u)\frac{du}{u}.$$

Since f is continuous at the origin, and because of Definition 6.2(II), it is straightforward to apply Leibniz's rule for parametric integrals and deduce that the first integral in the right-hand side defines a holomorphic function in $S(\tau, \omega(\mathbb{M}))$. Regarding the second integral, for u as before and by Definition 6.10 there exist $c_1, k_1 > 0$ such that

$$|f(u)| \leq c_1\big(h_{\mathbb{M}}(k_1/|u|)\big)^{-1}.$$

Also, for z such that $|\arg(z) - \tau| < \beta\pi/2$ we have that $u/z \in S_\beta$, and the property 6.2(III) provides us with constants $c_2, k_2 > 0$ such that

$$|e(u/z)| \leq c_2 h_{\mathbb{M}}(k_2|z|/|u|),$$

so that

$$\Big|\frac{1}{u}e(u/z)f(u)\Big| \leq \frac{c_1c_2}{|u|}\frac{h_{\mathbb{M}}(k_2|z|/|u|)}{h_{\mathbb{M}}(k_1/|u|)}.$$

Let $\rho(2) > 0$ be the constant appearing in (5) for $s = 2$, and consider $r := k_1/(\rho(2)k_2) > 0$. For any $z \in S(\tau, \beta, r)$ we have that $\rho(2)k_2|z| < k_1$, and from (5) and the monotonicity of $h_{\mathbb{M}}$ we deduce that

$$\Big|\frac{1}{u}e(u/z)f(u)\Big| \leq \frac{c_1c_2}{|u|}\frac{h^2_{\mathbb{M}}(\rho(2)k_2|z|/|u|)}{h_{\mathbb{M}}(k_1/|u|)} \leq \frac{c_1c_2}{|u|}h_{\mathbb{M}}(k_1/|u|).$$

By the very definition of $h_{\mathbb{M}}$ we have that $h_{\mathbb{M}}(k_1/|u|) < M_1k_1/|u|$, so the right-hand side of the last inequality is an integrable function of $|u|$ in (a, ∞), and again Leibniz's rule allows us to conclude the desired analyticity for the second integral.

Let $\sigma \in \mathbb{R}$ with $|\sigma - d| < \alpha\pi/2$. The map $T_{e,\sigma}f$ is a holomorphic function in a sectorial region $G_\sigma = G(\sigma, \omega(\mathbb{M}))$ which will overlap with G_τ whenever τ and σ are close enough. Since we know that

$$\lim_{t\to\infty} th_{\mathbb{M}}(1/t) = 0,$$

by Cauchy's residue theorem we easily deduce that $T_{e,\tau}f(z) \equiv T_{e,\sigma}f(z)$ whenever both maps are defined. Thus the family $\{T_{e,\tau}f\}_{\tau \text{ in } S}$ defines a holomorphic function T_ef in the union of the sectorial regions G_τ, which is indeed again a sectorial region $G = G(d, \alpha + \omega(\mathbb{M}))$. □

We now define the generalized Borel transforms.

Suppose $\omega(\mathbb{M}) < 2$, and let $G = G(d, \alpha)$ be a sectorial region with $\alpha > \omega(\mathbb{M})$, and $f : G \to \mathbb{C}$ be holomorphic in G and continuous at 0. For $\tau \in \mathbb{R}$ such that $|\tau - d| < (\alpha - \omega(\mathbb{M}))\pi/2$ we may consider a path $\delta_{\omega(\mathbb{M})}(\tau)$ in G like the ones used in the classical Borel transform, consisting of a segment from the origin to a point z_0 with $\arg(z_0) = \tau + \omega(\mathbb{M})(\pi + \varepsilon)/2$ (for some suitably small $\varepsilon \in (0, \pi)$), then the circular arc $|z| = |z_0|$ from z_0 to the point z_1 on the ray $\arg(z) = \tau - \omega(\mathbb{M})(\pi + \varepsilon)/2$ (traversed clockwise), and finally the segment from z_1 to the origin.

Given kernels e, E for $\mathbb{M}$-summability, we define the operator $T^-_{e,\tau}$ sending f to its *e-Borel transform in direction* τ, defined as

$$(T^-_{e,\tau}f)(u) := \frac{-1}{2\pi i}\int_{\delta_{\omega(\mathbb{M})}(\tau)} E(u/z)f(z)\frac{dz}{z}, \quad u \in S(\tau, \varepsilon_0), \quad \varepsilon_0 \text{ small enough.}$$

In case $\omega(\mathbb{M}) \geq 2$, choose $s > 0$ and a kernel $\tilde{e}$ for $\mathbb{M}^{(1/s)}$-summability as in Remark 6.3(iii), and let $T^{-}_{\tilde{e},\tau}$ be defined as before, where the kernel under the integral sign is the function $\tilde{E}$ given in (30). Then, if ϕ_s is the operator sending a function f to the function $f(z^s)$, we define $T^{-}_{e,\tau}$ by the identity

$$\phi_s \circ T^{-}_{e,\tau} = T^{-}_{\tilde{e},\tau} \circ \phi_s, \tag{33}$$

in the same way as in [2, p. 90].

Proposition 6.13 *For $G = G(d,\alpha)$ and $f : G \to \mathbb{C}$ as above, the family*

$$\{T^{-}_{e,\tau}f\}_{\tau},$$

where τ is a real number such that $|\tau - d| < (\alpha - \omega(\mathbb{M}))\pi/2$, defines a holomorphic function $T^{-}_{e}f$ in the sector $S = S(d, \alpha - \omega(\mathbb{M}))$. Moreover, $T^{-}_{e}f$ is of $\mathbb{M}$-growth in S.

Proof Due to the identity (33), it is clearly sufficient to prove our claim in the case $\omega(\mathbb{M}) < 2$. Since f is holomorphic in G and continuous at 0, for every $\tau \in \mathbb{R}$ such that $|\tau - d| < (\alpha - \omega(\mathbb{M}))\pi/2$ the condition in Definition 6.2(VI) implies that $T^{-}_{e,\tau}f$ is holomorphic in the sector $S(\tau, \varepsilon\omega(\mathbb{M})/\pi)$, where $\varepsilon > 0$ is the one entering in the definition of $\delta_{\omega(\mathbb{M})}(\tau)$, and it is small enough so that u/z stays in the sector $S(\pi, 2 - \omega(\mathbb{M}))$ as $u \in S(\tau, \varepsilon\omega(\mathbb{M})/\pi)$ and z runs over the two segments in $\delta_{\omega(\mathbb{M})}(\tau)$. Cauchy's theorem easily implies that the family $\{T^{-}_{e,\tau}f\}$, when $\tau \in \mathbb{R}$ and $|\tau - d| < (\alpha - \omega(\mathbb{M}))\pi/2$, defines a holomorphic function $T^{-}_{e}f$ in the sector $S = S(d, \alpha - \omega(\mathbb{M}))$. Let us finally check that $T^{-}_{e}f$ is of $\mathbb{M}$-growth in S. By compactness, it suffices to work on a proper unbounded subsector T of $S(\tau, \varepsilon\omega(\mathbb{M})/\pi)$, for τ and ε as before. Put $\delta_{\omega(\mathbb{M})}(\tau) = \delta_1 + \delta_2 + \delta_3$, where δ_1 and δ_3 are the aforementioned segments in directions, say, θ_1 and θ_3, and δ_2 is the circular arc with radius $r_2 > 0$. As f is continuous at the origin, there exists $M' > 0$ such that $|f(z)| \leq M'$ for every z in the trace of $\delta_{\omega(\mathbb{M})}(\tau)$. So, for every $u \in T$ and $j = 1, 3$ we have

$$\begin{aligned}\left|\frac{-1}{2\pi i}\int_{\delta_j} E(u/z)f(z)\frac{dz}{z}\right| &\leq \frac{M'}{2\pi}\int_0^{r_2}\frac{1}{t}|E(ue^{-i\theta_j}/t)|dt \\ &= \frac{M'}{2\pi}\int_0^{r_2/|u|}\frac{1}{s}|E(e^{i(\arg(u)-\theta_j)}/s)|ds,\end{aligned}$$

after the change of variable $t = |u|s$. According to the condition in Definition 6.2(VI), these expressions uniformly tend to 0 as u tends to infinity in T. On the other hand, the estimates for E in (28) allow us to write

$$\left|\frac{-1}{2\pi i}\int_{\delta_2} E(u/z)f(z)\frac{dz}{z}\right| \leq \frac{M'(\theta_1 - \theta_3)}{2\pi}\max_{|z|=r_2}|E(u/z)| \leq \frac{CM'(\theta_1 - \theta_3)}{2\pi h_{\mathbb{M}}(Kr_2/|u|)}.$$

So, this integral has the desired growth and the conclusion follows. □

In the next paragraph we follow the same ideas in [2, p. 87–88] in order to justify the forthcoming definition of formal Laplace and Borel transforms.

Let $\mathbb{M} = (M_p)_{p\in\mathbb{N}_0}$ be a strongly regular sequence, $S = S_\alpha$ and e a kernel for $\mathbb{M}$-summability. It is clear that for every $\lambda \in \mathbb{C}$ with $\Re(\lambda) \geq 0$, the function $f_\lambda(z) = z^\lambda$ belongs to the space $\mathcal{O}^{\mathbb{M}}(S)$. From Proposition 6.12, one can define $T_e f_\lambda(z)$ for every z in an appropriate sectorial region G. Moreover, for $z \in G$ and an adequate choice of $\tau \in \mathbb{R}$ one has

$$T_e f_\lambda(z) = \int_0^{\infty(\tau)} e\left(\frac{u}{z}\right) u^{\lambda-1} du.$$

In particular, for $z \in \mathcal{R}$ with $\arg(z) = \tau$, the change of variable $u/z = t$ turns the preceding integral into

$$T_e f_\lambda(z) = \int_0^{\infty} e(t) z^{\lambda-1} t^{\lambda-1} z dt = m_e(\lambda) z^\lambda. \tag{34}$$

Suppose we are given now an entire function $F(z) = \sum_{p\geq 0} a_p z^p$ of $\mathbb{M}$-growth in $\mathbb{C}$. Taking Proposition 5.9 into account, one may justify termwise integration to obtain that

$$T_e F(z) = \sum_{p=0}^{\infty} a_p m_e(p) z^p$$

whenever $|z|$ is small enough. In case $\omega(\mathbb{M}) < 2$, and particularizing this result for $F = E$ (the kernel function corresponding to e), we deduce that for every $z \neq 0$ and $w \neq 0$ such that $|z/w| < 1$ one has

$$\frac{w}{w-z} = \int_0^{\infty(\tau)} e(u/z) E(u/w) \frac{du}{u}, \tag{35}$$

a formula which remains valid as long as both sides are defined. Suppose now that f is holomorphic in a sectorial region $G(d, \alpha)$, with $\alpha > \omega(\mathbb{M})$, and continuous at the origin. By Propositions 6.13 and 6.12, $T_e T_e^- f$ is well defined; a change in the order of integration and the use of (35) prove that

$$T_e T_e^- f = f. \tag{36}$$

Finally, since f_λ (defined above) is continuous at the origin, it makes sense to compute

$$T_e^- f_\lambda(u) = \frac{-1}{2\pi i} \int_{\delta_{\omega(\mathbb{M})}(\tau)} E(u/z) z^{\lambda-1} dz.$$

Putting $u/z = t$, the integral is changed into

$$T_e^- f_\lambda(u) = \frac{u^\lambda}{2\pi i} \int_{\gamma_z} E(t) t^{-\lambda-1} dt,$$

for a corresponding path γ_z. However, Cauchy's theorem allows one to choose one and the same path of integration as long as z runs in a suitably small disk, and we deduce, by the identity principle, that $T_e^- f_\lambda(u)$ is a constant multiple of u^λ in $S(d, \alpha - \omega(\mathbb{M}))$. According to (34) and (36), we conclude that

$$T_e^- f_\lambda(u) = \frac{u^\lambda}{m_e(\lambda)}. \tag{37}$$

Observe that, taking into account (33), the same will be true if $\omega(\mathbb{M}) \geq 2$. Therefore, it is adequate to make the following definitions.

Definition 6.14 Given a strongly regular sequence $\mathbb{M}$, consider a kernel e for $\mathbb{M}$-summability. The formal e-Laplace transform $\hat{T}_e : \mathbb{C}[[z]] \to \mathbb{C}[[z]]$ is given by

$$\hat{T}_e\Big(\sum_{p=0}^{\infty} a_p z^p\Big) := \sum_{p=0}^{\infty} m_e(p) a_p z^p, \qquad \sum_{p=0}^{\infty} a_p z^p \in \mathbb{C}[[z]].$$

Accordingly, we define the formal e-Borel transform $\hat{T}_e^- : \mathbb{C}[[z]] \to \mathbb{C}[[z]]$ by

$$\hat{T}_e^-\Big(\sum_{p=0}^{\infty} a_p z^p\Big) := \sum_{p=0}^{\infty} \frac{a_p}{m_e(p)} z^p, \qquad \sum_{p=0}^{\infty} a_p z^p \in \mathbb{C}[[z]].$$

The operators $\hat{T}_e$ and $\hat{T}_e^-$ are inverse to each other.

The next result lets us know how these analytic and formal transforms interact with general asymptotic expansions. Given two sequences of positive real numbers $\mathbb{M} = (M_p)_{p\in\mathbb{N}_0}$ and $\mathbb{M}' = (M'_p)_{p\in\mathbb{N}_0}$, we consider the sequences $\mathbb{M}\cdot\mathbb{M}'$ and $\mathbb{M}'/\mathbb{M} = (M'_p/M_p)_{p\in\mathbb{N}_0}$. We note that if both sequences are strongly regular, $\mathbb{M}\cdot\mathbb{M}'$ is again strongly regular (see Proposition 2.18), but $\mathbb{M}'/\mathbb{M}$ might not be.

Theorem 6.15 *Suppose $\mathbb{M}$ is strongly regular and e is a kernel for $\mathbb{M}$-summability. For any sequence $\mathbb{M}'$ of positive real numbers the following hold:*

(i) *If $f \in \mathcal{O}^{\mathbb{M}}(S(d,\alpha))$ and $f \sim_{\mathbb{M}'} \hat{f}$, then $T_e f \sim_{\mathbb{M}\cdot\mathbb{M}'} \hat{T}_e \hat{f}$ in a sectorial region $G(d, \alpha + \omega(\mathbb{M}))$.*

(ii) *If $f \sim_{\mathbb{M}'} \hat{f}$ in a sectorial region $G(d,\alpha)$ with $\alpha > \omega(\mathbb{M})$, then $T_e^- f \sim_{\mathbb{M}'/\mathbb{M}} \hat{T}_e^- \hat{f}$ in the sector $S(d, \alpha - \omega(\mathbb{M}))$.*

Proof

(i) From Proposition 6.12 we know that $g := T_e f \in \mathcal{O}(G(d,\alpha+\omega(\mathbb{M})))$ for a sectorial region $G = G(d,\alpha+\omega(\mathbb{M}))$. Put $\hat{f} = \sum_{p=0}^{\infty} f_p u^p$. Given $\delta \in (0,\alpha)$, there exist $c,k>0$ such that for every $u \in S(d,\delta,1)$ and every $n \in \mathbb{N}$ one has

$$\Big|f(u) - \sum_{k=0}^{n-1} f_k u^k\Big| \le c k^n M'_n |u|^n.$$

Then, we deduce that $|f_n| \le c k^n M'_n$ for every n. Also, since f is of $\mathbb{M}$-growth, for every $u \in S(d,\delta)$ we have, by suitably enlarging the constants,

$$\Big|f(u) - \sum_{k=0}^{n-1} f_k u^k\Big| \le \frac{c_1 k_1^n M'_n |u|^n}{h_{\mathbb{M}}(k_1/|u|)}. \tag{38}$$

Observe that, by (34), for every $z \in G$ we have

$$g(z) - \sum_{k=0}^{n-1} m_e(k) f_k z^k = T_e\Big(f(u) - \sum_{k=0}^{n-1} f_k u^k\Big)(z).$$

So, given $\tau \in \mathbb{R}$ with $|\tau - d| < \alpha\pi/2$ and $z \in S(\tau,\beta)$ with $\beta \in (0,\omega(\mathbb{M}))$ and $|z|$ small enough, we have

$$g(z) - \sum_{k=0}^{n-1} m_e(k) f_k z^k = \int_0^{\infty(\tau)} e(u/z)\Big(f(u) - \sum_{k=0}^{n-1} f_k u^k\Big)\frac{du}{u}.$$

Since $u/z \in S_\beta$, by Definition 6.2(III) there exist $c_2, k_2 > 0$ such that

$$|e(u/z)| \le c_2 h_{\mathbb{M}}(k_2|z|/|u|),$$

and so, taking into account (38) and (5),

$$\begin{aligned}\Big|\frac{e(u/z)}{u}\Big(f(u) - \sum_{k=0}^{n-1} f_k u^k\Big)\Big| &\le c_1 c_2 k_1^n M'_n |u|^{n-1} \frac{h_{\mathbb{M}}(k_2|z|/|u|)}{h_{\mathbb{M}}(k_1/|u|)} \\ &\le c_1 c_2 k_1^n M'_n |u|^{n-1} \frac{h_{\mathbb{M}}^2(\rho(2)k_2|z|/|u|)}{h_{\mathbb{M}}(k_1/|u|)}.\end{aligned}$$

For $z \in S(\tau,\beta,k_1/(\rho(2)k_2))$ we have $\rho(2)k_2|z|/|u| < k_1/|u|$ and, $h_{\mathbb{M}}$ being increasing, we obtain that

$$\Big|\frac{e(u/z)}{u}\Big(f(u) - \sum_{k=0}^{n-1} f_k u^k\Big)\Big| \le c_1 c_2 k_1^n M'_n |u|^{n-1} h_{\mathbb{M}}(\rho(2)k_2|z|/|u|),$$

what implies that

$$\Big|g(z)-\sum_{k=0}^{n-1}m_e(k)f_kz^k\Big|\le c_1c_2k_1^nM'_n\int_0^\infty s^{n-1}h_{\mathbb{M}}(\rho(2)k_2|z|/s)ds.$$

Making the change of variable $s = |z|t$ and applying (19) leads to the conclusion.

(ii) From Proposition 6.13, $g := T_e^- f$ belongs to $\mathcal{O}(S(d,\alpha-\omega(\mathbb{M})))$. As in the proof of that Proposition, we limit ourselves to the case $\omega(\mathbb{M})<2$. It suffices to obtain estimates on a bounded subsector $T=S(\tau,\varepsilon\omega(\mathbb{M})/\pi,\rho)\ll S(d,\alpha-\omega(\mathbb{M}))$, for τ and ε entering in the definition of the path $\delta_{\omega(\mathbb{M})}(\tau)=\delta_1+\delta_2+\delta_3$ within $G(d,\alpha)$ (δ_1 and δ_3 are segments in directions θ_1 and θ_3, respectively, and δ_2 is a circular arc with radius $r>0$). If $\hat f=\sum_{n=0}^\infty f_nu^n$, there exist $c,k>0$ such that for every z in the trace of $\delta_{\omega(\mathbb{M})}(\tau)$ and every $n\in\mathbb{N}$ one has

$$|f(z)-\sum_{k=0}^{n-1}f_kz^k|\le ck^nM'_n|z|^n. \tag{39}$$

By (37), for every $u\in T$ we have

$$\begin{aligned}g(u)-\sum_{k=0}^{n-1}\frac{f_k}{m_e(k)}u^k&=T_e^-\big(f(z)-\sum_{k=0}^{n-1}f_kz^k\big)(u)\\&=\frac{-1}{2\pi i}\int_{\delta_{\omega(\mathbb{M})}(\tau)}E(u/z)\big(f(z)-\sum_{k=0}^{n-1}f_kz^k\big)\frac{dz}{z}\\&=\frac{-1}{2\pi i}\sum_{j=1}^{3}\int_{\delta_j}E(u/z)\big(f(z)-\sum_{k=0}^{n-1}f_kz^k\big)\frac{dz}{z}.\end{aligned} \tag{40}$$

By applying (39) and (28), we see that

$$\begin{aligned}\Big|\int_{\delta_2}E(u/z)\big(f(z)-\sum_{k=0}^{n-1}f_kz^k\big)\frac{dz}{z}\Big|&\le c(\theta_1-\theta_3)k^nM'_nr^n\max_{|z|=r}|E(u/z)|\\&\le\frac{cC(\theta_1-\theta_3)k^nM'_nr^n}{h_{\mathbb{M}}(Kr/|u|)}.\end{aligned}$$

So, for n large enough we may choose $r=|u|/(Km_n)$ and, since $h_{\mathbb{M}}(1/m_n)=M_n/m_n^n$, we deduce that

$$\Big|\int_{\delta_2}E(u/z)\big(f(z)-\sum_{k=0}^{n-1}f_kz^k\big)\frac{dz}{z}\Big|\le\frac{cC(\theta_1-\theta_3)k^nM'_n|u|^n}{K^nM_n}. \tag{41}$$

On the other hand, again (39) implies that for $j = 1, 3$,

$$\Big|\int_{\delta_j} E(u/z)\big(f(z) - \sum_{k=0}^{n-1} f_k z^k\big)\frac{dz}{z}\Big| \le ck^n M'_n \int_0^r t^n \frac{|E(ue^{-i\theta_j}/t)|}{t}dt$$

$$= ck^n M'_n |u|^n \int_0^{r/|u|} s^n \frac{|E(e^{i(\arg(u)-\theta_j)}/s)|}{s}ds,$$

after the change of variable $t = |u|s$. The same choice of r as before leads to

$$\Big|\int_{\delta_j} E(u/z)\big(f(z) - \sum_{k=0}^{n-1} f_k z^k\big)\frac{dz}{z}\Big| \le ck^n M'_n |u|^n \int_0^{\frac{1}{Km_n}} s^n \frac{|E(e^{i(\arg(u)-\theta_j)}/s)|}{s}ds$$

$$\le ck^n M'_n |u|^n \frac{1}{K^n m_n^n} \int_0^{1/(Km_n)} \frac{|E(e^{i(\arg(u)-\theta_j)}/s)|}{s}ds. \qquad (42)$$

Since $\lim_{n\to\infty} m_n = \infty$, the last integral admits an upper bound independent of u and n because of condition (VI) in Definition 6.2. According to (40), (41) and (42), and as $\mathbb{M}$ and $(m_p^p)_{p\in\mathbb{N}_0}$ are equivalent, the conclusion is reached. □

Remark 6.16 If in the previous statement we assume that $\mathbb{M} = \mathbb{M}' \cdot \mathbb{L}$, where $\mathbb{L}$ is (lc) and $\lim_{p\to\infty} \ell_p = \infty$, then for f as in part (ii) one can show $T_e^- f$ is entire and of $\mathbb{L}$-growth, according to Proposition 5.9.

We are ready for giving a definition of summability in a direction with respect to a kernel e of $\mathbb{M}$-summability. Let T_e be the corresponding Laplace operator, and recall that $\mathfrak{m}_e$ is strongly regular and equivalent to $\mathbb{M}$, so that, on one hand, $\Lambda_{\mathbb{M}} = \Lambda_{\mathfrak{m}_e}$ and, on the other hand, it makes sense to consider the space $\mathcal{O}^{\mathfrak{m}_e}(S)$ for any unbounded sector S and, moreover, $\mathcal{O}^{\mathfrak{m}_e}(S) = \mathcal{O}^{\mathbb{M}}(S)$ [see (6)].

Definition 6.17 We say $\hat{f} = \sum_{p\ge 0} \frac{f_p}{p!} z^p$ is *T_e-summable in direction $d \in \mathbb{R}$* if:

(i) $(f_p)_{p\in\mathbb{N}_0} \in \Lambda_{\mathfrak{m}_e}$, so that $g := \hat{T}_e^- \hat{f} = \sum_{p\ge 0} \frac{f_p}{p! m_e(p)} z^p$ converges in a disc, and
(ii) g admits analytic continuation in a sector $S = S(d, \varepsilon)$ for some $\varepsilon > 0$, and $g \in \mathcal{O}^{\mathfrak{m}_e}(S)$.

We state next the equivalence between $\mathbb{M}$-summability and T_e-summability in a direction, and provide a way to recover the $\mathbb{M}$-sum in a direction of a summable power series by means of the formal and analytic transforms previously introduced.

Theorem 6.18 *Given a strongly regular sequence* $\mathbb{M}$*, a direction* d *and a formal power series* $\hat{f} = \sum_{p\geq 0} \frac{f_p}{p!} z^p$*, the following are equivalent:*

(i) $\hat{f}$ *is* $\mathbb{M}$*-summable in direction* d*.*
(ii) *For every kernel* e *of* $\mathbb{M}$*-summability,* $\hat{f}$ *is* T_e*-summable in direction* d*.*
(iii) *For some kernel* e *of* $\mathbb{M}$*-summability,* $\hat{f}$ *is* T_e*-summable in direction* d*.*

In case any of the previous holds, we have (after analytic continuation)

$$\mathcal{S}_{\mathbb{M},d}\hat{f} = T_e(\hat{T}_e^{-}\hat{f})$$

for any kernel e *of* $\mathbb{M}$*-summability.*

Proof

(i) $\Longrightarrow$ (ii) Let $f = \mathcal{S}_{\mathbb{M},d}\hat{f}$, the $\mathbb{M}$-sum of $\hat{f}$ in direction d. Then $f \sim_{\mathbb{M}} \hat{f}$ in a sectorial region $G(d,\alpha)$ with $\alpha > \omega(\mathbb{M})$, and moreover $(f_p)_{p\in\mathbb{N}_0} \in \Lambda_{\mathbb{M}}$. If we put $\mathbb{M}' = (1)_{p\in\mathbb{N}_0}$ (the constant sequence whose terms are all equal to 1), item (ii) in Theorem 6.15 states that $g := T_e^{-}f \sim_{\mathbb{M}'} \hat{T}_e^{-}\hat{f}$, what implies that $\hat{T}_e^{-}\hat{f}$ converges to g in a disk, and g is, by Proposition 6.13, of $\mathbb{M}$-growth in a small unbounded sector around d, as we intended to prove.
(ii) $\Longrightarrow$ (iii) Trivial.
(iii) $\Longrightarrow$ (i) Since $g := \hat{T}_e^{-}\hat{f}$ converges in a disc and admits analytic continuation in a sector $S = S(d,\varepsilon)$ for some $\varepsilon > 0$, we have that $g \sim_{\mathbb{M}'} \hat{T}_e^{-}\hat{f}$ in S with $\mathbb{M}' = (1)_{p\in\mathbb{N}_0}$. Moreover, $g \in \mathcal{O}^{\mathfrak{m}_e}(S) = \mathcal{O}^{\mathbb{M}}(S)$, and due to (i) in Theorem 6.15, we obtain that the function $f := T_e g$ is holomorphic in a sectorial region of opening greater than $\pi\omega(\mathbb{M})$ and $f \sim_{\mathbb{M}} \hat{f}$ there, so we are done. □

Remark 6.19

(i) In case $\mathbb{M} = \mathbb{M}_{1/k}$, the summability methods described are just the classical k-summability and T_e-summability (in a direction) for kernels e of order $k > 0$, as defined by W. Balser.
(ii) If $\mathbb{M}$ and $\mathbb{M}'$ are equivalent strongly regular sequences, the respective families of kernels of summability coincide, as it is easily deduced from (6), hence the summability methods just introduced for $\mathbb{M}$ and $\mathbb{M}'$ are all the same, as well as the sums provided for every $\mathbb{M}$- (or equivalently, $\mathbb{M}'$-) summable series in a direction. Note that, by Remark 4.8, one also has $\omega(\mathbb{M}) = \omega(\mathbb{M}')$; however, it is important to note that for general (non-Gevrey) kernels the value $\omega(\mathbb{M})$ does not determine the equivalence class of strongly regular sequences, nor the summability methods, we are dealing with. This fact will be taken into account throughout the last section of this survey.
(iii) In particular, consider a kernel e, its moment function m_e and the strongly regular sequence of moments $\mathfrak{m}_e$, as in Remark 6.6. According to Definition 6.2, Remark 6.3(iii) and Remark 6.6, for every $s > 0$ one may

deduce that $e^{(s)}(z) := e(z^{1/s})/s$ is a kernel for $\mathfrak{m}_e^{(s)}$-summability (recall that $\mathfrak{m}_e^{(s)} = (m_e^s(n))_{n\in\mathbb{N}_0})$ with moment function $m_{e^{(s)}}(\lambda) = m_e(s\lambda)$ and sequence of moments $\mathfrak{m}_{e^{(s)}} = (m_e(sn))_{n\in\mathbb{N}_0}$, and consequently, $\mathfrak{m}_e^{(s)}$ and $\mathfrak{m}_{e^{(s)}}$ are equivalent (Proposition 5.8) and

$$\omega((m_e(sn))_{n\in\mathbb{N}_0}) = \omega(\mathfrak{m}_e^{(s)}) = s\omega(\mathfrak{m}_e) = s\omega((m_e(n))_{n\in\mathbb{N}_0}). \tag{43}$$

Moreover, by (6) and (29), there exist $A, B > 0$ such that for every $t \geq 0$ one has

$$\big(h_{\mathfrak{m}_e}(At)\big)^s = h_{\mathfrak{m}_e^{(s)}}(A^s t^s) \leq h_{\mathfrak{m}_{e^{(s)}}}(t^s) \leq h_{\mathfrak{m}_e^{(s)}}(B^s t^s) = \big(h_{\mathfrak{m}_e}(Bt)\big)^s. \tag{44}$$

The following properties are straightforward consequences of the results in this section. We will denote by $\mathbb{C}\{z\}_{\mathbb{M},d}$ the set of formal power series which are $\mathbb{M}$-summable in a direction $d \in \mathbb{R}$, where $\mathbb{M}$ is a strongly regular sequence such that $d(r)$ is a proximate order. A kernel for $\mathbb{M}$-summability will be denoted by e. As indicated before, the $\mathbb{M}$-sum in direction d of a formal power series $\hat{f}$ will be named $f = \mathcal{S}_{\mathbb{M},d}\hat{f}$.

Proposition 6.20

(i) $\mathbb{C}\{z\}_{\mathbb{M},d}$ *is a differential algebra, and the sum operator* $\mathcal{S}_{\mathbb{M},d}$ *respects the operations of addition, multiplication, derivation and inversion (whenever the latter is well defined).*
(ii) *If* $\hat{f}$ *is convergent, then it is* $\mathbb{M}$*-summable in every direction* d *and the sum coincides with* f *in the intersection of the corresponding domains.*
(iii) *If* $\hat{f} \in \mathbb{C}\{z\}_{\mathbb{M},d}$ *and* $\varepsilon > 0$ *is small enough, then* $\hat{f} \in \mathbb{C}\{z\}_{\mathbb{M},d'}$ *for every* d' *with* $|d - d'| < \varepsilon$ *and* $(\mathcal{S}_{\mathbb{M},d'}\hat{f})(z) = (\mathcal{S}_{\mathbb{M},d}\hat{f})(z)$ *for every* z *where both functions are defined.*
(iv) *If* $\hat{f} \in \mathbb{C}\{z\}_{\mathbb{M},d}$, $\varepsilon > 0$ *is small enough and* $\mathbb{M}'$ *is strongly regular,* $\mathbb{M} \ll \mathbb{M}'$ *and* $\omega(\mathbb{M}') \leq \omega(\mathbb{M}) + \epsilon$*, then* $\hat{f} \in \mathbb{C}\{z\}_{\mathbb{M}',d}$ *and* $(\mathcal{S}_{\mathbb{M}',d}\hat{f})(z) = (\mathcal{S}_{\mathbb{M},d}\hat{f})(z)$ *for every* z *where both functions are defined.*
(v) *Suppose* $\mathbb{M} \ll \mathbb{M}'$*, both inducing a proximate order, and* $\hat{f} \in \mathbb{C}\{z\}_{\mathbb{M},d} \cap \mathbb{C}\{z\}_{\mathbb{M}',d}$*. If either* $\omega(\mathbb{M}) < \omega(\mathbb{M}')$*, or* $\omega(\mathbb{M}) = \omega(\mathbb{M}')$*,* $\mathbb{M} \not\approx \mathbb{M}'$ *and* $\mathbb{M}'/\mathbb{M}$ *is (lc), then* $(\mathcal{S}_{\mathbb{M},d}\hat{f})(z) = (\mathcal{S}_{\mathbb{M}',d}\hat{f})(z)$ *for every* z *where both functions are defined.*

Example Consider $\mathbb{M}_{1,1} = \big(n!\prod_{m=0}^{n}\log(e+m)\big)_{n\in\mathbb{N}_0}$, which is strongly regular and induces a proximate order. So, we may take a kernel e for $\mathbb{M}_{1,1}$-summability and the sequence of moments $(m_e(n))_{n\in\mathbb{N}_0}$. Then, for any $k > 0$ the series $\hat{f} = \sum_{n=0}^{\infty} m_e(n)z^n$ is not k-summable in infinitely many directions. Indeed, for $k \geq 1$ it suffices to observe that the formal Borel transform of order k,

$$\sum_{n=0}^{\infty} \frac{m_e(n)}{\Gamma(1+n/k)} z^n,$$

does not converge. In case $k \in (0,1)$, the formal Borel transform of order k is entire and of exponential order $k/(1-k) > k$, and it has maximal type, what excludes infinitely many directions for k-summability.

However, note that $\hat{f}$ is $\mathbb{M}_{1,1}$-summable in every direction except $d = 0$ (mod 2π).

7 Applications

Following the idea of Balser and Yoshino [6], given a sequence of moments $\mathfrak{m} := (m(p))_{p\in\mathbb{N}_0}$ let us consider the operator $\partial_{\mathfrak{m},z}$, from $\mathbb{C}[[z]]$ into itself, given by

$$\partial_{\mathfrak{m},z}\left(\sum_{p\geq 0}\frac{f_p}{m(p)}z^p\right) = \sum_{p\geq 0}\frac{f_{p+1}}{m(p)}z^p.$$

Michalik [45] has studied the initial value problem for linear moment-partial differential equations of the form

$$P(\partial_{\mathfrak{m}_1,t},\partial_{\mathfrak{m}_2,z})u(t,z) = 0, \tag{45}$$

with given initial conditions

$$\partial^j_{\mathfrak{m}_1,t}u(0,z) = \varphi_j(z) \in \mathcal{O}(D), \quad j = 0,\dots,n-1, \tag{46}$$

for some $n \in \mathbb{N}$, and some neighborhood of the origin D, say $D(0,r)$ for some $r > 0$. Here, $P(\lambda,\xi) \in \mathbb{C}[[\lambda,\xi]]$ is a polynomial of degree n in the variable λ, and $\mathfrak{m}_1 = (m_1(p))_{p\in\mathbb{N}_0}$ and $\mathfrak{m}_2 = (m_2(p))_{p\in\mathbb{N}_0}$ are given moment sequences corresponding to kernels e_1 and e_2 of orders $k_1 > 0$ and $k_2 > 0$, respectively, as defined by W. Balser in [2]. In this last section we aim at stating analogous results to those in [45], now in the case when these kernels are associated with general strongly regular sequences (which might not be equivalent to Gevrey ones). So, our setting is as described in Remark 6.6. Although the class of linear moment-partial differential equations under study has been enlarged, the main ideas do not greatly differ from the ones in [45], so we will omit some proofs requiring only minor modifications with respect to the ones provided in that work.

The approach in [45] is based on the reduction of the initial problem (45),(46) into a finite number of problems which are easier to handle. For this purpose, we put

$$P(\lambda,\xi) = P_0(\xi)(\lambda-\lambda_1(\xi))^{n_1}\cdots(\lambda-\lambda_\ell(\xi))^{n_\ell}, \tag{47}$$

where $n_1,\dots,n_\ell \in \mathbb{N}$ with $n_1+\cdots+n_\ell = n$. For every $j = 1,\dots,\ell$, the function $\lambda_j(\xi)$ is an algebraic function, holomorphic for $|\xi| > R_0$, for some $R_0 > 0$, and with polynomial growth at infinity. The existence is proven of a normalized formal solution $\hat{u}$ to the main problem (45), (46), chosen so as to satisfy also the equation

$$(\partial_{\mathfrak{m}_1,t} - \lambda_1(\partial_{\mathfrak{m}_2,z}))^{n_1} \cdots (\partial_{\mathfrak{m}_1,t} - \lambda_\ell(\partial_{\mathfrak{m}_2,z}))^{n_\ell}\hat{u} = 0 \tag{48}$$

(the meaning of $\lambda_j(\partial_{\mathfrak{m}_2,z})$ to be specified). Indeed, Theorem 1 in [45] states that one can recover $\hat{u}$ as

$$\hat{u} = \sum_{\alpha=1}^{\ell}\sum_{\beta=1}^{n_\alpha} \hat{u}_{\alpha\beta}, \tag{49}$$

$\hat{u}_{\alpha\beta}$ being the formal solution of

$$\begin{cases} (\partial_{\mathfrak{m}_1,t} - \lambda_\alpha(\partial_{\mathfrak{m}_2,z}))^{\beta}\hat{u}_{\alpha\beta} = 0 \\ \partial^j_{\mathfrak{m}_1,t}\hat{u}_{\alpha\beta}(0,z) = 0, \quad j = 0,\dots,\beta-2 \\ \partial^{\beta-1}_{\mathfrak{m}_1,t}\hat{u}_{\alpha\beta}(0,z) = \lambda_\alpha^{\beta-1}(\partial_{\mathfrak{m}_2,z})\phi_{\alpha\beta}(z), \end{cases} \tag{50}$$

where $\phi_{\alpha\beta}(z) := \sum_{j=0}^{n-1} d_{\alpha\beta j}(\partial_{\mathfrak{m}_2,z})\phi_j(z) \in \mathcal{O}(D(0,r))$, and $d_{\alpha\beta j}(\xi)$ are holomorphic functions of polynomial growth at infinity for every α and β. One may easily check that the formal solution of (50) is given by

$$\hat{u}_{\alpha\beta}(t,z) = \sum_{j=\beta-1}^{\infty} \binom{j}{\beta-1} \frac{\lambda_\alpha^j(\partial_{\mathfrak{m}_2,z})\phi_{\alpha\beta}(z)}{m_1(j)} t^j. \tag{51}$$

We do not enter into details about this point, for the proof of this result is entirely analogous in our situation. We will focus our attention on the convergence of the formal solution, and also on the growth rate of its coefficients when it has null radius of convergence, but firstly we recall the meaning of the pseudodifferential operators $\lambda(\partial_{\mathfrak{m}_e,z})$ [like the ones appearing in (48), (50) and (51)], where $\lambda(\xi)$ is an element in the set $\{\lambda_j : j = 1,\dots,\ell\}$ and $\mathfrak{m}_e = (m_e(p))_{p\in\mathbb{N}_0}$ is the strongly regular sequence of moments of a kernel e with moment function $m_e(\lambda)$. Given $r > 0$, one can check (see Proposition 3 in [45]) that the differential operator $\partial_{\mathfrak{m}_e,z}$ is well-defined for any $\phi \in \mathcal{O}(D(0,r))$, and for $0 < \varepsilon < r$ and every $z \in D(0,\varepsilon)$ one has

$$\partial^n_{\mathfrak{m}_e,z}\phi(z) = \frac{1}{2\pi i}\oint_{|w|=\varepsilon} \phi(w)\int_0^{\infty(\theta)} \xi^n E(z\xi)\frac{e(w\xi)}{w\xi}d\xi dw,$$

for every $n \in \mathbb{N}_0$, where $\theta \in (-\arg(w) - \frac{\omega(\mathfrak{m}_e)\pi}{2}, -\arg(w) + \frac{\omega(\mathfrak{m}_e)\pi}{2})$ and E is the second kernel in Definition 6.2.

The previous expression inspires the definition of the pseudodifferential operator $\lambda(\partial_{\mathbb{m}_e,z})$ as

$$\lambda(\partial_{\mathbb{m}_e,z})\phi(z) := \frac{1}{2\pi i}\oint_{|w|=\varepsilon} \phi(w) \int_{\xi_0}^{\infty(\theta)} \lambda(\xi)E(\xi z)\frac{e(\xi w)}{\xi w} d\xi dw,$$

for every $\phi \in \mathcal{O}(D(0,r))$, where $\xi_0 = R_0 e^{i\theta}$ with suitably large $R_0 > 0$ and θ as before (see Definition 8 in [45]).

Next we study the growth rate of the formal solution of (50), given in (51). To this end, we need the following definition and lemmas.

Definition 7.1 Let $U \subseteq \mathbb{C}$ be a neighborhood of ∞, and $\Psi \in \mathcal{O}(U)$. The *pole order* $q \in \mathbb{Q}$ and the *leading term* $\psi \in \mathbb{C} \setminus \{0\}$ associated with Ψ are the elements satisfying

$$\lim_{z\to\infty} \frac{\Psi(z)}{z^q} = \psi,$$

if they exist.

Lemma 7.2 *Let* e, m_e *and* $\mathbb{m}_e$ *be as before,* $\lambda(\xi)$ *have pole order* q *and leading term* λ_0, *and let* $\phi \in \mathcal{O}(D(0,r))$. *There exist* $r_0, A, B > 0$ *such that*

$$\sup_{|z|<r_0} |\lambda(\partial_{\mathbb{m}_e,z})\phi(z)| \le |\lambda_0| A B^q m_e(q).$$

Proof One may choose $R_0 > 0$ such that $|\lambda(\xi)| \le 2|\lambda_0||\xi|^q$ for every ξ with $|\xi| \ge R_0$. Let $w \in \mathbb{C}$ with $0 < |w| = \varepsilon < r$, and put $\theta = -\arg(w)$ and $\xi_0 = R_0 e^{i\theta}$. One has

$$\left| \int_{\xi_0}^{\infty(\theta)} \lambda(\xi)E(\xi z)\frac{e(\xi w)}{\xi w} d\xi \right| \le 2|\lambda_0| \int_{R_0}^{\infty} s^q |E(se^{i\theta}z)| \frac{|e(se^{i\theta}w)|}{s\varepsilon} ds. \tag{52}$$

The properties of the kernel functions e and E stated in Definition 6.2, rephrased according to Remark 6.6, allow us to write

$$|E(se^{i\theta}z)e(se^{i\theta}w)| \le \frac{c_1}{h_{\mathbb{m}_e}\left(\frac{c_2}{s|z|}\right)} h_{\mathbb{m}_e}\left(\frac{c_3}{s\varepsilon}\right)$$

for some $c_1, c_2, c_3 > 0$ and for every $s \in [R_0, \infty)$. From (5) one has

$$\frac{c_1}{h_{\mathbb{m}_e}\left(\frac{c_2}{s|z|}\right)} h_{\mathbb{m}_e}\left(\frac{c_3}{s\varepsilon}\right) \le \frac{c_1 h^2_{\mathbb{m}_e}\left(\frac{\rho(2)c_3}{s\varepsilon}\right)}{h_{\mathbb{m}_e}\left(\frac{c_2}{s|z|}\right)}. \tag{53}$$

Let $r_0 > 0$ be such that $r_0 \leq c_2\varepsilon/(\rho(2)c_3)$, so that $\rho(2)c_3/(s\varepsilon) \leq c_2/(s|z|)$ for every $z \in D(0, r_0)$. For such z, the expression in the right-hand side of (53) is upper bounded by $c_1 h_{\mathbb{m}_e}\big(\rho(2)c_3/(s\varepsilon)\big)$, and one obtains that the last expression in (52) can be upper bounded by

$$2c_1|\lambda_0| \int_{R_0}^{\infty} \frac{s^{q-1}}{\varepsilon} h_{\mathbb{m}_e}\Big(\frac{\rho(2)c_3}{s\varepsilon}\Big) ds.$$

In turn, by the very definition of $h_{\mathbb{m}_e}$, the previous quantity is less than

$$2c_1|\lambda_0|(\rho(2)c_3)^{\lfloor q\rfloor+2} \frac{1}{\varepsilon^{\lfloor q\rfloor+3}} m_e(\lfloor q\rfloor + 2) \int_{R_0}^{\infty} \frac{1}{s^{3+\lfloor q\rfloor-q}} ds.$$

The last integral is easily seen to be bounded above by some constant independent of q. Moreover, the moderate growth property of $\mathbb{m}_e$ leads to an estimate of the form

$$|\lambda_0| A_0 B_0^{\lfloor q\rfloor} m_e(\lfloor q\rfloor).$$

Finally, we observe that the function $x \in [0, \infty) \to m_e(x)$ is continuous, strictly convex (since $m_e''(x) > 0$ for every $x > 0$) and $\lim_{x\to\infty} m_e(x) = \infty$, so it reaches its absolute minimum $m_e(x_0) > 0$ at a point $x_0 \geq 0$, and it is decreasing in $[0, x_0)$ (if $x_0 > 0$) and increasing in (x_0, ∞). So, we deduce that whenever $x_0 \leq x \leq y$ we have $m_e(x) \leq m_e(y)$, while if $0 \leq x < x_0$ and $x \leq y$, then $m_e(x)/m_e(y) \leq m_e(0)/m_e(x_0)$. In conclusion, there exists a constant $A_1 > 0$ such that $m_e(x) \leq A_1 m_e(y)$ whenever $0 \leq x \leq y$, and in particular, $m_e(\lfloor x\rfloor) \leq A_1 m_e(x)$ for every $x > 0$, what leads to the final estimate. □

For $j \in \mathbb{N}$, the function $\lambda^j(\xi)$ has pole order jq and leading term λ_0^j. So, an argument similar to the previous one provides the proof for the following result.

Corollary 7.3 *Let $j \in \mathbb{N}$. Under the assumptions of Lemma 7.2, one has*

$$\sup_{|z|<r_0} |\lambda^j(\partial_{\mathbb{m}_e,z})\phi(z)| \leq |\lambda_0|^j A B^{jq} m_e(qj)$$

for some $r_0, A, B > 0$.

Remark 7.4 The previous estimates, according to Remark 6.19(iii), could also be expressed as $|\lambda_0|^j A B_1^{jq} (m_e(j))^q$ for suitable $B_1 > 0$.

As indicated before [see (50) and (51)], a problem of the form

$$\begin{cases} (\partial_{\mathbb{m}_1,t} - \lambda(\partial_{\mathbb{m}_2,z}))^{\beta} \hat{u} = 0 \\ \partial^j_{\mathbb{m}_1,t} \hat{u}(0, z) = 0, \quad j = 0, \ldots, \beta - 2 \\ \partial^{\beta-1}_{\mathbb{m}_1,t} \hat{u}(0, z) = \lambda^{\beta-1}(\partial_{\mathbb{m}_2,z})\phi(z), \end{cases} \tag{54}$$

where $\beta \in \mathbb{N}$ and $\phi \in \mathcal{O}(D(0,r))$, has

$$\hat{u}(t,z) = \sum_{j=\beta-1}^{\infty} \binom{j}{\beta-1} \frac{\lambda^j(\partial_{\mathfrak{m}_2,z})\phi(z)}{m_1(j)} t^j =: \sum_{j=0}^{\infty} u_j(z)t^j \tag{55}$$

as its formal solution, and Corollary 7.3 allows us to claim that

$$\sup_{|z|<r_0} |u_j(z)| \le CD^j \frac{m_2(qj)}{m_1(j)},$$

for some $r_0, C, D > 0$ and for every $j \in \mathbb{N}_0$. Hence, convergence or divergence of $\hat{u}$ in some neighborhood of the origin is a consequence of the growth rate of the sequence $(\frac{m_2(qj)}{m_1(j)})_{j\ge 0}$. More precisely, one has

Corollary 7.5 *If*

$$\overline{\lim}_{j\to\infty} \left(\frac{m_2(qj)}{m_1(j)}\right)^{1/j} < \infty$$

(in other words, $\mathfrak{m}_2^{(q)} \ll \mathfrak{m}_1$*), then* $\hat{u}$ *in (55) defines a holomorphic function* $u(t,z)$ *on* $D_1 \times D(0,r_0)$ *for some neighborhood of the origin* $D_1 \subseteq \mathbb{C}$*, and* u *solves (54).*

We now turn our attention to the determination of sufficient conditions for $u(t,z)$ to admit analytic continuation in an unbounded sector with respect to the variable t and with adequate growth. We first need some notation, starting with $\mathcal{O}^{\mathfrak{m}_e}(S)$ (see Definition 6.10), where S is an unbounded sector in $\mathcal{R}$ and $\mathfrak{m}_e$ is a strongly regular sequence of moments for a kernel e.

Definition 7.6 We write $f \in \mathcal{O}^{\mathfrak{m}_e}(\hat{S})$ if $f \in \mathcal{O}^{\mathfrak{m}_e}(S) \cap \mathcal{O}(S \cup D)$ for some disc $D = D(0,r)$.

Let $D = D(0,r)$. We say $f(t,z)$, holomorphic in $S \times D$, belongs to $\mathcal{O}^{\mathfrak{m}_e}(S \times D)$ if for every $T \prec S$ and $r_1 \in (0,r)$ there exist $c, k > 0$ such that

$$\sup_{z\in D(0,r_1)} |f(t,z)| \le \frac{c}{h_{\mathfrak{m}_e}(k/|t|)}, \quad t \in T.$$

Analogously, we write $f \in \mathcal{O}^{\mathfrak{m}_e}(\hat{S} \times D)$ if $f \in \mathcal{O}((S \cup D_1) \times D) \cap \mathcal{O}^{\mathfrak{m}_e}(S \times D)$ for some disc D_1 around the origin. We also write $\mathcal{O}^{\mathfrak{m}_e}(\hat{S}(d))$ and $\mathcal{O}^{\mathfrak{m}_e}(\hat{S}(d) \times D)$ whenever the sector S is of the form $S(d,\varepsilon)$ for some inessential $\varepsilon > 0$.

From Proposition 5.9 we deduce the following result.

Corollary 7.7 *Suppose that there exists a strongly regular sequence of moments,* $\mathfrak{m}_e = (m_e(j))_{j\in\mathbb{N}_0}$*, and* $C, D > 0$ *such that*

$$m_2(qj)m_e(j) \le CD^j m_1(j)$$

for every $j \geq 0$. *Then,* $\hat{u}$ *in (55) defines a function* $u \in \mathcal{O}(\mathbb{C} \times D(0, r_0))$, *and one has*

$$\sup_{|z|<r_0} |u(t,z)| \leq \frac{c}{h_{\mathfrak{m}_e}(k/|t|)},$$

for some $c, k > 0$ *and for every* $t \in \mathbb{C}$.

Remark 7.8 In the particular case that $\mathfrak{m}_1 = \mathbb{M}_{1/k_1}$ and $\mathfrak{m}_2 = \mathbb{M}_{1/k_2}$ for some $k_1, k_2 > 0$ with $1/k_1 > q/k_2$, we would have that $\hat{u} \in \mathcal{O}(\mathbb{C} \times D(0, r_0))$, with exponential growth in the variable t of order $(\frac{1}{k_1} - q\frac{1}{k_2})^{-1}$, namely

$$\sup_{|z|<r_0} |u(t,z)| \leq Ce^{D|t|^{\frac{k_1 k_2}{k_2 - qk_1}}}, \quad t \in \mathbb{C},$$

for some $C, D > 0$, as stated in Proposition 5 of [45].

In order to go further in our study, by an argument entirely analogous to that in Lemma 4 in [45] one can prove that, under the assumptions of Corollary 7.5, the actual solution of (54) can be written in a neighborhood of $(0, 0)$ in the form

$$u(t,z) = \frac{t^{\beta-1}}{(\beta-1)!}\partial_t^{\beta-1}\frac{1}{2\pi i}\oint_{|w|=\varepsilon}\phi(w)\int_{\xi_0}^{\infty(\theta)} E_1(t\lambda(\xi))E_2(\xi z)\frac{e_2(\xi w)}{\xi w}d\xi dw, \tag{56}$$

with $\theta \in (-\arg(w) - \frac{\pi}{2}\omega(\mathfrak{m}_2), -\arg(w) + \frac{\pi}{2}\omega(\mathfrak{m}_2))$, and where E_1 and E_2 are the kernels corresponding to e_1 and e_2, respectively.

We are ready to relate the properties of analytic continuation and growth of the initial data with those of the solution. In these last results we assume the kernels e_1, E_1 have been constructed following the procedure in section "Flat Functions via Proximate Orders and Watson's Lemma".

Lemma 7.9 *Let* $q = \mu/\nu \in \mathbb{Q}$, *with* $\gcd(\mu, \nu) = 1$ *and* $\beta \geq 1$. *We assume the moment functions* $m_1(\lambda)$ *and* $m_2(\lambda)$ *are such that*

$$m_2(qj) \leq C_0 A_0^j m_1(j), \quad j \in \mathbb{N}_0, \tag{57}$$

and

$$m_1(j/q) \leq C_1 A_1^j m_2(j), \quad j \in \mathbb{N}_0, \tag{58}$$

for suitable $C_0, C_1, A_0, A_1 > 0$. *Let* $u(t,z)$ *be a solution of*

$$\begin{cases} (\partial_{\mathfrak{m}_1,t} - \lambda(\partial_{\mathfrak{m}_2,z}))^\beta u = 0 \\ \partial_{\mathfrak{m}_1,t}^j \hat{u}(0,z) = \phi_j(z) \in \mathcal{O}(D(0,r)), \quad j = 0, \ldots, \beta - 1, \end{cases} \tag{59}$$

for some $r > 0$. *If there exists a strongly regular sequence of moments* $\mathfrak{m} = (m(j))_{j\in\mathbb{N}_0}$ *such that:*

(i) *there exist* $C, A > 0$ *with*

$$m(j) \leq CA^j m_1(j), \quad j \in \mathbb{N}_0, \tag{60}$$

(ii) $\phi_j \in \mathcal{O}^{\mathfrak{m}^{(1/q)}}(\hat{S}((d + \arg(\lambda))/q + 2k\pi/\mu))$ *for every* $k = 0, \dots, \mu - 1$ *and* $j = 0, \dots, \beta - 1$, *and some* $d \in \mathbb{R}$,

then $u(t,z) \in \mathcal{O}^{\mathfrak{m}}(\hat{S}(d + 2n\pi/\nu) \times D(0,r))$ *for* $n = 0, \dots, \nu - 1$.

Remark 7.10 According to Remark 6.19(iii), the sequence $(m_2(qj))_{j\in\mathbb{N}_0}$ (respectively, $(m_1(j/q))_{j\in\mathbb{N}_0}$) is equivalent to $\mathfrak{m}_2^{(q)}$ (resp. to $\mathfrak{m}_1^{(1/q)}$). Together with this fact, the inequalities (57) and (58) amount to the equivalence of $\mathfrak{m}_2^{(q)}$ and $\mathfrak{m}_1$, and so, we deduce by (43) that

$$q\omega(\mathfrak{m}_2) = \omega(\mathfrak{m}_1). \tag{61}$$

Proof With the help of Lemma 3 in [45], which may be reproduced in our setting without modification, one can show that the general situation may always be taken into the case $\omega(\mathfrak{m}_1) < 2$, which will be the only one we consider. The principle of superposition of solutions allows us to reduce the study of (59) to that of some problems of the form (54), where $\lambda^{\beta-1}(\partial_{\mathfrak{m}_2,z})\phi$ in (54) turns out to be a function belonging to $\mathcal{O}^{\mathfrak{m}^{(1/q)}}(\hat{S}((d + \arg(\lambda))/q + 2k\pi/\mu))$ for every $k = 0, \dots, \mu - 1$ and $j = 0, \dots, \beta - 1$. Moreover, Corollary 7.5 and (57) guarantee the existence of a holomorphic solution $u(t,z)$ of (59), defined on some neighborhood of the origin in $\mathbb{C}^2$, which can be written in the form (56). Next, we claim that the function

$$t \mapsto \int_{\xi_0}^{\infty(\theta)} E_1(t\lambda(\xi))E_2(\xi z)\frac{e_2(\xi w)}{\xi w}d\xi, \tag{62}$$

which is holomorphic in $\{t \in \mathbb{C} : |t| \leq C_2|w|^q\}$ for some $C_2 > 0$, can be analytically continued to the set

$$\Omega = \{t \in \mathcal{R} : \arg(t) + 2k\pi + \arg(\lambda) \neq (\arg(w) + 2n\pi)q \text{ for every } k, n \in \mathbb{Z}\}.$$

Indeed, the equality (61) entails that, as long as $t \in \Omega$, one can replace θ in (62) by a direction $\tilde{\theta}$ such that

$$\arg(t) + 2k\pi + \arg(\lambda) + q\tilde{\theta} \in \big(\pi\omega(\mathfrak{m}_1)/2, 2\pi - \pi\omega(\mathfrak{m}_1)/2\big) \quad \text{for some } k \in \mathbb{Z}$$

and

$$\arg(w) + 2n\pi + \tilde{\theta} \in \big(-\pi\omega(\mathfrak{m}_2)/2, \pi\omega(\mathfrak{m}_2)/2\big) \quad \text{for some } n \in \mathbb{Z},$$

what makes the continuation possible by ensuring the adequate asymptotic behavior of the integrand as $\xi \to \infty$, $\arg(\xi) = \tilde{\theta}$. The rest of the proof, intended to estimate u, also follows the arguments in [45, Lemma 5], but estimates will be carefully given in order to highlight the techniques in this general situation. Suppose z is small relative to w. We deform the integration path $|w| = \varepsilon$ in order to write

$$u(t,z) = \frac{t^{\beta-1}}{(\beta-1)!}\partial_t^{\beta-1}\left(u_1(t,z) + u_2(t,z)\right),$$

with

$$u_1(t,z) = \sum_{k=0}^{\mu-1} \frac{1}{2\pi i} \oint_{\gamma_{2k}^R} \phi(w) \int_{\xi_0}^{\infty(\theta)} E_1(t\lambda(\xi))E_2(\xi z)\frac{e_2(\xi w)}{\xi w} d\xi dw,$$

and

$$u_2(t,z) = \sum_{k=0}^{\mu-1} \frac{1}{2\pi i} \oint_{\gamma_{2k+1}} \phi(w) \int_{\xi_0}^{\infty(\theta)} E_1(t\lambda(\xi))E_2(\xi z)\frac{e_2(\xi w)}{\xi w} d\xi dw.$$

Here, the path γ_{2k+1} is parameterized by

$$s \in I_{2k+1} := \left(\frac{d+\arg(\lambda)}{q} + \frac{2k\pi}{\mu} + \frac{\delta}{3}, \frac{d+\arg(\lambda)}{q} + \frac{2(k+1)\pi}{\mu} - \frac{\delta}{3}\right) \mapsto \varepsilon e^{is},$$

for some small enough $\delta > 0$. On the other hand, for large enough $R > 0$ the path γ_{2k}^R is $\gamma_{2k}^{R,-} + \gamma_{2k}^{R,1} - \gamma_{2k}^{R,+}$, where

$$\gamma_{2k}^{R,\star}(s) = s e^{i\left(\frac{d+\arg(\lambda)}{q} + \frac{2k\pi}{\mu} \star \frac{\delta}{3}\right)} = s e^{i\theta_\star}, \qquad \star \in \{-,+\}, \quad s \in [\varepsilon, R],$$

and

$$\gamma_{2k}^{R,1}(s) = Re^{is}, \qquad s \in \left(\frac{d+\arg(\lambda)}{q} - \frac{2k\pi}{\mu} - \frac{\delta}{3}, \frac{d+\arg(\lambda)}{q} + \frac{2k\pi}{\mu} + \frac{\delta}{3}\right).$$

We now give growth estimates for u_1 and u_2 in order to conclude the result. We first give bounds for $u_2(t,z)$. We take $k \in \{0,\ldots,\mu-1\}$. Let t be as above with $|t| \geq 1$, and consider ξ and w in the trace of the corresponding path defined by the path integrals in the definition of u_2. From the properties of the kernel functions in Definition 6.2, one has that

$$\left|E_1(t\lambda(\xi))E_2(\xi z)\frac{e_2(\xi w)}{\xi w}\right| \leq C_{11}|E_1(t\lambda(\xi))| \frac{h_{\mathfrak{m}_2}\left(\frac{C_{12}}{|\xi|\varepsilon}\right)}{h_{\mathfrak{m}_2}\left(\frac{C_{13}}{|\xi||z|}\right)|\xi|\varepsilon},$$

for some $C_{11}, C_{12}, C_{13} > 0$.

Taking into account (5) we have

$$J_1 := \left| \oint_{\gamma_{2k+1}} \phi(w) \int_{\xi_0}^{\infty(\theta)} E_1(t\lambda(\xi)) E_2(\xi z) \frac{e_2(\xi w)}{\xi w} d\xi dw \right|$$

$$\leq C_{11} \int_{s \in I_{2k+1}} |\phi(\varepsilon e^{is})| \int_{|\xi_0|}^{\infty} h_{\mathfrak{m}_2}\left(\frac{\rho(2)C_{12}}{|\xi|\varepsilon}\right) |E_1(t\lambda(|\xi|e^{i\theta}))| \frac{h_{\mathfrak{m}_2}\left(\frac{\rho(2)C_{12}}{|\xi|\varepsilon}\right)}{h_{\mathfrak{m}_2}\left(\frac{C_{13}}{|\xi||z|}\right)|\xi|\varepsilon} d|\xi| ds.$$

We assume z satisfies $|z| \leq C_{13}\varepsilon/(\rho(2)C_{12})$. This entails

$$h_{\mathfrak{m}_2}\left(\frac{\rho(2)C_{12}}{|\xi|\varepsilon}\right) \leq h_{\mathfrak{m}_2}\left(\frac{C_{13}}{|\xi||z|}\right).$$

By the careful choice of the direction θ above and because of Proposition 6.8 applied to E_1, we deduce there exists $\delta > 0$ such that the function $(t, |\xi|) \mapsto |E_1(t\lambda(|\xi|e^{i\theta}))|$ admits a maximum at a point, say $(t_1, |\xi_1|)$, as $(t, |\xi|)$ runs over $(S(d+2n\pi/\nu, \delta) \cap \{t : |t| \geq 1\}) \times [|\xi_0|, \infty)$. Then, for every such t and $|\xi| \geq |\xi_0|$, one easily obtains constants $C_{14}, C_{15} > 0$ such that

$$|E_1(t\lambda(|\xi|e^{i\theta}))| \leq |E_1(t\lambda(|\xi_1|e^{i\theta}))| \leq \frac{C_{14}}{h_{\mathfrak{m}_1}\left(\frac{C_{15}}{|t|}\right)}.$$

Moreover,

$$\int_{|\xi_0|}^{\infty} h_{\mathfrak{m}_2}\left(\frac{\rho(2)C_{12}}{|\xi|\varepsilon}\right) \frac{1}{|\xi|\varepsilon} d|\xi| < \infty,$$

so

$$J_1 \leq \frac{C_{16}}{h_{\mathfrak{m}_1}\left(\frac{C_{15}}{|t|}\right)} \int_{s \in I_{2k+1}} |\phi(\varepsilon e^{is})| \, ds.$$

Taking into account that

$$\sup_{|w|=\varepsilon} |\phi(w)| < \infty,$$

one concludes that $u_2 \in \mathcal{O}^{\mathfrak{m}_1}(\hat{S}(d + 2n\pi/\nu) \times D(0, r))$ for some $r > 0$ and for $n = 0, \ldots, \nu - 1$.

We now give estimates on $u_1(t,z)$. The inner integral in the definition of each term in the sum of u_1 can be upper bounded as before. We arrive at

$$J_2 := \left| \oint_{\gamma_{2k}^R} \phi(w) \int_{\xi_0}^{\infty(\theta)} E_1(t\lambda(\xi))E_2(\xi z)\frac{e_2(\xi w)}{\xi w} d\xi dw \right|$$

$$\leq C_{21} \int_{\varepsilon}^{R} \left(|\phi(se^{i\theta_+})| + |\phi(se^{i\theta_-})|\right) ds \frac{1}{h_{\mathfrak{m}_1}\left(\frac{C_{22}}{|t|}\right)} + C_{23} \int_{\theta_-}^{\theta_+} |\phi(Re^{i\theta})| d\theta \frac{1}{h_{\mathfrak{m}_1}\left(\frac{C_{22}}{|t|}\right)},$$

for some positive constants C_{21}, C_{22}, C_{23}. Since $\phi \in \mathcal{O}^{\mathfrak{m}^{(1/q)}}(\hat{S}((d + \arg(\lambda))/q + 2k\pi/\mu))$, it is straightforward to check that the previous expression can be upper bounded by

$$\frac{C_{24}}{h_{\mathfrak{m}^{(1/q)}}\left(\frac{C_{25}}{R}\right) h_{\mathfrak{m}_1}\left(\frac{C_{22}}{|t|}\right)},$$

for some $C_{24}, C_{25} > 0$. Cauchy's theorem allow us to choose R to be $R = |t|^{1/q}$. In addition to this, from property (29) one has

$$h_{\mathfrak{m}^{(1/q)}}\left(\frac{C_{25}}{|t|^{1/q}}\right) = \left(h_{\mathfrak{m}}\left(\frac{C_{25}^q}{|t|}\right)\right)^{1/q}.$$

If $0 < q \leq 1$, one can apply property (5) to obtain

$$\left(h_{\mathfrak{m}}\left(\frac{C_{25}^q}{|t|}\right)\right)^{1/q} \geq h_{\mathfrak{m}}\left(\frac{C_{25}^q}{\rho(1/q)|t|}\right),$$

and if $q \geq 1$, $h_{\mathfrak{m}}(s) \leq 1$ for all $s \in (0,\infty)$, so that

$$\left(h_{\mathfrak{m}}\left(\frac{C_{25}^q}{|t|}\right)\right)^{1/q} \geq h_{\mathfrak{m}}\left(\frac{C_{25}^q}{|t|}\right).$$

These facts entail, for some $C_{26} > 0$,

$$J_2 \leq \frac{C_{24}}{h_{\mathfrak{m}}\left(\frac{C_{26}}{|t|}\right) h_{\mathfrak{m}_1}\left(\frac{C_{22}}{|t|}\right)}.$$

From the hypothesis (60) we have, by (6), that $h_{\mathbb{m}}(v) \leq C h_{\mathbb{m}_1}(Av)$ for every $v > 0$, and so, putting $C_{27} = \min\{C_{22}/A, C_{26}\}$, one gets that

$$J_2 \leq \frac{CC_{24}}{h_{\mathbb{m}}\left(\frac{C_{26}}{|t|}\right) h_{\mathbb{m}}\left(\frac{C_{22}}{A|t|}\right)} \leq \frac{CC_{24}}{\left(h_{\mathbb{m}}\left(\frac{C_{27}}{|t|}\right)\right)^2} \leq \frac{CC_{24}}{h_{\mathbb{m}}\left(\frac{C_{27}}{\rho(2)|t|}\right)},$$

where (5) has been used in the last inequality. So, one obtains that $u_1 \in \mathcal{O}^{\mathbb{m}}(\hat{S}(d + 2n\pi/\nu) \times D(0,r))$ for some $r > 0$ and for $n = 0, \ldots, \nu - 1$, and the conclusion is immediate. □

Lemmas 6 and 7 in [45] can be easily rewritten in our context, and they lead us straightforward to the next result, an analogue of Theorem 3 in [45].

Theorem 7.11 *Let $q = \mu/\nu \in \mathbb{Q}$, with* $\gcd(\mu, \nu) = 1$. *Let $m_1(\lambda)$, $m_2(\lambda)$ and $\mathbb{m}$ be as in Lemma 7.9. If $u(t,z)$ is the solution of (54), then for every $d \in \mathbb{R}$ the following statements are equivalent:*

1. *$\phi \in \mathcal{O}^{\mathbb{m}^{(1/q)}}(\hat{S}((d + \arg(\lambda))/q + 2k\pi/\mu))$ for every $k = 0, \ldots, \mu - 1$.*
2. *$u \in \mathcal{O}^{\mathbb{m}}(\hat{S}(d + 2n\pi/\nu) \times D(0,r))$, for $n = 0, 1, \ldots, \nu - 1$.*

Although all the treatment of summability in this paper has been limited to complex valued functions, it can be extended without any difficulty to functions taking their values in a general complex Banach algebra. In particular, we may take this algebra to consist of the bounded holomorphic functions in a fixed neighborhood of the origin in the z plane with the norm of the supremum, and consider summability of formal power series in the t variable with such coefficients. The following definition is natural under this point of view.

Definition 7.12 Let $\hat{u}(t,z) = \sum_{j=0}^{\infty} u_j(z) t^j$ be a formal series with coefficients in $\mathcal{O}(D(0,r))$ for some $r > 0$ (independent of j), and let $\mathbb{m}_e = (m_e(j))_{j \in \mathbb{N}_0}$ be the strongly regular moment sequence of a kernel e. We say $\hat{u}$ is $\mathbb{m}_e$-summable in direction $d \in \mathbb{R}$ if

$$\hat{T}_e^{-}(t,z) = \sum_{j=0}^{\infty} \frac{u_j(z)}{m_e(j)} t^j \in \mathcal{O}^{\mathbb{m}_e}(\hat{S}(d) \times D(0,r)),$$

where $S(d)$ is an unbounded (small) sector bisected by d.

We are now able to establish a characterization of summability for the formal solutions of (54) and also for the initial problem (45), (46), under appropriate conditions regarding the moment functions involved.

Proposition 7.13 *Let $\hat{u}$ be a formal solution of (54). Let $q = \mu/\nu \in \mathbb{Q}$, with* $\gcd(\mu, \nu) = 1$, *and $d \in \mathbb{R}$. We assume a strongly regular sequence of moments $\mathbb{m} = (m(p))_{p \in \mathbb{N}_0}$ exists with*

$$m_2(qj) \leq C_0 A_0^j m(j) m_1(j), \quad j \in \mathbb{N}_0, \tag{63}$$

and

$$m(j/q)m_1(j/q) \leq C_1 A_1^j m_2(j), \quad j \in \mathbb{N}_0, \tag{64}$$

for suitable $C_0, C_1, A_0, A_1 > 0$. *Then,* $\hat{u}$ *is* $\mathfrak{m}$*-summable in direction* $d + 2n\pi/\nu$ *for* $n = 0, \ldots, \nu - 1$ *if, and only if,* $\phi \in \mathcal{O}^{\mathfrak{m}^{(1/q)}}(\hat{S}((d + \arg\lambda)/q + 2k\pi/\mu))$ *for* $k = 0, \ldots, \mu - 1$.

Remark 7.14 As in the previous Remark 7.10, the inequalities (63) and (64) amount to the equivalence of $\mathfrak{m}_2^{(q)}$ and $\mathfrak{m} \cdot \mathfrak{m}_1$ for some strongly regular sequence of moments $\mathfrak{m}$. The existence of such a sequence $\mathfrak{m}$, in turn, is equivalent to the fact that $q\omega(\mathfrak{m}_2) > \omega(\mathfrak{m}_1)$, as it may be proved by following the ideas in the work of Balser [2, Theorem 32]. One may note that in the Gevrey case treated by Michalik [45], in which the order (in the sense of W. Balser) of a kernel completely determines the summability method under consideration, one just imposes [instead of (63) and (64)] suitable conditions on the orders of the kernels e_1 and e_2. Moreover, in that situation the summability of $\hat{u}$ is characterized by the analytic continuation of ϕ with a (constant) order of exponential growth. As explained before [see Remark 6.19(ii)], this is not enough now, since these constants do not completely determine the summability method whose application is to be justified.

Proof Let $n \in \{0, \ldots, \nu - 1\}$. By Definition 7.12, $\hat{u}$ is $\mathfrak{m}$-summable in direction $d + 2n\pi/\nu$ if, and only if,

$$v(t,z) := \sum_{j=\beta-1}^{\infty} \binom{j}{\beta-1} \frac{\lambda^j (\partial_{\mathfrak{m}_2,z})\phi(z)}{m_1(j)m(j)} t^j \in \mathcal{O}^{\mathfrak{m}}(\hat{S}(d + 2n\pi/\nu) \times D(0,r)).$$

If we put $\tilde{\mathfrak{m}} = (\tilde{m}(p))_{p\geq 0}$, with $\tilde{m}(p) = m_1(p)m(p)$, then $\tilde{\mathfrak{m}}$ turns out to be a sequence of moments, as it may be deduced along the same lines as in the Gevrey case (see [2, Sect. 5.8]). One can observe that v turns out to be the solution of (54) when substituting $\mathfrak{m}_1$ by $\tilde{\mathfrak{m}}$. From Theorem 7.11, we know that $v \in \mathcal{O}^{\mathfrak{m}}(\hat{S}(d + 2n\pi/\nu) \times D(0,r))$ if, and only if, $\phi \in \mathcal{O}^{\mathfrak{m}^{(1/q)}}(\hat{S}((d + \arg\lambda)/q + 2k\pi/\mu))$ for $k = 0, \ldots, \mu - 1$, as desired. □

Finally, we consider the normalized formal solution for (45) given in (49). We make the following:

Assumption (A): There exists $q = \mu/\nu \in \mathbb{Q}$ with $gcd(\mu, \nu) = 1$ such that $P(\lambda, \xi)$ in (47) satisfies that

$$\lim_{z\to\infty} \frac{\lambda_\alpha(z)}{z^q} \in \mathbb{C} \setminus \{0\},$$

for every $\alpha = 1, \ldots, \ell$, i.e., $q \in \mathbb{Q}$ is the common order pole of λ_α for every $\alpha = 1, \ldots, \ell$.

The previous results lead to the main result of this last section.

Theorem 7.15 *Let $d \in \mathbb{R}$. Suppose a strongly regular sequence of moments $\mathfrak{m} = (m(p))_{p\in\mathbb{N}_0}$ exists such that (63) and (64) hold. Let $\hat{u}$ be the normalized formal solution of (45), (46). Then, under Assumption (A), $\hat{u}$ is $\mathfrak{m}$-summable in any direction of the form $d + 2n\pi/\nu$ for $n = 0, \ldots, \nu - 1$ if, and only if, $\phi \in \mathcal{O}^{\mathfrak{m}^{(1/q)}}(\hat{S}((d + \arg(\lambda_{\alpha\beta}))/q + 2k\pi/\mu))$ for every $k = 0, \ldots, \mu - 1$, every $\alpha = 1, \ldots, \ell$ and every $\beta = 1, \ldots, n_\alpha$.*

Acknowledgements This work is partially supported by the Spanish Ministry of Economy and Competitiveness under project MTM2012-31439.

References

1. W. Balser, *From Divergent Power Series to Analytic Functions. Theory and Application of Multisummable Power Series*. Lecture Notes in Mathematics, vol. 1582 (Springer, Berlin, 1994)
2. W. Balser, *Formal Power Series and Linear Systems of Meromorphic Ordinary Differential Equations* (Springer, Berlin 2000)
3. W. Balser, Multisummability of formal power series solutions of partial differential equations with constant coefficients. J. Differ. Equ. **201**, 63–74 (2004)
4. W. Balser, M. Miyake, Summability of formal solutions of certain partial differential equations. Acta Sci. Math. (Szeged) **65**, 543–551 (1999)
5. W. Balser, J. Mozo-Fernández, Multisummability of formal solutions of singular perturbation problems. J. Differ. Equ. **183**, 526–545 (2002)
6. W. Balser, M. Yoshino, Gevrey order of formal power series solutions of inhomogeneous partial differential equations with constant coefficients. Funkcialaj Ekvac. **53**, 411–434 (2010)
7. W. Balser, B.L.J. Braaksma, J.-P. Ramis, Y. Sibuya, Multisummability of formal power series solutions of linear ordinary differential equations. Asymptot. Anal. **5**, 27–45 (1991)
8. J. Bonet, R.W. Braun, R. Meise, B.A. Taylor, Whitney's extension theorem for nonquasi-analytic classes of ultradifferentiable functions. Stud. Math. **99**(2), 155–184 (1991)
9. B.L.J. Braaksma, Multisummability of formal power series solutions of nonlinear meromorphic differential equations. Ann. Inst. Fourier (Grenoble) **42**, 517–540 (1992)
10. B.L.J. Braaksma, B. Faber, G. Immink, Summation of formal solutions of a class of linear difference equations. Pac. J. Math. **195**(1), 35–65 (2000)
11. J. Bruna, An extension theorem of Whitney type for nonquasianalytic classes of functions. J. Lond. Math. Soc. II Ser. **22**, 495–505 (1980)
12. M. Canalis-Durand, J. Mozo-Fernández, R. Schäfke, Monomial summability and doubly singular differential equations. J. Differ. Equ. **233**, 485–511 (2007)
13. J.C. Canille, Desenvolvimento assintótico e introdução ao cálculo diferencial resurgente. 17 Colóquio Brasileiro de Matemática, IMPA (1989)
14. J. Chaumat, A.M. Chollet, Surjectivité de l'application restriction à un compact dans des classes de fonctions ultradifférentiables. Math. Ann. **298**(1), 7–40 (1994)
15. F. Galindo, J. Sanz, On strongly asymptotically developable functions and the Borel-Ritt theorem. Stud. Math. **133**(3), 231–248 (1999)
16. I.M. Gelfand, G.E. Shilov, *Generalized Functions, Vol. 2, Space of Fundamental and Generalized Functions* (Academic, New York, 1965)
17. A.A. Goldberg, I.V. Ostrovskii, *Value Distribution of Meromorphic Functions*, Translations of Mathematical Monographs, vol. 236 (American Mathematical Society, Providence, RI, 2008)

18. M. Hibino, Summability of formal solutions for singular first-order linear PDEs with holomorphic coefficients, in *Differential Equations and Exact WKB Analysis*. RIMS KôkyîVoku Bessatsu, B10 (Res. Inst. Math. Sci. (RIMS), Kyoto, 2008), pp. 47–62
19. A.S.B. Holland, *Introduction to the Theory of Entire Functions* (Academic, New York, 1973)
20. G.K. Immink, Exact asymptotics of nonlinear difference equations with levels 1 and 1^+. Ann. Fac. Sci. Toulouse T. XVII **2**, 309–356 (2008)
21. G.K. Immink, Accelero-summation of the formal solutions of nonlinear difference equations. Ann. Inst. Fourier (Grenoble) **61**(1), 1–51 (2011)
22. J.J. Jiménez-Garrido, J. Sanz, Strongly regular sequences and proximate orders. J. Math. Anal. Appl. **438**(2), 920–945 (2016)
23. J. Jiménez-Garrido, J. Sanz, G. Schindl, Log-convex sequences and nonzero proximate orders. J. Math. Anal. Appl. **448**(2), 1572–1599 (2017)
24. H. Komatsu, Ultradistributions, I: Structure theorems and a characterization. J. Fac. Sci. Univ. Tokyo Sect. IA Math. **20**, 25–105 (1973)
25. B.I. Korenbljum, Conditions of nontriviality of certain classes of functions analytic in a sector, and problems of quasianalyticity. Sov. Math. Dokl. **7** (1966), 232–236
26. A. Lastra, J. Sanz, Stieltjes moment problem in general Gelfand-Shilov spaces. Studia Math. **192**, 111–128 (2009)
27. A. Lastra, J. Sanz, Quasi-analyticity in Carleman ultraholomorphic classes. Ann. Inst. Fourier **60**, 1629–1648 (2010)
28. A. Lastra, J. Sanz, Extension operators in Carleman ultraholomorphic classes. J. Math. Anal. Appl. **372**(1), 287–305 (2010)
29. A. Lastra, S. Malek, J. Sanz, Continuous right inverses for the asymptotic Borel map in ultraholomorphic classes via a Laplace-type transform. J. Math. Anal. Appl. **396**, 724–740 (2012)
30. A. Lastra, S. Malek, J. Sanz, On Gevrey solutions of threefold singular nonlinear partial differential equations. J. Differ. Equ. **255**, 3205–3232 (2013)
31. A. Lastra, S. Malek, J. Sanz, Summability in general Carleman ultraholomorphic classes. J. Math. Anal. Appl. **430**, 1175–1206 (2015)
32. A. Lastra, S. Malek, J. Sanz, Strongly regular multi-level solutions of singularly perturbed linear partial differential equations. Results Math. **70**(3-4), 581–614 (2016)
33. B.J. Levin, *Distribution of Zeros of Entire Functions*. Translations of Mathematical Monographs, vol. 5 (American Mathematical Society, Providence, RI, 1980)
34. L.S. Maergoiz, Indicator diagram and generalized Borel-Laplace transforms for entire functions of a given proximate order. St. Petersburg Math. J. **12**(2), 191–232 (2001)
35. E. Maillet, Sur les séries divergentes et les équations différentielles. Ann. Éc. Norm. Sup. Paris **3**(20), 487–518 (1903)
36. S. Malek, On the summability of formal solutions of linear partial differential equations. J. Dyn. Control Syst. **11**(3), 389–403 (2005)
37. S. Malek, On the summability of formal solutions of nonlinear partial differential equations with schrinkings. J. Dyn. Control Syst. **13**(1), 1–13 (2007)
38. S. Malek, On Gevrey functional solutions of partial differential equations with Fuchsian and irregular singularities. J. Dyn. Control Syst. **15**(2), 277–305 (2009)
39. S. Malek, On singularly perturbed small step size difference-differential nonlinear PDEs. J. Differ. Equ. Appl. **20**(1), 118–168 (2014)
40. S. Mandelbrojt, *Séries adhérentes, régularisation des suites, applications*. Collection de monographies sur la théorie des fonctions (Gauthier-Villars, Paris, 1952)
41. J. Martinet, J.P. Ramis, Problèmes de modules pour des équations différentielles non linéaires du premier ordre. Publ. Math. Inst. Hautes Etudes Sci. **55**, 63–164 (1982)
42. J. Martinet, J.-P. Ramis, Elementary acceleration and multisummability. Ann. Inst. Henri Poincaré, Physique Theorique **54**, 331–401 (1991)
43. S. Michalik, Summability and fractional linear partial differential equations. J. Dyn. Control Syst. **16**(4), 557–584 (2010)

44. S. Michalik, On the multisummability of divergent solutions of linear partial differential equations with constant coefficients. J. Differ. Equ. **249**(3), 551–570 (2010)
45. S. Michalik, Analytic solutions of moment partial differential equations with constant coefficients. Funkcialaj Ekvac. **56**, 19–50 (2013)
46. S. Michalik, Summability of formal solutions of linear partial differential equations with divergent initial data. J. Math. Anal. Appl. **406**, 243–260 (2013)
47. S. Michalik, Analytic and summable solutions of inhomogeneous moment partial differential equations. Funkcialaj Ekvac (to appear). Available at http://arxiv.org/abs/1505.06724
48. S. Ouchi, Multisummability of formal solutions of some linear partial differential equations. J. Differ. Equ. **185**, 513–549 (2002)
49. H.-J. Petzsche, On E. Borel's theorem. Math. Ann. **282**(2), 299–313 (1988)
50. J.P. Ramis, Dévissage Gevrey. Asterisque **59–60**, 173–204 (1978)
51. J.P. Ramis, *Les séries k-sommables et leurs applications*. Lecture Notes in Physics, vol. 126 (Springer, Berlin, 1980)
52. J.-P. Ramis, Y. Sibuya, A new proof of multisummability of formal solutions of non linear meromorphic differential equations. Ann. Inst. Fourier (Grenoble) **44**, 811–848 (1994)
53. J. Sanz, Linear continuous extension operators for Gevrey classes on polysectors. Glasg. Math. J. **45**(2), 199–216 (2003)
54. J. Sanz, Flat functions in Carleman ultraholomorphic classes via proximate orders. J. Math. Anal. Appl. **415**, 623–643 (2014)
55. J. Schmets, M. Valdivia, Extension maps in ultradifferentiable and ultraholomorphic function spaces. Stud. Math. **143**(3), 221–250 (2000)
56. V. Thilliez, Extension Gevrey et rigidité dans un secteur. Stud. Math. **117**, 29–41 (1995)
57. V. Thilliez, Division by flat ultradifferentiable functions and sectorial extensions. Results Math. **44**, 169–188 (2003)
58. V. Thilliez, Smooth solutions of quasianalytic or ultraholomorphic equations. Monatsh. Math. **160**(4), 443–453 (2010)
59. J.C. Tougeron, An introduction to the theory of Gevrey expansions and to the Borel-Laplace transform with some applications, Toronto University (1990)
60. G. Valiron, *Théorie des Fonctions* (Masson et Cie., Paris, 1942)

WKB Analysis and Stokes Geometry of Differential Equations

Yoshitsugu Takei

Abstract In this article we survey the fundamental theory of the exact WKB analysis, that is, the WKB analysis based on the Borel resummation method. Starting with the exact WKB analysis for second order linear ordinary differential equations, we explain its application to the computation of monodromy groups of Fuchsian equations and its generalization to higher order equations. Some recent developments of the theory such as the exact WKB analysis for completely integrable systems are also briefly discussed.

Keywords Borel resummation • Completely integrable system • Connection formula • Exact WKB analysis • Monodromy group • New Stokes curve • Stokes geometry • Virtual turning point • Wall crossing formula • WKB solution

Mathematics Subject Classification (2000). Primary 34M60; Secondary 34E20, 34M40, 34M30, 34M35

1 Introduction

Since the very beginning of the quantum mechanics, the WKB (Wentzel-Kramers-Brillouin) approximation has been employed to obtain approximate eigenfunctions and solve the eigenvalue problems for Schrödinger equations. The (full-order) WKB approximations provide formal solutions (with respect to the Planck constant) of Schrödinger equations but, as they are divergent in almost all cases, they were not so often used in rigorous mathematical analysis. Around 1980, using the Borel resummed WKB solutions, Voros [36] successfully studied spectral functions of quartic oscillators and also Silverstone [33] discussed the WKB-type connection problem more rigorously. After their pioneering works, Pham, Delabaere and others (cf., e.g., [9, 11–13, 30]) have developed this new kind of WKB analysis (sometimes

Y. Takei (✉)
Department of Mathematical Sciences, Doshisha University, Kyotanabe, Kyoto 610-0321, Japan
e-mail: ytakei@mail.doshisha.ac.jp

G. Filipuk et al. (eds.), *Analytic, Algebraic and Geometric Aspects of Differential Equations*, Trends in Mathematics, DOI 10.1007/978-3-319-52842-7_5

called "exact WKB analysis" or "complex WKB analysis") based on the Borel resummation technique with the aid of Ecalle's theory of resurgent functions ([14–16], see also [32]). At present it turns out that the exact WKB analysis is very efficient not only for eigenvalue problems of Schrödinger equations but also for the global study of differential equations in the complex domain.

In this article, mainly using some concrete and illuminating examples, we explain the fundamental theory of the exact WKB analysis, its application to the global study of differential equations in the complex domain, and some recent developments of the theory.

The explanation will be done basically by following our monographs [27] and [25]. To be more specific, the article is organized as follows. We first discuss the exact WKB analysis for second order linear ordinary differential equations of Schrödinger type

$$\left(\frac{d^2}{dx^2} - \eta^2 Q(x)\right)\psi = 0, \tag{1}$$

where $Q(x)$ is a polynomial or a rational function and η denotes the inverse of the Planck constant (and hence a large parameter). Starting with the definition of WKB solutions, we introduce the Stokes geometry and explain the fundamental theorems of the exact WKB analysis, in particular, Voros' connection formula for Borel resummed WKB solutions, the most important result in the theory, in Sect. 2. Then, after illustrating an outline of the proof of the fundamental theorems in Sect. 3, we discuss its application to the computation of monodromy groups of Fuchsian equations (Sect. 4) and wall crossing formulas for WKB solutions with respect to the change of parameters contained in the equation (Sect. 5). In the latter part of the article, we consider generalizations of the exact WKB analysis to higher order linear ordinary differential equations of the form

$$\left(\frac{d^m}{dx^m} + \eta p_1(x)\frac{d^{m-1}}{dx^{m-1}} + \cdots + \eta^m p_m(x)\right)\psi = 0. \tag{2}$$

In Sect. 6 we discuss the problem of new Stokes curves pointed out by Berk-Nevins-Roberts [8], which is peculiar to higher order equations, and introduce the notion of virtual turning points with the help of the theory of microlocal analysis to treat new Stokes curves in a more intrinsic manner. Finally, in Sect. 7, we explain some recent developments of the theory such as the exact WKB analysis for completely integrable systems. These recent developments are also closely related to the problem of new Stokes curves and virtual turning points for higher order equations.

2 Exact WKB Analysis for Second Order Linear ODEs of Schrödinger Type

WKB Solutions

Let us first discuss the exact WKB analysis for second order linear ordinary differential equations of Schrödinger type

$$\left(\frac{d^2}{dx^2} - \eta^2 Q(x)\right)\psi = 0, \tag{3}$$

where $Q(x)$ is a polynomial or a rational function. Throughout this article η denotes a large parameter and is often assumed to be real and positive.

Definition 2.1 A WKB solution of (3) is a formal solution of the following form:

$$\psi(x,\eta) = \exp(\eta y_0(x)) \sum_{n=0}^{\infty} \psi_n(x)\eta^{-(n+\alpha)}, \tag{4}$$

where $y_0(x)$ and $\psi_n(x)$ are suitable analytic functions of x and $\alpha \geq 0$ is a constant.

In the case of (3) WKB solutions can be readily constructed in the following way: Assume that a solution of (3) has the form

$$\psi(x,\eta) = \exp \int^x S(x,\eta)dx, \tag{5}$$

then $S(x, \eta)$ should satisfy

$$S^2 + \frac{dS}{dx} = \eta^2 Q(x) \qquad (\text{"Riccati equation"}). \tag{6}$$

We further suppose that $S = S(x, \eta)$ can be expanded as $S = \eta S_{-1}(x) + S_0(x) + \eta^{-1}S_1(x) + \cdots$. It then follows from (6) that

$$(S_{-1})^2 = Q(x), \tag{7}$$

$$2S_{-1}S_{n+1} + \sum_{k=0}^{n} S_k S_{n-k} + \frac{dS_n}{dx} = 0 \qquad (n = -1, 0, 1, \ldots). \tag{8}$$

That is, once $S_{-1} = \pm\sqrt{Q(x)}$ is fixed, we obtain two solutions $S_{\pm}(x, \eta)$ of (6) in a recursive manner.

Remark 2.2 Let us denote $S_\pm$ as $S_\pm = \pm S_{\rm odd} + S_{\rm even}$, then the following relation is readily confirmed.

$$2S_{\rm odd}S_{\rm even} + \frac{dS_{\rm odd}}{dx} = 0, \quad \text{i.e.,} \quad S_{\rm even} = -\frac{1}{2}\frac{d}{dx}\log S_{\rm odd}. \tag{9}$$

Thus for Eq. (3) we obtain the following WKB solutions:

$$\begin{aligned}\psi_\pm(x,\eta) &= \frac{1}{\sqrt{S_{\rm odd}}}\exp\left(\pm\int_{x_0}^x S_{\rm odd}dx\right)\\ &= \exp\left(\pm\eta\int_{x_0}^x \sqrt{Q(x)}dx\right)\sum_{n=0}^{\infty}\psi_{\pm,n}(x)\eta^{-(n+1/2)},\end{aligned} \tag{10}$$

where x_0 is an arbitrarily chosen reference point.

Unfortunately WKB solutions are, in general, divergent. Concerning the behavior of $S_n(x)$ and $\psi_{\pm,n}(x)$ for large n, the following holds.

Proposition 2.3

(i) *Each $S_n(x)$ and $\psi_{\pm,n}(x)$ are holomorphic on*

$$U := \{x \in \mathbb{C} \mid Q(x) \text{ is holomorphic near } x \text{ and } Q(x) \neq 0\}. \tag{11}$$

(ii) *For any compact set K in U, there exist positive constants A_K and C_K satisfying*

$$|S_n(x)| \le A_K C_K^n n!, \quad |\psi_{\pm,n}(x)| \le A_K C_K^n n! \quad (x \in K) \tag{12}$$

for any n.

To give an analytic meaning to WKB solutions, we employ the Borel resummation technique (or the Borel-Laplace method) with respect to a large parameter η in the exact WKB analysis.

Definition 2.4 Let $\eta > 0$ be a large parameter. For an infinite series $f = \exp(\eta y_0)\sum_{n\ge0} f_n\eta^{-(n+\alpha)}$ ($\alpha > 0$, y_0, f_n : constants), we define

$$f_B(y) = \sum_{n=0}^{\infty}\frac{f_n}{\Gamma(n+\alpha)}(y+y_0)^{n+\alpha-1} \quad : \text{ Borel transform of } f, \tag{13}$$

$$F(\eta) = \int_{-y_0}^{\infty} e^{-y\eta}f_B(y)dy \quad : \text{ Borel sum of } f, \tag{14}$$

provided that they are well-defined. Here $\Gamma(s)$ denotes Euler's Γ-function and the integration path of (14) is taken to be parallel to the positive real axis.

See, e.g., [7, 10] for the details of the Borel-Laplace method. Here we only refer the following very fundamental properties of the Borel transform and the Borel sum.

Proposition 2.5

(i) *If f is convergent, then $(y+y_0)^{1-\alpha}f_B(y)$ is an entire function of exponential type. In this case the Borel sum $F(\eta)$ of f is well-defined for a sufficiently large $\eta > 0$ and coincides with the original f.*

(ii) *If f is Borel summable, that is,*

(a) $\sum \frac{f_n}{\Gamma(n+\alpha)}(y+y_0)^n$ *is convergent in a neighborhood of $y = -y_0$,*

(b) $f_B(y)$ *can be analytically continued along the integration path of the Borel sum, and*

(c) $\int_{-y_0}^{\infty} e^{-y\eta} f_B(y)dy$ *exists for a sufficiently large $\eta > 0$,*

then the following asymptotic formula holds :

$$\exp(-\eta y_0)\eta^{\alpha}F(\eta) \sim \sum_{n=0}^{\infty} f_n \eta^{-n} \qquad (\eta > 0,\ \eta \to \infty). \tag{15}$$

Proposition 2.6 *For $\psi(x,\eta) = \exp(\eta y_0(x)) \sum_{n\geq 0} \psi_n(x)\eta^{-(n+\alpha)}$ $(\alpha > 0, \alpha \notin \mathbb{Z})$ the following formulas hold :*

(i) $\left[\frac{\partial}{\partial x}\psi\right]_B = \frac{\partial}{\partial x}\psi_B.$

(ii) $\left[\eta^m \psi\right]_B = \left(\frac{\partial}{\partial y}\right)^m \psi_B \qquad (m = 1, 2, \ldots).$

(iii) $\left[\eta^{-m}\psi\right]_B = \frac{1}{(m-1)!}\int_{-y_0(x)}^{y} (y-y')^{m-1}\psi_B(x,y')dy' \ \left(=: \left(\frac{\partial}{\partial y}\right)^{-m}\psi_B \right).$

Furthermore, for $\psi = \sum \psi_n(x)\eta^{-(n+\alpha)}$ and $\varphi = \sum \varphi_n(x)\eta^{-(n+\beta)}$ $(\alpha, \beta > 0)$ we have

(iv) $\left[\varphi\psi\right]_B = \varphi_B * \psi_B := \int_0^y \varphi_B(x, y-y')\psi_B(x,y')dy'.$

WKB Solutions of the Airy Equation and Their Borel Transforms

To investigate properties of the Borel transform of WKB solutions, we consider WKB solutions of the Airy equation in this subsection.

Example (Airy Equation) Let us consider the Airy equation

$$\left(\frac{d^2}{dx^2} - \eta^2 x\right)\psi = 0 \tag{16}$$

and its WKB solutions normalized at $x = 0$

$$\psi_\pm = \frac{1}{\sqrt{S_{\mathrm{odd}}}} \exp\left(\pm\int_0^x S_{\mathrm{odd}}dx\right) = \exp(\pm\eta y_0(x)) \sum_{n=0}^{\infty} \psi_{\pm,n}(x)\eta^{-(n+1/2)}, \tag{17}$$

where $y_0(x) = \int_0^x \sqrt{x}dx = (2/3)x^{3/2}$. We compute the Borel transform of (17) explicitly.

By the recursion formulas (7)–(8) we easily find that each coefficient $S_n(x)$ of the formal power series solution of the Riccati equation associated with (16) has the form $S_n = c_n x^{-1-(3/2)n}$ with some constant c_n $(n = -1, 0, 1, \ldots)$. This implies that each coefficient $\psi_{\pm,n}(x)$ of (17) also has the form $\psi_{\pm,n} = d_{\pm,n}x^{-1/4-(3/2)n}$ with another constant $d_{\pm,n}$ $(n = 0, 1, 2, \ldots)$. Hence the Borel transform $\psi_{\pm,B}(x, y)$ can be written as

$$\psi_{\pm,B}(x,y) = \frac{1}{x}\sum_{n=0}^{\infty} \frac{d_{\pm,n}}{\Gamma(n+1/2)}\left(\frac{y}{x^{3/2}} \pm \frac{2}{3}\right)^{n-1/2} = \frac{1}{x}\phi_\pm(yx^{-3/2}) \tag{18}$$

with $\phi_\pm(t)$ being an analytic function of one variable $t = yx^{-3/2}$. On the other hand, since $\psi_\pm$ is a solution of (16), it follows from Proposition 2.6, (ii) that $\psi_{\pm,B}(x, y)$ should satisfy

$$\left(\frac{\partial^2}{\partial x^2} - x\frac{\partial^2}{\partial y^2}\right)\psi_{\pm,B}(x,y) = 0. \tag{19}$$

Consequently we obtain the following ODE for $\phi_\pm(t)$:

$$\left(\left(1 - \frac{9}{4}t^2\right)\frac{d^2}{dt^2} - \frac{27}{4}t\frac{d}{dt} - 2\right)\phi_\pm = 0, \tag{20}$$

or, employing a change of variable $s = 3t/4 + 1/2$,

$$\left(s(1-s)\frac{d^2}{ds^2} + \left(\frac{3}{2} - 3s\right)\frac{d}{ds} - \frac{8}{9}\right)\phi_\pm = 0. \tag{21}$$

Equation (21) is nothing but Gauss' hypergeometric equation (with the parameter $(\alpha, \beta, \gamma) = (4/3, 2/3, 3/2)$). Thus we have the following expression for $\psi_{\pm,B}(x, y)$:

$$\psi_{+,B}(x,y) = \frac{\sqrt{3}}{2\sqrt{\pi}}\frac{1}{x}s^{-1/2}F(5/6, 1/6, 1/2; s), \tag{22}$$

$$\psi_{-,B}(x,y) = \frac{\sqrt{3}}{2\sqrt{\pi}}\frac{1}{x}(1-s)^{-1/2}F(5/6, 1/6, 1/2; 1-s), \tag{23}$$

where $F(\alpha, \beta, \gamma; z)$ denotes Gauss' hypergeometric function and $s = 3yx^{-3/2}/4 + 1/2$.

Using this expression (22)–(23), we can deduce the following important properties of WKB solutions of the Airy equation.

Property (A) In addition to the reference point (singularity) $y = -y_0(x) = -(2/3)x^{3/2}$, $\psi_{+,B}(x,y)$ has a singularity also at $y = y_0(x) = (2/3)x^{3/2}$. This singularity is sometimes called a **"movable singularity"**, since its relative location with respect to the reference point moves according as x varies.

Property (B) The Borel sum $\Psi_+(x,\eta)$ is well-defined as long as $\Im(-y_0(x)) \neq \Im\, y_0(x)$, that is, provided that x does not belong to the set $\{\, \Im x^{3/2} = 0 \,\}$, whereas it is not defined on $\{\, \Im x^{3/2} = 0 \,\}$ where the movable singularity is located on the integration path of the Borel sum.

Property (C) The set $\{\, \Im x^{3/2} = 0 \,\}$ defined above consists of three half-lines emanating from the origin in x-plane. If we consider the analytic continuation of the Borel sum $\Psi_+(x,\eta)$ across one of them, say, the positive real axis, then $\Psi_+(x,\eta)$ becomes the sum of the two Laplace integrals of $\psi_{+,B}(x,y)$ along Γ_j $(j = 0, 1)$ described in Fig. 1.

Property (D) After the analytic continuation across the positive real axis, the Borel sum $\Psi_+(x,\eta)$ becomes the following

$$\Psi_+ \rightsquigarrow \Psi_+ + i\Psi_-, \tag{24}$$

that is, a Stokes phenomenon occurs for $\Psi_+(x,\eta)$ on the positive real axis. Formula (24) is often called the **"connection formula"** for $\Psi_+(x,\eta)$.

The connection formula (24) is a direct consequence of Property (C) and the following discontinuity formula for the Borel transform $\psi_{+,B}(x,y)$:

$$\Delta_{y=y_0(x)}\, \psi_{+,B}(x,y) = i\psi_{-,B}(x,y), \tag{25}$$

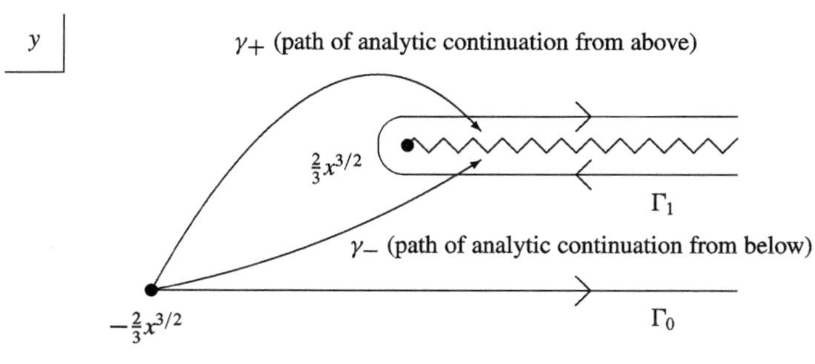

Fig. 1 Integration paths Γ_0 and Γ_1 (*wiggly lines* designate cuts to define a multi-valued analytic function $\psi_{+,B}(x,y)$)

where the discontinuity (or the **"alien derivative"** in the sense of Ecalle) of $\psi_{+,B}(x, y)$ is defined as follows:

$$\Delta_{y=y_0(x)}\, \psi_{+,B}(x, y) := (\gamma_+)_* \psi_{+,B}(x, y) - (\gamma_-)_* \psi_{+,B}(x, y), \tag{26}$$

where $(\gamma_\pm)_* \psi_{+,B}$ denotes the analytic continuation of $\psi_{+,B}$ along $\gamma_\pm$, that is, the discontinuity is the difference between the analytic continuations of $\psi_{+,B}(x, y)$ above the cut and below the cut (cf. Fig. 1). Note that in the case of the Airy equation the discontinuity formula (25) immediately follows from the expression (22)–(23) and Gauss' formula for hypergeometric functions:

$$\begin{aligned} s^{-1/2} F(5/6, 1/6, 1/2; s) &= \frac{1}{2}(1-s)^{-1/2} F(5/6, 1/6, 1/2; 1-s) \\ &\quad + \frac{1}{\sqrt{3}} F(4/3, 2/3, 3/2; 1-s) \end{aligned} \tag{27}$$

(cf. [17, p. 105, 2.9(1)–2.9(24) and p. 108, 2.10(1)]).

Stokes Geometry and Connection Formula

Taking into account Properties (A)–(D) for the Airy equation observed in the preceding subsection, we introduce the notion of turning points and Stokes curves for (3) as follows:

Definition 2.7

(i) A zero of $Q(x)$ is called a turning point of (3). In particular, a simple zero of $Q(x)$ is called a simple turning point of (3).
(ii) A Stokes curve of (3) is, by definition, an integral curve of the vector field (or, more precisely, the direction field) $\Im \sqrt{Q(x)}dx = 0$ emanating from a turning point, that is, a curve defined by

$$\Im \int_a^x \sqrt{Q(x)}dx = 0, \tag{28}$$

where a is a turning point of (3).

In the case of the Airy Eq. (16) the origin is the unique turning point (which is simple) and the Stokes curves are given by $\{\, \Im x^{3/2} = 0 \,\}$. Examples of the Stokes geometry for more complicated $Q(x)$ are shown in (Fig. 2).

In what follows we usually assume the following non-degenerate condition:

Condition (ND) There is no Stokes curve of (3) which connects two turning points. In other words, every Stokes curve of (3) emanating from a turning point flows into a singular point of $Q(x)$.

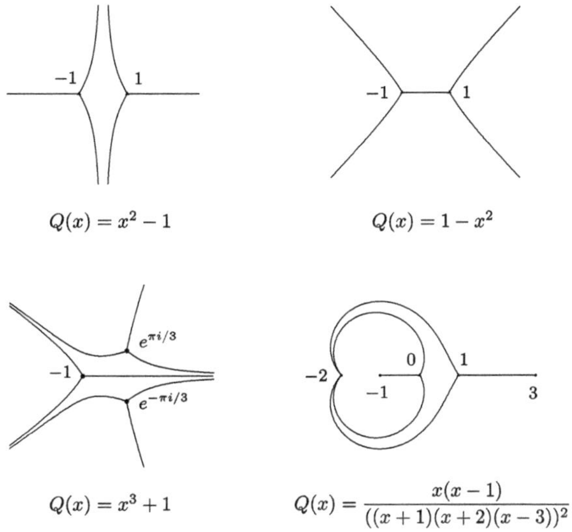

Fig. 2 Several examples of the Stokes geometry

Then, in parallel with the case of the Airy equation, the following fundamental theorems do hold for a second order Eq. (3) under Condition (ND).

Theorem 2.8 *Assume Condition (ND). Then WKB solutions $\psi_\pm(x,\eta)$ are Borel summable except on Stokes curves.*

Theorem 2.9 (Voros [36]) *Let $x = a$ be a simple turning point of* (3)*. Then, for WKB solutions*

$$\psi_\pm(x,\eta) = \frac{1}{\sqrt{S_{\rm odd}}}\exp\left(\pm\int_a^x S_{\rm odd}dx\right) \tag{29}$$

of (3) *normalized at $x = a$, the following properties hold in a neighborhood of $(x,y) = (a,0)$:*

(i) *$\psi_{\pm,B}(x,y)$ have singularities at $y = \pm y_0(x) = \pm\int_a^x \sqrt{Q(x)}dx$ (cf. Fig. 3).*
(ii)

$$\Delta_{y=y_0(x)}\,\psi_{+,B}(x,y) = i\psi_{-,B}(x,y), \quad \Delta_{y=-y_0(x)}\,\psi_{-,B}(x,y) = i\psi_{+,B}(x,y). \tag{30}$$

(iii) *On a Stokes curve Γ emanating from $x = a$ the following connection formula holds for the Borel sums $\Psi_\pm$ of $\psi_\pm$.*

Type (+) : *When $\Re\int_a^x \sqrt{Q(x)}dx > 0$ holds on Γ, that is, ψ_+ is dominant over ψ_- there, then*

$$\Psi_+ \rightsquigarrow \Psi_+ \pm i\Psi_-, \qquad \Psi_- \rightsquigarrow \Psi_-. \tag{31}$$

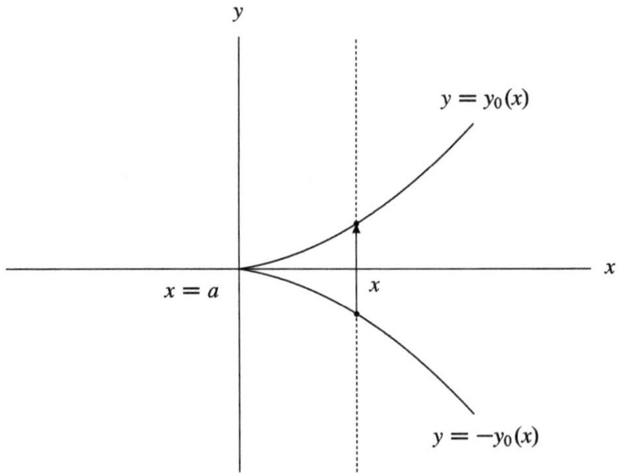

Fig. 3 Singularity locus of $\psi_{+,B}(x, y)$ near a simple turning point $x = a$

Type $(-)$ **:** *When* $\Re \int_a^x \sqrt{Q(x)}dx < 0$ *holds on* Γ*, that is,* ψ_- *is dominant over* ψ_+ *there, then*

$$\Psi_+ \rightsquigarrow \Psi_+, \qquad \Psi_- \rightsquigarrow \Psi_- \pm i\Psi_+. \tag{32}$$

Here the sign $\pm$ *depends on which direction one crosses the Stokes curve* Γ*. To be more precise, when one crosses* Γ *in an anticlockwise (resp., clockwise) manner viewed from* $x = a$*, we adopt the sign* $+$ *(resp.,* $-$*).*

3 Proof of the Fundamental Theorems

In this section we explain an outline of the proof of Theorems 2.8 and 2.9.

Outline of the Proof of Theorem 2.8

First, following the argument of Koike-Schäfke [28] and using in part an idea of Costin [10], we explain the proof of Theorem 2.8.

The central step is to verify the Borel summability of formal power series solutions of the Riccati equation

$$S^2 + \frac{dS}{dx} = \eta^2 Q(x). \tag{33}$$

For the sake of simplicity we assume $Q(x)$ is a polynomial and consider the Borel summability of $S(x,\eta) = S_+(x,\eta)$ only. We now write $S(x,\eta)$ as

$$S(x,\eta) = \eta S_{-1}(x) + S_0(x) + T(x,\eta), \qquad T(x,\eta) = \sum_{n=1}^{\infty} S_n(x)\eta^{-n}. \tag{34}$$

Then $T(x,\eta)$ satisfies

$$\frac{dT}{dx} + 2S_{-1}(\eta T - S_1) + 2S_0 T + T^2 = 0. \tag{35}$$

In view of Proposition 2.6, the Borel transform T_B of T satisfies

$$\frac{\partial T_B}{\partial x} + 2\sqrt{Q(x)}\,\frac{\partial T_B}{\partial y} + 2S_0(x)T_B + T_B * T_B = 0, \quad T_B(x,0) = S_1(x). \tag{36}$$

Since holomorphic solutions of (36) are unique near $y = 0$, it suffices to show the existence of a global holomorphic solution of (36) near the positive real axis $\mathbb{R}_+ = \{\, y \geq 0 \,\}$.

Definition 3.1 Let K be a compact subset of $U = \{\, x \in \mathbb{C} \mid Q(x) \neq 0 \,\}$. For a point $x_0 \in K$ we define

$$\Gamma_{x_0} := \{\, x \in \mathbb{C} \mid \Im \int_{x_0}^{x} \sqrt{Q(x)}dx = 0 \,\},$$

$$\Gamma_{x_0}^{(\pm)} := \{\, x \in \Gamma_{x_0} \mid \Re \int_{x_0}^{x} \sqrt{Q(x)}dx \gtrless 0 \,\} \cup \{\, x_0 \,\},$$

$$\widehat{K} := \bigcup_{x\in K} \Gamma_x, \quad \widehat{K}^{(\pm)} := \bigcup_{x\in K} \Gamma_x^{(\pm)}.$$

Theorem 3.2 *Let $Q(x)$ be a polynomial of degree d and K be a compact subset of $U = \{\, x \in \mathbb{C} \mid Q(x) \neq 0 \,\}$. If $\widehat{K}^{(-)}$ does not contain a turning point in its closure, the following hold :*

(i) *$T_B(x,y)$ is holomorphic in $\Omega := \widehat{K}^{(-)} \times \{\, y \mid \mathrm{dist}(y,\mathbb{R}_+) < \delta \,\}$ for a sufficiently small number $\delta > 0$.*

(ii) *There exist positive constants C_1 and C_2 that satisfy*

$$\left|T_B(x,y)\right| \leq \frac{C_1}{1+|x|^{d/2+2}}\, e^{C_2|y|} \quad in\ \Omega\,. \tag{37}$$

The proof of Theorem 3.2 consists of four steps.

Step 1. We employ the so-called Liouville transformation defined by

$$z(x) = \int_{x_0}^{x} \sqrt{Q(x)}\,dx. \tag{38}$$

Writing $T_B(x, y) = u(z, y)$, we find that $u(z, y)$ satisfies

$$\frac{\partial u}{\partial z} + 2\frac{\partial u}{\partial y} + 2A_1(z)u + A_2(z)u * u = 0, \qquad u(z, 0) = A_0(z), \tag{39}$$

where

$$A_0(z(x)) = S_1(x), \quad A_1(z(x)) = \frac{S_0(x)}{\sqrt{Q(x)}}, \quad A_2(z(x)) = \frac{1}{\sqrt{Q(x)}}. \tag{40}$$

Step 2. By using a linear change of variables $s = 2z - y, t = y$ and integrating (39) once with respect to the variable t, we can convert (39) into the following integral equation:

$$\begin{aligned} u(z, y) &= A_0\Big(z - \frac{y}{2}\Big) - \int_0^y A_1\Big(z - \frac{y-y'}{2}\Big)u\Big(z - \frac{y-y'}{2}, y'\Big)\,dy' \\ &\quad - \frac{1}{2}\int_0^y A_2\Big(z - \frac{y-y'}{2}\Big)(u * u)\Big(z - \frac{y-y'}{2}, y'\Big)\,dy'. \end{aligned} \tag{41}$$

Step 3. To discuss the existence of solutions of (41), we introduce the following domains:

$$\Omega^-(K, \delta) := \{ z \in \mathbb{C} \mid \mathrm{dist}\big(z, z(\widehat{K}^{(-)})\big) < \delta \}, \tag{42}$$

$$\begin{aligned} \mathcal{R} := \{ (z, y) \in \mathbb{C}^2 \mid \mathrm{dist}(y, \mathbb{R}_+) < \delta \text{ and the segment } [z, z - y/2] \\ \text{is contained in } \Omega^-(K, \delta) \}. \end{aligned} \tag{43}$$

Note that $\mathcal{R}$ is star-shaped with respect to the variable y (i.e., $(z, y) \in \mathcal{R}$ implies $(z, \theta y) \in \mathcal{R}$ for any $\theta \in [0, 1]$) and that

$$z(\widehat{K}^{(-)}) \times \{ y \in \mathbb{C} \mid \mathrm{dist}(y, \mathbb{R}_+) < \delta \} \subset \mathcal{R}$$

holds. Furthermore, if $\delta > 0$ is sufficiently small, we may assume that $A_j(z)$ ($j = 0, 1, 2$) are holomorphic and bounded in the closure of $\Omega^-(K, \delta)$.
In what follows we solve the integral Eq. (41) in $\mathcal{R}$.

Step 4. Let

$$\mathcal{O}_\lambda := \Big\{ u(z,y) \;\Big|\; u \text{ is holomorphic in } \mathcal{R} \text{ and}$$
$$\|u\|_\lambda := \sup_{(z,y)\in\mathcal{R}} \int_0^y |u(z,y')| e^{-\lambda|y'|}\, d|y'| < \infty \Big\}, \tag{44}$$

where $\lambda > 0$ is a parameter. Then we can prove

Proposition 3.3 *$\mathcal{O}_\lambda$ is a Banach algebra with respect to the convolution $*$, that is, $\mathcal{O}_\lambda$ is a Banach space and the following holds :*

$$\text{If } u, v \in \mathcal{O}_\lambda, \text{ then } u * v \in \mathcal{O}_\lambda \text{ and } \|u * v\|_\lambda \le \|u\|_\lambda \|v\|_\lambda. \tag{45}$$

Proposition 3.4 *Let $F(u)$ denote the right-hand side of* (41)*, that is,*

$$\begin{aligned}(F(u))(z,y) &= A_0\Big(z - \frac{y}{2}\Big) - \int_0^y A_1\Big(z - \frac{y-y'}{2}\Big) u\Big(z - \frac{y-y'}{2}, y'\Big)\, dy' \\ &\quad - \frac{1}{2}\int_0^y A_2\Big(z - \frac{y-y'}{2}\Big)(u*u)\Big(z - \frac{y-y'}{2}, y'\Big)\, dy'.\end{aligned} \tag{46}$$

Then $F(u)$ defines a contractive mapping from $\{ u \in \mathcal{O}_\lambda \mid \|u\|_\lambda \le 1 \}$ to itself when $\lambda > 0$ is sufficiently large.

Therefore the contractive mapping principle provides us with a (unique) holomorphic solution of (41) in $\mathcal{R} \supset z(\widehat{K}^{(-)}) \times \{ y \in \mathbb{C} \,|\, \mathrm{dist}(y, \mathbb{R}_+) < \delta \}$. The exponential estimate (37) for its solution can be obtained by repeating the above argument after replacing T by $\tilde{T} = (x - x_1)^{d/2+2} T$ ($x_1 \in \widehat{K}^{(-)}$) and further by using the boundedness of the norm $\|u\|_\lambda$ of the solution. Note that the use of the Banach algebra $\mathcal{O}_\lambda$ and the contractive mapping principle is essentially due to Costin [10], while successive approximation method is used in Koike-Schäfke [28].

Theorem 3.2 assures the Borel summability of formal power series solutions $S_\pm(x,\eta)$ of the Riccati equation. Once the Borel summability of $S_\pm(x,\eta)$ is established, then the Borel summability of WKB solutions

$$\psi_\pm(x,\eta) = \frac{1}{\sqrt{S_{\mathrm{odd}}}} \exp\Big(\pm \int_{x_0}^x S_{\mathrm{odd}} dx\Big)$$

of (3) can be confirmed by the following argument:

1. When the integration path from x_0 to x does not cross any Stokes curve, then the Borel summability of $\psi_\pm$ immediately follows from Theorem 3.2.

2. Even when the integration path crosses several Stokes curves, if we deform the integration path in such a way that

   at every crossing point $\hat{x}$ of the integration path with a Stokes curve Γ, we avoid the crossing with Γ and go to $x = \infty$ along one side of Γ and then return to $x = \hat{x}$ along the other side of Γ,

   then the Borel summability of $\psi_\pm$ is ensured by Theorem 3.2 also in this case. Note that Condition (ND) guarantees that such a deformation of the integration path is always possible.

This is an outline of the proof of Theorem 2.8. See [28] for more details of the discussion.

Outline of the Proof of Theorem 2.9

To prove Theorem 2.9, we make use of the transformation theory to the Airy equation developed in [1], [27, Chap. 2].

Let us consider

$$\left(\frac{d^2}{d\tilde{x}^2} - \eta^2 \tilde{Q}(\tilde{x})\right) \tilde{\psi} = 0 \tag{$\widetilde{47}$}$$

and apply a change of variable $x = x(\tilde{x})$ to ($\widetilde{47}$). If we further employ a change of unknown function $\tilde{\psi}(\tilde{x}) = (dx/d\tilde{x})^{-1/2}\psi(x(\tilde{x}))$, then ($\widetilde{47}$) is transformed to

$$\left(\frac{d^2}{dx^2} - \eta^2 Q(x)\right) \psi = 0 \tag{47}$$

with

$$\tilde{Q}(\tilde{x}) = \left(\frac{dx}{d\tilde{x}}\right)^2 Q(x(\tilde{x})) - \frac{1}{2}\eta^{-2}\{x;\tilde{x}\}. \tag{48}$$

Here $\{x;\tilde{x}\}$ stands for the Schwarzian derivative:

$$\{x;\tilde{x}\} = \frac{d^3x/d\tilde{x}^3}{dx/d\tilde{x}} - \frac{3}{2}\left(\frac{d^2x/d\tilde{x}^2}{dx/d\tilde{x}}\right)^2.$$

Taking this relation into account, we introduce the following terminology.

Definition 3.5 We say that ($\widetilde{47}$) is transformed (in the sense of exact WKB analysis) to (47) at $\tilde{x} = \tilde{x}_0$ if there exists an infinite series $x(\tilde{x},\eta) = \sum_{n\geq 0} x_n(\tilde{x})\eta^{-n}$ that satisfies the following conditions:

(i) $x_n(\tilde{x})$ is holomorphic in a fixed neighborhood $\tilde{U}$ of $\tilde{x} = \tilde{x}_0$ (i.e., $\tilde{U}$ is independent of n).
(ii) The following relation holds (as formal power series of η^{-1}):

$$\tilde{Q}(\tilde{x}) = \left(\frac{\partial x}{\partial \tilde{x}}(\tilde{x}, \eta)\right)^2 Q(x(\tilde{x}, \eta)) - \frac{1}{2}\eta^{-2}\{x(\tilde{x}, \eta); \tilde{x}\}. \tag{49}$$

Under this terminology we can prove the following

Theorem 3.6 *Let $\tilde{x} = \tilde{a}$ be a simple turning point of*

$$\left(\frac{d^2}{d\tilde{x}^2} - \eta^2 \tilde{Q}(\tilde{x})\right)\tilde{\psi} = 0, \tag{$\widetilde{47}$}$$

that is, $\tilde{a}$ is a simple zero of $\tilde{Q}(\tilde{x})$. Then at $\tilde{x} = \tilde{a}$ ($\widetilde{47}$) can be transformed (in the sense of exact WKB analysis) to the Airy equation

$$\left(\frac{d^2}{dx^2} - \eta^2 x\right)\psi = 0. \tag{50}$$

Theorem 3.6 is proved by constructing $x(\tilde{x}, \eta) = x_0(\tilde{x}) + x_1(\tilde{x})\eta^{-1} + \cdots$ that satisfies

$$\tilde{Q}(\tilde{x}) = \left(\frac{\partial x}{\partial \tilde{x}}(\tilde{x}, \eta)\right)^2 x(\tilde{x}, \eta) - \frac{1}{2}\eta^{-2}\{x(\tilde{x}, \eta); \tilde{x}\} \tag{51}$$

in a recursive manner. For example, the top order part $x_0(\tilde{x})$ is given by

$$x_0(\tilde{x}) = \left(\frac{3}{2}\int_{\tilde{a}}^{\tilde{x}} \sqrt{\tilde{Q}(\tilde{x})}\, d\tilde{x}\right)^{2/3} \tag{52}$$

and the higher order part $x_n(\tilde{x})$ is determined by solving a first order ODE of the form

$$\left(2x_0\left(\frac{dx_0}{d\tilde{x}}\right)^{-1}\frac{d}{d\tilde{x}} + 1\right)x_n = (\text{given}), \quad \text{i.e.,} \quad \left(2z\frac{d}{dz} + 1\right)x_n = (\text{given})$$

degree by degree. Here, for the sake of simplicity of notations, we use z to denote a new independent variable $z = x_0(\tilde{x})$. Note that $x_n(\tilde{x})$ identically vanishes for an odd integer n and also that $x_n(\tilde{x})$ satisfies the estimate

$$|x_n(\tilde{x})| \leq AC^n n! \tag{53}$$

for some positive constants $A, C > 0$ in a fixed neighborhood $\tilde{U}$ of $\tilde{x} = \tilde{a}$.

Thus Eq. $(\widetilde{47})$ is transformed to the Airy equation (50) near a simple turning point $\tilde{a}$ by the formal coordinate transformation $x = x(\tilde{x}, \eta)$. Furthermore, in the current situation we can verify the following relation between WKB solutions of $(\widetilde{47})$ and those of (50) in all orders of η^{-1}:

$$\tilde{\psi}_{\pm}(\tilde{x}, \eta) = \left(\frac{\partial x}{\partial \tilde{x}}\right)^{-1/2} \psi_{\pm}(x(\tilde{x}, \eta), \eta), \tag{54}$$

where $\tilde{\psi}_{\pm}$ and $\psi_{\pm}$ are WKB solutions of $(\widetilde{47})$ and (50) normalized at the turning points in question, respectively:

$$\tilde{\psi}_{\pm}(\tilde{x}, \eta) = \frac{1}{\sqrt{\tilde{S}_{\mathrm{odd}}}} \exp\left(\pm \int_{\tilde{a}}^{\tilde{x}} \tilde{S}_{\mathrm{odd}} d\tilde{x}\right), \tag{55}$$

$$\psi_{\pm}(x, \eta) = \frac{1}{\sqrt{S_{\mathrm{odd}}}} \exp\left(\pm \int_{0}^{x} S_{\mathrm{odd}} dx\right). \tag{56}$$

Theorem 2.9 is proved by considering the Borel transform of both sides of (54). As a matter of fact, the multiplication operator η^{-1} turns out to be an integral operator $(\partial/\partial y)^{-1}$ via the Borel transformation in view of Proposition 2.6. Thus, using the Taylor expansion, we find that the Borel transform of (54) is expressed as

$$\tilde{\psi}_{\pm,B}(\tilde{x}, y) =$$
$$\Big(\sum_{j\geq 0} \frac{\partial x_j}{\partial \tilde{x}} \big(\frac{\partial}{\partial y}\big)^{-j}\Big)^{-1/2} \sum_{n=0}^{\infty} \frac{1}{n!} \Big(\sum_{j\geq 1} x_j(\tilde{x}) \big(\frac{\partial}{\partial y}\big)^{-j}\Big)^{n} \Big(\frac{\partial^n}{\partial x^n} \psi_{\pm,B}\Big)(x_0(\tilde{x}), y). \tag{57}$$

As [6, Appendix C] shows, if we use $(x, y) = (x_0(\tilde{x}), y)$ as new independent variables instead of $(\tilde{x}, y)$, the right-hand side of (57) can be expressed also as

$$\int_{-y_0}^{y} K\big(x, y - y', \frac{\partial}{\partial x}\big) \psi_{\pm,B}(x, y')\, dy' =: L\big(x, \frac{\partial}{\partial x}, \big(\frac{\partial}{\partial y}\big)^{-1}\big) \psi_{\pm,B}(x, y) \tag{58}$$

with some integro-differential operator $L = L(x, \partial/\partial x, (\partial/\partial y)^{-1})$. Here y_0 is an arbitrarily chosen point (sufficiently close to 0) that fixes the action of $(\partial/\partial y)^{-1}$ as an integral operator.

Remark 3.7 For the sake of readers' reference we provide an explicit formula for the kernel $K(x, y - y', \partial/\partial x)$ of the operator $L(x, \partial/\partial x, (\partial/\partial y)^{-1})$. Replacing $\partial/\partial x$ and $\partial/\partial y$ by ξ and η, respectively, we first write the (total) symbol of L (i.e., the right-hand side of (57)) as

$$\sum_{j=0}^{\infty} \left(\sum_{k=0}^{\infty} a_{jk}(x) (\xi \eta^{-1})^k\right) \eta^{-j},$$

that is,

$$a_{j0}(x) = h_j(x), \qquad a_{jk}(x) = \sum_{\alpha+\beta=j} h_\alpha(x) f_{k,k+\beta}(x) \quad (k \geq 1),$$

where $h_j(x)$ and $f_{k,l}(x)$ are defined by

$$\left(\sum_{j\geq 0} \frac{\partial x_j}{\partial \tilde{x}}(\tilde{x})\left(\frac{\partial}{\partial y}\right)^{-j}\right)^{-1/2}\Bigg|_{\tilde{x}=x_0^{-1}(x)} = \sum_{j\geq 0} h_j(x)\eta^{-j},$$

$$\left(\sum_{n=0}^{\infty} \frac{1}{n!}\left(\sum_{j\geq 1} x_j(\tilde{x})\eta^{-j}\right)^n \xi^n\right)\Bigg|_{\tilde{x}=x_0^{-1}(x)} = 1 + \sum_{1\leq k\leq l} f_{k,l}(x)\xi^k\eta^{-l}.$$

Then $K(x, y-y', \xi)$ is explicitly given by

$$K(x, y-y', \xi) = \sum_{j,k=0}^{\infty} a_{jk}(x) \frac{(y-y')^{j+k-1}}{(j+k-1)!}\xi^k.$$

Thanks to the estimate (53), we find that the operator L defines what is called a **"microdifferential operator"** in the theory of microlocal analysis in a neighborhood of $(x, y) = (0, 0)$ (cf. [31]). As its consequence, it turns out that L does not change the location of singularities of the operand near $(0, 0)$. Hence, since the singular points of $\psi_{\pm,B}(x, y)$ are

$$y = \pm \int_0^x \sqrt{x}\, dx = \pm \frac{2}{3} x^{3/2}$$

(cf. Property (A) in section "WKB Solutions of the Airy Equation and Their Borel Transforms"), the singular points of $\tilde{\psi}_{\pm,B}(\tilde{x}, y)$ are also confined to

$$y = \pm \frac{2}{3} x^{3/2}\Big|_{x=x_0(\tilde{x})} = \pm \int_{\tilde{a}}^{\tilde{x}} \sqrt{\tilde{Q}(\tilde{x})}\, d\tilde{x} = \pm y_0(\tilde{x}).$$

Furthermore, the discontinuity formula (30) of $\tilde{\psi}_{\pm,B}(\tilde{x}, y)$ is also confirmed as

$$\begin{aligned} \Delta_{y=y_0(\tilde{x})}\, \tilde{\psi}_{+,B}(\tilde{x}, y) &= \Delta_{y=y_0(\tilde{x})}\, (L\psi_{+,B})\Big|_{x=x_0(\tilde{x})} \\ &= L\Delta_{y=2x^{3/2}/3}\, \psi_{+,B}\Big|_{x=x_0(\tilde{x})} \\ &= L(i\psi_{-,B})\Big|_{x=x_0(\tilde{x})} = i\tilde{\psi}_{-,B}(\tilde{x}, y). \end{aligned} \tag{59}$$

Thus we have verified Theorem 2.9, (i), (ii). For more details we refer the reader to [27, Chap. 2]. Finally, to verify Theorem 2.9, (iii), we need the Borel summability of the transformation series $x = x(\tilde{x}, \eta)$, which has been proved more recently by Kamimoto and Koike [26]. Thus the connection formula (31)–(32) for the WKB solutions (29) of (3) is also derived from that of the Airy equation. This completes the proof of Theorem 2.9.

4 Application: Computation of Monodromy Representations of Fuchsian Equations

In this section, as an application of the exact WKB analysis, let us compute the monodromy representations of second order equations of the form

$$\left(\frac{d^2}{dx^2} - \eta^2 Q(x)\right)\psi = 0 \tag{60}$$

with

$$Q(x) = \frac{F(x)}{G(x)^2} \qquad (F(x), G(x) : \text{polynomials}). \tag{61}$$

In what follows we assume that

$$\begin{aligned} \deg F &= 2g+2, \quad F(x) = (x-a_0)\cdots(x-a_{2g+1}), \\ \deg G &= g+2, \quad G(x) = (x-b_0)\cdots(x-b_{g+1}) \end{aligned}$$

for some non-negative integer $g \geq 0$ and that all a_j and b_k are mutually distinct. Then the set of singular points of (60) is given by

$$\mathcal{S} = \{\, b_0, \ldots, b_{g+1}, b_{g+2} = \infty \,\}$$

and all singular points become regular singular. Thus Eq. (60) is the so-called Fuchsian equation.

Remark 4.1 When $g = 0$ and $g = 1$, (60) is equivalent to Gauss' hypergeometric equation and Heun's equation, respectively.

For such a Fuchsian equation the monodromy representation is naturally defined: Take a base point $x_0 \in P^1(\mathbb{C}) \setminus \mathcal{S}$ and a fundamental system of solutions (ψ_0, ψ_1) around x_0. For any closed path γ in $P^1(\mathbb{C}) \setminus \mathcal{S}$ emanating from x_0 we consider analytic continuation of (ψ_0, ψ_1) along γ:

$$(\psi_0, \psi_1) \text{ near } x_0 \underset{\substack{\text{analytic continuation} \\ \text{along } \gamma}}{\rightsquigarrow} (\gamma_*\psi_0, \gamma_*\psi_1) = (\psi_0, \psi_1)\, {}^{\exists}A_\gamma, \tag{62}$$

where A_γ is an invertible 2×2 constant matrix. Then the monodromy representation of (60) is, by definition, the algebraic homomorphism

$$\pi_1(P^1(\mathbb{C}) \setminus \mathcal{S}, x_0) \ni [\gamma] \;\longmapsto\; A_\gamma \in GL_2(\mathbb{C}), \tag{63}$$

where π_1 designates the fundamental group.

From now on we compute the monodromy representation of (60) by making use of WKB solutions. Before doing the computation, we prepare one proposition which is concerned with the behavior of WKB solutions at a regular singular point.

Proposition 4.2 *At each regular singular point $x = b_k$, $S_{\mathrm{odd}}(x,\eta)$ has a pole of order* 1 *and its residue there is explicitly given by*

$$\operatorname*{Res}_{x=b_k} S_{\mathrm{odd}}(x,\eta) = c_k\eta\sqrt{1+\frac{1}{4c_k^2}\eta^{-2}}, \tag{64}$$

where

$$c_k = \operatorname*{Res}_{x=b_k}\sqrt{Q(x)} \quad (k=0,\ldots,g+1), \quad c_{g+2} = \operatorname*{Res}_{\zeta=0}\left(-\frac{\sqrt{Q(1/\zeta)}}{\zeta^2}\right). \tag{65}$$

We explain the computation by using the following example discussed in [27, Chap. 3].

Example Let us consider

$$\left(\frac{d^2}{dx^2} - \eta^2\frac{(x^2-9)(x^2-1/9)}{(x^3-\exp(i\pi/8))^2}\right)\psi = 0. \tag{66}$$

In this case $g = 1$ and we number turning points and regular singular points as follows:

$$\begin{aligned}&a_0 = -3, \quad a_1 = -1/3, \quad a_2 = 1/3, \quad a_3 = 3,\\ &b_0 = \exp(33i\pi/24), \;\; b_1 = \exp(i\pi/24), \;\; b_2 = \exp(17i\pi/24), \;\; b_3 = \infty.\end{aligned}$$

The Stokes geometry of (66) is described in Fig. 4. As is shown in Fig. 4, we take a base point x_0 between a_0 and a_1 and use (the Borel sums of) WKB solutions

$$\psi_\pm = \frac{1}{\sqrt{S_{\mathrm{odd}}}}\exp\left(\pm\int_{x_0}^{x} S_{\mathrm{odd}}dx\right) \tag{67}$$

as a fundamental system of solutions around x_0. In Fig. 4, for the later use, we draw (in blue) a path C_k $(0 \le k \le g+2)$ of analytic continuation which starts from x_0 and returns to x_0 after encircling a regular singular point b_k once in an anticlockwise manner, and also (in red) a path γ_j $(0 \le j \le 2g+1)$ which starts from x_0 and

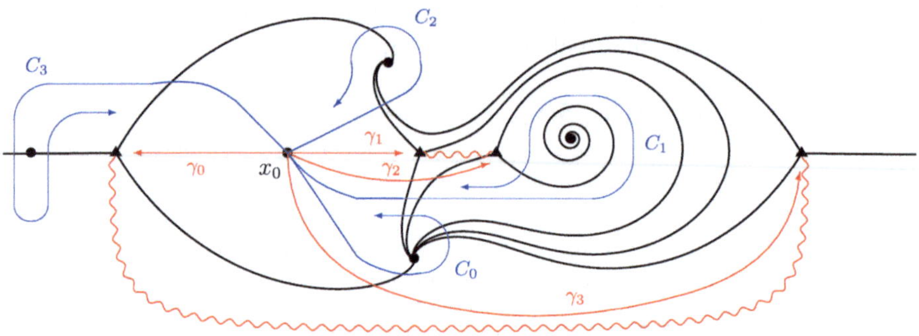

Fig. 4 Stokes geometry of Eq. (66) (*wiggly lines* designate cuts to define $\sqrt{Q(x)}$)

ends at a turning point a_j. Note that the branch of $\sqrt{Q(x)}$ is chosen here so that $\sqrt{Q(x)} \sim 1/x$ holds near $x = \infty$. This choice of the branch of $\sqrt{Q(x)}$ assures

$$\Re c_1, \Re c_2 > 0, \quad \Re c_0, \Re c_3 < 0$$

and hence it follows from Proposition 4.2 that on a Stokes curve flowing into b_1 or b_2 (resp., b_0 or b_3) the connection formula (32) of type $(-)$ (resp., (31) of type $(+)$) holds. In what follows we also use the following notations:

$$\nu_k^{\pm} := \exp\left(i\pi\left(1 \pm \sqrt{4c_k^2\eta^2 + 1}\right)\right) \quad (k = 0, \ldots, g+2), \tag{68}$$

$$u_j := \exp\left(2\int_{\gamma_j} S_{\mathrm{odd}}dx\right), \quad u_{jk} := u_j^{-1}u_k \quad (j, k = 0, \ldots, 2g+1). \tag{69}$$

Computation of monodromy matrices A_k along C_k

Since $\pi_1(P^1(\mathbb{C}) \setminus \mathcal{S}, x_0)$ is generated by C_k, it suffices to compute a monodromy matrix $A_k = A_{C_k}$ along C_k $(k = 0, 1, 2, 3)$. Let us first consider the computation of A_2 along C_2.

As is shown in Fig. 4, C_2 crosses three Stokes curves and at each crossing point a Stokes phenomenon described by Theorem 2.9 occurs. For example, at the first crossing point C_2 crosses a Stokes curve emanating from a turning point a_1. Note that on this Stokes curve the connection formula (32) of type $(-)$ holds, as was noted before. Since (32) is described in terms of the WKB solutions

$$\varphi_{\pm}^{(1)} = \frac{1}{\sqrt{S_{\mathrm{odd}}}} \exp\left(\pm\int_{a_1}^{x} S_{\mathrm{odd}}dx\right) \tag{70}$$

normalized at the turning point a_1 where the Stokes curve in question emanates, we factorize the WKB solutions (67) as

$$\psi_\pm = \varphi_\pm^{(1)} \exp\left(\pm \int_{\gamma_1} S_{\mathrm{odd}} dx\right). \tag{71}$$

For the WKB solutions $\varphi_\pm^{(1)}$ normalized at a_1 we have the connection formula (32). Hence for $\psi_\pm$ the following holds:

$$\Psi_+ \rightsquigarrow \Psi_+, \quad \Psi_- \rightsquigarrow \Psi_- - i\exp\left(-2\int_{\gamma_1} S_{\mathrm{odd}} dx\right)\Psi_+ = \Psi_- - iu_1^{-1}\Psi_+, \tag{72}$$

that is,

$$(\Psi_+, \Psi_-) \rightsquigarrow (\Psi_+, \Psi_-)\begin{pmatrix} 1 & -iu_1^{-1} \\ 0 & 1 \end{pmatrix}. \tag{73}$$

The Stokes phenomenon at the second crossing point can be similarly computed by using the WKB solutions

$$\varphi_\pm^{(3)} = \frac{1}{\sqrt{S_{\mathrm{odd}}}} \exp\left(\pm \int_{a_3}^{x} S_{\mathrm{odd}} dx\right) \tag{74}$$

normalized at a_3 and the factorization

$$\psi_\pm = \varphi_\pm^{(3)} \exp\left(\pm \int_{x_0}^{a_3} S_{\mathrm{odd}} dx\right). \tag{75}$$

However, in this case the integration path γ_{x_0,a_3} from x_0 to a_3 is not homotopic to γ_3; the closed path $\gamma_{x_0,a_3}(\gamma_3)^{-1}$ encircles two turning points b_0, b_1 and the cut connecting a_1 and a_2. Thus the factorization formula (75) reads as

$$\psi_\pm = \varphi_\pm^{(3)} \exp\left(\pm \int_{\gamma_3} S_{\mathrm{odd}} dx\right) \left(\nu_0^\pm\right)^{-1} \left(\nu_1^\pm\right)^{-1} \left(u_{21}\right)^{\pm 1} \tag{76}$$

and the Stokes phenomenon at the second crossing point is described by

$$(\Psi_+, \Psi_-) \rightsquigarrow (\Psi_+, \Psi_-)\begin{pmatrix} 1 & -iu_3^{-1}u_{21}^{-2}\dfrac{\nu_0^+\nu_1^+}{\nu_0^-\nu_1^-} \\ 0 & 1 \end{pmatrix}. \tag{77}$$

It is now clear how to compute the Stokes phenomenon at the third crossing point and consequently we obtain

$$A_2 = \begin{pmatrix} \nu_2^+ & 0 \\ 0 & \nu_2^- \end{pmatrix} \begin{pmatrix} 1 & -iu_0^{-1}\dfrac{\nu_2^-}{\nu_2^+} \\ 0 & 1 \end{pmatrix} \begin{pmatrix} 1 & -iu_3^{-1}u_{12}^2\dfrac{\nu_0^+\nu_1^+}{\nu_0^-\nu_1^-} \\ 0 & 1 \end{pmatrix} \begin{pmatrix} 1 & -iu_1^{-1} \\ 0 & 1 \end{pmatrix}. \tag{78}$$

Note that after the analytic continuation along C_2 WKB solutions $\psi_\pm$ given by (67) changes to

$$-\frac{1}{\sqrt{S_{\text{odd}}}}\exp\left(\pm\int_{C_2} S_{\text{odd}}dx\right)\exp\left(\pm\int_{x_0}^{x} S_{\text{odd}}dx\right) = \nu_2^\pm\psi_\pm,$$

as S_{odd} has a simple pole at $x = b_2$ and its residue there is given by (64) in view of Proposition 4.2; this is the reason why the left-most diagonal matrix in the right-hand side of (78) appears.

The computation of the other matrices A_k is the same as that of A_2. The result is as follows:

$$\begin{aligned} A_0 = &\begin{pmatrix} \nu_0^+ & 0 \\ 0 & \nu_0^- \end{pmatrix} \begin{pmatrix} 1 & 0 \\ -iu_1\dfrac{\nu_0^+}{\nu_0^-} & 1 \end{pmatrix} \begin{pmatrix} 1 & 0 \\ -iu_2\dfrac{\nu_0^+}{\nu_0^-} & 1 \end{pmatrix} \begin{pmatrix} 1 & 0 \\ -iu_2\dfrac{\nu_0^+\nu_1^+}{\nu_0^-\nu_1^-} & 1 \end{pmatrix} \\ &\times \begin{pmatrix} 1 & 0 \\ -iu_1u_{12}^2\dfrac{\nu_0^+\nu_1^+}{\nu_0^-\nu_1^-} & 1 \end{pmatrix} \begin{pmatrix} 1 & 0 \\ -iu_3 & 1 \end{pmatrix} \begin{pmatrix} 1 & 0 \\ -iu_0 & 1 \end{pmatrix}, \end{aligned} \tag{79}$$

$$\begin{aligned} A_1 = &\begin{pmatrix} \nu_1^+ & 0 \\ 0 & \nu_1^- \end{pmatrix} \begin{pmatrix} 1 & 0 \\ -iu_1\dfrac{\nu_1^+}{\nu_1^-} & 1 \end{pmatrix} \begin{pmatrix} 1 & 0 \\ -iu_2\dfrac{\nu_1^+}{\nu_1^-} & 1 \end{pmatrix} \\ &\times \begin{pmatrix} 1 & -iu_2^{-1}\dfrac{\nu_1^-}{\nu_1^+} \\ 0 & 1 \end{pmatrix} \begin{pmatrix} 1 & 0 \\ iu_2 & 1 \end{pmatrix} \begin{pmatrix} 1 & 0 \\ iu_1 & 1 \end{pmatrix}, \end{aligned} \tag{80}$$

$$\begin{aligned} A_3 = &\begin{pmatrix} \nu_3^+ & 0 \\ 0 & \nu_3^- \end{pmatrix} \begin{pmatrix} 1 & -iu_0^{-1}\dfrac{\nu_3^-}{\nu_3^+} \\ 0 & 1 \end{pmatrix} \begin{pmatrix} 1 & 0 \\ -iu_0\dfrac{\nu_3^+}{\nu_3^-} & 1 \end{pmatrix} \\ &\times \begin{pmatrix} 1 & 0 \\ -iu_3u_{21}^2\dfrac{\nu_0^-\nu_1^-\nu_2^-}{\nu_0^+\nu_1^+\nu_2^+} & 1 \end{pmatrix} \begin{pmatrix} 1 & iu_0^{-1} \\ 0 & 1 \end{pmatrix}. \end{aligned} \tag{81}$$

Finally, if we change a fundamental system of solutions as

$$\psi_{\pm} = \frac{1}{\sqrt{S_{\text{odd}}}} \exp\left(\pm \int_{x_0}^{x} S_{\text{odd}} dx \right) \longmapsto \tilde{\psi}_{\pm} = \exp\left(\mp \int_{\gamma_0} S_{\text{odd}} dx \right) \psi_{\pm}, \tag{82}$$

then we find that every A_k can be described solely by $\{\nu_k^{\pm}\}$ and $\{u_{jk}\}$ (or $\{\nu_k^{+}\}$ and $\{u_{01}, u_{12}\}$ thanks to Remark 4.3 below).

Remark 4.3 Among $\{\nu_k^{\pm}\}$ and $\{u_{jk}\}$ we have the following relations:

$$\nu_k^{+}\nu_k^{-} = 1 \quad (k = 0, 1, 2, 3),$$

$$\nu_0^{+}\nu_1^{+}\nu_2^{+}\nu_3^{+}u_{12}u_{30} = 1.$$

In conclusion we have

Theorem 4.4 *Every monodromy matrix can be described in terms of the following two kinds of quantities*:

(i) *Characteristic exponents* $\{\nu_k^{\pm}\}$ *at regular singular points* $\{b_k\}$.
(ii) *Contour integrals* $\{u_{jk}\}$ *of* S_{odd} *on the Riemann surface of* $\sqrt{Q(x)}$.

Remark 4.5 In the case of $g = 0$ Theorem 4.4 implies that the monodromy matrices of Gauss' hypergeometric equation are described solely in terms of characteristic exponents at its regular singular points. This is classically well-known result. Theorem 4.4 is its generalization to the case of $g \geq 1$. It shows the efficiency of the exact WKB analysis and, thanks to it, the role of the Riemann surface of $\sqrt{Q(x)}$ in the description of monodromic structure of (60) is clarified.

5 Voros Coefficients and Wall-Crossing Formulas

As we have seen so far, the connection formula (or Theorems 2.8 and 2.9) is very powerful to study the global behavior of solutions of second order ODEs. The most important analytic ingredient of the connection formula is the movable singular points $y = \pm y_0(x) = \pm \int_a^x \sqrt{Q(x)}dx$ of the Borel transform of WKB solutions. Here we should recall that, in applying the connection formula, we have assumed Condition (ND), that is, non-existence of Stokes curves connecting two turning points. In this section we consider the situation where this non-degeneracy condition (Condition (ND)) is violated. In such a degenerate situation another kind of singularities of the Borel transform of WKB solutions may play an important role. The study of this degenerate situation is also related to the so-called **"wall-crossing formula"** discussed by Gaiotto-Moore-Neitzke [18]. The discussion below can be regarded as a WKB-theoretic derivation of the wall-crossing formula.

Let us consider the problem by using a simple example.

Example (Weber Equation) We consider the Weber equation

$$\left(\frac{d^2}{dx^2} - \eta^2\left(c - \frac{x^2}{4}\right)\right)\psi = 0 \qquad (c > 0) \tag{83}$$

and its WKB solutions

$$\psi_\pm = \frac{1}{\sqrt{S_{\mathrm{odd}}}} \exp\left(\pm \int_{2\sqrt{c}}^{x} S_{\mathrm{odd}} dx\right). \tag{84}$$

Equation (83) has two simple turning points $x = \pm 2\sqrt{c}$ and they are connected by a Stokes curve (**"Stokes segment"**), as is shown in Fig. 2. In what follows we study the effect of this Stokes segment.

As $x = 2\sqrt{c}$ is a simple turning point, we can apply Theorem 2.9 to find that the singularity locus of $\psi_{+,B}(x,y)$ form a cusp and have two branches $y = \pm y_0(x)$ near $x = 2\sqrt{c}$, where

$$y_0(x) = \int_{2\sqrt{c}}^{x} \sqrt{c - \frac{x^2}{4}}\, dx$$

(cf. Fig. 3). These two branches of the singularity locus can be prolonged to $x = -2\sqrt{c}$ and they form again a cusp near $x = -2\sqrt{c}$. Repeating this process, we thus obtain Fig. 5 for the singularity locus of $\psi_{+,B}(x,y)$. They become a "ladder-like" set and, as

$$2\int_{-2\sqrt{c}}^{2\sqrt{c}} \sqrt{c - \frac{x^2}{4}}\, dx = 2\pi c$$

holds, the singularities of $\psi_{+,B}(x,y)$ for fixed x have the periodic structure with period $2\pi c$. Among them the singularities $-y_0(x) + 2\pi n c$ ($n \in \mathbb{Z}$) are often called **"fixed singularities"**, as their relative location with respect to the reference singularity $-y_0(x)$ does not depend on x.

The existence of fixed singularities can be rigorously confirmed by the following arguments.

First Approach We can verify the existence of fixed singularities by using the differential equation that $\psi_{\pm,B}(x,y)$ satisfy:

$$\left(\frac{\partial^2}{\partial x^2} - \left(c - \frac{x^2}{4}\right)\frac{\partial^2}{\partial y^2}\right)\psi_{\pm,B}(x,y) = 0. \tag{85}$$

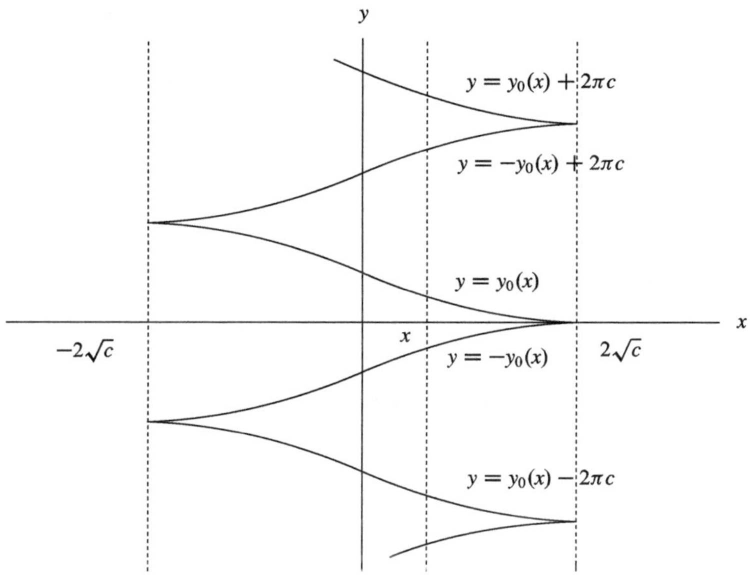

Fig. 5 Singularity locus of $\psi_{+,B}(x, y)$ for Eq. (83)

According to the general result for the propagation of singularities of solutions for linear partial differential equations established by the theory of microlocal analysis (cf. [31]), the singularities of solutions of (85) propagate along a bicharacteristic flow, that is, a Hamiltonian flow of the principal symbol of (85):

$$\begin{cases} \dot{x} = \dfrac{\partial p_B}{\partial \xi} = 2\xi, \\ \dot{y} = \dfrac{\partial p_B}{\partial \eta} = -2\Big(c - \dfrac{x^2}{4}\Big)\eta, \\ \dot{\xi} = -\dfrac{\partial p_B}{\partial x} = -\dfrac{x}{2}\eta^2, \\ \dot{\eta} = -\dfrac{\partial p_B}{\partial y} = 0, \end{cases} \tag{86}$$

where $\dot{x} = dx/dt$ etc. and $p_B(x, y, \xi, \eta) = \xi^2 - (c - x^2/4)\eta^2$ denotes the principal symbol of (85). A solution of (86) with the initial condition $(x(0), y(0), \xi(0), \eta(0)) = (2\sqrt{c}, 0, 0, 1)$ is explicitly given by

$$x = 2\sqrt{c}\cos t, \quad y = -c(t - \sin t \cos t), \quad \xi = -\sqrt{c}\sin t, \quad \eta = 1 \tag{87}$$

and its projection to the base space $\mathbb{C}^2_{(x,y)}$, i.e.,

$$x = 2\sqrt{c}\cos t, \quad y = -c(t - \sin t\cos t) \tag{88}$$

precisely describes the singularity locus of $\psi_{\pm,B}(x,y)$.

In this way the singularities of $\psi_{\pm,B}(x,y)$ can be analyzed by tracing the bicharacteristic flow of the Borel transformed Eq. (85).

Second Approach The second approach is more WKB-theoretic and provides us with more detailed information about the fixed singularities.

We start with the following factorization of $\psi_\pm(x,\eta)$.

$$\psi_\pm(x,\eta) = \psi_\pm^{(\infty)}(x,\eta)\exp\left(\pm\int_{2\sqrt{c}}^{\infty}(S_{\text{odd}} - \eta S_{-1})\,dx\right), \tag{89}$$

where

$$\psi_\pm^{(\infty)}(x,\eta) = \frac{1}{\sqrt{S_{\text{odd}}}}\exp\pm\left(\eta\int_{2\sqrt{c}}^{x} S_{-1}dx + \int_{\infty}^{x}(S_{\text{odd}} - \eta S_{-1})dx\right) \tag{90}$$

is a WKB solution normalized at $x = \infty$. Thanks to Theorem 3.2, we find that $\psi_\pm^{(\infty)}(x,\eta)$ is Borel summable near $\{x \in \mathbb{R} \mid x > 2\sqrt{c}\}$ and hence its Borel transform $\psi_{\pm,B}^{(\infty)}(x,y)$ has no singularities near

$$\{y \in \mathbb{C} \mid y = \mp y_0(x) + \rho, \ \ \rho > 0\}$$

for a fixed $x > 2\sqrt{c}$. On the other hand, the second factor or its exponent

$$V := \int_{2\sqrt{c}}^{\infty}(S_{\text{odd}} - \eta S_{-1})\,dx, \tag{91}$$

which is often called the **"Voros coefficient"**, has fixed singularities on the positive real axis. As a matter of fact, V has the following expression in terms of the Bernoulli numbers.

Proposition 5.1

$$2V = \sum_{n=1}^{\infty}\frac{2^{1-2n}-1}{2n(2n-1)}B_{2n}(ic\eta)^{1-2n}, \tag{92}$$

where B_{2n} stands for the Bernoulli numbers defined by

$$\frac{w}{e^w - 1} = 1 - \frac{w}{2} + \sum_{n=1}^{\infty}\frac{B_{2n}}{(2n)!}w^{2n}. \tag{93}$$

Proposition 5.1 is related to the shift operator with respect to the parameter c and the explicit form (92) of V is derived from the following difference equation for the solution $S_+ = S_+(x, c, \eta)$ of the Riccati equation associated with (83). (To clarify the dependence on the parameter c we here use the notation $S_+(x, c, \eta)$.)

$$S_+(x, c - \eta^{-1}i, \eta) - S_+(x, c, \eta) = \frac{d}{dx}\log\left(\eta^{-1}S_+(x, c, \eta) - i\frac{x}{2}\right). \tag{94}$$

Eq. (94) is an immediate consequence of the following commutation relation.

$$\left(\frac{d}{dx} - i\frac{x}{2}\eta\right)\left(\frac{d^2}{dx^2} - \eta^2(c - \frac{x^2}{4})\right) = \left(\frac{d^2}{dx^2} - \eta^2(c - \eta^{-1}i - \frac{x^2}{4})\right)\left(\frac{d}{dx} - i\frac{x}{2}\eta\right). \tag{95}$$

Using (94), we can verify that $2V$ satisfies a difference equation

$$\phi(\sigma + 1) - \phi(\sigma) = 1 + \log\left(1 + \frac{1}{2\sigma}\right) - (\sigma + 1)\log\left(1 + \frac{1}{\sigma}\right), \tag{96}$$

where $\sigma = ic\eta$. Since the right-hand side of (92) is the unique formal solution of the difference equation (96), we obtain Proposition 5.1.

It follows from Proposition 5.1 that

$$V_B(y) = \frac{1}{4y}\left(\frac{1}{e^{y/(2ic)} - 1} + \frac{1}{e^{y/(2ic)} + 1} - \frac{2ic}{y}\right), \tag{97}$$

which tells us that $V_B(y)$ has simple poles at $y = 2m\pi c$ ($m \in \mathbb{Z} \setminus \{0\}$) with residues $(-1)^{m-1}/(4\pi i m)$. This verifies V_B, and hence $\psi_{\pm,B}$ as well, has fixed singularities. Furthermore, as the Borel sum of the Voros coefficient can be computed explicitly by using (97) (in fact, Binet's formula implies the Borel sum of $2V$ is given by

$$\log\frac{\Gamma(ic\eta + 1/2)}{\sqrt{2\pi}} - ic\eta\left(\log(ic\eta) - 1\right)) \tag{98}$$

for $\arg c < 0$ (cf. [17, Sect. 1.9]) and

$$-\log\frac{\Gamma(-ic\eta + 1/2)}{\sqrt{2\pi}} - ic\eta\left(\log(ic\eta) - 1\right)) - \pi c\eta \tag{99}$$

for $\arg c > 0$, respectively), we obtain

Theorem 5.2 *Let $\Psi_+(x, \eta)$ (resp., $\widetilde{\Psi}_+(x, \eta)$) denote the Borel sum of $\psi_+(x, \eta)$ for $x > 2\sqrt{c}$ when* $\arg c < 0$ *(resp.,* $\arg c > 0$*). Then the following relation holds.*

$$\Psi_+ = (1 + \exp(-2\pi c\eta))^{-1/2}\,\widetilde{\Psi}_+(x, \eta). \tag{100}$$

Thus a kind of Stokes phenomena occurs with WKB solutions of the Weber equation (83) even when the parameter c varies (**"parametric Stokes phenomena"**). Formula (100) exactly coincides with the wall-crossing formula discussed by Gaiotto-Moore-Neitzke [18]. It has been analyzed from the viewpoint of the resurgent analysis by Pham and his collaborators (cf. [11, 13]). Note that, from the viewpoint of the resurgent analysis, (100) is equivalent to the following formula for the alien derivative (in the sense of Ecalle) of $\psi_{+,B}$ at the fixed singularities:

$$\Delta_{y=-y_0(x)+2m\pi c}\, \psi_{+,B}(x,y) = \frac{(-1)^m}{2m}\, \psi_{+,B}(x, y-2m\pi c). \tag{101}$$

For more details we refer the reader to [35].

6 Exact WKB Analysis for Higher Order ODEs

In this section we discuss generalization of the exact WKB analysis to higher order linear ordinary differential equations

$$P\psi = \left(\frac{d^m}{dx^m} + \eta p_1(x)\frac{d^{m-1}}{dx^{m-1}} + \cdots + \eta^m p_m(x)\right)\psi = 0. \tag{102}$$

Here $m \geq 3$ is an integer and η denotes a large parameter.

WKB Solutions, Stokes Geometry

Similarly to the case of second order equations, we can construct a WKB solution of (102) of the form

$$\psi_j(x,\eta) = \exp\left(\eta \int_{x_0}^{x} \zeta_j(x)dx\right) \sum_{n=0}^{\infty} \psi_{j,n}(x)\eta^{-(n+1/2)}, \tag{103}$$

where $\zeta_j(x)$ is a root of the characteristic equation of (102):

$$\zeta^m + p_1(x)\zeta^{m-1} + \cdots + p_m(x) = 0. \tag{104}$$

For details of the construction of WKB solutions we refer the reader to [3, 4].

Definition 6.1

(i) A point $x = a$ is said to be a turning point of (102) if (104) has a multiple root at $x = a$. In other words, a turning point is a zero of the discriminant of (104) in ζ. In particular, a simple zero of the discriminant is called a simple turning point of (102). When $\zeta_j(a) = \zeta_k(a)$ $(j \neq k)$ holds at $x = a$, the turning point $x = a$ is said to be of type (j, k).

(ii) A Stokes curve of type (j, k) of (102) is, by definition, a curve defined by

$$\Im \int_a^x (\zeta_j(x) - \zeta_k(x))dx = 0, \tag{105}$$

where $x = a$ is a turning point of type (j, k). Furthermore, if $\Re \int_a^x (\zeta_j(x) - \zeta_k(x))dx > 0$ holds in addition to (105), the Stokes curve is said to be of type $j > k$.

In parallel with Theorem 2.9 the following theorem holds also for WKB solutions of higher order equations.

Theorem 6.2 *Let $x = a$ be a simple turning point of* (102) *of type (j, k). Then, for suitably normalized WKB solutions ψ_j and ψ_k of* (102)*, the following properties hold in a neighborhood of $(x, y) = (a, 0)$:*

(i) *$\psi_{j,B}(x, y)$ and $\psi_{k,B}(x, y)$ are singular only along $\Gamma_j \cup \Gamma_k$, where*

$$\Gamma_j = \{ (x, y) \mid y = -\int_a^x \zeta_j(x)\, dx \}, \quad \Gamma_k = \{ (x, y) \mid y = -\int_a^x \zeta_k(x)\, dx \}. \tag{106}$$

(ii)

$$\Delta_{y=-\int_a^x \zeta_k(x)dx}\, \psi_{j,B}(x, y) = i\psi_{k,B}(x, y), \quad \Delta_{y=-\int_a^x \zeta_j(x)dx}\, \psi_{k,B}(x, y) = i\psi_{j,B}(x, y). \tag{107}$$

Theorem 6.2 is proved in the following manner: We first consider the factorization of the differential operator P to reduce the problem to that for second order equations, and then use transformation theory similar to Theorem 3.6. To be more specific, we prove the following two assertions.

Proposition 6.3 *In a neighborhood of a simple turning point $x = a$, we can find differential operators Q and R of order $(m - 2)$ and 2, respectively, that satisfy*

$$P = QR. \tag{108}$$

Here Q and R have the form

$$Q = \frac{d^{m-2}}{dx^{m-2}} + \eta q_1(x, \eta)\frac{d^{m-3}}{dx^{m-3}} + \cdots + \eta^{m-2} q_{m-2}(x, \eta), \tag{109}$$

$$R = \frac{d^2}{dx^2} + \eta r_1(x, \eta)\frac{d}{dx} + \eta^2 r_2(x, \eta), \tag{110}$$

where $q_j(x,\eta) = \sum_{n\ge 0} q_{j,n}(x)\eta^{-n}$ *and* $r_j(x,\eta) = \sum_{n\ge 0} r_{j,n}(x)\eta^{-n}$ *are formal power series in* η^{-1} *with holomorphic coefficients. Furthermore, the following conditions are also satisfied.*

$$\left(\zeta^{m-2} + q_{1,0}(x)\zeta^{m-3} + \cdots + q_{m-2,0}(x)\right)\Big|_{\zeta=\zeta_j(x)\ or\ \zeta_k(x)} \neq 0, \tag{111}$$

$$\zeta^2 + r_{1,0}(x)\zeta + r_{2,0}(x) = (\zeta - \zeta_j(x))(\zeta - \zeta_k(x)). \tag{112}$$

Proposition 6.4 *In a neighborhood of* $x = a$*, after the employment of the gauge transformation*

$$\psi \longmapsto \left(\exp\left(-\frac{1}{2}\eta\int_a^x r_1(x,\eta)\,dx\right)\right)\psi,$$

the second order differential equation $R\psi = 0$ *in Proposition 6.3 can be transformed (in the sense of exact WKB analysis) to the Airy equation.*

For more detailed explanation see [3, 4, 25].

BNR Equation: Appearance of New Stokes Curves

Theorem 6.2 asserts that, as far as the local theory near a simple turning point is concerned, the behavior of Borel resummed WKB solutions of higher order equations is the same as that of second order equations. However, the global behavior is completely different, as Berk et al. [8] pointed out by using the following example.

Example (BNR Equation)

$$\left(\frac{d^3}{dx^3} + 3\eta^2\frac{d}{dx} + 2ix\eta^3\right)\psi = 0. \tag{113}$$

The characteristic equation of (113) is $\zeta^3 + 3\zeta + 2ix = 0$. Considering its discriminant, we find that (113) has two turning points $x = \pm 1$. Figure 6 indicates the configuration of Stokes curves of (113).

As Fig. 6 shows, there exist crossing points of Stokes curves for (113). Such crossing points cause the following serious difficulty: We consider the analytic continuation of the Borel sum of a WKB solution $\Psi_3(x,\eta)$ near a crossing point x_* of Stokes curves. Assuming the Borel summability, we can expect that a Stokes phenomenon of the form

$$\Psi_3 \rightsquigarrow \Psi_3 + \alpha\Psi_2 \quad (\text{resp., } \Psi_2 \rightsquigarrow \Psi_2 + \beta\Psi_1)$$

Fig. 6 Stokes curves of the BNR equation (113)

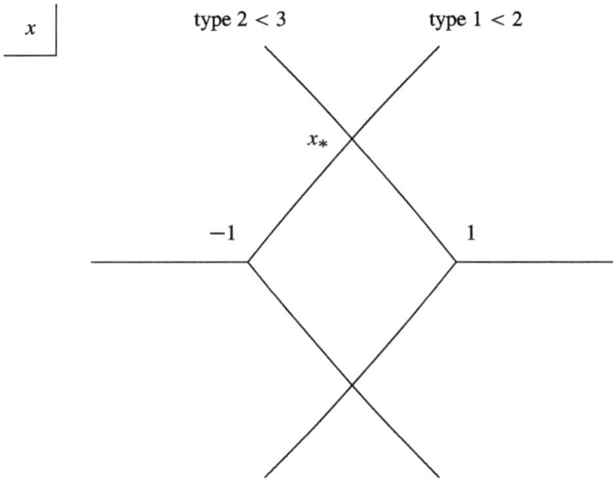

Fig. 7 Paths of analytic continuation near x_*

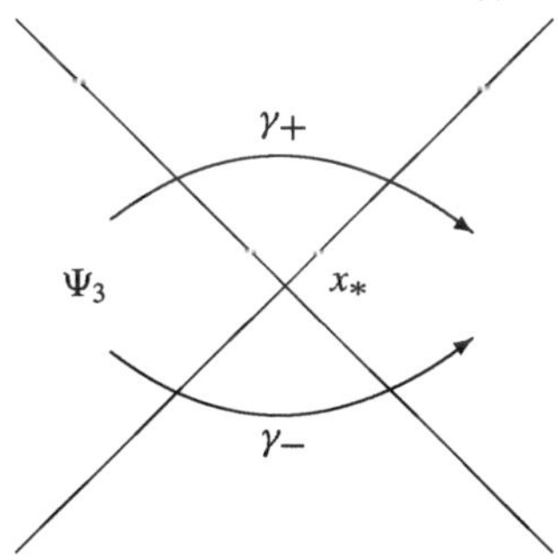

with some suitable constant α (resp., β) occurs on a Stokes curve of type $2 < 3$ (resp., of type $1 < 2$), in view of Theorem 6.2. Hence, by the analytic continuation along γ_+ (cf. Fig. 7) Ψ_3 should be changed to $\Psi_3 + \alpha(\Psi_2 + \beta\Psi_1)$, whereas by the analytic continuation along γ_- Ψ_3 should become $\Psi_3 + \alpha\Psi_2$. This is a contradiction if $\alpha\beta \neq 0$, since Eq. (113) does not have any singularity near x_*!

To resolve this paradoxical problem Berk et al. [8] proposed to introduce a **"new Stokes curve"** of type $1 < 3$ that emanates from x_* and tends to ∞ (cf. Fig. 8). As a matter of fact, if a Stokes phenomenon of the form

$$\Psi_3 \rightsquigarrow \Psi_3 - \alpha\beta\Psi_1$$

occurs on it, the contradiction disappears. Berk et al. confirmed the existence of a new Stokes curve by investigating an integral representation of solutions of (113) through the steepest descent method ([8], see also [34]).

If we consider the structure of singularities of $\psi_{3,B}(x,y)$ in y-plane (the so-called Borel plane) near x_*, we find that at x_* three relevant singular points $-y_j(x) := -\int^x \zeta_j(x)dx$ ($j = 1,2,3$) of $\psi_{3,B}(x,y)$ have the same imaginary part. From the singularity structure the new Stokes curve is characterized as a curve where the two distant singular points $-y_3(x)$ and $-y_1(x)$ have the same imaginary part. Note that on the upper portion of the new Stokes curve $-y_1(x)$ is visible from $-y_3(x)$, whereas it is not visible on the lower portion, as is indicated in Fig. 8. In this way a new Stokes curve is also related to the singularities of the Borel transform $\psi_{j,B}(x,y)$ of WKB solutions. Since the sheet structure of the Riemann surface of $\psi_{j,B}(x,y)$ is complicated, a new Stokes curve may become inert on some portion of it.

Virtual Turning Points

The Borel transform $\psi_{j,B}(x,y)$ of WKB solutions of the BNR equation (113) satisfies

$$\left(\frac{\partial^3}{\partial x^3} + 3\frac{\partial^3}{\partial x \partial y^2} + 2ix\frac{\partial^3}{\partial y^3}\right)\psi_{j,B}(x,y) = 0. \tag{114}$$

Since the new Stokes curve of the BNR equation is related to the singularities of $\psi_{j,B}(x,y)$, let us investigate the bicharacteristic flow of (114) to understand the singularity structure of $\psi_{j,B}(x,y)$ and a new Stokes curve more thoroughly.

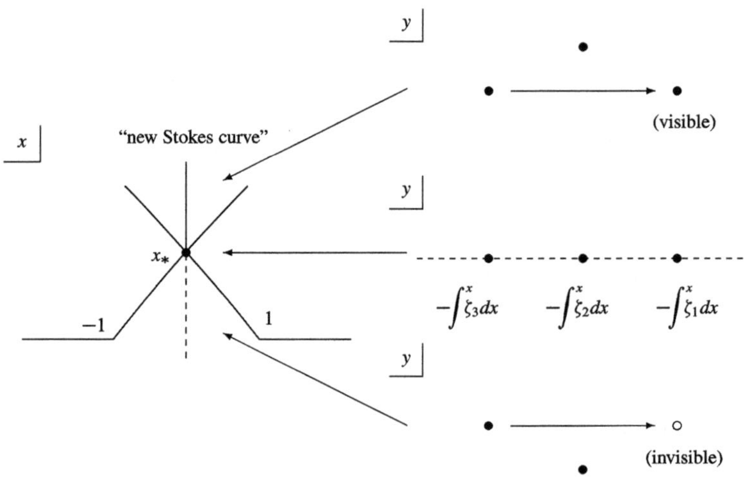

Fig. 8 New Stokes curve passing through x_* and singularities of $\psi_{3,B}(x,y)$ in the Borel plane near x_*

The bicharacteristic flow of (114) is defined by

$$\begin{cases} \dot{x} = \dfrac{\partial p_B}{\partial \xi} = 3\xi^2 + 3\eta^2, \\ \dot{y} = \dfrac{\partial p_B}{\partial \eta} = 6\xi\eta + 6ix\eta^2, \\ \dot{\xi} = -\dfrac{\partial p_B}{\partial x} = -2i\eta^3, \\ \dot{\eta} = -\dfrac{\partial p_B}{\partial y} = 0, \end{cases} \tag{115}$$

where $p_B(x, y, \xi, \eta) = \xi^3 + 3\xi\eta^2 + 2ix\eta^3$ is the principal symbol of (114). A solution of (115) with the initial condition $(x(0), y(0), \xi(0), \eta(0)) = (1, 0, -i, 1)$ (note that $\xi(0) = -i$ is a double root of $\zeta^3 + 3\zeta + 2ix = 0$ at $x = 1$) is given by

$$\begin{aligned} x &= -4t^3 - 6t^2 + 1 = -(2t + 1)(2t^2 + 2t - 1), \\ y &= -6it^4 - 12it^3 - 6it^2 = -6it^2(t + 1)^2, \\ \xi &= -2it - i, \quad \eta = 1. \end{aligned} \tag{116}$$

Hence its projection to the base space $\mathbb{C}^2_{(x,y)}$, which describes the singularities of $\psi_{j,B}(x, y)$, becomes as is visualized in Fig. 9. Near the simple turning points $x = \pm 1$ two branches of singularities coalesce and form a cuspidal singularity. We also observe that, in addition to these turning points, the singularities make a self-intersection point at $(x, y) = (0, -3i/2)$ and two branches intersect there. In fact, if we regard this self-intersection point (to be more precise, its x-component $x = 0$) as a new kind of turning points and add a Stokes curve

$$\Im \int_0^x \big(\zeta_1(x) - \zeta_3(x)\big)\, dx = 0$$

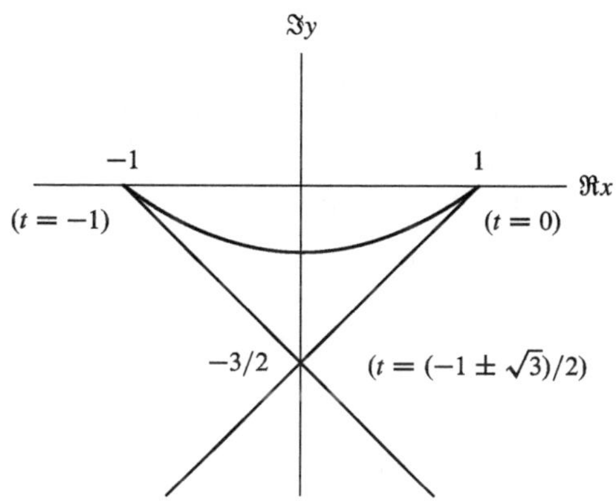

Fig. 9 Bicharacteristic curve of (114)

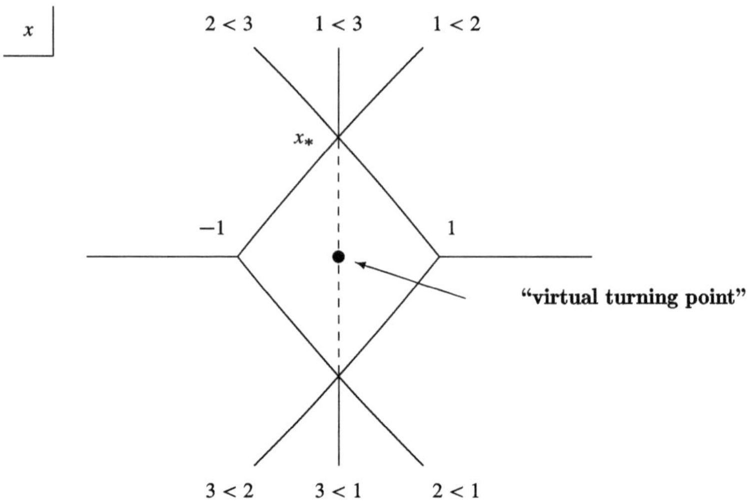

Fig. 10 Complete Stokes geometry of the BNR equation (113). (A *dotted line* indicates the inert portion of a new Stokes curve)

emanating from this point to Fig. 6 (i.e., the original configuration of Stokes curves of (113)), we obtain Fig. 10. Thus we can re-obtain a new Stokes curve of the BNR equation.

This consideration naturally leads to the following

Definition 6.5 Let

$$P_B\psi_B = \left(\frac{\partial^m}{\partial x^m} + p_1(x)\frac{\partial^m}{\partial x^{m-1}\partial y} + \cdots + p_m(x)\frac{\partial^m}{\partial y^m}\right)\psi_B = 0 \tag{117}$$

be the Borel transformed equation of (102) and

$$p_B(x, y, \xi, \eta) = \xi^m + p_1(x)\xi^{m-1}\eta + \cdots + p_m(x)\eta^m \tag{118}$$

its principal symbol. Then we call the x-component of a self-intersection point of a bicharacteristic curve of (117) a virtual turning point of (102). Here a bicharacteristic curve of (117) means the projection of a bicharacteristic flow of (117) onto the base space $\mathbb{C}^2_{(x,y)}$.

We can verify that each singularity of $\psi_{j,B}(x, y)$ (or, equivalently, a bicharacteristic curve) is locally described by $y = -\int^x \zeta_k(x)dx$ and at a virtual turning point two branches of singularities of $\psi_{j,B}(x, y)$ (for example, $y = -\int^x \zeta_k(x)dx$ and $y = -\int^x \zeta_l(x)dx$ with $k \neq l$) cross by its definition. Thus we can naturally define a Stokes curve emanating from a virtual turning point (concretely by $\Im \int^x(\zeta_k(x) - \zeta_l(x))dx = 0$ in the above situation).

Remark 6.6 A virtual turning point was first introduced in [2] under the name of "new turning point".

Remark 6.7 A crossing point of Stokes curves is highly dependent on the way how the Borel resummation is performed (for example, it heavily depends on $\arg \eta$), whereas the definition of a virtual turning point is independent of the way of resummation. In this sense a virtual turning point is related to the operator P in (102) more intrinsically than a new Stokes curve.

Once the definition of virtual turning points is provided, we obtain the following recipe for finding the proper Stokes geometry of a higher order equation (102).

Recipe 6.8

(a) Draw all Stokes curves that emanate from turning points defined in Definition 6.1.
(b) Draw the new Stokes curve that emanates from a virtual turning point.
(c) As the portion of a new Stokes curve near a virtual turning point is inert, we draw the new Stokes curve in (b) by a dotted line until it hits a crossing point of (new) Stokes curves.
(d) When the new Stokes curve in (b) is of type $j > l$ and it hits a crossing point of a (new) Stokes curve of type $j > k$ and that of type $k > l$, we use a solid line to draw the portion of the new Stokes curve in (b) after passing over the crossing point.

Practically speaking, Recipe 6.8 is powerful enough to discuss the Stokes geometry (including new Stokes curves) of a higher order equation (102). However, it is not complete: As there exist in general infinitely many virtual turning points due to the existence of fixed singularities, it is necessary to exclude redundant virtual turning points.

Example

$$\left(\frac{d^3}{dx^3} - 6(x+1)\eta^2\frac{d}{dx} + (4x+2i)\eta^3\right)\psi = 0. \tag{119}$$

As discussed in [2, Example 2.5], (119) has infinitely many virtual turning points. Among them at most three are non-redundant and Fig. 11 describes the complete Stokes geometry of (119). (In Fig. 11 wiggly lines designate cuts to define a root $\zeta_j(x)$ of the characteristic equation associated with (119) as a single-valued function.)

At present no complete criterion is available for the determination of redundant virtual turning points. This problem is also related to the Borel summability of WKB solutions of a higher order equation (102). To establish the Borel summability of WKB solutions of a higher order equation is still an open problem.

For more detailed explanation of virtual turning points and new Stokes curves including the connection formula on it, we refer the reader to [25].

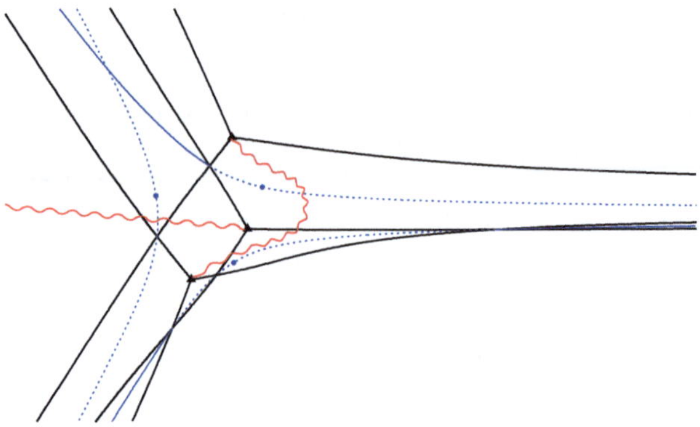

Fig. 11 Complete Stokes geometry of Eq. (119)

Remark 6.9 Honda made a detailed study on the efficiency of Recipe 6.8 and gave a satisfactory answer to the finiteness of non-redundant virtual turning points. See [23] and [24] for his study.

Remark 6.10 Recently the Stokes geometry of higher order equations including new Stokes curves (but no virtual turning points) is investigated under the name of "spectral networks" also by Gaiotto-Moore-Neitzke [19, 20].

7 Some Recent Developments

In the final section we briefly discuss two recent developments of the exact WKB analysis, both of which are related to the problem of new Stokes curves and virtual turning points for higher order ODEs.

Borel Summability of Formal Solutions of Inhomogeneous Second Order Equations

Let us first discuss the Borel summability of formal solutions of the following inhomogeneous second order equations:

$$P\psi = \left(\frac{d^2}{dx^2} + \eta p(x)\frac{d}{dx} + \eta^2 q(x)\right)\psi = F(x), \tag{120}$$

where $p(x), q(x), F(x)$ are assumed to be polynomials for the sake of simplicity. As one can easily confirm, Eq. (120) has a unique formal power series solution of the form

$$\widehat{\psi} = \eta^{-2}\psi_2(x) + \eta^{-3}\psi_3(x) + \cdots, \tag{121}$$

whose coefficients $\psi_n(x)$ $(n = 2, 3, \ldots)$ are determined by the recursive relation

$$q\psi_2 = F, \quad q\psi_3 + p\psi_2' = 0, \quad q\psi_n + p\psi_{n-1}' + \psi_{n-2}'' = 0 \ (n \geq 4).$$

In [29] we showed the following result for the Borel summability of $\widehat{\psi}$:

Theorem 7.1 *Let $\zeta_\pm(x)$ be the roots of $\zeta^2 + p(x)\zeta + q(x) = 0$, i.e., $\zeta_\pm(x) = (-p \pm \sqrt{p^2 - 4q})/2$.*

Case I (The Case Where $p(x) \equiv 0$) *Assume $p(x) \equiv 0$. Then, if a curve $\Im \int_{x_0}^x \sqrt{-q(x)}\, dx = 0$ passing through x_0 does not flow into a turning point (i.e., x_0 is not located on any Stokes curve of $P\psi = 0$), then $\widehat{\psi}$ is Borel summable at $x = x_0$.*

Case II (The Case Where $p(x) \not\equiv 0$) *Suppose that the following three conditions are satisfied:*

(i) *x_0 is not located on any Stokes curve of $P\psi = 0$, that is, a curve $\Im \int_{x_0}^x (\zeta_+ - \zeta_-)\, dx = \Im \int_{x_0}^x \sqrt{p^2 - 4q(x)}\, dx = 0$ passing through x_0 does not flow into a turning point.*
(ii) *A curve $\Gamma_\pm^0$ defined by*

$$\Im \int_{x_0}^x (-\zeta_\mp)\, dx = \Im \int_{x_0}^x \frac{p \pm \sqrt{p^2 - 4q}}{2}\, dx = 0, \tag{122}$$

that is, a steepest descent path of $\Re \int_{x_0}^x (-\zeta_\mp)\, dx$ passing through x_0, can be extended to $x = \infty$.
(iii) *When $\Gamma_\pm^0$ crosses a Stokes curve of $P\psi = 0$ of type $\pm > \mp$ at $x = x_1$, then a bifurcated path $\Gamma_\mp^1$ emanating from $x = x_1$ defined by*

$$\Im \int_{x_1}^x (-\zeta_\pm)\, dx = 0 \tag{123}$$

is also extensible to $x = \infty$.

If these three conditions (i)–(iii) *are satisfied, then $\widehat{\psi}$ is Borel summable at $x = x_0$.*

Remark 7.2 In Case II, if $\Gamma_\pm^0$ and/or $\Gamma_\mp^1$ cross other Stokes curves of $P\psi = 0$, we further consider additional bifurcated paths emanating from crossing points similarly defined as in (iii) and impose their extensibility to $x = \infty$.

An inhomogeneous second order equation can be thought of as a special case of a third order homogeneous equation. (As a matter of fact, (120) can be written as $(d/dx)(F(x)^{-1}P\psi) = 0$.) Hence, Theorem 7.1 suggests the difficulty for characterizing the Borel summability of formal solutions of higher order equations.

WKB Analysis for Completely Integrable Systems

The following equation is a variant of the BNR equation.

$$\left(\frac{d^3}{dx^3} + \frac{c}{2}\eta^2\frac{d}{dx} + \frac{x}{4}\eta^3\right)\psi = 0, \tag{124}$$

where a parameter c is introduced into the coefficient of the first order term. Note that the Pearcey integral

$$\int \exp\bigl(\eta(t^4 + ct^2 + xt)\bigr)\, dt$$

gives a particular solution of (124) (for an appropriate choice of the integration path). If we regard the parameter c as a new independent variable, we find the Pearcey integral satisfies the following system of differential equations in two variables $(x_1, x_2) = (x, c)$:

$$\begin{cases} \left(\dfrac{\partial^3}{\partial x_1^3} + \dfrac{x_2}{2}\eta^2\dfrac{\partial}{\partial x_1} + \dfrac{x_1}{4}\eta^3\right)\psi = 0, \\ \left(\eta\dfrac{\partial}{\partial x_2} - \dfrac{\partial^2}{\partial x_1^2}\right)\psi = 0. \end{cases} \tag{125}$$

Equation (125) can be written also in the form of completely integrable systems of first order equations as follows:

$$\begin{cases} \eta^{-1}\dfrac{\partial}{\partial x_1}\Psi = P(x)\Psi, & P = \begin{pmatrix} 0 & 1 & 0 \\ 0 & 0 & 1 \\ -x_1/4 & -x_2/2 & 0 \end{pmatrix}, \\ \eta^{-1}\dfrac{\partial}{\partial x_2}\Psi = Q(x)\Psi, & Q = P^2 + \dfrac{x_2}{3} - \dfrac{\eta^{-1}}{4}\begin{pmatrix} 0 & 0 & 0 \\ 0 & 0 & 0 \\ 1 & 0 & 0 \end{pmatrix}, \end{cases} \tag{126}$$

where Ψ is an unknown 3-vector. Here the complete integrability condition of (126) means

$$\eta^{-1}\frac{\partial P}{\partial x_2} - \eta^{-1}\frac{\partial Q}{\partial x_1} + (PQ - QP) = 0. \tag{127}$$

Equation (126) (or, equivalently, (125)) is called the **"Pearcey system"** as the Pearcey integral gives a particular solution of it. In what follows, as the first step toward the exact WKB theory for completely integrable systems (or holonomic systems in more recent terminologies), we discuss the exact WKB analysis of the Pearcey system.

For such a holonomic system with a large parameter η, making use of the complete integrability condition (127), we can construct a WKB solution of the form

$$\psi^{(j)} = \exp \int^x \omega^{(j)} \tag{128}$$

($j = 1, 2, 3$ in the case of the Pearcey system), where $\omega^{(j)} = S^{(j)}dx_1 + T^{(j)}dx_2$ is an infinite series of the closed 1-form:

$$S^{(j)} = \eta S^{(j)}_{-1} + S^{(j)}_0 + \eta^{-1} S^{(j)}_{-1} + \cdots, \quad T^{(j)} = \eta T^{(j)}_{-1} + T^{(j)}_0 + \eta^{-1} T^{(j)}_{-1} + \cdots. \tag{129}$$

As a matter of fact, the top order part $\omega^{(j)}_{-1} = S^{(j)}_{-1}dx_1 + T^{(j)}_{-1}dx_2$ is determined by some algebraic equations (for example, $(S^{(j)}_{-1}, T^{(j)}_{-1})$ satisfies

$$(S^{(j)}_{-1})^3 + \frac{x_2}{2} S^{(j)}_{-1} + \frac{x_1}{4} = 0, \quad T^{(j)}_{-1} = (S^{(j)}_{-1})^2$$

in the case of the Pearcey system) and, once $\omega^{(j)}_{-1}$ is fixed, the higher order parts $\omega^{(j)}_n = S^{(j)}_n dx_1 + T^{(j)}_n dx_2$ ($n \geq 0$) are uniquely determined in a recursive manner.

Turning points and Stokes surfaces are defined in parallel with Definition 6.1.

Definition 7.3

(i) A point $a = (a_1, a_2) \in \mathbb{C}^2$ is said to be a turning point of type (j, k) if

$$(\omega^{(j)}_{-1} - \omega^{(k)}_{-1})\Big|_{x=a} = 0, \quad \text{i.e.,} \quad S^{(j)}_{-1}(a) = S^{(k)}_{-1}(a) \text{ and } T^{(j)}_{-1}(a) = T^{(k)}_{-1}(a) \tag{130}$$

hold for some (j, k) with $j \neq k$.

(ii) A Stokes surface of type (j, k) is, by definition, a real hypersurface defined by

$$\Im \int_a^x (\omega^{(j)}_{-1} - \omega^{(k)}_{-1}) = 0, \tag{131}$$

where $a = (a_1, a_2)$ is a turning point of type (j, k).

For example, by straightforward computations we find that the set of turning points of the Pearcey system is explicitly given by

$$\Lambda = \{ (x_1, x_2) \in \mathbb{C}^2 \mid 27x_1^2 + 8x_2^3 = 0 \}. \tag{132}$$

Note also that, by the definition, the section of the Stokes surfaces of the Pearcey system (125) with $x_2 = c$ contains the Stokes curves of the BNR equation (124).

The following result of Hirose impressively shows an advantage of considering the Pearcey system instead of considering the BNR equation.

Theorem 7.4 (Hirose [21]) *The Stokes surface of the Pearcey system* (125) *contains not only the Stokes curves of the BNR equation* (124) *but also its new Stokes curves in their section with* $x_2 = c$.

The reason why the Stokes surface of (125) contains the new Stokes curves of (124) is the following: If we change the parameter c (or x_2) in (124), we encounter the degenerate configuration where a Stokes curve emanating from a turning point hits another turning point with different type at some value of c. As was discussed in [5], the role of a new Stokes curve and that of an ordinary Stokes curve are interchanged through such a degenerate configuration. Otherwise stated, by the change of the parameter c a new Stokes curve of (124) is continuously deformed to an ordinary Stokes curve. Thus, in $\mathbb{C}^2_{(x_1,x_2)}$ new Stokes curves and ordinary Stokes curves are connected and hence the Stokes surface of (125) inevitably contains the new Stokes curves of (124).

Theorem 7.4 strongly suggests that the exact WKB analysis for holonomic systems or completely integrable systems may play an important role also in the analysis of new Stokes curves for higher order ODEs.

Another peculiar feature of the Pearcey system is that the set Λ of its turning points has a unique cuspidal singularity at the origin $(x_1, x_2) = (0, 0)$. At this cuspidal singularity two turning points with different types coalesce. Furthermore, the virtual turning point of the BNR equation (124) also coalesces there.

For this cuspidal singular point of the Pearcey system Hirose proves the following intriguing result.

Theorem 7.5 (Hirose [22]) *Under some genericity condition every completely integrable system of two independent variables can be transformed (in the sense of exact WKB analysis) to the Pearcey system at a cuspidal singularity of the set of turning points.*

For the proof of Theorem 7.5 see [22].

To clarify the implication of Theorem 7.5, let us consider, for example, the following holonomic system:

$$\begin{cases} \left(\dfrac{\partial^3}{\partial x_1^3} + \dfrac{2}{3}x_2\eta\dfrac{\partial^2}{\partial x_1^2} + \dfrac{1}{3}x_1\eta^2\dfrac{\partial}{\partial x_1} - \dfrac{\alpha}{3}\eta^3\right)\psi = 0, \\ \left(\eta\dfrac{\partial}{\partial x_2} - \dfrac{\partial^2}{\partial x_1^2}\right)\psi = 0. \end{cases} \tag{133}$$

Equation (133) is the so-called "(1, 4)-hypergeometric system", an example of (confluent) hypergeometric systems of two variables. Note that, when x_2 is fixed, (133) is equivalent to (119) discussed in section "6".

By a straightforward computation we confirm that the set of turning points of (133) has three cuspidal singular points. Then Theorem 7.5 tells us that at each cuspidal singular point the $(1, 4)$-hypergeometric system (133) can be transformed to the Pearcey system. In particular, there exists a virtual turning point that coalesces with two ordinary turning points with different types at each cuspidal singularity. These three virtual turning points are all non-redundant and play an important role in describing the complete Stokes geometry of the higher order ODE (119) discussed in section "Virtual Turning Points". In this way the cuspidal singularity of the set of turning points of a completely integrable system is closely related to the problem of (non-)redundant virtual turning points of a higher order ODE.

These two results of Hirose brings a new insight to the problem of new Stokes curves and virtual turning points for higher order ODEs. It is the future problem to develop the exact WKB analysis for completely integrable systems in a systematic manner.

Acknowledgements The author would like to thank Dr. Shinji Sasaki for his kind help in preparing some figures contained in this article. This research is supported by JSPS KAKENHI Grant No. 26287015 and No. 24340026.

References

1. T. Aoki, T. Kawai, Y. Takei, The Bender-Wu analysis and the Voros theory, in *Special Functions*, eds. M. Kashiwara and T. Miwa (Springer, Berlin, 1991), pp. 1–29
2. T. Aoki, T. Kawai, Y. Takei, New turning points in the exact WKB analysis for higher order ordinary differential equations, in *Analyse algébrique des perturbations singulières. I*, ed. L. Boutet de Monvel (Hermann, Paris, 1994), pp. 69–84
3. T. Aoki, T. Kawai, T. Koike, Y. Takei, On the exact WKB analysis of operators admitting infinitely many phases. Adv. Math. **181**, 165–189 (2004)
4. T. Aoki, T. Kawai, T. Koike, Y. Takei, On the exact WKB analysis of microdifferential operators of WKB type. Ann. Inst. Fourier (Grenoble) **54**, 1393–1421 (2004)
5. T. Aoki, T. Kawai, S. Sasaki, A. Shudo, Y. Takei, Virtual turning points and bifurcation of Stokes curves. J. Phys. A **38**, 3317–3336 (2005)
6. T. Aoki, T. Kawai, Y. Takei, The Bender-Wu analysis and the Voros theory. II, in *Adv. Stud. Pure Math.*, Vol. 54, eds. T. Miwa, A. Matsuo, T. Nakashima and Y. Saito) (The Mathemetical Society of Japan, Tokyo, 2009), pp. 19–94
7. W. Balser, *From Divergent Power Series to Analytic Functions*, Vol. 1582, Lecture Notes in Mathematics (Springer, Berlin, 1994)
8. H.L. Berk, W.M. Nevins, K.V. Roberts, New Stokes' line in WKB theory. J. Math. Phys. **23**, 988–1002 (1982)
9. B. Candelpergher, J.C. Nosmas, F. Pham, *Approche de la résurgence* (Hermann, Paris, 1993)
10. O. Costin, *Asymptotics and Borel Summability*, Monographs and Surveys in Pure and Applied Mathematics, vol. 141 (CRC Press, Boca Raton, 2009)
11. E. Delabaere, H. Dillinger, F. Pham, Résurgence de Voros et périodes des courbes hyperelliptiques. Ann. Inst. Fourier (Grenoble) **43**, 163–199 (1993)
12. E. Delabaere, H. Dillinger, F. Pham, Exact semi-classical expansions for one dimensional quantum oscillators. J. Math. Phys. **38**, 6126–6184 (1997)
13. E. Delabaere, F. Pham, Resurgent methods in semi-classical asymptotics. Ann. Inst. H. Poincaré **71**, 1–94 (1999)

14. J. Ecalle, Les fonctions résurgentes. Tome I, II, III. Publ. Math. d'Orsay, Univ. Paris-Sud, 1981 (Tome I, II), 1985 (Tome III)
15. J. Ecalle, Cinq applications des fonctions résurgentes. Publ. Math. d'Orsay, 84T62, Univ. Paris-Sud, 1984
16. J. Ecalle, Weighted products and parametric resurgence, in *Analyse algébrique des perturbations singulières*. I, ed. by L. Boutet de Monvel (Hermann, Paris, 1994), pp. 7–49
17. A. Erdélyi et al., *Higher Transcendental Functions* (Reprint Edition), vol. I (McGraw-Hill, New York, 1981)
18. D. Gaiotto, G.W. Moore, A. Neitzke, Wall-crossing, Hitchin systems, and the WKB approximation. Adv. Math. **234**, 239–403 (2013)
19. D. Gaiotto, G.W. Moore, A. Neitzke, Spectral networks. Ann. Henri Poincaré **14**, 1643–1731 (2013)
20. D. Gaiotto, G.W. Moore, A. Neitzke, Spectral networks and snakes. Ann. Henri Poincaré **15**, 61–141 (2014)
21. S. Hirose, On the stokes geometry for the Pearcey system and the $(1, 4)$ hypergeometric system, RIMS Kôkyûroku Bessatsu, vol. B40, pp. 243–292 (2013)
22. S. Hirose, On a WKB theoretic transformation for a completely integrable system near a degenerate point where two turning points coalesce. Publ. RIMS **50**, 19–84 (2014)
23. N. Honda, On the stokes geometry of the Noumi-Yamada system, RIMS Kôkyûroku Bessatsu, vol. B2, pp. 45–72 (2007)
24. N. Honda, The geometric structure of a virtual turning point and the model of the stokes geometry, RIMS Kôkyûroku Bessatsu, vol. B10, pp. 63–113 (2008)
25. N. Honda, T. Kawai, Y. Takei, *Virtual Turning Points*. Springer Briefs in Mathematical Physics, vol. 4 (Springer, Berlin, 2015)
26. S. Kamimoto, T. Koike, On the Borel summability of WKB theoretic transformation series. Preprint (RIMS-1726) (2011), available at http://www.kurims.kyoto-u.ac.jp/preprint/preprint_y2011.html. RIMS Kôkyûroku Bessatsu (to appear)
27. T. Kawai, Y. Takei, *Algebraic Analysis of Singular Perturbation Theory*. Translations of Mathematical Monographs, vol. 227 (American Mathematical Society, Providence, RI, 2005)
28. T. Koike, R. Schäfke, On the Borel summability of WKB solutions of Schrödinger equations with polynomial potentials and its applications. In preparation (The completion of the paper is delayed. A preliminary version can be obtained through the contact with Tatsuya Koike at Kobe University)
29. T. Koike, Y. Takei, Exact WKB analysis of second-order non-homogeneous linear ordinary differential equations, RIMS Kôkyûroku Bessatsu, vol. B40, pp. 293–312 (2013)
30. F. Pham, Resurgence, quantized canonical transformations, and multi-instanton expansions, in *Algebraic Analysis*, ed. by M. Kashiwara, T. Kawai, vol. II (Academic Press, New York, 1988), pp. 699–726
31. M. Sato, T. Kawai, M. Kashiwara, *Microfunctions and Pseudo-Differential Equations*. Lecture Notes in Mathematics, vol. 287 (Springer, Berlin, 1973), pp. 265–529
32. D. Sauzin, Resurgent functions and splitting problems, RIMS Kôkyûroku, vol. 1493, pp. 48–117 (2006)
33. H.J. Silverstone, JWKB connection-formula problem revisited via Borel summation. Phys. Rev. Lett. **55**, 2523–2526 (1985)
34. Y. Takei, Exact WKB analysis, and exact steepest descent method. Sugaku Expositions **20**, 169–189 (2007)
35. Y. Takei, Sato's conjecture for the Weber equation and transformation theory for Schrödinger equations with a merging pair of turning points, RIMS Kôkyûroku Bessatsu, vol. B10, pp. 205–224 (2008)
36. A. Voros, The return of the quartic oscillator. The complex WKB method. Ann. Inst. H. Poincaré **39**, 211–338 (1983)

Part II
Research Papers

Transcendental Meromorphic Solutions of P_{34} and Small Targets

Ewa Ciechanowicz and Galina Filipuk

Abstract The equation P_{34} is closely related to the well-known second Painlevé equation P_2 and some properties of its meromorphic solutions are similar to those of the solutions of P_4. We discuss various value distribution properties of solutions of P_{34}, including growth, the second main theorem and behaviour towards small target functions.

Keywords Defect • Meromorphic function • Painlevé equation • Small target

Mathematics Subject Classification (2000). Primary 30D35; Secondary 34M05, 34M55

The paper is organized as follows. We begin with a short overview of elementary notions of value distribution theory. Then we give some facts about value distribution of transcendental solutions of the Painlevé equations and present results concerning the equation P_{34}. First we estimate growth of its meromorphic solutions and then present the specific form of the second main theorem for solutions of this equation. Finally, we discuss small target functions of meromorphic solutions of P_{34}.

E. Ciechanowicz
Faculty of Mathematics and Physics, University of Szczecin, ul. Wielkopolska 15, 70-451 Szczecin, Poland
e-mail: ewa.ciechanowicz@usz.edu.pl

G. Filipuk (✉)
Department of Mathematics, Informatics and Mechanics, University of Warsaw, Banacha 2, 02-097 Warsaw, Poland

Institute of Mathematics of the Polish Academy of Sciences, Sniadeckich str. 8, 00-956 Warsaw, Poland
e-mail: filipuk@mimuw.edu.pl

G. Filipuk et al. (eds.), *Analytic, Algebraic and Geometric Aspects of Differential Equations*, Trends in Mathematics, DOI 10.1007/978-3-319-52842-7_6

1 Preliminaries

Let us start with the basic definitions and results of the Nevanlinna theory. We apply the following standard notations (see: [9, 12, 14]). Let

$$N(r,f) = \int_0^r (n(t,f) - n(0,f))\frac{dt}{t} + n(0,f)\log r$$

be the integrated counting function, where $n(r,f)$ counts the number of poles, including multiplicity, in the disk of radius r, and let

$$\overline{N}(r,f) = \int_0^r (\overline{n}(t,f) - \overline{n}(0,f))\frac{dt}{t} + \overline{n}(0,f)\log r$$

count poles of f ignoring multiplicity. Then

$$N(r,a,f) := N(r,\frac{1}{f-a}),\quad \overline{N}(r,a,f) := \overline{N}(r,\frac{1}{f-a})\quad (a \in \mathbb{C})$$

count a-points with and without multiplicity respectively. We also put

$$N_1(r,a,f) := N(r,a,f) - \overline{N}(r,a,f).$$

Next,

$$m(r,f) = m(r,\infty,f) := \frac{1}{2\pi}\int_0^{2\pi} \log^+ |f(re^{i\phi})|d\phi,\quad \log^+ x = \max(0,\log x),$$

is the mean proximity function and

$$m(r,a,f) := m(r,\frac{1}{f-a})$$

denotes the mean proximity to $a \in \mathbb{C}$. Then $T(r,f) := m(r,f) + N(r,f)$ is called *the characteristic function of* f.

The following theorems are known as the first and the second main theorems of Nevanlinna theory.

Theorem 1.1 ([14]) *For any function f meromorphic in the disc $|z| < R \leq \infty$ the equality*

$$m(r,a,f) + N(r,a,f) = T(r,f) + \phi(r,a)$$

holds for each $a \in \overline{\mathbb{C}}$, where $|\phi(r,a)| \leq \log^+ |a| + |\log|c|| + \log 2$ and c is the first non-vanishing coefficient of the Laurent expansion of $f - a$ at zero.

Theorem 1.2 ([14]) *Let $\{a_k\}_{\nu=1}^{k}$ be a finite set of pairwise distinct complex numbers and let f be a meromorphic function. The inequality*

$$m(r,f) + \sum_{\nu=1}^{k} m(r,a_\nu,f) \le 2T(r,f) - (N(r,\frac{1}{f'}) + 2N(r,f) - N(r,f'))$$
$$+O(\log T(r,f) + \log r)$$

holds true for all $r \to \infty$, possibly except for r in a set E of finite linear measure.

In the standard way we define $\delta(a,f)$, the defect of f at a value $a \in \overline{\mathbb{C}}$,

$$\delta(a,f) = \liminf_{r\to\infty} \frac{m(r,a,f)}{T(r,f)} = 1 - \limsup_{r\to\infty} \frac{N(r,a,f)}{T(r,f)}.$$

If $\delta(a,f) > 0$ we say that the value a is defective (in the sense of Nevanlinna). As a result of the first and the second main theorems of Nevanlinna, the set $E_N(f)$ of defective values of a meromorphic function f is at most countable and the following relations are true:

$$0 \le \delta(a,f) \le 1, \qquad \sum_{a\in\overline{\mathbb{C}}} \delta(a,f) \le 2.$$

The order of a meromorphic function f is defined by

$$\varrho(f) := \limsup_{r\to\infty} \frac{\log T(r,f)}{\log r}.$$

We say that $\phi : (0,+\infty) \to \mathbb{R}$ is $S(r,f)$ if

$$\phi(r) = o(T(r,f)) \quad (r \to \infty,\ r \notin E),$$

where E is a set of finite linear measure. For meromorphic functions a, f, we say that *a is small with respect to f* if $T(r,a) = S(r,f)$. Meromorphic functions small with respect to a meromorphic function f are sometimes called *small targets of f*. The Nevanlinna functions for small targets are defined similarly as for constants. Namely, if f is a meromorphic function and a is a meromorphic small target of f, then we put

$$N(r,a,f) := N(r,\frac{1}{f-a}), \quad m(r,a,f) := m(r,\frac{1}{f-a}), \quad \delta(a,f) = \delta(\infty,\frac{1}{f-a}).$$

Attempts to prove the second main theorem for small target functions were initiated by Nevanlinna and his theorem on three small functions. The most important results belong to Yang, who applied the notion of spread [30], Chuang, with the estimate for entire functions [1], Osgood, who applied algebraic methods

[16], Frank and Weissenborn, who applied a Wronskian to obtain the result for rational small targets [4], and Steinmetz with his result for small targets in general [25]. The exact analogue of the second main theorem was finally obtained by Yamanoi in 2004.

Theorem 1.3 ([29]) *Let f be a nonconstant meromorphic function on $\mathbb{C}$ and let $a_1, \ldots, a_q$ be distinct small meromorphic functions of f. Then we have the second main theorem,*

$$(k-2)T(r,f) \leq \sum_{\nu=1}^{k} \overline{N}(r, a_\nu, f) + S(r,f),$$

and the defect relation,

$$\sum_{a \in \mathcal{S}(f)} (\delta(a,f) + \theta(a,f)) \leq 2,$$

where $\theta(a,f) := \liminf_{r\to\infty} \frac{\overline{N}(r,1/(f-a))}{T(r,f)}$ *and $S(f)$ denotes the set of all small meromorphic functions of f.*

2 Main Results

The second Painlevé equation is given by

$$f'' = 2f^3 + zf + \alpha, \tag{P_2}$$

where α is an arbitrary complex parameter and $f = f(z)$. The solutions of this equation are meromorphic functions in the sense that every local solution has a continuation to a function meromorphic in $\mathbb{C}$. For recent proofs see: Hinkkanen and Laine [10], Steinmetz [26] and Shimomura [23]. The solutions are also known to be of finite order [20, 22, 27]. The deficiencies and ramification indices of transcendental solutions have been estimated both in case of P_2 and other Painlevé equations. The estimates of deficiencies for P_2 are due to Schubart [17] and Schubart and Wittich [18]. In case of transcendental meromorphic solutions of P_2 the inequality in the second main theorem becomes an asymptotic equality (see: [7, 12]). There are also estimates concerning defects from small target functions [19, 21].

In general, the Nevanlinna theory has wide applications in the theory of both linear and nonlinear ordinary differential equations in the complex plane [12]. There are also attempts to study the global behaviour of solutions of nonlinear differential equations (e.g., [8]).

In this paper we are particularly interested in equation P_{34}, also called equation XXXIV in [11, Chap. 14]. Equation P_{34} is the second order nonlinear ordinary differential equation of the form

$$f'' = \frac{(f')^2}{2f} + Bf(2f - z) - \frac{A}{2f}, \tag{1}$$

where A and B are fixed complex parameters. We shall omit the case $B = 0$ due to the integrability of the equation (the solution is polynomial).

Equation (1) is related to the second Painlevé equation P_2 in the following way (see: [5]). Equation P_2 admits a Hamiltonian formulation [15]. If

$$H = 1/2f^2 - (g^2 + z/2)f - (\alpha + 1/2)g$$

is a Hamiltonian of the system

$$\frac{dg}{dz} = \frac{\partial H}{\partial f}, \quad \frac{df}{dz} = -\frac{\partial H}{\partial g},$$

then by eliminating the function

$$f = (z + 2g^2 + 2g')/2, \tag{2}$$

the function g satisfies P_2. By eliminating

$$g = (2f' - 2\alpha - 1)/(4f) \tag{3}$$

between these equations, it is easy to show that the function f satisfies Eq. (1) with $A = (2\alpha + 1)^2/4$, $B = 1$. Clearly, since solutions of P_2 are meromorphic in $\mathbb{C}$, then the solutions of $P_{34}(A, 1)$ are also meromorphic.

Equation P_{34} has a scaling symmetry: if $f(z)$ is a solution of P_{34} with the parameters A, 1, then $B^{-1/3}f(B^{1/3}z)$ is a solution of P_{34} with the parameters A and B. Thus it follows from the relationship with P_2 that for any choice of the parameters A, B the solutions of $P_{34}(A, B)$ are meromorphic and of finite order of growth.

Theorem 2.1 *The solutions of the equation P_{34} are meromorphic functions of finite order of growth.*

The order of a solution of the second Painlevé equation is less or equal to 3 (see: [20, 22, 27]). Thus we may formulate the following statement.

Corollary 2.2 *If f is a solution of P_{34}, then $\varrho(f) \leq 3$.*

Let us now present some value distribution properties of the solutions of P_{34}. First, we reformulate the results concerning the distribution of a-points ($a \in \overline{\mathbb{C}}$) of a transcendental solution of (1) which we proved in [2]. Here, by Theorem 2.1, we take into account the finiteness of order of meromorphic solutions of the equation.

Theorem 2.3 ([2]) *Transcendental meromorphic solutions of $P_{34}(A, B)$ satisfy the following conditions:*

1. $m(r,f) = O(\log r)$ *and* $\delta(\infty,f) = 0$;
2. $m(r, 1/(f-a)) = O(\log r)$ *and* $\delta(a,f) = 0$ *for all* $a \in \mathbb{C} \setminus \{0\}$;
3. *if* $A \neq 0$, *then* $m(r, 1/f) = O(\log r)$ *and* $\delta(0,f) = 0$;
4. *if* $A = 0$, *then* $m(r, 1/f) \leq \frac{1}{2}T(r,f) + O(\log r)$ *and* $\delta(0,f) \leq \frac{1}{2}$.

From point 1 of Theorem 2.3 we easily get the following corollary.

Corollary 2.4 *The equation P_{34} does not admit transcendental entire solutions.*

We remark that an arbitrary solution of $P_{34}(A, B)$, $B \neq 0$, has double poles. Around a pole $z = z_0$ we have the following Laurent series expansion with $\xi = z - z_0$:

$$f(z) = \frac{2}{B\xi^2} + \frac{z_0}{3} + \frac{\xi}{2} + a_2\xi^2 + \frac{Bz_0\xi^3}{18} + \dots, \tag{4}$$

where a_2 is arbitrary. Around a zero $z = z_0$ we have

$$f(z) = \varepsilon\sqrt{A}(z - z_0) + a_2(z - z_0)^2 + \dots, \quad \varepsilon^2 = 1, \tag{5}$$

where a_2 is an arbitrary constant.

The second main theorem for transcendental meromorphic solutions of P_{34}, similarly to the second and fourth Painlevé equations, has the form of an asymptotic equality.

Theorem 2.5 *Let f be an arbitrary transcendental meromorphic solution of $P_{34}(A, B)$, $A, B \in \mathbb{C}$. Then*

$$m(r, 1/f) + N(r, 1/f') + N_1(r,f) = 2T(r,f) + O(\log r) \qquad (r \to \infty).$$

If, in addition, $A \neq 0$, then

$$N(r, 1/f') + N_1(r,f) = 2T(r,f) + O(\log r) \qquad (r \to \infty).$$

The following result gives estimates of defects of a transcendental meromorphic solution of P_{34} with respect to small target functions.

Theorem 2.6 *Let a and f be arbitrary transcendental meromorphic solutions of $P_{34}(A, B)$, $A, B \in \mathbb{C}$, such that $T(r, a) = S(r,f)$. Then*

$$m(r, \frac{1}{f-a}) \leq \frac{1}{2}T(r,f) + O(\log r + T(r,a)) \qquad \text{and} \qquad \delta(a,f) \leq \frac{1}{2}.$$

If a does not solve P_{34} and $T(r, a) = S(r,f)$, where f is an arbitrary transcendental meromorphic solutions of $P_{34}(A, B)$, then

$$m(r, \frac{1}{f-a}) = O(\log r + T(r,a)) \qquad \text{and} \qquad \delta(a,f) = 0.$$

3 Auxiliary Results

Before proving the main results, we formulate several necessary well-known theorems. The lemma on the logarithmic derivative is as follows.

Lemma 3.1 ([6, Chap. 3]) *If f is a meromorphic function, then*

$$m(r, \frac{f'}{f}) = O(\log rT(r,f)) \qquad (r \to \infty)$$

possibly apart from r in a set $E \subset (0, \infty)$ of finite linear measure. For functions of finite order the following equality holds:

$$m(r, \frac{f'}{f}) = O(\log r) \qquad (r \to \infty).$$

Lemma 3.2 ([28]) *Let f be a meromorphic function in $\mathbb{C}$ of order ϱ. Then its derivative f' is also of order ϱ.*

In the proofs of the main results we shall use the following version of the original lemma by Clunie (see: [3]).

Lemma 3.3 ([7, App. B], [12, Chap. 2]) *Let f be a transcendental meromorphic function of finite order such that*

$$f^{p+1} = Q(z,f), \; p \in \mathbb{N},$$

where $Q(z, u)$ is a polynomial in u and its derivatives with meromorphic coefficients b_μ ($\mu \in M$). If the total degree of $Q(z, u)$ as a polynomial in u and its derivatives does not exceed p, then

$$m(r,f) = O(\sum_{\mu \in M} m(r, b_\mu)) + O(\log r) \qquad (r \to \infty).$$

Let us now recall the well-known result of A.Z. Mohon'ko and V.D. Mohon'ko.

Lemma 3.4 ([13], [7, App. B]) *Let $F(z, u)$ be a polynomial in u and its derivatives with meromorphic coefficients b_μ ($\mu \in M$). Suppose that f is a transcendental meromorphic function of finite order such that $F(z,f) = 0$ and let $c \in \mathbb{C}$. If $F(z, c) \not\equiv 0$, then*

$$m(r, \frac{1}{f-c}) = O(\sum_{\mu \in M} T(r, b_\mu)) + O(\log r) \qquad (r \to \infty).$$

Lemma 3.5 ([6, Chap. 1]) *Let f be a meromorphic function and $f_1(z) = f(Az)$, where $A \neq 0$. Then*

$$m(r,f_1) = m(|A|r,f),$$
$$N(r,f_1) = N(|A|r,f) - n(r,0,f)\ln|A|,$$
$$T(r,f_1) = T(|A|r,f) - n(r,0,f)\ln|A|,$$

where $n(r, 0, f)$ is the multiplicity of a pole at zero.

Lemma 3.6 ([7, Chap. 3]) *Let f be an arbitrary transcendental meromorphic function satisfying*

$$m(r,f') = O(\log r) \quad (r \to \infty) \qquad \text{and} \qquad m(r, \frac{f}{f'}) = O(\log r) \quad (r \to \infty).$$

Then

$$m(r, 1/f) + N(r, 1/f') = N(r,f') + O(\log r).$$

4 Proofs of the Main Results

Proof of Theorem 2.1

Let g be a transcendental meromorphic solution of $P_2(\alpha)$ ($\alpha \in \mathbb{C}$). By (2), the function

$$f = g^2 + g' + \frac{z}{2}$$

is then a transcendental meromorphic solution of $P_{34}(A, 1)$, where

$$A = \frac{(2\alpha + 1)^2}{4}.$$

Applying elementary properties of the characteristic function of the sum and product of functions, properties of the counting function, and the estimate of growth of rational functions, we get

$$\begin{aligned} T(r,f) &\le T(r,g^2) + T(r,g') + T(r,\frac{z}{2}) + O(1) \\ &\le 2T(r,g) + m(r,g') + N(r,g') + O(\log r) \\ &\le 2T(r,g) + m(r,g) + m(r,\frac{g'}{g}) + 2N(r,g) + O(\log r). \end{aligned}$$

Now, as g is of finite order, by Lemma 3.1 we obtain

$$T(r,f) \le 4T(r,g) + O(\log r) \qquad (r \to \infty).$$

It follows that the order of f is also finite and fulfills the inequality

$$\varrho(f) \le \varrho(g). \tag{6}$$

By (3) and (2), each solution f of the equation $P_{34}(A,1)$ is of the form

$$f = g^2 + g' + \frac{z}{2},$$

with g being a solution of P_2. Thus all solutions of $P_{34}(A,1)$ fulfill the inequality (6) for a certain solution g of P_2.

Let us now consider the equation $P_{34}(A,B)$. If $f(z)$ is a solution of this equation, then, by the scaling symmetry, $f_1(z) = B^{1/3}f(B^{-1/3}z)$ is a solution of $P_{34}(A,1)$. Thus f_1 is of finite order. By Lemma 3.5

$$\varrho(f) = \varrho(f_1)$$

and the function f is also of finite order.

Proof of Theorem 2.5

For the sake of completeness we begin with showing the points 1 and 3 of Theorem 2.3.

1. Let f be a transcendental meromorphic solution of the equation $P_{34}(A,B)$. We write the equation in the form

$$f^{p+1} = Q(z,f),$$

where $p = 2$ and

$$Q(z,f) = \frac{1}{2B}ff'' - \frac{1}{4B}(f')^2 + \frac{1}{2}zf^2 + \frac{A}{4B}.$$

We can see that the coefficients of $Q(z,f)$ are constant or polynomial and the degree of $Q(z,f)$ with respect to f and its derivatives equals 2. Thus, since f is of finite order, the conditions of Theorem 3.3 are fulfilled, so we obtain

$$m(r,f) = O(\log r) \qquad (r \to \infty).$$

3. We put

$$F(z,f) := 2ff'' - (f')^2 - 4Bf^3 + 2Bzf^2 + A. \tag{7}$$

The equality $F(z,0) \equiv 0$ is possible only if $A = 0$. The coefficients of $F(z,f)$ are constant or polynomial. Thus, if $A \neq 0$, then by Lemma 3.4 we get

$$m(r,\frac{1}{f}) = O(\log r) \qquad (r \to \infty).$$

The proof of Theorem 2.5 now follows similar lines as the proof of Theorem 11.3 in [7]. Let $A, B \in \mathbb{C}$ be fixed and let us put the equation $P_{34}(A,B)$ in the form $F(z,f) \equiv 0$, where $F(z,f)$ is as defined in (7).

Computing $\frac{F'}{ff'}$, we obtain

$$\frac{f}{f'} = 6f - \frac{1}{B}\frac{f'''}{f'} - 2z.$$

Applying the properties of the mean proximity function we get

$$\begin{aligned} m(r,\frac{f}{f'}) &\leq m(r,f) + m(r,\frac{f'''}{f'}) + O(\log r) \\ &\leq m(r,f) + m(r,\frac{f'''}{f''}) + m(r,\frac{f''}{f'}) + O(\log r). \end{aligned}$$

As f is of finite order, it follows from Lemma 3.2 that f' and f'' are of finite order as well. Applying point 1 of Theorem 2.3 to f and Lemma 3.1 to f' and f'' we get,

$$m(r,\frac{f}{f'}) = O(\log r) \qquad (r \to \infty).$$

Moreover, by point 1 of Theorem 2.3 and Lemma 3.1 again,

$$m(r,f') \leq m(r,\frac{f'}{f}) + m(r,f) = O(\log r).$$

Thus, if f is a transcendental meromorphic solution of $P_{34}(A,B)$, the conditions of Lemma 3.6 are fulfilled, so we have

$$m(r,1/f) + N(r,1/f') = N(r,f') + O(\log r).$$

By the expansion (4) a solution f of P_{34} has only double poles, so

$$N_1(r,f) = N(r,f) - \overline{N}(r,f) = \frac{1}{2}N(r,f),$$

$$N(r,f') = \frac{3}{2}N(r,f).$$

This way we obtain the equality

$$\begin{aligned} m(r,1/f) + N(r,1/f') + N_1(r,f) &= N(r,f') + N_1(r,f) + O(\log r) \\ &= 2T(r,f) + O(\log r) \qquad (r \to \infty). \end{aligned}$$

It remains to notice that, if $A \neq 0$, by point 3 of Theorem 2.3 for a transcendental solution of $P_{34}(A,B)$

$$m(r,\frac{1}{f}) = O(\log r) \qquad (r \to \infty).$$

Proof of Theorem 2.6

The proof follows similar lines as in [21] (see also [24]). Let f be a transcendental meromorphic solution of $P_{34}(A,B)$ and let a be a meromorphic function such that $T(r,a) = S(r,f)$. We put $g(z) := f(z) - a(z)$. Then, as the equality (1) holds for f, we have for g

$$\begin{aligned} g(z)^3 = &\frac{1}{4B}\{2(g(z)+a(z))g''(z) - g'(z)^2 - 2a'(z)g'(z) + (2Bz - 12Ba(z))g(z)^2 \\ &+ (2a''(z) - 12Ba^2(z) + 4Bza(z))g(z) + F(z,a)\}, \end{aligned} \tag{8}$$

where F is defined as in (7).

By elementary properties of the mean proximity function and the fact that a is of finite order,

$$m(r, a^n) = O(m(r, a)), \qquad m(r, a^{(n)}) = O(m(r, a)) + O(\log r) \qquad (r \to \infty).$$

If we put

$$\begin{aligned} Q(z, g) =& \frac{1}{4B}\{2(g(z) + a(z))g''(z) - g'(z)^2 - 2a'(z)g'(z) + (2Bz - 12Ba(z))g(z)^2 \\ &+ (2a''(z) - 12Ba^2(z) + 4Bza(z))g(z) + F(z, a)\}, \end{aligned}$$

then by Lemma 3.3 we obtain

$$m(r, f - a) = m(r, g) = O(\log r + m(r, a)). \tag{9}$$

Let us now assume that $F(z, a) \not\equiv 0$. Then if we put a constant $c = 0$ in place of $g(z)$ we can see that c does not solve Eq. (8). Thus, by Lemma 3.4, we get

$$m(r, \frac{1}{f - a}) = m(r, \frac{1}{g}) = O(\log r + T(r, a)) \qquad (r \to \infty). \tag{10}$$

Let now $F(z, a) \equiv 0$. It means that $a(z)$ is a solution of P_{34}. In this case we consider an auxiliary function

$$G(z) := \frac{(g'(z) + a'(z))^2}{(g(z) + a(z))} - 2Bg(z)^2 + 2B(z - 2a(z))g(z) + \frac{Ag(z)}{a(z)(g(z) + a(z))} - \frac{a'(z)^2}{a(z)}.$$

Then we differentiate and apply the fact that both $g(z) + a(z)$ and $a(z)$ are solutions of (1) to obtain

$$G'(z) = 2Bg(z). \tag{11}$$

We put

$$H(z) := \frac{G(z)}{g(z)}. \tag{12}$$

Then we replace $(g'(z) + a(z))^2/(g(z) + a(z))$ and $a''(z)$ using (1) and get

$$H(z) = 2\frac{g''(z)}{g(z)} - 6Bg(z) - 12Ba(z) + 4Bz.$$

By elementary properties of the mean proximity function,

$$m(r, H) \leq m(r, \frac{g''}{g}) + m(r, g) + m(r, a) + O(\log r).$$

As f is of finite order and a is small with respect to f, also g is of finite order. It follows from Lemma 3.1, that

$$m(r, \frac{g''}{g}) \leq m(r, \frac{g''}{g'}) + m(r, \frac{g'}{g}) = O(\log r) \qquad (r \to \infty).$$

Moreover, by (9) $m(r, g) = O(\log r + m(r, a))$. This way we obtain

$$m(r, H) = O(\log r + m(r, a)) \qquad (r \to \infty). \tag{13}$$

From the expansion (4) we know that all the poles of f and a, as the solutions of P_{34}, are double, so all the poles of g are also double. Also the expansion (4) shows, the points where both f and a have poles are not the poles of g. Thus we have poles of g belonging to two sets:

$$P = \{z : f(z) = \infty, \ a(z) \neq \infty\} \qquad \text{or} \qquad P' = \{z : f(z) \neq \infty, \ a(z) = \infty\}.$$

We put

$$N_P(r, g) := \int_0^r [n_P(t, g) - n_P(0, g)] \frac{dt}{t} + n_P(0, g) \log r,$$

where $n_P(t, g)$ is the number of poles of g in $\{z : \ |z| \leq t\} \cap P$, counted with multiplicity and $n_P(0, g)$ is the multiplicity of a pole of g at zero, if $0 \in P$. We also put

$$N_{P'}(r, g) := \int_0^r [n_{P'}(t, g) - n_{P'}(0, g)] \frac{dt}{t} + n_{P'}(0, g) \log r,$$

where $n_{P'}(t, g)$ is the number of poles of g in $\{z : \ |z| \leq t\} \cap P'$, counted with multiplicity and $n_{P'}(0, f)$ is the multiplicity of a pole at zero, if $0 \in P'$. It follows that

$$N(r, g) = N_P(r, g) + N_{P'}(r, g) \leq N_P(r, g) + T(r, a). \tag{14}$$

Moreover,

$$\frac{g'(z)}{g(z)} H(z) + H'(z) = \frac{G'(z)}{g(z)} = 2B.$$

Thus, if z_0 is a pole of g and $z_0 \in P$, then H is analytic at z_0 with the expansion

$$H(z) = -2B(z - z_0) + \sum_{n=2}^{\infty} b_n(z - z_0)^n \quad (b_n \in \mathbb{C}).$$

Notice that, by definition (12), if $H(z) \equiv 0$, then also $G(z) \equiv 0$. Thus, by (11), it means that $g(z) \equiv 0$, which contradicts the assumption that g is transcendental. Then, as the zeros of H at the poles of g are simple,

$$\frac{1}{2}N_P(r, g) \le N(r, \frac{1}{H}) \le T(r, H) + O(1). \tag{15}$$

By definition (12), the poles of H may appear only at the poles of G or the zeros of g. From (11) we can see that the poles of G may appear only at the poles of g. As H has zeros at the poles of g, we need only to take into account the poles of H at the zeros of g. The function g has a zero at a point z_0 in three cases: if $f(z_0) = a(z_0) \neq 0, \infty$, if $f(z_0) = a(z_0) = 0$ or, by the expansion (4), if both f and a have a pole at z_0. We put

$$S := \{z \in \mathbb{C} : f(z) = a(z) = c,\ c \in \mathbb{C} \setminus \{0\}\},$$

$$S' := \{z \in \mathbb{C} : f(z) = a(z) = 0\},$$

$$S'' := \{z \in \mathbb{C} : f(z) = a(z) = \infty\}.$$

We also put

$$N_S(r, \frac{1}{g}) := \int_0^r [n_S(t, \frac{1}{g}) - n_S(0, \frac{1}{g})]\frac{dt}{t} + n_S(0, \frac{1}{g}) \log r,$$

where $n_P(t, \frac{1}{g})$ is the number of zeros of g in $\{z : |z| \le t\} \cap S$, counted with multiplicity and $n_S(0, \frac{1}{g})$ is the multiplicity of a zero of g at zero, if $0 \in S$. Similarly, we define $N_{S'}(r, \frac{1}{g})$ and $N_{S''}(r, \frac{1}{g})$. This way we have

$$N(r, \frac{1}{g}) = N_S(r, \frac{1}{g}) + N_{S'}(r, \frac{1}{g}) + N_{S''}(r, \frac{1}{g}).$$

It is easy to notice that by (4) and (5),

$$N_{S'}(r, \frac{1}{g}) = O(T(r, a)) \quad \text{and} \quad N_{S''}(r, \frac{1}{g}) = O(T(r, a)).$$

Thus we have

$$N(r, \frac{1}{g}) = N_S(r, \frac{1}{g}) + O(T(r, a)).$$

As f and a are solutions of P_{34}, by simple computations we get that unless $f \equiv a$, which cannot be as $T(r, a) = S(r, f)$, the zeros of $f - a$ in S must be simple. Indeed, assuming that $f(z_0) = a(z_0) = c \neq 0$ and $f'(z_0) = a'(z_0)$, by (1) we get

$$2cf''(z_0) = (f'(z_0))^2 + 4Bc^3 - 2Bz_0c^2 - A$$

and

$$2ca''(z_0) = (a'(z_0))^2 + 4Bc^3 - 2Bz_0c - A.$$

Subtracting the equations we get

$$2c(f''(z_0) - a''(z_0)) = (f'(z_0))^2 - (a'(z_0))^2,$$

which means that, by our assumption, also $f''(z_0) = a''(z_0)$. Next, we differentiate (1) to get

$$2ff''' = 12Bf^2f' - 4Bzff' - 2Bf^2.$$

By assumption, at z_0 we have

$$2cf'''(z_0) = 12Bc^2f'(z_0) - 4Bz_0cf'(z_0) - 2Bc^2$$

and

$$2ca'''(z_0) = 12Bc^2a'(z_0) - 4Bz_0ca'(z_0) - 2Bc^2,$$

which, after subtracting the equations, shows that $f'''(z_0) = a'''(z_0)$. The equality of the higher order derivatives of f and a at z_0 follows in a similar way. Thus $f \equiv a$, which is a contradiction. Hence the zeros of $f - a$ in S are indeed simple.

By the definition of G and (11), if $z_0 \in S$, then G is analytic at z_0. Then the pole of H at z_0 must be simple. Thus, applying the first main theorem, we have

$$N(r, H) \leq N_S(r, \frac{1}{g}) + O(T(r, a)) \leq T(r, g) - m(r, \frac{1}{g}) + O(T(r, a)). \tag{16}$$

It follows from (15) and then (16) that

$$\begin{aligned}\frac{1}{2}N_P(r,g) &\leq m(r,H) + N(r,H) + O(1)\\ &\leq m(r,H) + T(r,g) - m(r,\frac{1}{g}) + O(T(r,a)).\end{aligned}$$

By (14) we have

$$\begin{aligned}\frac{1}{2}N(r,g) &\leq \frac{1}{2}N_P(r,g) + \frac{1}{2}T(r,a)\\ &\leq m(r,H) + T(r,g) - m(r,\frac{1}{g}) + O(T(r,a)).\end{aligned}$$

Thus, we get

$$\begin{aligned}m(r,\frac{1}{g}) &\leq T(r,g) - \frac{1}{2}N(r,g) + m(r,H) + O(T(r,a))\\ &= \frac{1}{2}T(r,g) + \frac{1}{2}m(r,g) + m(r,H) + O(T(r,a)).\end{aligned}$$

Applying (9), (13) and the fact that $g = f - a$ leads to

$$m(r,\frac{1}{f-a}) \leq \frac{1}{2}T(r,f) + O(\log r + T(r,a)).$$

Together with (10) this completes the proof.

Acknowledgements Galina Filipuk acknowledges the support of the National Science grant 2011/03/B/ST1/00330.

References

1. C.T. Chuang, Une généralisation d'une inégalité de Nevanlinna. Sci. Sinica **13**, 887–895 (1964)
2. E. Ciechanowicz, G. Filipuk, Meromorphic solutions of $P_{4,34}$ and their value distribution. Ann. Acad. Sci. Fenn. Math. **41**, 617–638 (2016)
3. J. Clunie, On integral and meromorphic functions. J. Lond. Math. Soc. **37**, 17–27 (1962)
4. G. Frank, G. Weissenborn, Rational deficient functions of meromorphic functions. Bull. Lond. Math. Soc. **18**, 29–33 (1986)
5. B. Gambier, Sur les équations différentielles du second ordre et du premier degré dont l'intégrale générale est à points critiques fixés. Acta Math. **33**, 1–55 (1909)
6. A.A. Gol'dberg, I.V. Ostrowskii, *Distribution of Values of Meromorphic Functions* (Nauka, Moscow, 1970). (Russian)
7. V. Gromak, I. Laine, S. Shimomura, *Painlevé Differential Equations in the Complex Plane*. De Gruyter Studies in Mathematics, vol. 28 (Walter de Gruyter, Berlin, 2002)

8. R. Halburd, T. Kecker, Local and global finite branching of solutions of ordinary differential equations. Reports and Studies in Forestry and Natural Sciences, vol. 14, University of Eastern Finland, 2014, pp. 57–78
9. W.K. Hayman, *Meromorphic Functions* (Clarendon, Oxford, 1964)
10. A. Hinkkanen, I. Laine, Solutions of the first and second Painlevé equations are meromorphic. Journal d'Analyse Math. **79**, 345–377 (1999)
11. E.L. Ince, *Ordinary Differential Equations* (Longmans Green, London, 1927)
12. I. Laine, *Nevanlinna Theory and Complex Differential Equations* (De Gruyter, Berlin, New York, 1993)
13. A.Z. Mohon'ko, V.D. Mohon'ko, Estimates of the Nevanlinna characteristics of certain classes of meromorphic functions, and their applications to differential equations. Sibirsk. Mat. Zh. **15**, 1305–1322 (1974). (Russian)
14. R. Nevanlinna, *Eindeutige Analytische Funktionen* (Springer, Berlin, 1970)
15. K. Okamoto, Studies on the Painlevé equations. III. Second and fourth Painlevé equations, P_{II} and P_{IV}. Math. Ann. **275**, 221–255 (1986)
16. C.F. Osgood, A number-theoretic-differential equations approach to generalizing Nevanlinna theory. Indian J. Math. **23**, 1–15 (1981)
17. H. Schubart, Zur Wertverteilung der Painlevéschen Transzendenten. Arch. Math. (Basel) **7**, 284–290 (1956)
18. H. Schubart, H. Wittich, Über die Lösungen der ersten beiden Painlevéschen Differentialgleichungen. Math. Z. **66**, 364–370 (1957)
19. S. Shimomura, Value distribution of Painlevé transcendents of the first and the second kind. J. Anal. Math. **82**, 333–346 (2000)
20. S. Shimomura, The first, the second and the fourth Painlevé transcendents are of finite order. Proc. Jpn. Acad. Ser. A **77**, 42–45 (2001)
21. S. Shimomura, On deficiencies of small functions for Painlevé transcendents of the fourth kind. Ann. Acad. Sci. Fenn. Ser A I Math. **27**, 109–120 (2002)
22. S. Shimomura, Growth of the first, the second and the fourth Painlevé transcendents. Math. Proc. Camb. Philos. Soc. **134**(2), 259–269 (2003)
23. S. Shimomura, Proofs of the Painlevé property for all Painlevé equations. Jpn. J. Math. **29**, 159–179 (2003)
24. N. Steinmetz, Zur Wertverteilung der Lösungen der vierten Painlevéschen Differentialgleichung. Math. Z. **181**, 553–561 (1982)
25. N. Steinmetz, Eine Verallgemainerung des zweiten Nevanlinnaschen Hauptsatzes. J. Reine Angew. Math. **368**, 134–141 (1986)
26. N. Steinmetz, On Painlevé equations I, II and IV. J. d'Analyse Math. **82**, 363–377 (2000)
27. N. Steinmetz, Value distribution of the Painlevé transcendents. Israel J. Math. **128**, 29–52 (2002)
28. J.M. Whittaker, The order of the derivative of a meromorphic function. J. Lond. Math. Soc. **11**, 82–87 (1936)
29. K. Yamanoi, The second main theorem for small functions and related problems. Acta Math. **192**, 225–294 (2004)
30. L. Yang, Deficient functions of meromorphic functions. Sci. Sinica **24**, 1179–1189 (1981)

Towards the Convergence of Generalized Power Series Solutions of Algebraic ODEs

Renat R. Gontsov and Irina V. Goryuchkina

Abstract The aim of this work is to provide another proof of the sufficient condition of the convergence of a generalized power series (with complex power exponents) formally satisfying an algebraic (polynomial) ordinary differential equation. This proof is based on the implicit mapping theorem for Banach spaces rather than on the majorant method used in our previous proof. We also discuss some examples of a such type formal solutions of Painlevé equations.

Keywords Algebraic ODE • Convergence • Formal solution • Generalized power series

Mathematics Subject Classification (2000). Primary 34M25; Secondary 34A25

1 Introduction

Let us consider an ordinary differential equation (ODE)

$$F(z, u, \delta u, \ldots, \delta^m u) = 0 \tag{1}$$

of order m with respect to the unknown u, where $F(z, u_0, u_1, \ldots, u_m) \not\equiv 0$ is a polynomial of $m+2$ variables, $\delta = z\frac{d}{dz}$.

R.R. Gontsov (✉)
Institute for Information Transmission Problems of RAS, Bolshoy Karetny per. 19, Moscow 127994, Russia
e-mail: gontsovrr@gmail.com

I.V. Goryuchkina
Keldysh Institute of Applied Mathematics of RAS, Miusskaya sq. 4, Moscow 125047, Russia
e-mail: igoryuchkina@gmail.com

G. Filipuk et al. (eds.), *Analytic, Algebraic and Geometric Aspects of Differential Equations*, Trends in Mathematics, DOI 10.1007/978-3-319-52842-7_7

In the paper we study *generalized* power series solutions of (1) of the form

$$\varphi = \sum_{n=0}^{\infty} c_n z^{s_n}, \qquad c_n \in \mathbb{C}, \qquad s_n \in \mathbb{C}, \tag{2}$$

with the power exponents satisfying conditions

$$0 \leqslant \operatorname{Re} s_0 \leqslant \operatorname{Re} s_1 \leqslant \ldots, \qquad \lim_{n\to\infty} \operatorname{Re} s_n = +\infty$$

(the latter, in particular, implies that a set of exponents having a fixed real part is finite).

Note that substituting the series (2) into Eq. (1) makes sense, as only a finite number of terms in φ contribute to any term of the form cz^s in the expansion of $F(z, \Phi) = F(z, \varphi, \delta\varphi, \ldots, \delta^m\varphi)$ in powers of z. Indeed, $\delta^j\varphi = \sum_{n=0}^{\infty} c_n s_n^j z^{s_n}$, and an equation $s = s_{n_0} + s_{n_1} + \ldots + s_{n_l}$ has a finite number of solutions $(s_{n_0}, s_{n_1}, \ldots, s_{n_l})$, since $0 \leqslant \operatorname{Re} s_n \to +\infty$. Furthermore, for any integer N an inequality $\operatorname{Re}(s_{n_0} + s_{n_1} + \ldots + s_{n_l}) \leqslant N$ has also a finite number of solutions, so that powers of z in the expansion of $F(z, \Phi)$ can be ordered by the increasing of real parts. Thus, one may correctly define the notion of a formal solution of (1) in the form of a generalized power series. In particular, the Painlevé III, V, VI equations are known to have such formal solutions (see [2, 5, 7, 8, 10–13]). Their convergence in sectorial domains near zero is also proved in some of those papers. Here we are interested in convergence for an equation of the general form (1).

There is the following sufficient condition [4] of the convergence of a generalized power series solution of (1).

Theorem 1.1 *Let the generalized power series* (2) *formally satisfy Eq.* (1), $\frac{\partial F}{\partial u_m}(z, \Phi) \neq 0$, *and for each* $i = 0, 1, \ldots, m$ *one have*

$$\frac{\partial F}{\partial u_i}(z, \Phi) = A_i z^{\lambda} + B_i z^{\lambda_i} + \ldots, \qquad \operatorname{Re}\lambda_i > \operatorname{Re}\lambda, \quad A_m \neq 0. \tag{3}$$

Then for any sector S of sufficiently small radius with the vertex at the origin and of the opening less than 2π, the series φ converges uniformly in S.

In this paper we propose a shorter proof of Theorem 1.1 based on the implicit mapping theorem, whereas in the original proof in [4] we used the majorant method. In the case of integer powers $s_n = n \in \mathbb{Z}_+$, Theorem 1.1 was obtained by Malgrange [9], and in the case of real powers $s_n \in \mathbb{R}$ this theorem was formulated in a somewhat different form in [1, Theorem 3.4].

2 Auxiliary Lemmas

The proof of Theorem 1.1 is preceded by some auxiliary lemmas which have been proved in [4].

Lemma 2.1 *Under the assumptions of Theorem 1.1, there exists an integer* $\mu' \geqslant 0$ *such that for any integer* $\mu \geqslant \mu'$ *satisfying* $\mathrm{Re}(s_{\mu+1} - s_\mu) > 0$, *a transformation*

$$u = \sum_{n=0}^{\mu} c_n z^{s_n} + z^{s_\mu} v \tag{4}$$

reduces Eq. (1) *to an equation of the form*

$$L(\delta)v + N(z, v, \delta v, \ldots, \delta^m v) = 0, \tag{5}$$

where

- *L is a polynomial of degree m,*
- $L(s) \neq 0$ *for any s with* $\mathrm{Re}\, s > 0$, *and*
- *N is a finite linear combination of monomials of the form*

$$z^\beta v^{q_0} (\delta v)^{q_1} \ldots (\delta^m v)^{q_m}, \qquad \beta \in \mathbb{C},\ \mathrm{Re}\, \beta > 0,\ q_i \in \mathbb{Z}_+.$$

As follows from the form of the transformation (4), the reduced Eq. (5) has a generalized power series solution $\psi = \sum_{n=\mu+1}^{\infty} c_n z^{s_n - s_\mu}$. The second auxiliary lemma describes a structure of the set of power exponents $s_n - s_\mu \in \mathbb{C}$ of this series.

Let us define an additive semi-group Γ generated by a (finite) set of power exponents of the variable z containing in $N(z, v, \delta v, \ldots, \delta^m v)$, and let $r_1, \ldots, r_l$ be generators of this semi-group, that is,

$$\Gamma = \{m_1 r_1 + \ldots + m_l r_l \mid m_i \in \mathbb{Z}_+, \sum_{i=1}^{l} m_i > 0\}, \qquad \mathrm{Re}\, r_i > 0.$$

Lemma 2.2 *All the numbers* $s_n - s_\mu$, $n \geqslant \mu + 1$, *belong to the semi-group* Γ.

We may assume that the generators $r_1, \ldots, r_l$ of Γ are linearly independent over $\mathbb{Z}$. This is provided by the following lemma.

Lemma 2.3 *There are complex numbers* $\rho_1, \ldots, \rho_\tau$ *linearly independent over* $\mathbb{Z}$, *such that all* $\mathrm{Re}\, \rho_i > 0$, *and an additive semi-group* Γ' *generated by them contains the above semi-group* Γ *generated by* $r_1, \ldots, r_l$.

3 Proof of Theorem 1.1

For the simplicity of exposition we assume that the semi-group Γ is generated by two numbers:

$$\Gamma = \{m_1r_1 + m_2r_2 \mid m_1, m_2 \in \mathbb{Z}_+,\ m_1 + m_2 > 0\}, \qquad \operatorname{Re} r_1,\ \operatorname{Re} r_2 > 0.$$

In the case of an arbitrary number l of generators all constructions are analogous, only multivariate Taylor series in l rather than in two variables are involved.

We should establish the convergence of the generalized power series

$$\psi = \sum_{n=\mu+1}^{\infty} c_n z^{s_n - s_\mu},$$

which satisfies the equality

$$L(\delta)\psi + N(z, \psi, \delta\psi, \dots, \delta^m\psi) = 0. \tag{6}$$

According to Lemma 2.2, all the exponents $s_n - s_\mu$ belong to the semi-group Γ:

$$s_n - s_\mu = m_1r_1 + m_2r_2, \qquad (m_1, m_2) \in M \subseteq \mathbb{Z}_+^2 \setminus \{0\},$$

for some set M such that the map $n \mapsto (m_1, m_2)$ is a bijection from $\mathbb{N} \setminus \{1, \dots, \mu\}$ to M. Hence,

$$\psi = \sum_{(m_1,m_2)\in M} c_{m_1,m_2} z^{m_1r_1+m_2r_2} = \sum_{(m_1,m_2)\in\mathbb{Z}_+^2\setminus\{0\}} c_{m_1,m_2} z^{m_1r_1+m_2r_2}$$

(in the last series one puts $c_{m_1,m_2} = 0$, if $(m_1, m_2) \notin M$).

Now we define a natural linear map $\sigma : \mathbb{C}[[z^\Gamma]] \to \mathbb{C}[[z_1, z_2]]_*$ from the $\mathbb{C}$-algebra of generalized power series with exponents in Γ to the $\mathbb{C}$-algebra of Taylor series in two variables without a constant term,

$$\sigma : \sum_{\gamma=m_1r_1+m_2r_2\in\Gamma} a_\gamma z^\gamma \mapsto \sum_{\gamma=m_1r_1+m_2r_2\in\Gamma} a_\gamma z_1^{m_1} z_2^{m_2}.$$

As follows from the linear independence of the generators r_1, r_2 over $\mathbb{Z}$,

$$\sigma(\eta_1\eta_2) = \sigma(\eta_1)\sigma(\eta_2) \qquad \forall \eta_1, \eta_2 \in \mathbb{C}[[z^\Gamma]],$$

hence σ is an isomorphism. The differentiation $\delta : \mathbb{C}[[z^\Gamma]] \to \mathbb{C}[[z^\Gamma]]$ naturally induces a linear bijective map Δ of $\mathbb{C}[[z_1, z_2]]_*$ to itself,

$$\Delta : \sum_{\gamma\in\Gamma} a_\gamma z_1^{m_1} z_2^{m_2} \mapsto \sum_{\gamma\in\Gamma} \gamma\, a_\gamma z_1^{m_1} z_2^{m_2},$$

which clearly satisfies $\Delta \circ \sigma = \sigma \circ \delta$, so that the following commutative diagram holds:

$$\begin{array}{ccc} \mathbb{C}[[z^\Gamma]] & \xrightarrow{\delta} & \mathbb{C}[[z^\Gamma]] \\ \downarrow \sigma & & \downarrow \sigma \\ \mathbb{C}[[z_1, z_2]]_* & \xrightarrow{\Delta} & \mathbb{C}[[z_1, z_2]]_* \end{array}$$

Thus we have the representation

$$\tilde{\psi} = \sigma(\psi) = \sum_{\gamma\in\Gamma} c_\gamma z_1^{m_1} z_2^{m_2}$$

of the formal solution ψ of (5) by a multivariate Taylor series, where $c_\gamma = c_{m_1,m_2}$ for every $\gamma = m_1 r_1 + m_2 r_2$. Now we apply the map σ to the both sides of the equality (6) and obtain a relation for $\tilde{\psi}$:

$$L(\Delta)\tilde{\psi} + \widetilde{N}(z_1, z_2, \tilde{\psi}, \Delta\tilde{\psi}, \dots, \Delta^m\tilde{\psi}) = 0, \tag{7}$$

where $\widetilde{N}(z_1, z_2, u_0, \dots, u_m)$ is a polynomial such that $\widetilde{N}(0, 0, u_0, \dots, u_m) \equiv 0$.

We conclude the proof of Theorem 1.1 establishing the convergence of the bivariate Taylor series $\tilde{\psi}$, which represents the generalized power series ψ and satisfies the relation (7). We use the dilatation method based on the implicit mapping theorem for Banach spaces. This was originally used by Malgrange [9] for proving Theorem 1.1 in the case of integer powers $s_n = n \in \mathbb{Z}_+$.

Let us define the following Banach spaces H^j of (formal) Taylor series in two variables without a constant term:

$$H^j = \Big\{\eta = \sum_{\gamma\in\Gamma} a_\gamma z_1^{m_1} z_2^{m_2} \mid \sum_{\gamma\in\Gamma} |\gamma|^j\, |a_\gamma| < +\infty\Big\}, \qquad j = 0, 1, \dots, m,$$

with the norm

$$\|\eta\|_j = \sum_{\gamma\in\Gamma} |\gamma|^j\, |a_\gamma| = \|\Delta^j \eta\|_0.$$

(The completeness of each H^j is checked in a way similar to that how one checks the completeness of the space l_2; see, for example, [3, Chap. 6, Sect. 4].) One clearly has $H^m \subset H^{m-1} \ldots \subset H^0$ and

$$\Delta : H^j \to H^{j-1}, \qquad j = 1, \ldots, m,$$

are continuous linear mappings.

We recall below the implicit mapping theorem for Banach spaces (see [3, Theorem 10.2.1]).

Let $\mathcal{E}$, $\mathcal{F}$, $\mathcal{G}$ be Banach spaces, A an open subset of the direct product $\mathcal{E} \times \mathcal{F}$, and $h : A \to \mathcal{G}$ a continuously differentiable mapping. Consider a point $(x_0, y_0) \in A$ such that $h(x_0, y_0) = 0$ and $\frac{\partial h}{\partial y}(x_0, y_0)$ is a bijective linear mapping from $\mathcal{F}$ to $\mathcal{G}$.

Then there are a neighbourhood $U_0 \subset \mathcal{E}$ of the point x_0 and a unique continuous mapping $g : U_0 \to \mathcal{F}$ such that $g(x_0) = y_0$, $(x, g(x)) \in A$, and $h(x, g(x)) = 0$ for any $x \in U_0$.

We will apply this theorem to the Banach spaces $\mathbb{C}$, H^m, H^0, and to the mapping $h : \mathbb{C} \times H^m \to H^0$ defined by

$$h : (\lambda, \eta) \mapsto L(\Delta)\eta + \widetilde{N}(\lambda z_1, \lambda z_2,\ \eta, \Delta\eta, \ldots, \Delta^m \eta),$$

with L and $\widetilde{N}$ coming from (7). This mapping is continuously differentiable, moreover $h(0,0) = 0$, and $\frac{\partial h}{\partial \eta}(0,0) = L(\Delta)$ is a bijective linear mapping from H^m to H^0. Indeed,

$$L(\Delta) : a_\gamma z_1^{m_1} z_2^{m_2} \mapsto a_\gamma L(\gamma) z_1^{m_1} z_2^{m_2} \ (= 0 \Longleftrightarrow a_\gamma = 0),$$

therefore $\ker L(\Delta) = \{0\}$ (recall that $L(\gamma) \neq 0$ for any γ with $\operatorname{Re}\gamma > 0$). In the same time, if $\sum_{\gamma \in \Gamma} a_\gamma z_1^{m_1} z_2^{m_2} \in H^0$, then $\sum_{\gamma \in \Gamma} (a_\gamma / L(\gamma))\, z_1^{m_1} z_2^{m_2} \in H^m$, that is, the image of $L(\Delta)$ coincides with H^0.

Hence, by the implicit mapping theorem, there are a real number $\rho > 0$ and $\eta_\rho \in H^m$ such that

$$L(\Delta)\eta_\rho + \widetilde{N}(\rho z_1, \rho z_2,\ \eta_\rho, \Delta\eta_\rho \ldots, \Delta^m \eta_\rho) = 0.$$

Making the change of variables $(z_1, z_2) \mapsto (\frac{z_1}{\rho}, \frac{z_2}{\rho})$, which induces an automorphism $\eta(z_1, z_2) \mapsto \eta(\frac{z_1}{\rho}, \frac{z_2}{\rho})$ of $\mathbb{C}[[z_1, z_2]]_*$ commuting with Δ, one can easily see that the above relation implies that the power series $\eta_\rho(\frac{z_1}{\rho}, \frac{z_2}{\rho})$ satisfies the same equality (7) as $\tilde{\psi} = \sum_{\gamma \in \Gamma} c_\gamma z_1^{m_1} z_2^{m_2}$ does. Hence, these two series coincide (the coefficients of a series satisfying (7) are determined uniquely by this equality) and $\tilde{\psi}$ has a non-zero radius of convergence. This implies (substitute $z_1 = z^{r_1}$, $z_2 = z^{r_2}$ remembering that $\operatorname{Re} r_1, \operatorname{Re} r_2 > 0$) the convergence of the series

$$\sum_{\gamma \in \Gamma} c_\gamma z^\gamma = \sum_{n=\mu+1}^{\infty} c_n z^{s_n - s_\mu}$$

for any z from a sector S of sufficiently small radius with the vertex at the origin and of the opening less than 2π, whence Theorem 1.1 follows.

4 Examples

As was mentioned in Introduction, for the Painlevé III, V, VI equations, their generalized power series solutions of the form (2) converge in some sectorial domains near the origin. This is proved in each case mainly by using a kind of majorant series. Here we give examples of such formal solutions and illustrate how Theorem 1.1 can be applied to prove their convergence.

Let us consider the Painlevé III equation with the parameters $a = b = 0$, $c = d = 1$:

$$\frac{d^2u}{dz^2} = \frac{1}{u}\left(\frac{du}{dz}\right)^2 - \frac{1}{z}\frac{du}{dz} + u^3 + \frac{1}{u}.$$

Rewritten in the form (1), this becomes

$$u\,\delta^2 u - (\delta u)^2 - z^2u^4 - z^2 = 0 \quad \text{or} \quad F(z, u, \delta u, \delta^2 u) = 0, \tag{8}$$

where

$$F(z, u_0, u_1, u_2) = u_0u_2 - u_1^2 - z^2(u_0^4 + 1).$$

As known [5, 12], Eq. (8) has a two-parametric family of formal solutions

$$\varphi = c_r z^r + \sum_{s\in \mathrm{K}} c_s z^s, \tag{9}$$

where $c_r \neq 0$ is an arbitrary complex number and r is any complex number with $-1 \leqslant \operatorname{Re} r \leqslant 1$. The other coefficients c_s are determined uniquely by c_r, and the set K of power exponents is of the form

$$\mathrm{K} = \{r + m_1(1 - r) + m_2(1 + r) \mid m_1, m_2 \in \mathbb{Z}_+, m_1 + m_2 > 0\}.$$

Denote $r = \rho + i\sigma$. There are two essentially different types of formal solutions in the family above.

1) Solutions with $\rho \in (-1, 1)$. For any solution of such type there is only a finite number of exponents $s = r + m_1(1 - r) + m_2(1 + r)$ with a fixed real part $\operatorname{Re} s = \rho + m_1(1-\rho) + m_2(1+\rho)$, since $1-\rho$ and $1+\rho$ are positive. Therefore, such solutions are of the form (2), and we will apply Theorem 1.1 to prove their convergence.

2) Solutions with $\rho = \pm 1$. For any solution of such type there are infinitely many exponents $s = r + m_1(1-r) + m_2(1+r)$ with a fixed real part $\mathrm{Re}\, s = \rho + m_1(1-\rho) + m_2(1+\rho)$, since the latter depends only on one of the two indexes m_1, m_2. Therefore, such solutions are not generalized power series of the form (2), and Theorem 1.1 cannot be used for studying their convergence. In fact, such series diverge along some rays coming to the origin, which will be explained below.

Let us consider a formal solution φ of the **first type**. To prove the convergence of φ in sectors of small radius, it is sufficient to find the partial derivatives $\frac{\partial F}{\partial u_0}$, $\frac{\partial F}{\partial u_1}$, $\frac{\partial F}{\partial u_2}$ along φ and verify the assumption (3). Note that in this case $\mathrm{Re}\, s > \rho$ for any $s \in \mathrm{K}$. One has

$$\frac{\partial F}{\partial u_0} = u_2 - 4z^2 u_0^3, \qquad \frac{\partial F}{\partial u_1} = -2u_1, \qquad \frac{\partial F}{\partial u_2} = u_0.$$

Hence,

$$\frac{\partial F}{\partial u_2}(z, \Phi) = \varphi = c_r z^r + \sum_{s \in \mathrm{K}} c_s z^s, \qquad c_r \neq 0, \quad \mathrm{Re}\, s > \rho \;\; \forall s \in \mathrm{K},$$

$$\frac{\partial F}{\partial u_1}(z, \Phi) = -2\,\delta\varphi = -2r\, c_r z^r - \sum_{s \in \mathrm{K}} 2s\, c_s z^s, \;\; \mathrm{Re}\, s > \rho \;\; \forall s \in \mathrm{K},$$

$$\frac{\partial F}{\partial u_0}(z, \Phi) = \delta^2 \varphi - 4z^2 \varphi^3.$$

To prove the convergence of φ, it is remaining to find first terms of the expansion of $\frac{\partial F}{\partial u_0}(z, \Phi)$. Since

$$\delta^2 \varphi = r^2 c_r z^r + \sum_{s \in \mathrm{K}} s^2 c_s z^s, \qquad \mathrm{Re}\, s > \rho \;\; \forall s \in \mathrm{K},$$

$$z^2 \varphi^3 = c_r^3 z^{3r+2} + \ldots, \qquad 3\rho + 2 > \rho \; (\text{as } -1 < \rho < 1),$$

one finally has

$$\frac{\partial F}{\partial u_0}(z, \Phi) = r^2 c_r z^r + B_0 z^{\lambda_0} + \ldots, \qquad \mathrm{Re}\, \lambda_0 > \rho,$$

whence the convergence follows.

Now we consider a formal solution φ of the **second type**. Let $\rho = -1$. Then φ can be written in the form

$$\varphi = \sum_{(m_1,m_2)\in\mathbb{Z}_+^2} c_{m_1,m_2} z^{r+m_1(1-r)+m_2(1+r)} =$$

$$= \sum_{m_1=0}^{\infty} z^{2m_1-1} \sum_{m_2=0}^{\infty} c_{m_1,m_2} z^{i\sigma(1-m_1+m_2)} = \sum_{l=0}^{\infty} z^{2l-1} y_l(z), \tag{10}$$

where

$$y_l(z) = z^{i\sigma(1-l)} \sum_{m=0}^{\infty} c_{l,m} z^{i\sigma m}, \qquad l = 0, 1, \ldots .$$

Applying the technique of the Newton–Bruno polygon (see [1]), one can check that the first term $z^{-1}y_0(z)$ of the series (10) is a solution of the truncated equation

$$u\,\delta^2 u - (\delta u)^2 - z^2 u^4 = 0. \tag{11}$$

A general solution of (11) has the form

$$u = z^{-1} \frac{4c_1^2}{(c_2/z^{c_1}) - 4c_1^2(z^{c_1}/c_2)},$$

where $c_1, c_2 \in \mathbb{C}$, $c_2 \neq 0$, are arbitrary constants. This solution coincides with the first term $z^{-1}y_0(z)$ of the series (10) only if $\operatorname{Re} c_1 = 0$. Let $c_1 = i\mu$, $\mu \in \mathbb{R} \setminus \{0\}$. Then

$$u = z^{-1} \frac{-4\mu^2}{(c_2/z^{i\mu}) + 4\mu^2(z^{i\mu}/c_2)},$$

which has expansions

$$z^{-1}\left(-\frac{4\mu^2}{c_2}\right) z^{i\mu} \sum_{m=0}^{\infty} (-1)^m \left(\frac{2\mu}{c_2}\right)^{2m} z^{2i\mu m}, \qquad |z^{i\mu}| < |c_2/2\mu|, \tag{12}$$

$$z^{-1}(-c_2) z^{-i\mu} \sum_{m=0}^{\infty} (-1)^m \left(\frac{c_2}{2\mu}\right)^{2m} z^{-2i\mu m}, \qquad |z^{i\mu}| > |c_2/2\mu|. \tag{13}$$

These expansions coincide with

$$z^{-1}y_0(z) = z^{-1} z^{i\sigma} \sum_{m=0}^{\infty} c_{0,m} z^{i\sigma m}, \qquad c_{0,0} = c_r \neq 0,$$

if one puts $\mu = \sigma$, $c_2 = -4\sigma^2/c_{0,0}$ for (12), and $\mu = -\sigma$, $c_2 = -c_{0,0}$ for (13).

Thus, the first term $z^{-1}y_0(z)$ of the formal solution (10) converges to the function

$$u = z^{-1}\frac{4\sigma^2}{c_{0,0}z^{i\sigma} + (4\sigma^2/c_{0,0})z^{-i\sigma}} \tag{14}$$

in sectors contained in the domain $\{|z^{i\sigma}| < |2\sigma/c_{0,0}|\}$ with the boundary ray

$$\{|z^{i\sigma}| = |2\sigma/c_{0,0}|\} = \{\arg z = (-1/\sigma)\ln|2\sigma/c_{0,0}|\}.$$

The poles of the function (14) accumulate to the origin along this ray. For example, if $c_{0,0} = 2\sigma$, then

$$u = z^{-1}\frac{2\sigma}{z^{i\sigma} + z^{-i\sigma}} = z^{-1}\frac{\sigma}{\cos(\sigma\ln z)},$$

whose poles $z_k = e^{(\pi+2\pi k)/2\sigma}$ accumulate to the origin along the positive real axis. In this case the series $z^{-1}y_0(z)$ cannot converge in a whole sector containing this ray.

Concerning the formal solution φ, in general the points where it diverges do not coincide with the poles of the solution (14) of the truncated equation, but we guess that they are asymptotically distributed near those poles (something similar holds for the pole distribution of some Painlevé VI transcendents near its critical point, see [6]). This means that the formal solution φ (of the second type) converges not in any sector of sufficiently small radius and opening less than 2π, but convergence depends on the bisecting direction of a sector. Such solutions are "of measure null", as they form a (real) three-parametric subfamily in the (real) four-parametric family (9) of formal solutions of (8), whereas "most" solutions (9) converge in any sector near the origin (which could correspond to the accumulation of poles along spirals around the origin).

Acknowledgements We would like to thank the referee for pointing us a minor error in the statement of Lemma 1 of [4]. Indeed, in the statement of this lemma we have written "for any integer $\mu \geqslant \mu'$" and missed an additional assumption "$\mathrm{Re}(s_{\mu+1} - s_\mu) > 0$" which was used in its proof. Taking this into consideration, in Lemma 2.1 we have written a more precise "for any integer $\mu \geqslant \mu'$ satisfying $\mathrm{Re}(s_{\mu+1} - s_\mu) > 0$". The research is supported by the Russian Foundation for Basic Research (projects no. RFBR 14-01-00346, RFBR-CNRS 16-51-150005).

References

1. A.D. Bruno, Asymptotic behaviour and expansions of solutions of an ordinary differential equation. Russ. Math. Surv. **59**(3), 429–480 (2004)
2. A.D. Bruno, I.V. Goryuchkina, Asymptotic expansions of the solutions of the sixth Painlevé equation. Trans. Mosc. Math. Soc. **71**, 1–104 (2010)
3. J. Dieudonné, *Foundations of Modern Analysis* (Academic, New York, 1960)

4. R.R. Gontsov, I.V. Goryuchkina, On the convergence of generalized power series satisfying an algebraic ODE. Asymptot. Anal. **93**, 311–325 (2015)
5. A.V. Gridnev, Power expansions of solutions to the modified third Painlevé equation in a neighborhood of zero. J. Math. Sci. **145**, 5180–5187 (2007)
6. D. Guzzetti, Pole distribution of PVI transcendents close to a critical point. Physica D **241**, 2188–2203 (2012)
7. D. Guzzetti, A review on the sixth Painlevé equation. Constr. Approx. **41**, 495–527 (2015)
8. H. Kimura, The construction of a general solution of a Hamiltonian system with regular type singularity and its application to Painlevé equations. Ann. Mat. Pura Appl. **134**, 363–392 (1983)
9. B. Malgrange, Sur le théorème de Maillet. Asymptot. Anal. **2**, 1–4 (1989)
10. A. Parusnikova, Asymptotic expansions of solutions to the fifth Painlevé equation in neighbourhoods of singular and nonsingular points of the equation, in *Formal and Analytic Solutions of Differential and Difference Equations*. Banach Center Publications, vol. 97, pp. 113–124 (Polish Academy of Sciences, Warszawa, 2012)
11. S. Shimomura, Series expansions of Painlevé transcendents in the neighbourhood of a fixed singular point. Funk. Ekvac. **25**, 185–197; Supplement, 363–371 (1982).
12. S. Shimomura, A family of solutions of a nonlinear ordinary differential equation and its application to Painlevé equations (III), (V) and (VI). J. Math. Soc. Jpn. **39**, 649–662 (1987)
13. K. Takano, Reduction for Painlevé equations at the fixed singular points of the first kind. Funk. Ekvac. **29**, 99–119 (1986)

Connection Problem for Regular Holonomic Systems in Several Variables

Yoshishige Haraoka

Abstract We formulate the connection problem for regular holonomic systems in several variables on the basis of local monodromies. As examples, we solve the connection problem for Appell's hypergeometric functions F_1 and F_2.

Keywords Appell's hypergeometric functions • Local monodromy • Singular locus

Mathematics Subject Classification (2000). Primary 34M40; Secondary 33C65

1 Introduction

Connection problem for differential equations is a part of the global analysis, and has decisive meanings both for theories and for applications. In this paper, we study an intrinsic formulation of the connection problem for regular holonomic systems in several variables.

So far, connection problem for ordinary differential equations has been intensively studied, and there are many works. On the other hand, as far as I know, connection problem for holonomic systems in several variables has not yet been studied so intensively even in regular holonomic case. In my knowledge, for Appell's hypergeometric functions, connection problem for F_2 is studied by Takayama [10], Sekiguchi [9] and Kato [5], and connection problem for F_4 is studied by Kato [4]. For Heckman-Opdam hypergeometric function, Opdam [7] first solved a connection problem, and then Oshima-Shimeno [8] solved it in a different way.

In extending the Katz theory of Fuchsian ordinary differential equations to regular holonomic systems, we noticed that the definition of the local monodromy is basic, and that the local monodromy is attached to each irreducible component of the singular locus, not to a point [3]. Then it is natural to formulate the connection

Y. Haraoka (✉)
Department of Mathematics, Kumamoto University, Kumamoto 860-8555, Japan
e-mail: haraoka@kumamoto-u.ac.jp

G. Filipuk et al. (eds.), *Analytic, Algebraic and Geometric Aspects of Differential Equations*, Trends in Mathematics, DOI 10.1007/978-3-319-52842-7_8

problem as a connection between irreducible components of the singular locus. We shall proceed in this way.

On this formulation, we solve connection problems for Appell's hypergeometric functions F_1 and F_2 by using integral representations of solutions.

2 Connection Problem for Fuchsian Ordinary Differential Equations

We formulate the connection problem for Fuchsian ordinary differential equations in order to clarify the idea of the formulation for regular holonomic case. The connection problem for Fuchsian ordinary differential equations is usually formulated as a problem of obtaining linear relations among the sets of solutions each of which is specialized by characteristic exponents at a regular singular point. Note that the characteristic exponents correspond to the eigenvalues of the local monodromies. Then in order to formulate the connection problem, we first recall the definition of the local monodromy.

Let $S = \{a_0, a_1, \ldots, a_p\}$ be a finite subset of the projective line $\mathbb{P}^1$, and set $X = \mathbb{P}^1 \setminus S$. Take any point $b \in X$ as a base point. Take a point $a_j \in S$. Let K be a circle centered at a_j with so small radius that the other points in S are in outside K. We regard K as a loop with positive direction and with a base point $c \in K$. Take any curve L in X from b to c. We call the loop LKL^{-1} a *monodromy* or a $(+1)$-*loop* for a_j. In this article we use the latter in order to avoid the confusion with monodromy representations. By the definition, any two $(+1)$-loops for a_j are conjugate in $\pi_1(X, b)$. Let

$$\rho : \pi_1(X, b) \to \mathrm{GL}(V)$$

be an anti-representation on a vector space V. Then, for a $(+1)$-loop γ for a_j, the conjugacy class $[\rho(\gamma)]$ is independent of the choice of a $(+1)$-loop, and hence is uniquely determined by a_j. We call the conjugacy class the *local monodromy* at a_j.

Now we proceed to formulate the connection problem. We consider a Fuchsian ordinary differential equation (E) of rank n with the set of the regular singular points $S = \{a_0, a_1, \ldots, a_p\}$. For each a_j, we choose a simply connected domain U_j in X near a_j and a point $b_j \in U_j$. Let V_j be the vector space of the solutions of (E) on U_j. Take a $(+1)$-loop γ_j for a_j with base point b_j. Then we have the local monodromy action

$$\gamma_{j*} : V_j \to V_j,$$

which is given by the analytic continuation along γ_j. By this action, we can decompose V_j into a direct sum

$$V_j = \bigoplus_{\alpha} V_j^{\alpha},$$

where α denotes an eigenvalue of γ_{j*} and V_j^α the generalized eigenspace for the eigenvalue α. For each j, α, we denote by

$$\pi_j^\alpha : V_j \to V_j^\alpha$$

the projection onto the direct sum component V_j^α. Note that, if γ_{j*} is semi-simple, all V_j^α are eigenspaces. In general, we have a filtration

$$V_j^{\alpha,0} \subset V_j^{\alpha,1} \subset \cdots \subset V_j^\alpha,$$

where each $V_j^{\alpha,r}$ is stable under the action of γ_{j*}. Each $V_j^{\alpha,r}$ consists of solutions containing $(\log(x - a_j))^k$ with $k \leq r$. We call this filtration logarithmic filtration.

For each pair (j, k) of indices, we take a path μ_{jk} in X from b_j to b_k. Analytic continuations of solutions in V_j along μ_{jk} are contained in V_k. Then, for any α and for any solution $u \in V_j^\alpha$, we have the decomposition

$$\mu_{jk*}u = \sum_\beta \pi_k^\beta (\mu_{jk*}u)$$

according to the direct sum decomposition of V_k. The connection problem is to describe this decomposition. If $\dim V_j^\alpha = 1$, this decomposition is intrinsic because u is unique up to scalar multiplication. When $\dim V_j^\alpha > 1$, we are interested in

$$\dim \pi_k^\beta \left(\mu_{jk*}(V_j^\alpha)\right),$$

which does not depend on the choice of a basis of V_j^α. More precise version of the connection problem is to describe the subspace $\pi_k^\beta \left(\mu_{jk*} V_j^{\alpha,r}\right) \cap V_k^{\beta,s}$. If we fix a basis for every V_j^α compatible with the logarithmic filtration, the connection problem is reduced to describing the connection coefficients.

3 Formulation of the Connection Problem

Let $\varphi(x)$ be a reduced polynomial in n variables. We consider a hypersurface $S = \{x \in \mathbb{C}^n \mid \varphi(x) = 0\} \subset \mathbb{C}^n$. We may take any compactification of $\mathbb{C}^n$, and in this article we take $\mathbb{P}^n$ for simplicity. Let H_∞ be the hyperplane at infinity in $\mathbb{P}^n$. We set $\bar{S} = S \cup H_\infty$, and

$$X = \mathbb{C}^n \setminus S = \mathbb{P}^n \setminus \bar{S}.$$

We consider a completely integrable system (H) on X with logarithmic singularities along $\bar{S}$.

Compared with the formulation of the connection problem for Fuchsian ordinary differential equations, which we have recalled in the previous section, we should solve the following two questions to formulate the connection problem for the present case.

(i) What is a local monodromy action?
(ii) How can we take a path corresponding to μ_{jk}?

We consider the first question. Let

$$S = \bigcup_j S_j$$

be the irreducible decomposition of S. Take an irreducible component S_j. Let $p \in S_j$ be a point which is regular in S, i.e. $\operatorname{grad}\varphi(p) \neq 0$. We take a complex line H passing through p in general position with respect to S. We fix a base point $b \in X$. Take a $(+1)$-loop K for p in H, and a path L in X from b to the base point of K. We call the loop LKL^{-1} a $(+1)$-*loop* for S_j. The following fact is basic.

Proposition 3.1 *Any two* $(+1)$*-loops for* S_j *are conjugate in* $\pi_1(X, b)$*.*

This proposition can be shown by noting $S_j \cap S^\circ$ is path connected, where S° denotes the set of regular points in S. Thanks to this proposition, we can define local monodromies in a similar way as in the ODE case. Let

$$\rho : \pi_1(X, b) \to \mathrm{GL}(V)$$

be an anti-representation on a vector space V. For a $(+1)$-loop γ for S_j, the conjugacy class $[\rho(\gamma)]$ is uniquely determined by S_j. We call the conjugacy class the *local monodromy* at S_j. Thus, each irreducible component of the singular locus plays a similar role as a singular point of an ordinary differential equation. Accordingly, it is natural to formulate the connection problem as to describe a linear relation between solution spaces at two irreducible components, not at two points. The local monodromy action is given by a $(+1)$-loop for each irreducible component S_j.

Now we find that the question (ii) has a simple solution after solving the question (i). Since we understand that the problem is between two irreducible components, and since two irreducible components intersect, there is no need to take a new path which connect two irreducible components.

Thus we can formulate the connection problem in the following way. Let S_j, S_k be two irreducible components. We take a simply connected domain U_{jk} in X near an intersection point of S_j and S_k. We denote by V_{jk} the vector space of the solutions of (H) on U_{jk}. Take $(+1)$-loops γ_j and γ_k for S_j and S_k, respectively, with a common base point b' in U_{jk}. Then we decompose V_{jk} into direct sums in two ways. By the local monodromy action γ_{j*}, we have

$$V_{jk} = \bigoplus_\alpha V_j^\alpha,$$

and by the local monodromy action γ_{k*}, we have

$$V_{jk} = \bigoplus_{\beta} V_k^\beta,$$

where V_j^α (resp. V_k^β) denotes the generalized eigenspace of γ_{j*} (resp. γ_{k*}) for the eigenvalue α (resp. β). According to these decompositions, we have two sets of projections

$$\pi_j^\alpha : V_{jk} \to V_j^\alpha,$$
$$\pi_k^\beta : V_{jk} \to V_k^\beta.$$

If γ_{j*} or γ_{k*} is not semi-simple, we may consider logarithmic filtrations as well. Then the connection problem is to describe the decomposition

$$u = \sum_{\beta} \pi_k^\beta(u)$$

for each α and each $u \in V_j^\alpha$. When $\dim V_j^\alpha > 1$, we are interested in the intrinsic quantity

$$\dim \pi_k^\beta \left(V_j^\alpha\right)$$

for each α and each β.

To consider our connection problem, the following result is fundamental.

Theorem 3.2 (Gérard [2], Yoshida-Takano [11]) *Let $p \in S$ be an intersection point of irreducible components of S at which the intersecting irreducible components are normal crossing. By a local coordinate change, we can send p to the origin and the irreducible components passing through p to $x_1 = 0, x_2 = 0, \ldots, x_\nu = 0$ with $2 \le \nu \le n$. Then there exists a local fundamental matrix solution*

$$\mathcal{U}(x) = F(x) \prod_{j=1}^{\nu} x_j{}^{L_j} \prod_{j=1}^{\nu} x_j{}^{B_j},$$

where $F(x)$ is a matrix function invertible and holomorphic at $x = 0$, $L_1, L_2, \ldots, L_\nu$ diagonal matrices with non-negative integers in the diagonal entries, and $B_1, B_2, \ldots, B_\nu$ mutually commutative constant matrices.

The fundamental matrix solution $\mathcal{U}(x)$ in the theorem gives bases of generalized eigenspaces of the local monodromy actions at $x_1 = 0, x_2 = 0, \ldots, x_\nu = 0$ in common. Thus, for a normally crossing point p of S, the connection problem between the irreducible components passing through p is reduced to the problem of choosing bases of generalized eigenspaces for each irreducible component from a common set of solutions. We call this problem *trivialization*. Note that, in this case, we have

$$\pi_k^{\beta}\left(V_j^{\alpha}\right) = V_j^{\alpha} \cap V_k^{\beta}.$$

On the other hand, if two irreducible components intersect at a non-normally crossing point of S, we should solve a "usual" connection problem.

4 Connection Problem for Appell's Hypergeometric Series

Appell's F_1

As an easiest and illustrative example, we consider the connection problem for Appell's hypergeometric series F_1. Appell's $F_1(a, b, b', c; x, y)$ is a power series in two variables x, y with parameters a, b, b', c, and satisfies a system of partial differential equations. We can transform the system to a Pfaffian system

$$du = \left(A_0 \frac{dx}{x} + A_1 \frac{dx}{x-1} + B_0 \frac{dy}{y} + B_1 \frac{dy}{y-1} + C \frac{d(x-y)}{x-y} \right) u, \tag{1}$$

where A_0, A_1, B_0, B_1, C are 3×3 constant matrices satisfying

$$\begin{aligned} A_0 &\sim \text{diag}[0, 0, b' - c + 1],\ A_1 \sim \text{diag}[0, 0, c - a - b - 1], \\ B_0 &\sim \text{diag}[0, 0, b - c + 1],\ B_1 \sim \text{diag}[0, 0, c - a - b' - 1], \\ C &\sim \text{diag}[0, 0, -b - b'],\ -A_0 - A_1 - B_0 - B_1 - C \sim \text{diag}[a, a, b + b']. \end{aligned}$$

For simplicity, in this paper we assume that there is no linear relation with coefficient in $\mathbb{Z}$ among a, b, b', c and 1.

We consider the connection problem of the Pfaffian system (1). We see that the singular locus is $\{x = 0\} \cup \{x = 1\} \cup \{y = 0\} \cup \{y = 1\} \cup \{x = y\} \cup H_\infty$, and that the eigenvalues of each residue matrix given above are characteristic exponents at the corresponding irreducible component of the singular locus. We see that $(0, 1)$ and $(1, 0)$ are normally crossing points, and $(0, 0)$ and $(1, 1)$ are non-normally crossing points (Fig. 1).

It is known that any solution of the system (1) can be given by an integral

$$u(x, y) = \int_\Delta t^{\lambda_1}(1-t)^{\lambda_2}(1-xt)^{\lambda_3}(1-yt)^{\lambda_4}\, \vec{\eta},$$

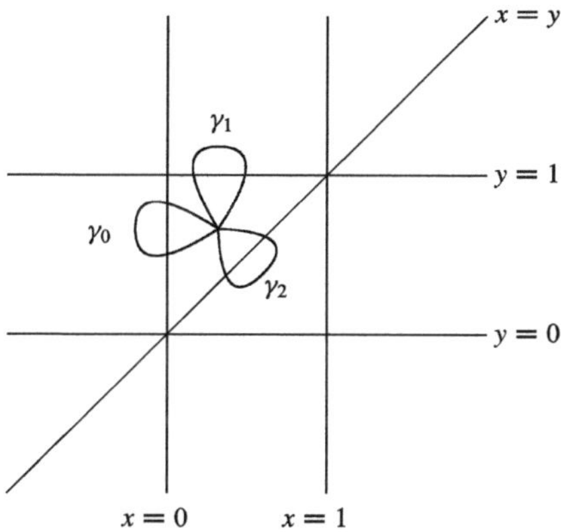

Fig. 1 Singular locus for F_1

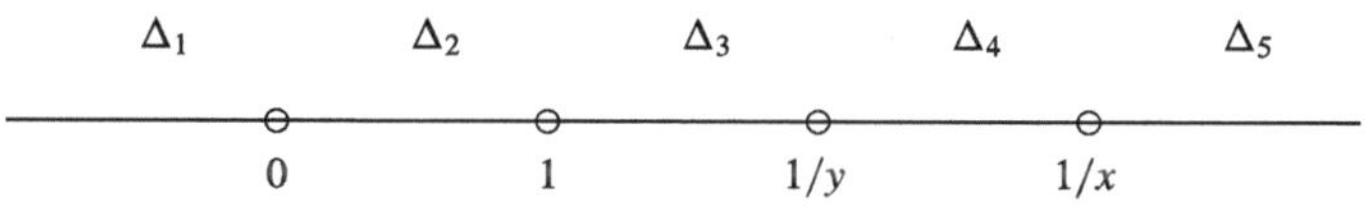

Fig. 2 Twisted cycles for F_1

where

$$\vec{\eta} = {}^t\left(\frac{dt}{t}, \frac{bt}{1-xt}\,dt, \frac{b't}{1-yt}\,dt\right),$$

and $\lambda_1 = a, \lambda_2 = c - a - 1, \lambda_3 = -b, \lambda_4 = -b'$. The path Δ of integration is understood as a twisted cycle on the space $\mathbb{C}_t \setminus \{0, 1, 1/x, 1/y\}$. We denote the solution given by the integral over Δ also by Δ.

We shall consider the connection problems between $\{x = 0\}$ and $\{y = 1\}$, and between $\{x = 0\}$ and $\{x = y\}$.

1. Trivialization at $(0, 1)$.

Since the irreducible components $\{x = 0\}$ and $\{y = 1\}$ intersect at the normally crossing point $(0, 1)$, the connection problem between them is trivialization. We consider in a simply connected region $U = \{(x, y) \in \mathbb{R}^2 \mid 0 < x < y < 1\}$ in $\mathbb{R}^2$. Then the branch points in the integrand are located on the real axis in t-plane as described in Fig. 2. We define twisted cycles Δ_j $(1 \le j \le 5)$ by simple chains connecting two adjacent branch points (Fig. 2).

Let V be the vector space of solutions of (1) on U. Since the characteristic exponents at $x = 0$ is 0 of multiplicity 2 and $b' - c + 1$ of multiplicity free, the eigenvalues of the local monodromy at $\{x = 0\}$ are 1 of multiplicity 2 and $e^{2\pi i(b'-c+1)} =: \alpha_0$ of multiplicity free. Similarly, the eigenvalues of the local monodromy at $\{y = 1\}$ are 1 of multiplicity 2 and $e^{2\pi i(c-a-b'-1)} =: \beta_1$ of multiplicity free. We take a base point $q = (x_0, y_0)$ in U. We take a $(+1)$-loop

γ_0 for $\{x = 0\}$ by a simple circle in the complex line $y = y_0$ with the base point q which encircles $x = 0$ in the positive direction, and a $(+1)$-loop γ_1 for $\{y = 1\}$ by a simple circle in the complex line $x = x_0$ with the same base point q which encircles $y = 1$ in the positive direction (Fig. 1). We shall decompose V into direct sums by the actions of γ_{0*} and γ_{1*}.

By a usual argument, we obtain eigenfunctions for the local monodromy actions γ_{0*} and γ_{1*} by choosing appropriate twisted cycles. Then we have the direct sum decompositions

$$\begin{aligned} V &= V^1_{x=0} \oplus V^{\alpha_0}_{x=0} = \langle \Delta_2, \Delta_3 \rangle \oplus \langle \Delta_5 \rangle, \\ V &= V^1_{y=1} \oplus V^{\beta_1}_{y=1} = \langle \Delta_1, \Delta_5 \rangle \oplus \langle \Delta_3 \rangle. \end{aligned} \tag{2}$$

Thanks to Theorem 3.2, we can choose a basis of V which constitutes the eigenspaces for γ_{0*} and the eigenspaces for γ_{1*}. In (2), Δ_3 and Δ_5 are common, while Δ_2 and Δ_1 are not. Then we want to find another eigenfunction contained in both $V^1_{x=0}$ and $V^1_{y=1}$. We consider the projective line $\mathbb{P}^1$ in t such that the real axis is the equator. Then the twisted cycles $\Delta_1, \dots, \Delta_5$ are on the equator. We take as another twisted cycle a circle on $\mathbb{P}^1$ with base point 0 such that 1 and $1/y$ are in the inside and $1/x$ is in the outside. We denote it by Δ_0. Then as illustrated in Fig. 3, Δ_0 can be written as a linear combination of Δ_2 and Δ_3, and also as a linear combination of Δ_1 and Δ_5. Thus we can replace Δ_2 in $V^1_{x=0}$ and Δ_1 in $V^1_{y=1}$ by Δ_0 to obtain

$$\begin{aligned} V &= V^1_{x=0} \oplus V^{\alpha_0}_{x=0} = \langle \Delta_0, \Delta_3 \rangle \oplus \langle \Delta_5 \rangle, \\ V &= V^1_{y=1} \oplus V^{\beta_1}_{y=1} = \langle \Delta_0, \Delta_5 \rangle \oplus \langle \Delta_3 \rangle. \end{aligned} \tag{3}$$

This completes the trivialization.

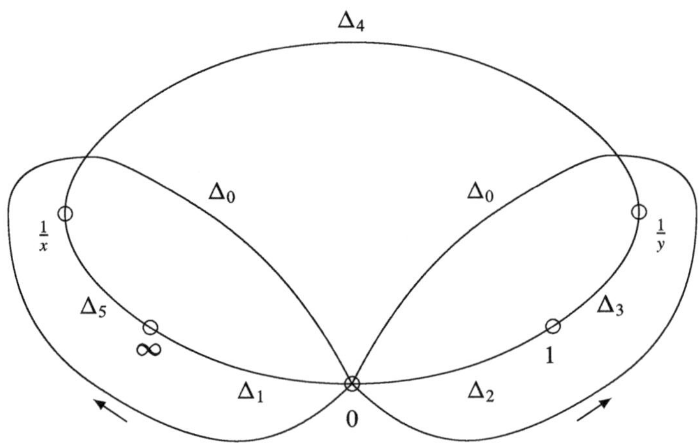

Fig. 3 Twisted cycle Δ_0

2. Connection between $\{x = 0\}$ and $\{x = y\}$.

The irreducible components $\{x = 0\}$ and $\{x = y\}$ intersect at non-normally crossing point $(0,0)$. We take the same region U, and consider the vector space V of solutions on U. V is decomposed into a direct sum of the eigenspaces for the action of γ_{0*}. We use bases of the eigenspaces as in (2). Take a $(+1)$-loop γ_2 for $x = y$ as in Fig. 1. The eigenvalues of the action of γ_{2*} are 1 of multiplicity 2 and $e^{2\pi i(-b-b')} =: \gamma$ of multiplicity free. Then we have the direct sum decomposition

$$V = V^1_{x=y} \oplus V^\gamma_{x=y} = \langle \Delta_1, \Delta_2 \rangle \oplus \langle \Delta_4 \rangle.$$

Hence the connection problem between $x = 0$ and $x = y$ is to obtain linear relations among two sets $\{\Delta_2, \Delta_3, \Delta_5\}$ and $\{\Delta_1, \Delta_2, \Delta_4\}$.

We determine the branches on Δ_j by the standard loading, which is a way of assigning branches by using a real structure [6]. By using Cauchy's theorem, we have the following two relations among the twisted cycles:

$$\begin{aligned} &\Delta_1 + f_1\Delta_2 + f_{12}\Delta_3 + f_{124}\Delta_4 + f_{1234}\Delta_5 = 0, \\ &\Delta_1 + f_1^{-1}\Delta_2 + f_{12}^{-1}\Delta_3 + f_{124}^{-1}\Delta_4 + f_{1234}^{-1}\Delta_5 = 0, \end{aligned} \tag{4}$$

where $f_j = e^{\pi i \lambda_j}$ $(1 \le j \le 4)$ and $f_{jk\ldots} = f_j f_k \cdots$. Solving the relations (4), we get

$$\Delta_3 = \frac{f_{1234} - f_{1234}^{-1}}{f_{34}^{-1} - f_{34}}\Delta_1 + \frac{f_{234} - f_{234}^{-1}}{f_{34}^{-1} - f_{34}}\Delta_2 + \frac{f_3 - f_3^{-1}}{f_{34}^{-1} - f_{34}}\Delta_4,$$

$$\Delta_5 = \frac{f_{12} - f_{12}^{-1}}{f_{34} - f_{34}^{-1}}\Delta_1 + \frac{f_2 - f_2^{-1}}{f_{34} - f_{34}^{-1}}\Delta_2 - \frac{f_4 - f_4^{-1}}{f_{34} - f_{34}^{-1}}\Delta_4.$$

By using the equality $f_j - f_j^{-1} = 2i \sin \pi\lambda_j$, the above relations are written as

$$\begin{aligned} \Delta_3 &= \frac{\sin\pi(b+b'-c)}{\sin\pi(b+b')}\Delta_1 + \frac{\sin\pi(a+b+b'-c)}{\sin\pi(b+b')}\Delta_2 - \frac{\sin\pi b}{\sin\pi(b+b')}\Delta_4, \\ \Delta_5 &= \frac{\sin\pi c}{\sin\pi(b+b')}\Delta_1 + \frac{\sin\pi(c-a)}{\sin\pi(b+b')}\Delta_2 - \frac{\sin\pi b'}{\sin\pi(b+b')}\Delta_4. \end{aligned} \tag{5}$$

Owing to the assumption on a, b, b', c, we see that no coefficient vanishes.

Appell's F_2

Appell's $F_2(a,b,b',c,c';x,y)$ is also a power series in two variables, and satisfies a system of partial differential equations, which can be transformed to a Pfaffian system

$$du = \left(A_0 \frac{dx}{x} + A_1 \frac{dx}{x-1} + B_0 \frac{dy}{y} + B_1 \frac{dy}{y-1} + C \frac{d(x+y)}{x+y-1} \right) u, \tag{6}$$

where A_0, A_1, B_0, B_1, C are 4×4 constant matrices satisfying

$$\begin{aligned} A_0 &\sim \operatorname{diag}[0,0,1-c,1-c],\ A_1 \sim \operatorname{diag}[0,0,0,c-a-b+b'-1], \\ B_0 &\sim \operatorname{diag}[0,0,1-c',1-c'],\ B_1 \sim \operatorname{diag}[0,0,0,c'-a+b-b'-1], \\ C &\sim \operatorname{diag}[0,0,0,c+c'-a-b-b'-2], \\ -A_0 - A_1 - B_0 - B_1 - C &\sim \operatorname{diag}[a,a,a,b+b']. \end{aligned}$$

We assume that there is no linear relation with coefficients in $\mathbb{Z}$ among a, b, b', c, c' and 1. The irreducible components of the singular locus are $\{x = 0\}, \{x = 1\}, \{y = 0\}, \{y = 1\}, \{x + y - 1 = 0\}$ and H_∞.

It is known that any solution of the system (6) can be given by an integral

$$u(x,y) = \int_\Delta s^{\lambda_1} t^{\lambda_2} (1-s)^{\lambda_3} (1-t)^{\lambda_4} (1-xs-yt)^{\lambda_5} \vec{\eta},$$

where $\vec{\eta}$ is a 4-vector of twisted cocycles and $\lambda_1 = b-1, \lambda_2 = b'-1, \lambda_3 = c-b-1, \lambda_4 = c'-b'-1, \lambda_5 = -a$. By using this integral representation, we shall consider connection problems between $\{x = 0\}$ and $\{y = 0\}$, between $\{x = 1\}$ and $\{y = 1\}$, and between $\{x = 0\}$ and $\{y = 1\}$.

1. Trivialization at $(0, 0)$.

The intersection point of the irreducible components $\{x = 0\}$ and $\{y = 0\}$ is a normal crossing point of the singular locus, and hence we consider a trivialization. Let U_1 be the simply connected region $\{(x,y) \in \mathbb{R}^2 \mid 0 < x < 1-y < 1\}$ in $\mathbb{R}^2$, and V_1 the vector space of solutions of (6) on U_1. For $(x,y) \in U_1$, the configuration of twisted cycles for the integral representation is as in Fig. 4.

In a similar way as in F_1's case, we have

$$\begin{aligned} V^1_{x=0} &= \langle \Delta_4, \Delta_8 \rangle,\ V^{\alpha_0}_{x=0} = \langle \Delta_{10}, \Delta_{14} \rangle, \\ V^1_{y=0} &= \langle \Delta_8, \Delta_{10} \rangle,\ V^{\beta_0}_{y=0} = \langle \Delta_2, \Delta_4 \rangle, \end{aligned}$$

where we set $e^{2\pi i(1-c)} = \alpha_0, e^{2\pi i(1-c')} = \beta_0$. We need to find a twisted cycle which is a common element of $V^{\alpha_0}_{x=0}$ and $V^{\beta_0}_{y=0}$. For $\rho > 1$, we have the sections

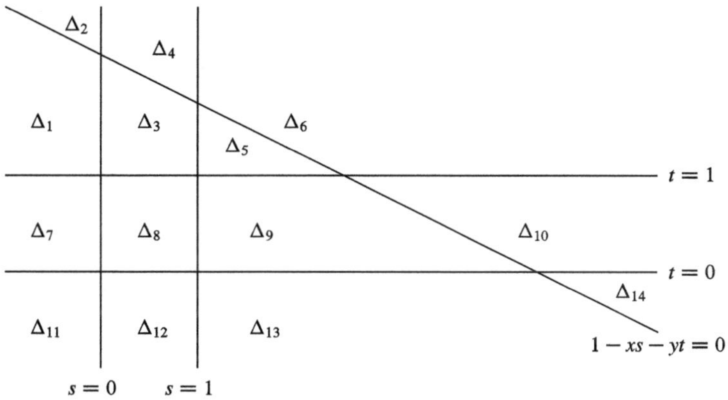

Fig. 4 Configuration of twisted cycles for F_2 on U_1

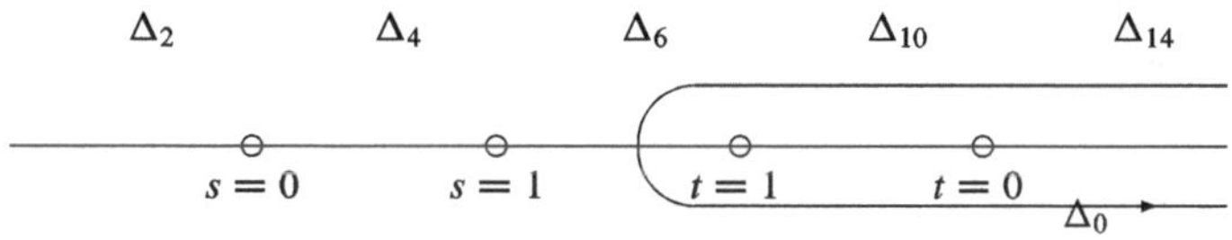

Fig. 5 Twisted cycles on the line ℓ_ρ

of $\Delta_2, \Delta_4, \Delta_6, \Delta_{10}, \Delta_{14}$ on the line $\ell_\rho : xs + yt = \rho$ as in Fig. 5. Define a twisted cycle Δ_0 whose section by ℓ_ρ is given by Fig. 5. Then Δ_0 is a linear combination of Δ_{10} and Δ_{14}, and also a linear combination of Δ_2 and Δ_4.

By using Δ_0, we obtain the trivialization

$$\begin{aligned} V_1 &= V^1_{x=0} \oplus V^{\alpha_0}_{x=0} = \langle \Delta_4, \Delta_8 \rangle \oplus \langle \Delta_0, \Delta_{10} \rangle \\ &= V^1_{y=0} \oplus V^{\beta_0}_{y=0} = \langle \Delta_8, \Delta_{10} \rangle \oplus \langle \Delta_0, \Delta_4 \rangle. \end{aligned} \tag{7}$$

2. Trivialization at $(1, 1)$.

To consider the connection between $\{x = 1\}$ and $\{y = 1\}$, we take a simply connected region $U_2 = \{(x, y) \in \mathbb{R}^2 \mid 0 < 1 - y < x < 1\}$, and the vector space V_2 of solutions of (6) on U_2. For $(x, y) \in U_2$, the configuration of twisted cycles is as in Fig. 6. Set $e^{2\pi i(c-a-b+b'-1)} = \alpha_1$ and $e^{2\pi i(c'-a+b-b'-1)} = \beta_1$. Then we have the trivialization

$$\begin{aligned} V_2 &= V^1_{x=1} \oplus V^{\alpha_1}_{x=1} = \langle \Delta'_3, \Delta'_6, \Delta'_{11} \rangle \oplus \langle \Delta'_9 \rangle \\ &= V^1_{y=1} \oplus V^{\beta_1}_{y=1} = \langle \Delta'_6, \Delta'_9, \Delta'_{11} \rangle \oplus \langle \Delta'_3 \rangle. \end{aligned} \tag{8}$$

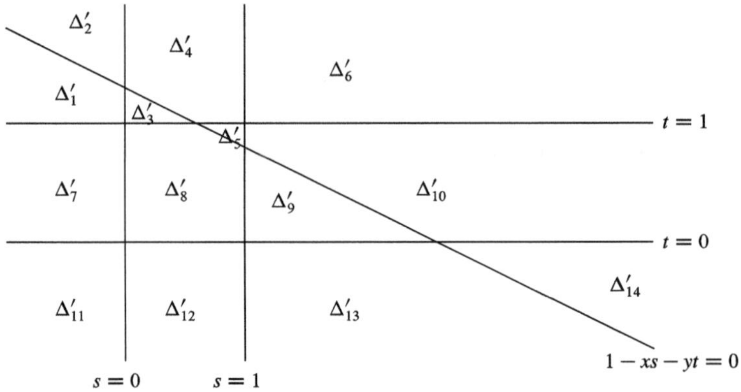

Fig. 6 Configuration of twisted cycles for F_2 on U_2

3. Connection between $\{x = 0\}$ and $\{y = 1\}$.

We combine the above two results to get the connection between $\{x = 0\}$ and $\{y = 1\}$. For the purpose, we need a path from U_1 to U_2. Take a point $(x_0, y_0) \in U_1$, and define a path μ starting from (x_0, y_0) by

$$x(\theta) = 1 - y_0 - re^{-i\theta}, \ y(\theta) = y_0 \quad (\theta \in [0, \pi]),$$

where $r = 1 - y_0 - x_0 > 0$. Note that the end point $(x(\pi), y(\pi))$ is in U_2. The branch on each Δ_j, Δ'_k is assumed to be defined by the standard loading. Then we have

$$\begin{aligned} \mu_* \Delta_4 &= \Delta'_4 + f_4^{-1} \Delta'_5, \\ \mu_* \Delta_8 &= \Delta'_8 + f_5^{-1} \Delta'_5, \\ \mu_* \Delta_{10} &= \Delta'_{10} + f_3^{-1} \Delta'_5, \\ \mu_* \Delta_{14} &= \Delta'_{14}, \end{aligned}$$

where $f_j = e^{\pi i \lambda_j}$ $(1 \leq j \leq 5)$. We can express $\Delta'_4, \Delta'_5, \Delta'_8, \Delta'_{10}, \Delta'_{14}$ in terms of $\Delta'_3, \Delta'_6, \Delta'_9$ and Δ'_{11} by the way given in [1]. We finally obtain the solution of the connection problem:

$$\begin{aligned} \mu_* \Delta_4 &= c_{4,3} \Delta'_3 + c_{4,6} \Delta'_6 + c_{4,9} \Delta'_9 + c_{4,11} \Delta'_{11}, \\ \mu_* \Delta_8 &= c_{8,3} \Delta'_3 + c_{8,6} \Delta'_6 + c_{8,9} \Delta'_9 + c_{8,11} \Delta'_{11}, \\ \mu_* \Delta_{10} &= c_{10,3} \Delta'_3 + c_{10,6} \Delta'_6 + c_{10,9} \Delta'_9 + c_{10,11} \Delta'_{11}, \\ \mu_* \Delta_0 &= c_{0,3} \Delta'_3 + c_{0,6} \Delta'_6 + c_{0,9} \Delta'_9 + c_{0,11} \Delta'_{11}, \end{aligned} \tag{9}$$

where the connection coefficients $c_{j,k}$ are given as follows:

$$
\begin{aligned}
c_{4,3} &= -\frac{f_1 \sin \pi(c' - b')}{\sin \pi(c' - a + b - b')}, & c_{4,9} &= -\frac{\sin \pi b'}{f_4 \sin \pi(c - a - b + b')}, \\
c_{8,3} &= \frac{f_1 \sin \pi a}{\sin \pi(c' - a + b - b')}, & c_{8,9} &= \frac{f_2 \sin \pi a}{\sin \pi(c - a - b + b')}, \\
c_{10,3} &= -\frac{\sin \pi b}{f_3 \sin \pi(c' - a + b - b')}, & c_{10,9} &= \frac{f_2 \sin \pi(c - b)}{\sin \pi(c - a + b - b')}, \\
c_{0,3} &= \frac{2 i f_2 \sin \pi b \sin \pi(c' - b')}{f_3 \sin \pi(c' - a + b - b')}, & c_{0,9} &= \frac{2 i f_2 \sin \pi b' \sin \pi(c - b)}{\sin \pi(c - a - b + b')},
\end{aligned}
\tag{10}
$$

and

$$
\begin{aligned}
c_{4,6} &= \frac{(f_4^2 - 1)(1 - f_{35}^2 - f_{1345}^2 + f_{13455}^2 - f_{123455}^2 + f_{1233455}^2)}{f_3 f_4^2 (f_5^2 - 1)(f_{235}^2 - 1)(f_{145}^2 - 1)}, \\
c_{4,11} &= -\frac{f_5 (f_2^2 - 1)(1 - f_1^2 + f_{14}^2 - f_{145}^2 - f_{12345}^2 + f_{1123455}^2)}{f_{124} (f_5^2 - 1)(f_{235}^2 - 1)(f_{145}^2 - 1)}, \\
c_{8,6} &= \frac{1 - f_{1345}^2 - f_{2345}^2 - f_{123455}^2 + f_{1233455}^2 + f_{1234455}^2}{f_{345} (f_{235}^2 - 1)(f_{145}^2 - 1)}, \\
c_{8,11} &= \frac{f_1^2 + f_2^2 - f_{12}^2 - f_{1235}^2 - f_{1245}^2 + f_{112234455}^2}{f_{12} (f_{235}^2 - 1)(f_{145}^2 - 1)}, \\
c_{10,6} &= \frac{(f_3^2 - 1)(1 - f_{45}^2 - f_{2345}^2 + f_{23455}^2 - f_{123455}^2 + f_{1234455}^2)}{f_3^2 f_4 (f_5^2 - 1)(f_{235}^2 - 1)(f_{145}^2 - 1)}, \\
c_{10,11} &= -\frac{f_5 (f_1^2 - 1)(1 - f_2^2 + f_{23}^2 - f_{235}^2 - f_{12345}^2 + f_{1223455}^2)}{f_{123} (f_5^2 - 1)(f_{235}^2 - 1)(f_{145}^2 - 1)}, \\
c_{0,6} &= \frac{f_2 (f_3^2 - 1)(f_4^2 - 1)(1 - f_{35}^2 - f_{45}^2 + f_{13455}^2 + f_{23455}^2 - f_{123455}^2)}{f_3^2 f_4 (f_5^2 - 1)(f_{235}^2 - 1)(f_{145}^2 - 1)}, \\
c_{0,11} &= -\frac{f_5 (f_1^2 - 1)(f_2^2 - 1)(1 - f_3^2 - f_4^2 + f_{1345}^2 + f_{2345}^2 - f_{123455}^2)}{f_{13} (f_5^2 - 1)(f_{235}^2 - 1)(f_{145}^2 - 1)}.
\end{aligned}
\tag{11}
$$

By using the explicit expression (10), we obtain $c_{4,3} c_{8,3} c_{10,3} c_{0,3} \neq 0$ from the assumption, which implies

$$
\dim \pi_{y=1}^{\beta_1} \left(\mu_* V_{x=0}^{1} \right) = \dim \pi_{y=1}^{\beta_1} \left(\mu_* V_{x=0}^{\alpha_0} \right) = 1.
$$

Although the expression (11) is complicated, we can also derive a similar conclusion. In fact, we have

$$\begin{vmatrix} c_{4,6} & c_{4,9} \\ c_{8,6} & c_{8,9} \end{vmatrix} = \frac{\sin \pi c' \sin \pi (c + c' - a - b')}{f_{45} \sin \pi (c - a - b + b') \sin \pi (c' - a + b - b')},$$

which does not vanish by the assumption. This implies

$$\operatorname{rank} \begin{pmatrix} c_{4,6} & c_{4,9} & c_{4,11} \\ c_{8,6} & c_{8,9} & c_{8,11} \end{pmatrix} = 2,$$

and hence

$$\dim \pi^1_{y=1} \left(\mu_* V^1_{x=0} \right) = 2.$$

Since we have the trivializations (7) and (8) at $(0, 0)$ and $(1, 1)$, the connection formula (9) can be also regarded as a connection between $\{x = 0\}$ and $\{x = 1\}$, or between $\{y = 0\}$ and $\{x = 1\}$, or between $\{y = 0\}$ and $\{y = 1\}$.

Acknowledgements This work is supported by the JSPS grants-in-aid for scientific research B, No. 15H03628.

References

1. K. Aomoto, On the structure of integrals of power products of linear functions. Sci. Pap. Coll. Gen. Educ. Univ. Tokyo **27**, 49–61 (1977)
2. R. Gérard, Théorie de Fuchs sur une variété analytique complexe. J. Math. Pures Appl. **47**, 321–404 (1968)
3. Y. Haraoka, *Linear Differential Equations in the Complex Domain* (Sugaku Shobo, Tokyo, 2015) (in Japanese)
4. M. Kato, Connection formulas for Appell's system F_4 and some applications. Funkcial. Ekvac. **38**, 243–266 (1995)
5. M. Kato, Connection formulas and irreducibility conditions for Appell's F_2. Kyushu J. Math. **66**, 325–363 (2012)
6. K. Mimachi, Intersection numbers for twisted cycles and the connection problem associated with the generalized hypergeometric function ${}_{n+1}F_n$. Int. Math. Res. Not. **2011**, 1757–1781 (2011)
7. E.M. Opdam, An analogue of the Gauss summation formula for hypergeometric functions related to root systems. Math. Z. **212**, 313–336 (1993)
8. T. Oshima, N. Shimeno, Heckman-Opdam hypergeometric functions and their specializations. RIMS Kôkyûuroku Bessatsu **B20**, 129–162 (2010)
9. J. Sekiguchi, Appell's hypergeometric function $F_2(a, b, b'; c, c'; x, y)$ and the blowing up space of $\mathbb{P}^2$. RIMS Kôkyûuroku **773**, 66–77 (1991)
10. N. Takayama, Boundary values and connection formulas of holonomic systems with regular singularities, in *Special Differential Equations, Proceedings of the Taniguchi Workshop*, pp. 125–149 (1991)
11. M. Yoshida, K. Takano, On a linear system of Pfaffian equations with regular singular points. Funkcial. Ekvac. **19**, 175–189 (1976)

On k-Summability of Formal Solutions for Certain Higher Order Partial Differential Operators with Polynomial Coefficients

Kunio Ichinobe

Abstract We study the k-summability of divergent formal solutions to the Cauchy problem for a class of linear partial differential operators of higher order with respect to t which have polynomial coefficients in t. We obtain a sufficient condition for the k-summability of formal solutions in terms of a global analyticity and a proper exponential growth estimate of the Cauchy data.

Keywords Cauchy problem • k-summability • Power series solutions

Mathematics Subject Classification (2000). Primary 35C10; Secondary 35G10

1 Result

Let us consider the following linear partial differential operator L with polynomial coefficients of t

$$L = \partial_t^M - P^M(t, \partial_t, \partial_x), \quad P^M(t, \partial_t, \partial_x) = \sum_{\substack{1 \le j \le M, \\ \alpha \,:\, \text{finite}}} a_{\alpha j}(t) \partial_t^{M-j} \partial_x^{\alpha}, \tag{1}$$

where $t, x \in \mathbb{C}$, $M \geq 2$ and $a_{\alpha j}(t) \in \mathbb{C}[t]$ for all α and j.

We consider the following Cauchy problem

$$\begin{cases} LU(t,x) = 0, \\ \partial_t^n U(0,x) = 0 \ (0 \le n \le M-2), \\ \partial_t^{M-1} U(0,x) = \varphi(x) \in \mathcal{O}_x, \end{cases} \tag{CP}$$

K. Ichinobe (✉)
Department of Mathematics Education, Aichi University of Education, 1 Hirosawa, Igaya, Kariya City, Aichi Prefecture 448-8542, Japan
e-mail: ichinobe@auecc.aichi-edu.ac.jp

G. Filipuk et al. (eds.), *Analytic, Algebraic and Geometric Aspects of Differential Equations*, Trends in Mathematics, DOI 10.1007/978-3-319-52842-7_9

where $\mathcal{O}_x$ denotes the set of holomorphic functions at $x = 0$. This Cauchy problem has a unique formal power series solution of the form

$$\hat{U}(t,x) = \sum_{n \geq M-1} U_n(x)\frac{t^n}{n!}, \quad U_{M-1}(x) = \varphi(x). \tag{2}$$

Throughout this paper, we assume

$$\alpha_* - M := \max\{\alpha - j; a_{\alpha j}(t) \not\equiv 0\} > 0. \tag{A-1}$$

Since this formal solution is divergent in general under this assumption, we shall study k-summability of the divergent solution under some conditions. We shall explain the conditions by using the Newton polygon, which is defined as follows.

We define a domain $\mathrm{N}(\alpha, j)$ by

$$\mathrm{N}(\alpha,j) := \{(x,y);\ x \leq M - j + \alpha,\ y \geq i(\alpha,j) - M + j\} \subset \mathbb{R}^2,$$

for $a_{\alpha j}(t) \not\equiv 0$, where $i(\alpha,j) = O(a_{\alpha j}(t))$ which denotes the order of zero of $a_{\alpha j}(t)$ at $t = 0$, and $\mathrm{N}(\alpha,j) := \phi$ for $a_{\alpha j}(t) \equiv 0$. Then the Newton polygon $\mathrm{N}(L)$ is defined by

$$\mathrm{N}(L) := \mathrm{Ch}\left\{\mathrm{N}(0,0) \cup \bigcup_{\alpha,j:\text{finite}} \mathrm{N}(\alpha,j)\right\}, \tag{3}$$

where $\mathrm{Ch}\{\cdots\}$ denotes the convex hull of the set $\mathrm{N}(0,0) \cup \cup_{\alpha,j}\mathrm{N}(\alpha,j)$ in $\mathbb{R}^2$. Here $\mathrm{N}(0,0) := \{(x,y);\ x \leq M,\ y \geq -M\}$.

We assume that

$\mathrm{N}(L)$ has only one side of a positive slope with $(M,-M)$ and $(\alpha_*, i(\alpha_*, M))$. (A-2)

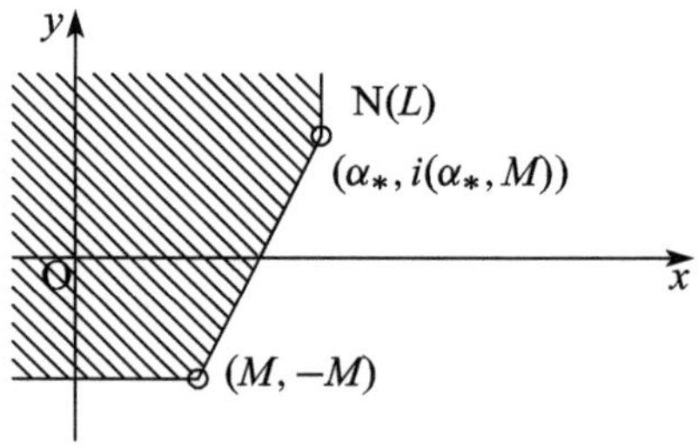

We put $i_* := i(\alpha_*, M)$. Then, we assume that for any (α, j) with $a_{\alpha j}(t) \not\equiv 0$ we have

$$\frac{\alpha_*}{i_* + M} \geq \frac{\alpha}{i(\alpha,j) + j}. \tag{A-3}$$

We put $\frac{\alpha_*}{i_*+M} = \frac{p}{q}$, which is called the modified order of the operator L, where p and q are relatively prime numbers.

Moreover, for any α $(0 \leq \alpha \leq \alpha_*)$ and j $(1 \leq j \leq M)$, we assume that

$$a_{\alpha j}(t) = \sum_{\substack{i(\alpha,j) \leq i \leq [i_* j/M], \\ i+j \in q\mathbb{N}}} a_i^{(\alpha,j)} t^i, \tag{A-4}$$

where $[\xi]$ denotes the integer part of $\xi \in \mathbb{R}$ and $\mathbb{N}$ denotes the set of non-negative integers. Especially, we have

$$a_{\alpha_* M}(t) = a_{i_*}^{(\alpha_*,M)} t^{i_*}. \tag{4}$$

In order to state our result, we define a characteristic equation for L with respect to z by

$$\sum_J a_{\frac{i_* j}{M}}^{(\frac{\alpha_* j}{M},j)} z^{\alpha_* j/M} - 1 = 0, \tag{5}$$

where $J = \{j \in \mathbb{N}; 1 \leq j \leq M,\ i_* j/M, \alpha_* j/M \in \mathbb{N}\}$. Let z_n $(n = 1, 2, \cdots, \alpha_*)$ be the roots of the characteristic equation.

Finally, we prepare the notation $S(d, \beta, \rho)$. For $d \in \mathbb{R}$, $\beta > 0$ and ρ $(0 < \rho \leq \infty)$, we define a sector $S = S(d, \beta, \rho)$ by

$$S(d, \beta, \rho) := \left\{ t \in \mathbb{C}; |d - \arg t| < \frac{\beta}{2}, 0 < |t| < \rho \right\}, \tag{6}$$

where d, β and ρ are called the direction, the opening angle and the radius of S, respectively. We write $S(d, \beta, \infty) = S(d, \beta)$ for short.

Under these preparations, our result is stated as follows.

Theorem 1.1 *Let*

$$k = \frac{i_* + M}{\alpha_* - M}. \tag{7}$$

Let $d \in \mathbb{R}$ be fixed and $d_n := qd/p - \arg z_n$ for $1 \leq n \leq \alpha_*$. We assume that the Cauchy data $\varphi(x)$ can be analytically continued in $\cup_{n=1}^{\alpha_*} S(d_n, \varepsilon)$ for some $\varepsilon > 0$, and has the exponential growth estimate of order at most $\alpha_*/(\alpha_* - M)$ there, that is,

$$|\varphi(x)| \leq C \exp\left(\delta |x|^{\alpha_*/(\alpha_* - M)}\right) \tag{8}$$

by some positive constants C and δ. Then under the assumptions (A-1)–(A-4), *the formal solution $\hat{U}(t, x)$ of* (CP) *is k-summable in d direction.*

We may assume that $k \geq 1$ without loss of generality by a change of variable, e.g. $t^{1/(\alpha_*-M)} = \tau$. This assumption is only needed for the analysis of convolution equations (see section "Proof of Proposition 7.2").

The k-summability of divergent formal solutions of non-Kowalevskian equations like heat equation has been developed by many mathematicians. Especially there are many studies for partial differential equation with constant coefficients (e.g. [8] for the heat equation, [11] for the operator $\partial_t^p - \partial_x^q, (p < q)$, [2] for general equations, [9] for moment partial differential equations). But, there are not many study for equations with variable coefficients yet. In the papers [4] and [5], we treated the equations whose coefficients are monomial of t. In the paper [7], we treated the first order equations with respect to t, whose coefficients are polynomials of t and modified order is equal to 1. In this paper, we consider the higher order equations with respect to t, whose coefficients are polynomials of t and modified order is general. The paper [6] treated the first order equations with respect to t whose coefficients are polynomials of t and modified order is general. Our theorem is a generalization of results in [4–7].

The paper consists of the following contents. We give a review of k-summability in Sect. 2. In Sect. 3, we give a construction of formal solutions of Cauchy problem (CP) by employing the method of successive approximation. We give a result of Gevrey order of formal solutions and its simple proof in Sect. 4. In Sect. 5, we introduce the moment series associated with formal solutions and the important result for k-summability of moment series is given and in Sect. 6 we will give a simple proof of Theorem 1.1. Section 7 is devoted to a proof of the result for k-summability of moment series given in Sect. 5.

2 Review of k-Summability

In this section, we give some notation and definitions in the way of Ramis or Balser (cf. Balser [1] for detail).

Let $k > 0$, $S = S(d, \beta)$ and $B(\sigma) := \{x \in \mathbb{C}; |x| \leq \sigma\}$. Let $v(t,x) \in \mathcal{O}(S \times B(\sigma))$ which means that $v(t,x)$ is holomorphic in $S \times B(\sigma)$. Then we define that $v(t,x) \in \mathrm{Exp}_t^k(S \times B(\sigma))$ if, for any closed subsector S' of S, there exist some positive constants C and δ such that

$$\max_{|x|\leq\sigma} |v(t,x)| \leq Ce^{\delta|t|^k}, \quad t \in S'. \tag{9}$$

For $k > 0$, we define that $\hat{v}(t,x) = \sum_{n=0}^{\infty} v_n(x)t^n \in \mathcal{O}_x[[t]]_{1/k}$ (we say $\hat{v}(t,x)$ is a formal power series of Gevrey order $1/k$) if $v_n(x)$ are holomorphic on a common closed disk $B(\sigma)$ for some $\sigma > 0$ and there exist some positive constants C and K such that for any n,

$$\max_{|x|\leq\sigma} |v_n(x)| \leq CK^n\Gamma\left(1 + \frac{n}{k}\right). \tag{10}$$

Here when $v_n(x) \equiv v_n$ (constants) for all n, we use the notation $\mathbb{C}[[t]]_{1/k}$ instead of $\mathcal{O}_x[[t]]_{1/k}$. In the following, we use the similar notation.

Let $k > 0$, $\hat{v}(t,x) = \sum_{n=0}^{\infty} v_n(x)t^n \in \mathcal{O}_x[[t]]_{1/k}$ and $v(t,x)$ be an analytic function on $S(d,\beta,\rho) \times B(\sigma)$. Then we define that

$$v(t,x) \cong_k \hat{v}(t,x) \quad \text{in } S = S(d,\beta,\rho), \tag{11}$$

if for any closed subsector S' of S, there exist some positive constants C and K such that for any $N \geq 1$, we have

$$\max_{|x|\leq\sigma} \left| v(t,x) - \sum_{n=0}^{N-1} v_n(x)t^n \right| \leq CK^N |t|^N \Gamma\left(1 + \frac{N}{k}\right), \quad t \in S'. \tag{12}$$

For $k > 0$, $d \in \mathbb{R}$ and $\hat{v}(t,x) \in \mathcal{O}_x[[t]]_{1/k}$, we say that $\hat{v}(t,x)$ is *k-summable* in d direction, and denote it by $\hat{v}(t,x) \in \mathcal{O}_x\{t\}_{k,d}$, if there exist a sector $S = S(d,\beta,\rho)$ with $\beta > \pi/k$ and an analytic function $v(t,x)$ on $S \times B(\sigma)$ such that $v(t,x) \cong_k \hat{v}(t,x)$ in S.

We remark that the function $v(t,x)$ above for a *k-summable* $\hat{v}(t,x)$ is unique if it exists. Therefore such a function $v(t,x)$ is called the *k-sum* of $\hat{v}(t,x)$ in d direction.

3 Construction of a Formal Solution

Decomposition of P^M

We give a decomposition of operator $P^M(t,\partial_t,\partial_x)$.

For $\ell \geq 0$, we define

$$K_\ell := \left\{ (i,j,\alpha)\ ;\ \ell = \frac{p}{q}(i+j) - \alpha,\ a_i^{(\alpha,j)} \neq 0 \right\}$$

and we put

$$P_\ell^M(t,\partial_t,\partial_x) := \sum_{(i,j,\alpha)\in K_\ell} a_i^{(\alpha,j)} t^i \partial_t^{M-j} \partial_x^\alpha.$$

In this case, we have

$$K_\ell = \left\{ (i,j,\alpha); 1 \leq j \leq M,\ i(\alpha,j) \leq i \leq \frac{i_* j}{M},\ i+j \in q\mathbb{N},\ 0 \leq \alpha = \frac{p}{q}(i+j) - \ell \right\},$$

and

$$P^M(t, \partial_t, \partial_x) = \sum_{\ell=0}^{\alpha_*} P_\ell^M(t, \partial_t, \partial_x) \tag{13}$$

because of $\ell = \frac{p}{q}(i+j) - \alpha \le \frac{p}{q}(i_* + M) = \alpha_*$.

The Successive Approximation Solutions

By employing the decomposition of P^M, we consider the following Cauchy problems for $\nu \ge 0$

$$\begin{cases} \partial_t^M u_\nu(t,x) = \displaystyle\sum_{\ell=0}^{\min\{\alpha_*,\nu\}} P_\ell^M u_{\nu-\ell}(t,x), \\ \partial_t^n u_\nu(0,x) = 0 \ (0 \le n \le M-2), \\ \partial_t^{M-1} u_\nu(0,x) = \varphi(x) \ (\nu = 0), \quad = 0 \ (\nu \ge 1). \end{cases} \tag{E_ν}$$

For each ν, the Cauchy problem (E_ν) has a unique formal solution of the form

$$\hat{u}_\nu(t,x) = \sum_{n\ge 0} u_{\nu,n}(x) \frac{t^n}{n!}. \tag{14}$$

Then $\hat{U}(t,x) = \sum_{\nu \ge 0} \hat{u}_\nu(t,x)$ is the formal solution of the Cauchy problem (CP).

Construction of a Formal Solution $\hat{u}_\nu$

We give a exact construction of the formal solution $\hat{u}_\nu(t,x)$.

Lemma 3.1 *For each ν, we have*

$$u_{\nu,n}(x) = A_\nu(n) \varphi^{\left(\frac{p}{q}(n-M+1)-\nu\right)}(x) \quad \left(\frac{p}{q}(n-M+1) - \nu \in \mathbb{N}\right)$$

and $u_{\nu,n}(x) \equiv 0$ *(*$p(n-M+1)/q-\nu \notin \mathbb{N}$*). Here* $\{A_\nu(n)\}$ *satisfy the following recurrence formula*

$$\begin{cases} A_\nu(n+M) = \displaystyle\sum_{\ell=0}^{\min\{\alpha_*,\nu\}} \sum_{K_\ell} a_i^{(\alpha,j)}[n]_i A_{\nu-\ell}(n+M-i-j) \quad (n \geq 0), \\ A_\nu(n) = 0 \quad (n < M-1), \\ A_\nu(M-1) = 1 \ (\nu = 0), \quad = 0 \ (\nu \geq 1), \end{cases} \tag{R_ν}$$

where the notation $[n]_i$ *is defined by*

$$[n]_i = \begin{cases} n(n-1)\cdots(n-i+1) & i \geq 1, \\ 1 & i = 0. \end{cases}$$

4 Gevrey Order of Formal Solution $\hat{U}$

We give the Gevrey order of formal solution $\hat{U}(t,x)$ by employing a result of Gevrey order of formal solutions $\hat{u}_\nu(t,x)$ without proof (cf. [4, 5, 7] for detail).

Proposition 4.1 *We assume* $\varphi \in \mathcal{O}_x$*. Then for each* ν*, we have* $\hat{u}_\nu(t,x) \in \mathcal{O}_x[[t]]_{1/k}$*,* $k = (i_* + M)/(\alpha_* - M)$*. More exactly, we have*

$$\max_{|x| \leq \sigma} \left| \frac{u_{\nu,n}(x)}{n!} \right| \leq \frac{AB^{\nu+n}}{\nu!} \Gamma\left(1 + \frac{n}{k}\right) \tag{15}$$

with some positive constants A, B *and* σ*.*

By Proposition 4.1 we see that $\hat{U}(t,x) \in \mathcal{O}_x[[t]]_{1/k}$. In fact, we put

$$\hat{U}(t,x) = \sum_{\nu=0}^{\infty} \hat{u}_\nu(t,x) = \sum_\nu \sum_n \frac{u_{\nu,n}(x)}{n!} t^n = \sum_n \frac{\sum_\nu u_{\nu,n}(x)}{n!} t^n =: \sum_n \frac{U_n(x)}{n!} t^n.$$

Then we have

$$\begin{aligned} \max_{|x| \leq \sigma} \left| \frac{U_n(x)}{n!} \right| &\leq \sum_\nu \max_{|x| \leq \sigma} \left| \frac{u_{\nu,n}(x)}{n!} \right| \leq \sum_\nu \frac{AB^{\nu+n}}{\nu!} \Gamma\left(1 + \frac{n}{k}\right) \\ &= AB^n \Gamma\left(1 + \frac{n}{k}\right) \sum_\nu \frac{B^\nu}{\nu!} \leq Ae^B B^n \Gamma\left(1 + \frac{n}{k}\right). \end{aligned}$$

5 Preliminaries for Proof of Theorem 1.1

We prepare some results which are employed for proving Theorem 1.1.

First, we give an important lemma for the summability theory (cf. [1, 8]).

Lemma 5.1 *Let* $k > 0$*,* $d \in \mathbb{R}$ *and* $\hat{v}(t,x) = \sum_{n=0}^{\infty} v_n(x)t^n \in \mathcal{O}_x[[t]]_{1/k}$*. Then the following statements are equivalent:*

i) $\hat{v}(t,x) \in \mathcal{O}_x\{t\}_{k,d}$*.*

ii) We put

$$v_B(s,x) = (\hat{\mathcal{B}}_k\hat{v})(s,x) := \sum_{n=0}^{\infty} \frac{v_n(x)}{\Gamma(1+n/k)} s^n, \tag{16}$$

which is called a formal k-Borel transform of $\hat{v}(t,x)$*, that is convergent in a neighborhood of* $(s,x) = (0,0)$*. Then* $v_B(s,x) \in \mathrm{Exp}_s^k(S(d,\varepsilon) \times B(\sigma))$ *for some* $\varepsilon > 0$ *and* $\sigma > 0$*.*

Now, we introduce three formal series. For $\nu \geq 0$, we define

$$\hat{f}_\nu(t) := \sum_{n\geq 0, \frac{p}{q}(n-M+1)-\nu \in \mathbb{N}} A_\nu(n)t^n = \sum\nolimits_{n\geq 0}^{(\nu)} A_\nu(n)t^n, \tag{17}$$

which are the generating functions of $\{A_\nu(n)\}$, and

$$\hat{g}_\nu(t) := \sum\nolimits_{n\geq 0}^{(\nu)} A_\nu(n) \frac{\left(\frac{p}{q}(n-M+1)-\nu\right)!}{n!} t^n \in \mathbb{C}[[t]]_{1/k}, \tag{18}$$

$$\hat{h}_\nu(t) := \sum\nolimits_{n\geq 0}^{(\nu)} A_\nu(n) \frac{\left(\frac{p}{q}(n-M+1)\right)!}{n!} t^n \in \mathbb{C}[[t]]_{1/k}, \tag{19}$$

which are called moment series of $\hat{f}_\nu$. Here, $\sum^{(\nu)}$ denotes the sum over $n \geq 0$ satisfying $p(n-M+1)/q - \nu \in \mathbb{N}$. We note that we can find $\hat{g}_\nu(t)$ in $\hat{u}_\nu(t,x)$ by the formal use of the Cauchy integral formula.

$$\begin{aligned}\hat{u}_\nu(t,x) &= \sum\nolimits_{n\geq 0}^{(\nu)} A_\nu(n)\varphi^{\left(\frac{p}{q}(n-M+1)-\nu\right)}(x)\frac{t^n}{n!} \\ &= \frac{1}{2\pi i}\oint \varphi(x+\zeta)\zeta^{\frac{p}{q}(M-1)+\nu-1}\hat{g}_\nu\left(\frac{t}{\zeta^{p/q}}\right)d\zeta.\end{aligned}$$

Moreover, we have a formal relationship between $\hat{g}_\nu$ and $\hat{h}_\nu$. For $\nu \geq 1$,

$$\hat{g}_\nu(t) = \frac{1}{\Gamma(\nu)} \int_0^1 \tau^{-\frac{p}{q}(M-1)-\nu}(1-\tau)^{\nu-1}\hat{h}_\nu(\tau^{p/q}t)d\tau \tag{20}$$

and $\hat{g}_0(t) = \hat{h}_0(t)$ when $\nu = 0$.

Next, we prepare a lemma for the summability of $\hat{h}_\nu(t)$, whose proof will be given in Sect. 7. Let s_n $(n = 1, 2, \ldots, i_* + M)$ be the roots of the following equation

$$\sum_J a_{\frac{i_* j}{M}}^{(\frac{\alpha_* j}{M}, j)} s^{\frac{i_* + M}{M} j} - 1 = 0. \tag{21}$$

This equation is obtained by substituting $z = s^{q/p}$ into the characteristic equation (5). Then we have the following lemma.

Lemma 5.2 *Let $k = (i_* + M)/(\alpha_* - M)$ and $h_{\nu B}(s) = (\hat{\mathcal{B}}_k \hat{h}_\nu)(s)$. Then we obtain $h_{\nu B}(s) \in \mathrm{Exp}_s^k(S(\theta, \varepsilon_0))$, where θ satisfies*

$$\theta \not\equiv \arg s_n \pmod{2\pi} \quad (n = 1, 2, \ldots, i_* + M) \tag{22}$$

and $\varepsilon_0 > 0$. Exactly, we obtain the following estimates for $\nu \geq 0$

$$|h_{\nu B}(s)| \leq CK^\nu \exp(\delta |s|^k), \quad s \in S(\theta, \varepsilon_0), \tag{23}$$

where positive constants C, K and δ are independent of ν.

We remark that Lemma 5.2 means that $\hat{h}_\nu(t) \in \mathbb{C}\{t\}_{k,\theta}$.

6 Proof of Theorem 1.1

By employing Lemmas 5.1 and 5.2, we obtain the following result which means that $\hat{u}_\nu(t,x) \in \mathcal{O}_x\{t\}_{k,d}$ whose proof is omitted (cf. [7–11] for detail).

Proposition 6.1 *Let $d \in \mathbb{R}$ be a fixed and put $u_{\nu B}(s,x) = (\hat{\mathcal{B}}_k \hat{u}_\nu)(s,x)$. Then if the Cauchy data $\varphi(x)$ satisfies the same assumptions as in Theorem 1.1, for all ν, we have*

$$\max_{|x| \leq \sigma} |u_{\nu B}(s,x)| \leq C \frac{K^\nu}{\nu!} \exp\left(\delta |s|^k\right), \quad s \in S(d, \varepsilon) \tag{24}$$

by some positive constants C, K, δ and σ.

We can prove Theorem 1.1 by using this proposition.

Proof Let $\hat{U}(t,x) = \sum_{\nu\geq 0} \hat{u}_\nu(t,x)$ be the formal solution of original Cauchy problem (CP). We finish the proof of Theorem 1.1 by showing that $U_B(s,x) = (\hat{\mathcal{B}}_k\hat{U})(s,x) = \sum_{\nu\geq 0} u_{\nu B}(s,x) \in \mathrm{Exp}_s^k(S(d,\varepsilon)\times B(\sigma))$. In fact, we obtain the desired estimate of $U_B(s,x)$

$$\max_{|x|\leq\sigma} |U_B(s,x)| \leq \sum_{\nu\geq 0} \max_{|x|\leq\sigma} |u_{\nu B}(s,x)| \leq C\exp(\delta|s|^k)\sum_{\nu\geq 0}\frac{K^\nu}{\nu!} = Ce^K\exp(\delta|s|^k).$$

□

7 Proof of Lemma 5.2

We shall give a proof of Lemma 5.2. For the purpose, we will obtain the differential equations of $\hat{h}_\nu$ and their convolution equations. After that, we will prove Lemma 5.2 by employing the method of successive approximation for the convolution equations.

Differential Equations of $\hat{h}_\nu$

We recall that

$$\hat{h}_\nu(t) = \sum\nolimits_{n\geq 0}^{(\nu)} A_\nu(n)\frac{\left(\frac{p}{q}(n-M+1)\right)!}{n!}t^n,$$

where $\sum^{(\nu)}$ denotes the sum over $n\geq 0$ satisfying $p(n-M+1)/q-\nu\in\mathbb{N}$. For $n\geq 0$ satisfying $p(n-M+1)/q\in\mathbb{N}$, we put

$$m(n) := \frac{\left(\frac{p}{q}(n-M+1)\right)!}{n!}. \tag{25}$$

For $n+1, n+1-i-j\in q\mathbb{N}$, we have by using formula $n! = [n]_i(n-i)!$

$$\frac{m(n+M)}{m(n+M-i-j)} = \frac{\left[\frac{p}{q}(n+1)\right]_{\frac{p}{q}(i+j)}}{[n+M]_{i+j}}.$$

By multiplying both sides of (R_ν) by $[n+M]_M m(n+M)t^{n+M}$ and taking sum $\sum_n'$ over $n \geq 0$ satisfying $p(n+1)/q - \nu \in \mathbb{N}$, we get

$$\sum_{n\geq 0}'[n+M]_M A_\nu(n+M)m(n+M)t^{n+M} = \sum_{\ell=0}^{\min\{\alpha_*,\nu\}}\sum_{K_\ell} a_i^{(\alpha,j)}$$

$$\times\sum_{n\geq 0}'[n+M]_M[n]_i\frac{\left[\frac{p}{q}(n+1)\right]_{\frac{p}{q}(i+j)}}{[n+M]_{i+j}}A_{\nu-\ell}(n+M-i-j)m(n+M-i-j)t^{n+M}.$$

Here we have $[n+M]_M[n]_i/[n+M]_{i+j} = [n+M-i-j]_{M-j}$ and by noticing $\ell = p(i+j)/q-\alpha$, $A_{\nu-\ell}(n) = 0$ if $p(n-M+1)/q-(\nu-\ell) \notin \mathbb{N}$ or $n < M-1$, we have $\sum_{n\geq 0}' A_{\nu-\ell}(n+M-i-j) = \sum_{n\geq 0}^{(\nu-\ell)} A_{\nu-\ell}(n)$. Finally, we obtain a differential equation of $\hat{h}_\nu(t)$. For $\nu \geq 0$, we have

$$[\delta_t]_M(\hat{h}_\nu(t) - c_\nu t^{M-1}) \tag{26}$$
$$= \sum_{\ell-0}^{\min\{\alpha_*,\nu\}}\sum_{K_\ell} a_i^{(\alpha,j)} t^{i+j}[\delta_t]_{M-j}\left[\frac{p}{q}(\delta_t - M + i + j + 1)\right]_{\frac{p}{q}(i+j)} \hat{h}_{\nu-\ell}(t),$$

where $\delta_t = t(d/dt)$ denotes the Euler operator and $c_0 = 1/(M-1)!$, $c_\nu = 0\ (\nu \geq 1)$, and $K_\ell - \{(l,j,\alpha);\ 1 \leq j \leq M, i(\alpha,j) \leq i \leq \frac{i*j}{M}, i+j \in q\mathbb{N}, 0 \leq \alpha = p(i+j)/q-\ell\}$.

A Canonical Form for Differential Equation of $\hat{h}_\nu$

We shall reduce the differential equation of $\hat{h}_\nu(t)$ to a certain canonical form.

Lemma 7.1 *Let $a, b \in \mathbb{R}$ and $k > 0$. Then for $n \in \mathbb{N}$, we have*

$$[a\delta_t + b + n]_n = \sum_{m=0}^{n} d_{n,m}^{[a,b]} t^{-km}(t^k\delta_t)^m, \tag{27}$$

where $d_{0,0}^{[a,b]} = 1$, and for $n \geq 1$

$$d_{n,m}^{[a,b]} = a d_{n-1,m-1}^{[a,b]} + (b+n-akm)d_{n-1,m}^{[a,b]}, \quad 0 \leq m \leq n$$

with $d_{n-1,-1}^{[a,b]} = d_{n-1,n}^{[a,b]} = 0$. Then we have $d_{n,n}^{[a,b]} = a^n$ and $d_{n,0}^{[a,b]} = [b+n]_n$.

By multiplying both sides of (26) by t^{kM} and using this lemma, we can rewrite (26) into the following form for each $\nu \geq 0$

$$\sum_{m=0}^{M} d_{M,m}^{[1,-M]} t^{k(M-m)} (t^k \delta_t)^m (\hat{h}_\nu(t) - c_\nu t^{M-1})$$
$$= \sum_{\ell=0}^{\min\{\alpha_*,\nu\}} \sum_{K_\ell} a_i^{(\alpha,j)} t^{i+j+kM} \sum_{m=0}^{\frac{p}{q}(i+j)+M-j} \mathcal{D}_{m,i,j} t^{-km} (t^k \delta_t)^m \hat{h}_{\nu-\ell}(t),$$

where $\mathcal{D}_{m,i,j} = \sum_{n_1+n_2=m} d_{\frac{p}{q}(i+j),n_1}^{[\frac{p}{q},-\frac{p}{q}(M-1)]} \times d_{M-j,n_2}^{[1,-kn_1-M+j]}$. In fact, we calculate as follows;

$$[\delta_t]_{M-j} \left[\frac{p}{q}(\delta_t - M + i + j + 1) \right]_{\frac{p}{q}(i+j)} = \sum_{n_1=0}^{\frac{p}{q}(i+j)} d_{\frac{p}{q}(i+j),n_1}^{[\frac{p}{q},-\frac{p}{q}(M-1)]} t^{-kn_1} [\delta_t - kn_1]_{M-j} (t^k \delta_t)^{n_1}.$$

Since we have

$$\sum_{K_\ell} \sum_{0 \leq m \leq \frac{p}{q}(i+j)+M-j} = \sum_{1 \leq j \leq M} \sum_{0 \leq m \leq \frac{\alpha_*}{M} j+M-j} \sum_{\substack{\max\{i(\alpha,j),\, \frac{q}{p}(m-M+j)-j\} \leq i \leq \frac{i_*}{M} j \\ \alpha = \frac{p}{q}(i+j)-\ell}},$$

we put for ℓ, j and m with $0 \leq \ell \leq \alpha_*$, $1 \leq j \leq M$ and $0 \leq m \leq \alpha_* j/M + M - j$

$$A_{jm}^{[\ell]}(t) := \sum_{\substack{\max\{i(\alpha,j),\, \frac{q}{p}(m-M+j)-j\} \leq i \leq \frac{i_*}{M} j \\ \alpha = \frac{p}{q}(i+j)-\ell}} a_i^{(\alpha,j)} \mathcal{D}_{m,i,j} t^{i+j+k(M-m)}.$$

Then we notice for any ℓ and j that when $0 \leq m < \alpha_* j/M + M - j$, we have $O(A_{jm}^{[\ell]}(t)) > 0$ and when $m = \alpha_* j/M + M - j =: m_j \in \mathbb{N}$, we have

$$A_{j,m_j}^{[\ell]}(t) \equiv a_{\frac{i_* j}{M}}^{(\frac{\alpha_* j}{M} - \ell, j)} \left(\frac{p}{q} \right)^{\frac{\alpha_* j}{M}} =: A_{[j]}^{[\ell]}.$$

Because by noticing $k = (i_* + M)/(\alpha_* - M)$ and $q/p = (i_* + M)/\alpha_*$, we have

$$i + j + k(M-m) \geq \tfrac{q}{p}(m - M + j) + k(M-m) = \tfrac{q}{p} j + (M-m)(k - \tfrac{q}{p})$$
$$= \tfrac{q}{p} j + (M-m) \tfrac{q}{p} \tfrac{M}{\alpha_* - M} = \tfrac{q}{p} \tfrac{M}{\alpha_* - M} \left\{ \tfrac{\alpha_*}{M} j + M - j - m \right\} \geq 0.$$

Therefore we can obtain the following canonical differential equation of $\hat{h}_\nu$

$$\left[(t^k\delta_t)^M - \sum_J A_{[j]}^{[0]}(t^k\delta_t)^{m_j}\right]\hat{h}_\nu(t) = c'_\nu t^{M-1+kM} \tag{28}$$

$$+ \sum_{\ell=1}^{\min\{\alpha_*.\nu\}} \sum_J A_{[j]}^{[\ell]}(t^k\delta_t)^{m_j}\hat{h}_{\nu-\ell}(t) + \sum_{\ell=0}^{\min\{\alpha_*,\nu\}} \sum_{0\le m<\alpha_*} \mathcal{A}_m^{[\ell]}(t)(t^k\delta_t)^m\hat{h}_{\nu-\ell}(t),$$

where $m_j = \frac{\alpha_* j}{M} + M - j$, $J = \{j \in \mathbb{N}; 1 \le j \le M, \frac{i_* j}{M}, \frac{\alpha_* j}{M} \in \mathbb{N}\}$ and $c'_\nu = \sum_{m=0}^M d_{M,m}^{[1,-M]} k^m[m-1+\frac{M-1}{k}]_m$. Here for $0 \le m < \alpha_*$ and $1 \le \ell \le \alpha_*$, we put

$$\mathcal{A}_m^{[\ell]}(t) := \sum_{\frac{m-M}{\alpha_*-M}M<j\le M, j\in\mathbb{N}} A_{jm}^{[\ell]}(t),$$

and when $\ell = 0$, we put

$$\mathcal{A}_m^{[0]}(t) = \begin{cases} \sum_{\frac{m-M}{\alpha_*-M}M<j\le M, j\in\mathbb{N}} A_{jm}^{[0]}(t) - d_{M,m}^{[1,-M]} t^{k(M-m)} & (0 \le m \le M-1), \\ \sum_{\frac{m-M}{\alpha_*-M}M<j\le M, j\in\mathbb{N}} A_{jm}^{[0]}(t) & (M < m < \alpha_*). \end{cases}$$

Convolution Equations

We shall obtain the convolution equations by operating the Borel transform to the canonical differential equations which are obtained in the previous subsection.

After operating the formal k-Borel transform to Eq. (28) and differentiating the both sides, we put $w_\nu(s) = D_s h_{\nu B}(s)$ or $h_{\nu B}(s) = D_s^{-1} w_\nu(s)$ for each ν, where $D_s = d/ds$ and $D_s^{-1} = \int_0^s$. Then by noticing $\hat{\mathcal{B}}_k(t^k\delta_t) = kD_s^{-1}s^k D_s$, the convolution equation for $w_\nu(s)$ is given by the following expression

$$\left[k^M s^{kM} - \sum_J A_{[j]}^{[0]} k^{m_j} s^{km_j}\right] w_\nu(s) = \sum_{\ell=1}^{\min\{\alpha_*,\nu\}} \sum_J A_{[j]}^{[\ell]} k^{m_j} s^{km_j} w_{\nu-\ell}(s) \tag{29}$$

$$+ \tilde{c}_\nu s^{M-2+kM} + \sum_{\ell=0}^{\min\{\alpha_*,\nu\}} \sum_{0\le m<\alpha_*} k^m D_s\left(\mathcal{A}_{mB}^{[\ell]}(s) *_k D_s^{-1} s^{km} w_{\nu-\ell}(s)\right),$$

where $\tilde{c}_\nu = c'_\nu k/\Gamma(M + (M-1)/k)$ and $\mathcal{A}_{mB}^{[\ell]}(s) = (\hat{\mathcal{B}}_k \mathcal{A}_m^{[\ell]})(s)$ for $0 \le m < \alpha_*$.

Here the k-convolution $a(s) *_k b(s)$ with $a(0) = b(0) = 0$ is defined by the following integral

$$(a *_k b)(s) = \int_0^s a\left((s^k - u^k)^{1/k}\right) \frac{d}{du} b(u) du. \tag{30}$$

We remark that if $a(0) = b(0) = 0$, the convolution is commutative. Note that this formula is same with that in [1, Sect. 5.3] although the expression is a little different from it.

We put

$$A_*(s) := k^M s^{kM} - \sum_J A_{[j]}^{[0]} k^{m_j} s^{km_j} = k^M s^{kM} \left[1 - \sum_J A_{[j]}^{[0]} k^{\frac{\alpha * j}{M} - j} s^{\frac{i_* + M}{M} j} \right]$$

and

$$T_\nu(w_\nu)(s) := \frac{1}{A_*(s)} \left(\text{the right hand side of (29)} \right).$$

Then we remark that for all ν, $T_\nu : \mathbb{C}[[s]] \to \mathbb{C}[[s]]$, where $\mathbb{C}[[s]]$ denotes the set of formal power series. Therefore for each ν, the function $w_\nu(s) = D_s h_{\nu B}(s) = D_s \sum_{n \geq 0}^{(\nu)} A_\nu(n) m(n) s^n / \Gamma(1 + n/k)$ is a unique holomorphic solution of the convolution equation (29) in $|s| < \sigma_*$ for some positive σ_* (cf. [5] and [6] for detail). We have to remark that $s = 0$ is a removable singularity for solutions of the convolution equations $w = T_\nu(w)$. Moreover, we have to remark that for each ν, the solution w_ν may be continued analytically on $S(\theta, \varepsilon_0)$ with $\theta \not\equiv \arg s_n$ (mod 2π) with $n = 1, 2, \ldots, i_* + M$ and $\varepsilon_0 > 0$, because the roots of $A_*(s)$ are the only singular points of the analytic convolution equations $w = T_\nu(w)$.

Proof of Lemma 5.2

We shall show that $w_\nu(s)$ has the exponential growth estimate of order at most k in a sector with infinite radius. This is same as $h_{\nu B}(s)$'s having the exponential growth estimate of order at most k in the sector. For obtaining the estimate of w_ν, we consider the convolution equation $w = T_\nu(w)$ on $S_1 := \{s \in S(\theta, \varepsilon_0); |s| \geq \sigma_*/2\}$ (cf. [3]).

Let $s_0 \in S_1$ with $|s_0| = \sigma_*/2$. We modify the operator T_ν by $\tilde{T}_\nu$ on S_1 by replacing in $w = T_\nu(w)$ the convolutions $b *_k a$ by $b \,\tilde{*}\, a$, where

$$(b \,\tilde{*}\, a)(s) := \int_{s_0}^s b\left((s^k - u^k)^{1/k}\right) \frac{d}{du} a(u) du, \quad s \in S_1.$$

Then we have the convolution equation $w_\nu = \tilde{T}_\nu(w_\nu)$ on S_1, where

$$\tilde{T}_\nu(w_\nu) = F_\nu(s) + \frac{1}{A_*(s)} \sum_{\ell=1}^{\min\{\alpha_*,\nu\}} \sum_J A_{[j]}^{[\ell]} k^{m_j} s^{km_j} w_{\nu-\ell}(s)$$
$$+\frac{1}{A_*(s)} \sum_{\ell=0}^{\min\{\alpha_*,\nu\}} \sum_{m=0}^{\alpha_*-1} k^m D_s \left(D_s^{-1} s^{km} w_{\nu-\ell} \tilde{*} \mathcal{A}_{mB}^{[\ell]}\right)(s),$$

$$F_\nu(s) = \frac{1}{A_*(s)} \left[\tilde{c}_\nu s^{kM+M-2} + \sum_{\ell=0}^{\min\{\alpha_*,\nu\}} \sum_{m=0}^{\alpha_*-1} k^m \int_0^{s_0} \mathcal{A}_{mB}^{[\ell]}\left((s^k - u^k)^{\frac{1}{k}}\right) u^{km} w_{\nu-\ell}(u) du \right].$$

We assume that for $s \in S_1$,

$$\begin{aligned} &\left|\frac{1}{A_*(s)}\right| \leq \frac{B_1}{(1+|s|^{i_*+M})|s|^{kM}}, \quad \left|D_s \mathcal{A}_{mB}^{[\ell]}(s)\right| \leq B_2 |s|^{i_*+M+k(M-m)-1} \\ &|c_\nu|, \left|A_{[j]}^{[\ell]} k^{m_j}\right| \leq B_2, \quad \frac{|s|^\mu}{1+|s|^{i_*+M}} \leq B_3 \end{aligned} \tag{31}$$

with some positive constants B_1, B_2 and B_3 satisfying $B := B_1 B_2 B_3 \geq 1$ and $\mu \leq i_* + M$ for $0 \leq m < \alpha_*$, $0 \leq \ell \leq \alpha_*$ $\nu \geq 0$ and $j \in J$.

Let $S_* := \{s \in \mathbb{C}; \theta - \varepsilon_0/2 + \gamma \leq \arg(s^k - s_0^k)^{\frac{1}{k}} \leq \theta + \varepsilon_0/2 - \gamma, |s| \geq \sigma_*/2\} \subset S_1$ for $\gamma > 0$. Then we notice that $F_\nu(s)$ are bounded above by B_0 in S_* for some $B_0 > 0$, because $w_{\nu-\ell}(s)$ are bounded on $|s| \leqq |s_0|_*$.

By employing the method of successive approximation for the convolution equation $w = \tilde{T}_\nu(w)$ on S_1, we obtain the desired exponential estimate for w_ν.

The case $\nu = 0$. We define the functions $\{w_{0,n}(s)\}$ by the following

$$\begin{cases} w_{0,0}(s) = F_0(s), \\ w_{0,n}(s) = \dfrac{1}{A_*(s)} \displaystyle\sum_{0 \leq m < \alpha_*} k^m D_s \left(D_s^{-1} s^{km} w_{0,n-1} \tilde{*} \mathcal{A}_{mB}^{[0]}\right)(s) \ (n \geq 1). \end{cases}$$

The case $\nu \geq 1$. We define the functions $\{w_{\nu,n}(s)\}$ by the following

$$\begin{cases} w_{\nu,0}(s) = F_\nu(s) + \dfrac{1}{A_*(s)} \displaystyle\sum_{\ell=1}^{\min\{\alpha_*,\nu\}} \sum_J A_{[j]}^{[\ell]} k^{m_j} s^{km_j} w_{\nu-\ell,0}(s), \\ w_{\nu,n}(s) = \dfrac{1}{A_*(s)} \left[\displaystyle\sum_{\ell=1}^{\min\{\alpha_*,\nu\}} \sum_J A_{[j]}^{[\ell]} k^{m_j} s^{km_j} w_{\nu-\ell,n}(s) \right. \\ \left. + \displaystyle\sum_{\ell=0}^{\min\{\alpha_*,\nu\}} \sum_{0 \leq m < \alpha_*} k^m D_s \left(D_s^{-1} s^{km} w_{\nu-\ell,n-1} \tilde{*} \mathcal{A}_{mB}^{[\ell]}\right)(s) \right] \ (n \geq 1). \end{cases} \tag{32}$$

We remark that from the above construction for each ν, $W_\nu(s) = \sum_{n\geq 0} w_{\nu,n}(s)$ is a convergent power series solution of the convolution equation $w_\nu(s) = T_\nu(w_\nu)(s)$ in a neighborhood of the origin, and therefore $W_\nu(s)$ coincides with $w_\nu(s) = D_s\sum_{n\geq 0}^{(\nu)} A_\nu(n)m(n)s^n/\Gamma(1+n/k)$.

Now, under the assumptions (31), the following proposition holds.

Proposition 7.2 *For each ν, we have for all n*

$$|w_{\nu,n}(s)| \leq C_\nu K^n \frac{|s|^n}{\Gamma\left(\frac{n+1}{k}\right)}, \quad s \in S_*, \tag{33}$$

where $C_0 = B_0\Gamma(1/k)$ $C_\nu = \beta C_{\nu-1}$ with $\beta \geq 2(1+BM)$ and K is a positive constant such that

$$4B\Gamma(1+1/k)\Gamma(\alpha_*)\sum_{0\leq m<\alpha_*} k^m \leq K. \tag{34}$$

This proposition means that $w_\nu(s) = \sum w_{\nu,n}(s)$ has the exponential growth estimate of order at most k in $S(\theta,\varepsilon_0)$ by changing the position of s_0. Therefore the estimate of $h_{\nu B}(s)$ in Lemma 5.2 follows from this proposition immediately.

Proof of Proposition 7.2

We shall prove Proposition 7.2 by induction on ν and n.

The Case $\nu = 0$ When $n = 0$, it is trivial from the estimate $|F_0(s)| \leq B_0$ for $s \in S_*$.

When $n \geq 1$, we have for $s \in S_*$

$$|w_{0,n}(s)| \leq \left|\frac{1}{A_*(s)}\right| \sum_{m=0}^{\alpha_*-1} k^m \left|D_s\left(D_s^{-1}s^{km}w_{0,n-1} \tilde{*} \mathcal{A}_{mB}^{[\ell]}\right)(s)\right| =: \left|\frac{1}{A_*(s)}\right| \sum_{m=0}^{\alpha_*-1} k^m I_m.$$

Since

$$\begin{aligned} I_m &= \left|D_s \int_{s_0}^{s} D_s^{-1}(s^{km}w_{0,n-1}(s))\big|_{s=(s^k-u^k)^{\frac{1}{k}}} D_u\mathcal{A}_{mB}^{[\ell]}(u)du\right| \\ &= \left|s^{k-1}\int_{s_0}^{s} w_{0,n-1}\left((s^k-u^k)^{\frac{1}{k}}\right)(s^k-u^k)^{m+\frac{1}{k}-1}D_u\mathcal{A}_{mB}^{[\ell]}(u)du\right|, \end{aligned}$$

we get from (31) and the assumption of induction

$$
\begin{aligned}
I_m &\leq B_2 C_0 \frac{K^{n-1}}{\Gamma(n/k)} |s|^{k-1} \int_{s_0}^{s} |s^k - u^k|^{\frac{n-1}{k}+m+\frac{1}{k}-1} |u|^{i_*+M+k(M-m)-1} |du| \\
&\leq B_2 C_0 \frac{K^{n-1}}{\Gamma(n/k)} |s|^{n+(i_*+M-1)+kM} \int_0^1 (1-t^k)^{\frac{n}{k}+m-1} t^{i_*+M+k(M-m)-1} dt.
\end{aligned}
$$

Here, we use the following formulas.

Lemma 7.3 *Let $p, q > 0$, $k \geq 1$. Then we have*

$$
\int_0^1 (1-t^k)^{p-1} t^{q-1} dt = \frac{1}{k} \frac{\Gamma(p)\Gamma(q/k)}{\Gamma(p+q/k)}, \qquad \frac{1}{\Gamma(n/k)} \leq \frac{\frac{n}{k}\Gamma(1/k)}{\Gamma(\frac{n+1}{k})}.
$$

Then we have

$$
|w_{0,n}(s)| \leq C_0 K^{n-1} \frac{|s|^n}{\Gamma((n+1)/k)} B \frac{\Gamma(1/k)}{k} \Gamma(\alpha_*) \sum_{0 \leq m < \alpha_*} k^m.
$$

Therefore from the condition (34) of K, we obtain the desired estimate of $w_{0,n}(s)$.

The Case $\nu \geq 1$ We assume that the estimates (33) hold up to $\nu - 1$.

When $n = 0$, by noticing $km_j = i_* + M + kM$, we have for $s \in S_*$

$$
|w_{\nu,0}(s)| \leq B_0 + B \sum_{\ell=1}^{\nu} \sum_J C_{\nu-\ell} \frac{1}{\Gamma(1/k)} \leq C_0 \frac{1}{\Gamma(1/k)} + BM \frac{1}{\Gamma(1/k)} \sum_{\ell=1}^{\nu} C_{\nu-\ell}.
$$

From $C_\nu = \beta C_{\nu-1}$, $\beta \geq 2(1+BM) \geq 4$ and $BM/\beta \leq 1/2$, we have

$$
C_0 = \frac{C_\nu}{\beta^\nu} \leq \frac{C_\nu}{4}, \quad BM \sum_{\ell=1}^{\nu} C_{\nu-\ell} = BM \sum_{\ell=1}^{\nu} \frac{C_\nu}{\beta^\ell} = \frac{C_\nu}{2} \sum_{\ell=1}^{\nu} \frac{1}{4^{\ell-1}} \leq \frac{2}{3} C_\nu.
$$

Therefore we obtain the desired estimate of $w_{\nu,0}(s)$.

When $n \geq 1$, from the expression (32) we put $w_{\nu,n}(s) =: J_1(s) + J_2(s)$.

Similarly with the estimate of $w_{\nu,0}(s)$, we have

$$
J_1 \leq \frac{2}{3} C_\nu K^n \frac{|s|^n}{\Gamma((n+1)/k)}.
$$

Similarly with the estimates of $w_{0,n}(s)$ and $w_{\nu,0}(s)$, we have

$$J_2 \le \sum_{\ell=0}^{\nu} C_{\nu-\ell} \frac{K^{n-1}|s|^n}{\Gamma((n+1)/k)} B\Gamma(1+1/k)\Gamma(\alpha_*) \sum_{m=0}^{\alpha_*-1} k^m$$
$$\le \frac{1}{4} \frac{K^n|s|^n}{\Gamma((n+1)/k)} \sum_{\ell=0}^{\nu} C_{\nu-\ell} \le \frac{1}{3} C_\nu \frac{K^n|s|^n}{\Gamma((n+1)/k)}.$$

Therefore we obtain the desired estimate of $w_{\nu,n}(s)$.

Under the above observations, the proof of Proposition 7.2 is completed.

Acknowledgements The author is most grateful to the referees for their careful reading and kind comments and suggestions to improve this paper. The author is partially supported by the Grant-in-Aid for Scientific Research No. 15K04898 of Japan Society for the Promotion of Science.

References

1. W. Balser, *From Divergent Power Series to Analytic Functions*. Springer Lecture Notes in Mathematics, No. 1582 (Springer, Berlin, 1994)
2. W. Balser, Summability of power-series solutions of partial differential equations with constant coefficients. J. Math. Sci. **124**(4), 5085–5097 (2004)
3. B.L.J. Braaksma, Multisummability and stokes multipliers of linear meromorphic differential equations. J. Differ. Equ. **92**(1), 45–75 (1991)
4. K. Ichinobe, Summability of formal solution of Cauchy problem for some PDE with variable coefficients. Recent development of micro-local analysis for the theory of asymptotic analysis. RIMS Kôkyûroku Bessatsu **B40**, 081–094 (2013)
5. K. Ichinobe, On a k-summability of formal solutions for a class of partial differential operators with time dependent coefficients. J. Differ. Equ. **257**(8), 3048–3070 (2014)
6. K. Ichinobe, k-summability of formal solutions for certain partial differential equations and the method of successive approximation. RIMS Kôkyûroku (2020, to appear)
7. K. Ichinobe, M. Miyake, On k-summability of formal solutions for certain partial differential operators with polynomial coefficients. Opusc. Math. **35**(5), 625–653 (2015)
8. D. Lutz, M. Miyake, R. Schäfke, On the Borel summability of divergent solutions of the heat equation. Nagoya Math. J. **154**, 1–29 (1999)
9. S. Michalik, Analytic solutions of moment partial differential equations with constant coefficients. Funkcial. Ekvac **56**, 19–50 (2013)
10. S. Michalik, On the summability of formal solutions to some linear partial differential equations. Manuscript (2009) (available on the web page https://www.ma.utexas.edu/mp_arc/c/09/09-96.pdf)
11. M. Miyake, Borel summability of divergent solutions of the Cauchy problem to non-Kowalevskian equations, in *Partial Differential Equations and Their Applications (Wuhan, 1999)* (World Scientific Publishing, River Edge, 1999), pp. 225–239

On Stokes Phenomena for the Alternate Discrete PI Equation

Nalini Joshi and Yoshitsugu Takei

Abstract The alternate discrete PI equation (or alt-dPI) arises from the continuous second Painlevé equation (PII) through Bäcklund transformations. In this announcement paper we consider Stokes phenomena for (alt-dPI) from the viewpoint of exact WKB analysis. After constructing transseries solutions and defining the Stokes geometry of (alt-dPI), we derive explicit connection formulas that describe Stokes phenomena for transseries solutions of (alt-dPI) on its Stokes curves. The derivation is based on the computation of Stokes multipliers of the Lax pair associated with (PII) and (alt-dPI). The detailed proof and computations will be discussed in our forthcoming paper.

Keywords Connection formula • Discrete Painlevé equation • Exact WKB analysis • Lax pair • Stokes geometry • Stokes phenomenon • Transseries solution

Mathematics Subject Classification (2000). Primary 34M60; Secondary 34E20, 34M40, 34M55, 33E17, 34K25, 34K26, 39A45

1 Introduction

As is well-known, solutions of the second Painlevé equation

$$\frac{d^2u}{dz^2} = 2u^3 + zu + c \qquad \text{(PII)}$$

N. Joshi
The University of Sydney, NSW 2006, Australia
e-mail: nalini.joshi@sydney.edu.au

Y. Takei (✉)
Department of Mathematical Sciences, Doshisha University, Kyotanabe, Kyoto 610-0321, Japan
e-mail: ytakei@mail.doshisha.ac.jp

G. Filipuk et al. (eds.), *Analytic, Algebraic and Geometric Aspects of Differential Equations*, Trends in Mathematics, DOI 10.1007/978-3-319-52842-7_10

admit a Bäcklund transformation, that is, if u is a solution of (PII), then $\overline{u}$ (resp., $\underline{u}$) defined by

$$\overline{u} = -u - \frac{c+1/2}{u^2 + du/dz + z/2} \quad \left(\text{resp., } \underline{u} = -u - \frac{c-1/2}{u^2 - du/dz + z/2}\right) \tag{1}$$

satisfies the same Eq. (PII) with the parameter c being shifted by 1 (resp., -1). Hence, regarding $\overline{u}$ (resp., $\underline{u}$) as a shift of u by 1 (resp., -1) with respect to the parameter c

$$\overline{u} = u\Big|_{c \mapsto c+1} \quad \left(\text{resp., } \underline{u} = u\Big|_{c \mapsto c-1}\right) \tag{2}$$

and eliminating du/dz from the defining Eq. (1) of $\overline{u}$ and $\underline{u}$, we obtain a discrete Painlevé equation known as (alt-dPI):

$$\frac{c+1/2}{\overline{u}+u} + \frac{c-1/2}{\underline{u}+u} + 2u^2 + z = 0. \tag{alt-dPI}$$

Note that the roles of the variables z and c are interchanged here from the original Eq. (PII), that is, the independent variable of (alt-dPI) is c and z is just a parameter there. Now the purpose of this paper is to make an asymptotic study of solutions of (alt-dPI) for an (arbitrarily) fixed z, in particular, to discuss Stokes phenomena for (alt-dPI) for fixed z from the viewpoint of exact WKB analysis; To be more specific, we will present explicit connection formulas that describe Stokes phenomena for formal transseries solutions of (alt-dPI).

The discrete Painlevé equation (alt-dPI) first appeared in [10] as a contiguity relation of (PII). Then Fokas et al. [7] discussed (alt-dPI) and (alt-dPII) together with their continuum limits. Later Ramani et al. [17] considered coalescence limits of contiguity relations of the sixth Painlevé equation (PVI) and showed how (alt-dPI) arises from this approach. However, as far as the authors know, studies of solutions of discrete Painlevé equations are focused mainly on their special solutions and there are not so many known results about Stokes phenomena for discrete Painlevé equations. One exception is a joint paper [12] of the first author with Lustri. In [12] Stokes phenomena for the discrete PI equation were investigated from the viewpoint of exponential asymptotics or the hyperasymptotic analysis (cf. [4]).

In this paper we employ the exact WKB analysis to analyze Stokes phenomena for (alt-dPI). To this aim we first introduce a large parameter η (i.e., inverse of the semi-classical parameter) into the equations by scaling of variables

$$u = \eta^{1/3}\lambda, \quad z = \eta^{2/3}t, \quad c = \eta\zeta. \tag{3}$$

This scaling of variables transforms (PII), (1), (2) and (alt-dPI) into

$$\eta^{-2}\frac{d^2\lambda}{dt^2} = 2\lambda^3 + t\lambda + \zeta, \tag{PII}$$

$$\overline{\lambda} = \lambda\Big|_{\zeta\mapsto\zeta+\eta^{-1}} = -\lambda - \frac{\zeta+\eta^{-1}/2}{\lambda^2+\eta^{-1}(d\lambda/dt)+t/2}, \tag{4}$$

$$\left(\text{resp.}, \underline{\lambda} = \lambda\Big|_{\zeta\mapsto\zeta-\eta^{-1}} = -\lambda - \frac{\zeta-\eta^{-1}/2}{\lambda^2-\eta^{-1}(d\lambda/dt)+t/2}, \right) \tag{5}$$

$$\frac{\zeta+\eta^{-1}/2}{\overline{\lambda}+\lambda} + \frac{\zeta-\eta^{-1}/2}{\underline{\lambda}+\lambda} + 2\lambda^2 + t = 0. \tag{alt-dPI}$$

(To the scaled equations, for the sake of simplicity, we attach the same symbols (PII) and (alt-dPI) as to the original ones. We hope there will be no fear of confusions.) The exact WKB analysis, i.e., WKB analysis based on the Borel resummation technique, was initiated by Silverstone [18] and Voros [22] for one-dimensional stationary Schrödinger equations and later developed by the French school and the Japanese school ([5, 6, 15] and references cited therein). The exact WKB analysis was generalized also to the continuous Painlevé equations with a large parameter in a series of papers [2, 13, 14, 19] of the second author with Aoki and Kawai and, as its consequence, the connection formula for Ablowitz-Segur's connection problem for the second Painlevé equation (PII) was reobtained by the exact WKB approach [20]. We should also mention that, by using the exact WKB analysis, Iwaki studied in [9], which is closely related to this paper, some Stokes phenomena (called "parametric Stokes phenomena") for solutions of (PII) that are observed when the parameter ζ (or c) changes. In this paper, extending the exact WKB analysis to discrete equations with a large parameter, we explicitly discuss Stokes phenomena for (alt-dPI). Here we only make an announcement of our results. The details will be discussed in our forthcoming paper.

The plan of the paper is as follows: In Sect. 2 we construct formal transseries solutions of (alt-dPI) that will be used for the description of Stokes phenomena. Then we define the Stokes geometry, that is, turning points and Stokes curves of (alt-dPI) in Sect. 3. Finally, in Sect. 4, we present explicit connection formulas describing Stokes phenomena for the transseries solutions of (alt-dPI) constructed in Sect. 2.

2 Formal Solutions of (alt-dPI)

The most important ingredients of the exact WKB analysis for differential equations are formal solutions (e.g., WKB solutions in the case of one-dimensional stationary Schrödinger equations), Stokes geometry (i.e., turning points and Stokes curves) and connection formulas that describe Stokes phenomena on Stokes curves. The situation is the same also for the discrete Painlevé equation (alt-dPI). Let us start with the construction of formal solutions of (alt-dPI).

Formal Power Series Solution

We first note that, replacing the shift operators $\overline{\lambda}$ and $\underline{\lambda}$ by $\sum_{n\geq 0} \frac{\eta^{-n}}{n!}\frac{\partial^n \lambda}{\partial \zeta^n}$ and $\sum_{n\geq 0}(-1)^n \frac{\eta^{-n}}{n!}\frac{\partial^n \lambda}{\partial \zeta^n}$, respectively, we can regard (alt-dPI) as an ∞-order differential equation of WKB type. Then, in parallel to the case of continuous Painlevé equations (cf. [13]), we can readily construct the following formal power series solution of (alt-dPI):

$$\lambda^{(0)} = \lambda_0(\zeta) + \eta^{-1}\lambda_1(\zeta) + \eta^{-2}\lambda_2(\zeta) + \cdots, \tag{6}$$

where the top order term $\lambda_0(\zeta)$ satisfies

$$2\lambda_0^3 + t\lambda_0 + \zeta = 0 \tag{7}$$

and the lower order terms $\lambda_j(\zeta)$ $(j \geq 1)$ are uniquely determined in a recursive manner. Here, although each $\lambda_j(\zeta)$ depends also on t, we do not specify the dependence of $\lambda_j(\zeta)$ on t as we are considering (alt-dPI) with keeping t fixed.

For the formal power series solution the following holds.

Proposition 2.1 *The formal power series solution* $\lambda^{(0)}$ *of* (alt-dPI) *coincides with the formal power series solution of* (PII).

Proposition 2.1 suggests that we should consider not only (alt-dPI) but also (PII) simultaneously or, equivalently, we should consider a system of differential equations

$$\begin{cases} \eta^{-2}\dfrac{d^2\lambda}{dt^2} = 2\lambda^3 + t\lambda + \zeta, \\ \overline{\lambda}\left(\stackrel{\text{def}}{=} \displaystyle\sum_{n=0}^{\infty} \frac{\eta^{-n}}{n!}\frac{\partial^n \lambda}{\partial \zeta^n}\right) = -\lambda - \dfrac{\zeta + \eta^{-1}/2}{\lambda^2 + \eta^{-1}(d\lambda/dt) + t/2}. \end{cases} \tag{8}$$

In what follows we deal with this system (8) to construct formal solutions of (alt-dPI).

Transseries Solution

It is not possible to describe Stokes phenomena solely in terms of formal power series solutions as they are almost unique. To describe Stokes phenomena we need more general solutions, called transseries solutions, which contain free parameters.

Transseries solutions of (8) are solutions of the following form:

$$\lambda = \lambda^{(0)} + \lambda^{(1)} + \cdots , \tag{9}$$

where $\lambda^{(0)}$ is a formal power series solution and $\lambda^{(1)} + \cdots$ are assumed to be exponentially small compared to $\lambda^{(0)}$. Then the subleading term $\lambda^{(1)}$ should satisfy the Fréchet derivative (i.e., variational equation) of (8) along $\lambda^{(0)}$, that is, $\lambda^{(1)}$ satisfies

$$\begin{cases} \eta^{-2}\dfrac{d^2}{dt^2}\lambda^{(1)} = \left(6(\lambda^{(0)})^2 + t\right)\lambda^{(1)}, \\ \overline{\lambda^{(1)}} = \lambda^{(1)} + \left(\zeta + \dfrac{\eta^{-1}}{2}\right)\dfrac{2\lambda^{(0)}\lambda^{(1)} + \eta^{-1}(d\lambda^{(1)}/dt)}{((\lambda^{(0)})^2 + \eta^{-1}(d\lambda^{(0)}/dt) + t/2)^2}. \end{cases} \tag{10}$$

In particular, $\lambda^{(1)}$ is a WKB solution of this linear system (10). Furthermore, the remainder part (i.e., lower order terms $\lambda^{(k)}$ $(k \geq 2)$ of $\lambda^{(0)} + \lambda^{(1)} + \lambda^{(2)} + \cdots$) are recursively determined. Thus we obtain

Proposition 2.2 *For a given infinite series α specified by* (14) *below the system* (8) *has the following transseries solution:*

$$\lambda(t, \zeta, \eta; \alpha) = \lambda^{(0)} + \eta^{-1/2}\alpha\,\lambda^{(1)} + (\eta^{-1/2}\alpha)^2\,\lambda^{(2)} + \cdots , \tag{11}$$

where $\lambda^{(0)}$ is a formal power series solution and $\lambda^{(k)}$ $(k \geq 1)$ is of the form

$$\exp\left(k\eta \int_{(t_0,\zeta_0)}^{(t,\zeta)} \omega_{-1}\right) \sum_{n=0}^{\infty} \eta^{-n}\lambda_n^{(k)}(t, \zeta). \tag{12}$$

Here $\omega_{-1} = S_{-1}(t, \zeta)\,dt + Z_{-1}(t, \zeta)\,d\zeta$ is a closed 1*-form explicitly given by*

$$S_{-1} = \sqrt{6\lambda_0^2 + t}, \quad Z_{-1} = \cosh^{-1}\left(\frac{8\lambda_0^3 - \zeta}{\zeta}\right), \tag{13}$$

and α is an infinite series of the form

$$\alpha = \sum_{l=0}^{\infty} \alpha_l e^{2\pi i l \eta \zeta} \qquad (\alpha_l \in \mathbb{C}). \tag{14}$$

Remark 2.3 (i) $\overrightarrow{\alpha} = (\alpha_0, \alpha_1, \alpha_2, \ldots)$ gives a set of free parameters of a transseries solution of (8).

(ii) If we fix t and regard $\lambda(t, \zeta, \eta; \alpha)$ as a formal series depending on the variable ζ, then we obtain a formal solution of (alt-dPI).

3 Stokes Geometry of (alt-dPI)

We next consider the Stokes geometry of (alt-dPI). In what follows, as we are interested in the analysis of (alt-dPI), we let t be fixed.

The transseries solution of (alt-dPI) is provided by (11) (with t being fixed). There the phase factor of $\lambda^{(1)}$ (in the ζ-direction) is given by

$$Z_{-1,(\pm,l)}(\zeta) = \operatorname{Cosh}^{-1}\left(\frac{8\lambda_0^3 - \zeta}{\zeta}\right) + 2\pi i l$$

$$= \operatorname{Log}\left(\frac{8\lambda_0^3 - \zeta}{\zeta} \pm \frac{4\lambda_0^2}{\zeta}\sqrt{6\lambda_0^2 + t}\right) + 2\pi i l. \tag{15}$$

Here we use the suffix $(\pm, l)$ ($l \in \mathbb{Z}_{\geq 0}$) to specify the branch of $Z_{-1}(\zeta)$. The Stokes geometry, that is, turning points and Stokes curves of (alt-dPI) are defined by this phase factor $Z_{-1,(\pm,l)}(\zeta)$ as follows:

Definition 3.1

(i) A point $\zeta = \widehat{\zeta}$ is said to be a turning point of (alt-dPI) if there exist two suffices $(*, l) \neq (*', l')$ for which

$$Z_{-1,(*,l)}(\widehat{\zeta}) = Z_{-1,(*',l')}(\widehat{\zeta}) \tag{16}$$

holds.

(ii) A Stokes curve of (alt-dPI) is defined by

$$\Im \int_{\widehat{\zeta}}^{\zeta} (Z_{-1,(*,l)} - Z_{-1,(*',l')})d\zeta = 0, \tag{17}$$

where $\widehat{\zeta}$ is a turning point at which (16) is satisfied.

Example Let t be fixed at $t = e^{\pi i/6}$. Then the Stokes geometry of (alt-dPI) (for $t = e^{\pi i/6}$) is given by Fig. 1. Note that, as turning points and Stokes curves are defined through the top order term λ_0 of the formal power series solution and λ_0 is an algebraic function satisfying (7), Fig. 1 is drawn on the Riemann surface of λ_0. To be more precise, Fig. 1 is drawn on λ_0-plane since λ_0 itself gives a global coordinate of the Riemann surface.

When $t = e^{\pi i/6}$, there exist seven turning points. Among them p_1 and p_2 are turning points where the relation $Z_{-1,(+,l)}(p_k) = Z_{-1,(-,l)}(p_k)$ holds. Such a turning

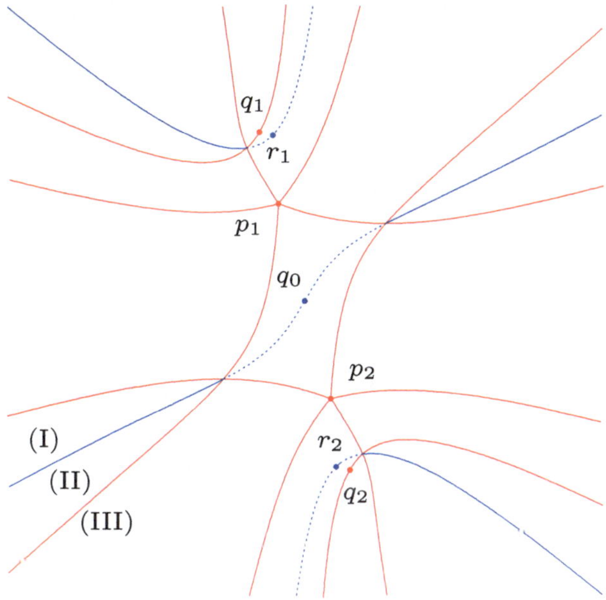

Fig. 1 Stokes geometry of (alt-dPI) for $t = e^{\pi i/6}$

point is said to be of type $((+, l), (-, l))$. Similarly, q_k $(k = 0, 1, 2)$ is a turning point of type $((+, l), (+, l'))$ and r_k $(k = 1, 2)$ is a turning point of type $((+, l), (-, l+2))$.

Remark 3.2 As is easily seen in Fig. 1, there are several crossing points of Stokes curves for (alt-dPI). This is because (alt-dPI) is considered to be an ∞-order differential equation. As a matter of fact, such crossing points of Stokes curves often appear for higher order ordinary differential equations with a large parameter. Thus some Stokes curves of Fig. 1 can be regarded as nonlinear analogue of "new Stokes curves" emanating from "virtual turning points" introduced by Berk et al. [3] and Aoki et al. [1], respectively. We refer the reader to [8] for more details of new Stokes curves and virtual turning points.

4 Stokes Phenomena for (alt-dPI)

In the preceding section we defined and presented an example (Fig. 1) of the Stokes geometry of (alt-dPI). On each Stokes curve a Stokes phenomenon is expected to occur with transseries solutions of (alt-dPI). In this section, using linear differential-difference equations ("Lax pair") associated with (PII) and (alt-dPI), we analyze such Stokes phenomena and seek for connection formulas describing them in an explicit manner.

Lax Pair Associated with (PII) and (alt-dPI) and Its Stokes Geometry

The following system of linear differential-difference equations are associated with (PII) and (alt-dPI), that is, (PII) and (alt-dPI) describe its compatibility condition [11, 16]. The system (18)–(20) is often called the "Lax pair".

$$\left(\eta^{-2}\frac{\partial^2}{\partial x^2} - Q_{\mathrm{II}}\right)\psi = 0, \tag{18}$$

$$\eta^{-1}\frac{\partial\psi}{\partial t} = A_{\mathrm{II}}\,\eta^{-1}\frac{\partial\psi}{\partial x} - \frac{\eta^{-1}}{2}\frac{\partial A_{\mathrm{II}}}{\partial x}\psi, \tag{19}$$

$$\overline{\psi} = g_{\mathrm{II}}\,\eta^{-1}\frac{\partial\psi}{\partial x} + f_{\mathrm{II}}\,\psi. \tag{20}$$

Here the explicit form of the coefficient functions Q_{II}, A_{II}, f_{II} and g_{II} is given as follows:

$$Q_{\mathrm{II}} = x^4 + tx^2 + 2\zeta x + \nu^2 - (\lambda^4 + t\lambda^2 + 2\zeta\lambda) - \eta^{-1}\frac{\nu}{x-\lambda} + \eta^{-2}\frac{3}{4(x-\lambda)^2}, \tag{21}$$

$$A_{\mathrm{II}} = \frac{1}{2(x-\lambda)}, \tag{22}$$

$$g_{\mathrm{II}} = \left((2\nu + 2\lambda^2 + t)(x-\lambda)(x-\overline{\lambda})\right)^{-1/2}, \tag{23}$$

$$f_{\mathrm{II}} = \left(x^2 - \lambda^2 - \nu + \eta^{-1}\frac{1}{2(x-\lambda)}\right) g_{\mathrm{II}}. \tag{24}$$

We use this system (18)–(20), especially the first Eq. (18) in the x-variable, to analyze the Stokes phenomena for (alt-dPI).

The following properties of the Lax pair play an important role in our discussion:

Fundamental properties of the Lax pair

(i) *Suppose that analytic solutions of* (10)*, i.e., solutions of the simultaneous equations* (PII) *and* (alt-dPI) *are substituted into the coefficients of the Lax pair* (18)–(20)*. Then the Stokes multipliers of* (18) *become analytic functions of* (t, ζ)*. In particular, they are independent of* t *thanks to the isomonodromic property.*

(ii) *Stokes multipliers of* (18) *can be explicitly computed by applying the exact WKB analysis to* (18)*. The computation is based (hence heavily depends) on the Stokes geometry of* (18)*.*

(iii) *On each Stokes curve of* (alt-dPI) *some degenerate configuration is observed for the Stokes geometry of* (18). *That is, if* ζ, *or more precisely a point* $\lambda_0(\zeta)$ *on the Riemann surface of* λ_0 *corresponding to* ζ, *is located on a Stokes curve of* (alt-dPI), *there exist two turning points of* (18) *that are connected by a Stokes curve of* (18).

Among these three properties the third one is crucially important. To see the relationship between the Stokes geometry of (alt-dPI) and that of (18) more concretely, let us take three regions (I), (II) and (III) on the Riemann surface of λ_0 specified in Fig. 1 and observe the configuration of Stokes curves of (18) when $\lambda_0(\zeta)$ belongs to each Region (*J*) ($J = \mathrm{I}, \mathrm{II}, \mathrm{III}$). In what follows we take a turning point as an endpoint (t_0, ζ_0) of the integral in the definition of a transseries solution $\lambda(t, \zeta, \eta; \alpha)$ and substitute $\lambda(t, \zeta, \eta; \alpha)$ thus normalized into the coefficients of the Lax pair (18)–(20).

Example Figure 2 shows the Stokes geometry of (18) when $\lambda_0(\zeta)$ belongs to each Region (*J*) ($J = \mathrm{I}, \mathrm{II}, \mathrm{III}$). Equation (18) has one double turning point at $x = \lambda_0$ and two simple turning points at a_1 and a_2. As is easily surmised from comparison between (I) and (II) of Fig. 2, two simple turning points a_1 and a_2 are connected by a Stokes curve in the transition from Region (I) to Region (II). This degenerate configuration is observed exactly on a Stokes curve of (alt-dPI) separating two Regions (I) and (II) in Fig. 1. Similarly, in the transition from Region (II) to Region

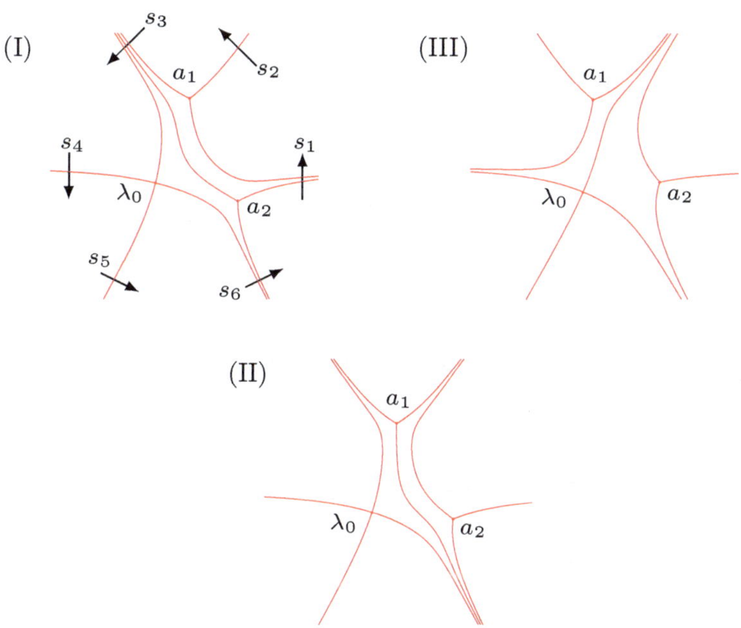

Fig. 2 Stokes geometry of (18) in Region (*J*) ($J = \mathrm{I}, \mathrm{II}, \mathrm{III}$)

(III) a double turning point λ_0 and a simple turning point a_1 are connected on a Stokes curve of (alt-dPI) separating Regions (II) and (III).

Stokes Multipliers of (18)

Applying the exact WKB analysis to the linear equation (18) or, more specifically, using the connection formulas for second order linear ordinary differential equations in view of the configuration of Stokes curves given by Fig. 2, we can compute the Stokes multipliers of (18). Such computation of Stokes multipliers through the exact WKB analysis was done in [19] for the linear equation associated with the first Painlevé equation and in [9] for that associated with the second Painlevé equation. Adjusting the computations in [19] and [9] to the current situation, we obtain the following explicit formulas for Stokes multipliers of (18). Note that in the case of (18) there exist six Stokes multipliers s_k $(k = 1, \ldots, 6)$ corresponding to six Stokes directions along which Stokes curves asymptotically tend to an irregular singular point $x = \infty$ (cf. Fig. 2, (I)). Since the computations depend on the Stokes geometry, the expressions for s_j differ according as $\lambda_0(\zeta)$ belongs to Region (I), (II) or (III). Thus in the following formulas we use the notation $s_k^{(J)}$ $(k = 1, \ldots, 6,$ $J = \mathrm{I}, \mathrm{II}, \mathrm{III})$ to denote the Stokes multipliers when $\lambda_0(\zeta)$ belongs to Region (J).

Stokes multipliers of (18) when $\lambda_0(\zeta)$ belongs to Region (I)

$$\begin{cases} s_1^{(\mathrm{I})} = ie^{V}(1 + e^{2\pi i\eta\zeta}), \\ s_2^{(\mathrm{I})} = ie^{-V}e^{-2\pi i\eta\zeta}, \\ s_3^{(\mathrm{I})} = ie^{V}(1 + e^{-2\pi i\eta\zeta}), \\ s_4^{(\mathrm{I})} = -2\sqrt{\pi}\alpha_{\mathrm{I}}e^{-V}, \\ s_5^{(\mathrm{I})} = 0, \\ s_6^{(\mathrm{I})} = (2\sqrt{\pi}\alpha_{\mathrm{I}} + i)e^{-V}. \end{cases} \tag{25}$$

Stokes multipliers of (18) when $\lambda_0(\zeta)$ belongs to Region (II)

$$\begin{cases} s_1^{(\mathrm{II})} = ie^{V}, \\ s_2^{(\mathrm{II})} = ie^{-V}(1 + e^{-2\pi i\eta\zeta}), \\ s_3^{(\mathrm{II})} = ie^{V}e^{-2\pi i\eta\zeta}, \\ s_4^{(\mathrm{II})} = -2\sqrt{\pi}\alpha_{\mathrm{II}}e^{-V}, \\ s_5^{(\mathrm{II})} = 0, \\ s_6^{(\mathrm{II})} = (2\sqrt{\pi}\alpha_{\mathrm{II}} + i(1 + e^{2\pi i\eta\zeta}))e^{-V}. \end{cases} \tag{26}$$

Stokes multipliers of (18) when $\lambda_0(\zeta)$ belongs to Region (III)

$$\begin{cases} s_1^{(\mathrm{III})} = ie^V, \\ s_2^{(\mathrm{III})} = ie^{-V}(1+e^{-2\pi i\eta\zeta}), \\ s_3^{(\mathrm{III})} = ie^V e^{-2\pi i\eta\zeta}, \\ s_4^{(\mathrm{III})} = (-2\sqrt{\pi}\alpha_{\mathrm{III}} + ie^{2\pi i\eta\zeta})e^{-V}, \\ s_5^{(\mathrm{III})} = 0, \\ s_6^{(\mathrm{III})} = (2\sqrt{\pi}\alpha_{\mathrm{III}} + i)e^{-V}. \end{cases} \tag{27}$$

In these formulas α_J ($J = \mathrm{I}, \mathrm{II}, \mathrm{III}$) denotes the free parameter of a transseries solution given by (14) when $\lambda_0(\zeta)$ belongs to Region (J), and V designates the following formal series:

$$V = \sum_{n=1}^{\infty} \frac{2^{1-2n}-1}{2n(2n-1)} B_{2n}(\eta\zeta)^{1-2n}, \tag{28}$$

where B_{2n} stands for the Bernoulli number defined by

$$\frac{w}{e^w - 1} = 1 - \frac{w}{2} + \sum_{n=1}^{\infty} \frac{B_{2n}}{(2n)!} w^{2n}. \tag{29}$$

The formal series V is called the "Voros coefficient", which appears from the comparison between WKB solutions of (18) normalized at a simple turning point and those normalized at $x = \infty$. For more details see [9]. (See also [21].)

Connection Formula for (alt-dPI)

Let us assume that the two transseries solutions $\lambda(t, \zeta, \eta; \alpha_J)$ in Region (J) and $\lambda(t, \zeta, \eta; \alpha_{J+1})$ in Region ($J+1$) should define the same analytic solution of (alt-dPI). Then, thanks to the fundamental property (i), $s_k^{(J)} = s_k^{(J)}(\alpha_J)$ and $s_k^{(J+1)} = s_k^{(J+1)}(\alpha_{J+1})$ should be the same analytic function of ζ. This gives a constraint on α_J and α_{J+1}:

$$s_k^{(J)}(\alpha_J) = s_k^{(J+1)}(\alpha_{J+1}) \quad (k = 1, 2, \ldots, 6). \tag{30}$$

The relation (30) describes the Stokes phenomena for transseries solutions of (alt-dPI). Making use of the explicit formulas (25)–(27) for $s_k^{(J)}$, we thus obtain the following connection formula for (alt-dPI).

Connection formula for (alt-dPI)

Suppose that the transseries solutions $\lambda(t,\zeta,\eta;\alpha_J)$ in Region $(J)(J = \mathrm{I},\mathrm{II},\mathrm{III})$ should define the same analytic solution of (alt-dPI). Then the following relations hold among the free parameters α_J.

$$\alpha_{\mathrm{II}} = \alpha_{\mathrm{I}}(1 + e^{2\pi i\eta\zeta}), \tag{31}$$

$$\alpha_{\mathrm{III}} = \alpha_{\mathrm{II}} + \frac{i}{2\sqrt{\pi}} e^{2\pi i\eta\zeta}. \tag{32}$$

Remark 4.1 In terms of $\overrightarrow{\alpha_J} = (\alpha_0^{(J)}, \alpha_1^{(J)}, \alpha_2^{(J)}, \ldots)$ (cf. Remark 2.3), the formulas (31) and (32) can be expressed also as

$$\alpha_l^{(\mathrm{II})} = \alpha_l^{(\mathrm{I})} + \alpha_{l-1}^{(\mathrm{I})}, \tag{33}$$

$$\alpha_l^{(\mathrm{III})} = \alpha_l^{(\mathrm{II})} + \frac{i}{2\sqrt{\pi}} \delta_{l1} \tag{34}$$

$(l = 0, 1, 2, \ldots)$, where δ_{jk} denotes Kronecker's delta and $\alpha_{-1}^{(\mathrm{I})} = 0$.

Remark 4.2 The formula (32) immediately follows from comparison between (26) and (27). On the other hand, derivation of (31) is not so straightforward, since a Stokes phenomenon for the Voros coefficient V also occurs on a Stokes curve in question. Making comparison between (25) and (26) and taking the effect of a Stokes phenomenon of V into account, we obtain (31).

We finally remark that the relation (32) between α_{II} and α_{III} is the same as the connection formula for Stokes phenomena of the continuous first Painlevé equation (PI) discussed in [19]. Similarly, the relation (31) between α_{I} and α_{II} is the same as the connection formula for parametric Stokes phenomena of the continuous second Painlevé equation (PII) studied by Iwaki [9]. Thus both Stokes phenomena of (PI) type and those of (PII) type occur with the discrete Painlevé equation (alt-dPI).

Acknowledgements The authors would like to express their sincere gratitude to Dr. Shinji Sasaki, who kindly helps us in preparing the figures used in the paper. These figures of Stokes curves are very useful and play an important role in our deriving various explicit formulas.

The research of Nalini Joshi is supported by an ARC Georgina Sweet Australian Laureate Fellowship FL120100094 and the research of Yoshitsugu Takei is supported by JSPS KAKENHI Grant No. 26287015 and No. 24340026.

References

1. T. Aoki, T. Kawai, Y. Takei, New turning points in the exact WKB analysis for higher-order ordinary differential equations, in *Analyse algébrique des perturbations singulières*, I (Hermann, Paris, 1994), pp. 69–84
2. T. Aoki, T. Kawai, Y. Takei, WKB analysis of Painlevé transcendents with a large parameter. II, in *Structure of Solutions of Differential Equations* (World Scientific Publishing, Singapore, 1996), pp. 1–49
3. H.L. Berk, W.M. Nevins, K.V. Roberts, New Stokes' line in WKB theory. J. Math. Phys. **23**, 988–1002 (1982)
4. M.V. Berry, C.J. Howls, Hyperasymptotics. Proc. R. Soc. Lond. Ser. A **430**, 653–668 (1990)
5. B. Candelpergher, J.C. Nosmas, F. Pham, *Approche de la résurgence* (Hermann, Paris, 1993)
6. E. Delabaere, F. Pham, Resurgent methods in semi-classical asymptotics. Ann. Inst. H. Poincaré **71**, 1–94 (1999)
7. A.S. Fokas, B. Grammaticos, A. Ramani, From continuous to discrete Painlevé equations. J. Math. Anal. **180**, 342–360 (1993)
8. N. Honda, T. Kawai, Y. Takei, *Virtual Turning Points*. Springer Briefs in Mathematical Physics, vol. 4 (Springer, Tokyo, 2015)
9. K. Iwaki, Parametric Stokes phenomenon for the second Painlevé equation. Funkcial. Ekvac. **57**, 173–243 (2014)
10. M. Jimbo, T. Miwa, Monodromy preserving deformation of linear ordinary differential equations with rational coefficients. II. Phys. D **2**, 407–448 (1981)
11. M. Jimbo, T. Miwa, K. Ueno, Monodromy preserving deformation of linear ordinary differential equations with rational coefficients. I. Phys. D **2**, 306–352 (1981)
12. N. Joshi, C.J. Lustri, Stokes phenomena in discrete Painlevé I. Proc. A. **471**, 20140874, 22 pp. (2015)
13. T. Kawai, Y. Takei, WKB analysis of Painlevé transcendents with a large parameter. I. Adv. Math. **118**, 1–33 (1996)
14. T. Kawai, Y. Takei, WKB analysis of Painlevé transcendents with a large parameter. III. Adv. Math. **134**, 178–218 (1998)
15. T. Kawai, Y. Takei, *Algebraic Analysis of Singular Perturbation Theory*. Translations of Mathematical Monographs, vol. 227 (American Mathematical Society, Providence, RI, 2005)
16. K. Okamoto, Isomonodromic deformation and Painlevé equations, and the Garnier systems. J. Fac. Sci. Univ. Tokyo Sect. IA **33**, 575–618 (1986)
17. A. Ramani, B. Grammaticos, T. Tamizhmani, On the alternate discrete Painlevé equations and related systems, in *Séminaires et Congrès*, vol. 14 (Soc. Math. France, Paris, 2006), pp. 205–215
18. H.J. Silverstone, JWKB connection-formula problem revisited via Borel summation. Phys. Rev. Lett. **55**, 2523–2526 (1985)
19. Y. Takei, An explicit description of the connection formula for the first Painlevé equation, in *Toward the Exact WKB Analysis of Differential Equations, Linear or Non-linear* (Kyoto University Press, Kyoto, 2000), pp. 271–296
20. Y. Takei, On an exact WKB approach to Ablowitz-Segur's connection problem for the second Painlevé equation. ANZIAM J. **44**, 111–119 (2002)
21. Y. Takei, Sato's conjecture for the Weber equation and transformation theory for Schrödinger equations with a merging pair of turning points. RIMS Kôkyûroku Bessatsu **B10**, 205–224 (2008)
22. A. Voros, The return of the quartic oscillator. The complex WKB method. Ann. Inst. H. Poincaré **39**, 211–338 (1983)

Flat Structures and Algebraic Solutions to Painlevé VI Equation

Mitsuo Kato, Toshiyuki Mano, and Jiro Sekiguchi

Abstract The aim of this paper is first to formulate the definition of Frobenius manifolds and its generalization. Then we study the algebraic solutions to Painlevé VI obtained by Dubrovin-Mazzocco related with the reflection group of type H_3.

Keywords Flat structure • Painlevé VI equation • Potential vector field

Mathematics Subject Classification (2000). Primary 34M56; Secondary 35N10, 32S25.

1 Introduction

In this paper, we mainly treat the three algebraic solutions to Painlevé VI obtained by Dubrovin-Mazzocco [4] and related topics. In the first half, we explain the definition of potentials of Frobenius manifolds and that of potential vector fields of flat structure which is a generalization of Frobenius manifold. In our formulation of flat structure, systems of differential equations of Okubo type play an important role. In the second half, we construct examples of systems of Okubo type related with algebraic solutions to Painlevé VI called the icosahedron, great icosahedron and great dodecahedron solutions in [4].

M. Kato
Department of Mathematics, College of Educations, University of Ryukyus, Nishihara-cho, Okinawa 903-0213, Japan
e-mail: mkato@edu.u-ryukyu.ac.jp

T. Mano
Faculty of Sciences, Department of Mathematical Sciences, University of Ryukyus, Nishihara-cho, Okinawa 903-0213, Japan
e-mail: tmano@math.u-ryukyu.ac.jp

J. Sekiguchi (✉)
Faculty of Engineering, Department of Mathematics, Tokyo University of Agriculture and Technology, Koganei, Tokyo 184-8588, Japan
e-mail: sekiguti@cc.tuat.ac.jp

G. Filipuk et al. (eds.), *Analytic, Algebraic and Geometric Aspects of Differential Equations*, Trends in Mathematics, DOI 10.1007/978-3-319-52842-7_11

2 Free Divisors

Let $h(x) = h(x_1, x_2, \ldots, x_n)$ be a holomorphic function defined in a neighbourhood U of the origin 0 in $\mathbf{C}^n$. Assume that $h(x)$ is reduced.

Definition 2.1 (cf. K. Saito [11]) $h(x) = 0$ is a free divisor if there is an $n \times n$ matrix $(v_{ij}(x))$ whose entries are holomorphic functions satisfying the following conditions (i), (ii).

(i) Put $V_i = \sum_{j=1}^{n} v_{ij}(x)\partial_j$ $(i = 1, 2, \ldots, n)$. Then $(V_i h)/h$ is holomorphic.
(ii) $\det(v_{ij}(x)) = g(x)h(x)$ for a holomorphic function $g(x)$ such that $g(x) \neq 0$ on U.

Example 1 Let V be a real vector space and let W be an irreducible reflection group generated by reflections of V. Put $V_c = V \otimes_{\mathbf{R}} \mathbf{C}$ and $S = \mathbf{C}[V_c]$ is the polynomial ring over V_c. Then W acts on S. Put $R = S^W$ which consists of W-invariant polynomials. Then it is known that R is also a polynomial ring. Taking generators $x_1, x_2, \ldots, x_n$ of R, we have $R = \mathbf{C}[x_1, x_2, \ldots, x_n]$. We may take each x_j as a homogeneous polynomial. Let d_j be the degree of x_j. We may also assume that $0 < d_1 \leq d_2 \leq \cdots \leq d_n$. Then it is known that $d_1 = 2 < d_2$, $d_{n-1} < d_n$ and that $d_j + d_{n-j+1} = d_1 + d_n$ for all j. A basic anti-invariant is a non-zero homogeneous polynomial $D \in S$ such that $D^s = -D$ for any reflection $s \in W$ and that its degree is minimal among such polynomials. Then there exists a basic anti-invariant D up to a non-zero constant and D^2 is W-invariant. As a consequence, D^2 is regarded as a polynomial of $x_1, x_2, \ldots, x_n$. We denote by $\Delta_W(x_1, x_2, \ldots, x_n)$ the polynomial obtained in this manner.

Basic results obtained by K. Saito [11, 12] are

(1) As a polynomial of x_n,

$$\Delta_W = c_0 x_n^n + c_1(x')x_n^{n-1} + \cdots + c_{n-1}(x')x_n + c_n(x'),$$

where each c_j is a polynomial of $x' = (x_1, \ldots, x_{n-1})$ and c_0 is a non-zero constant.
(2) $\Delta_W = 0$ is free.

3 Potentials of Frobenius Manifolds

Let $F(x) = F(x_1, x_2, \ldots, x_n)$ be a function of the form

$$F = \begin{cases} \frac{1}{2}x_1 x_n^2 + \sum_{j=1}^{m-1} x_{j+1}x_{2m-j+1}x_n + \frac{1}{2}x_{m+1}^2 x_n + F_0(x') & (n = 2m+1), \\ \frac{1}{2}x_1 x_n^2 + \sum_{j=1}^{m-1} x_{j+1}x_{2m-j}x_n + F_0(x') & (n = 2m), \end{cases}$$

where $x' = (x_1, \ldots, x_{n-1})$. Assume that $F(x)$ is weighted homogeneous. This means that there are non-zero numbers $w_1, w_2, \ldots, w_n, d$ such that $\sum_{j=1}^n w_j x_j \partial_j F = dF$. We put $E = \sum_{j=1}^n w_j x_j \partial_j$. E is called an Euler vector field. We also assume that each w_j is a rational number and that[1] $0 < w_1 < w_2 < \cdots < w_n = 1$. Using F, we define $g_j = \partial_{n-j+1} F$ $(j = 1, 2, \ldots, n)$ and put $P = (g_1, g_2, \ldots, g_n)$. By differentiating P, we obtain n-vectors $\partial_j P$ and construct a matrix $C = (C_{ij})$ by

$$C = \begin{pmatrix} \partial_1 P \\ \partial_2 P \\ \vdots \\ \partial_n P \end{pmatrix}. \tag{1}$$

In particular $C_{ij} = \partial_i g_j$ and $\partial_n P = (x_1, x_2, \ldots, x_n)$. Moreover we put $\tilde{B}^{(k)} = \partial_k C$. It follows from the definition that $(\tilde{B}^{(k)})_{ij} = \partial_k \partial_i g_j$. As a consequence, $(\tilde{B}^{(k)})_{ij} = (\tilde{B}^{(i)})_{kj}$. It is clear from the definition that $C - x_n I_n$ is independent of x_n. This implies that $\tilde{B}^{(n)} = I_n$. We need the matrix $T = EC$ for later consideration.

Definition 3.1 The function F is a potential if $\tilde{B}^{(p)}\tilde{B}^{(q)} = \tilde{B}^{(q)}\tilde{B}^{(p)}$ for all p, q and $x = (x_1, x_2, \ldots, x_n)$ is called a flat coordinate.

Remark 3.2 The matrix entries of $\tilde{B}^{(p)}\tilde{B}^{(q)} = \tilde{B}^{(q)}\tilde{B}^{(p)}$ are expressed as non-linear differential equations for F. The collection of such differential equations is called a WDVV(=Witten-Dijkgraaf-Verlinde-Verlinde) equation. For this reason, a potential is a solution of WDVV equation.

We put $h = \det T$. It follows from the definition that

$$h = x_n^n + c_1(x')x_n^{n-1} + \cdots + c_{n-1}(x')x_n + c_n(x'),$$

where each $c_j(x')$ is a function of $x' = (x_1, \ldots, x_{n-1})$. An important property of h is shown by C. Sabbah:

Proposition 3.3 ([10]) *The hypersurface of* $\mathbf{C}^n$ *defined by* $\det(T) = 0$ *is free.*

Example 2 (Continuation of Example 1) The following result is due to Dubrovin [3]. (See also [14], which treated a prototype of Frobenius manifold.)

(3) Taking the basic invariants $x_1, x_2, \ldots, x_n$ appropriately, we find that there is a potential $F = F(x_1, x_2, \ldots, x_n)$ such that the polynomial Δ_W is identified with $\det(T)$ up to a non-zero constant and basic logarithmic vector fields defined in Definition 2.1 are obtained by T.

We explain the statement (3) more precisely. In the definition of a potential one needs the numbers $w_1, \ldots, w_n$. We start with taking these numbers so that

[1] Sometimes we treat the case where some of $w_1, \ldots, w_n$ are equal. In this paper we use the notation $\partial_j = \partial_{x_j}$ for the sake of simplicity.

$w_j = d_j/d_n$, where d_j are degrees of basic invariants. The next step is to find a polynomial solution $F(x)$ of the WDVV equation corresponding to these numbers. The third step is to define matrices C and T as before. Finally we obtain $\Delta_W = \det T$.

Relating with real reflection groups of rank three (cf. [3]), there are three kinds of polynomial potentials:

$$A_3 \text{ case} : F = \frac{x_1x_3^2 + x_2^2x_3}{2} + \frac{x_1^2x_2^2}{4} + \frac{x_1^5}{60},$$

$$B_3 \text{ case} : F = \frac{x_1x_3^2 + x_2^2x_3}{2} + \frac{x_1x_2^3}{6} + \frac{x_1^3x_2^2}{6} + \frac{x_1^7}{210},$$

$$H_3 \text{ case} : F = \frac{x_1x_3^2 + x_2^2x_3}{2} + \frac{x_1^2x_2^3}{6} + \frac{x_1^5x_2^2}{20} + \frac{x_1^{11}}{3960}.$$

Remark 3.4 The three polynomials above were already obtained by K. Saito and T. Yano at the end of 1970s (unpublished).

Remark 3.5 Dubrovin [3] classified polynomial potentials of three variables. It is interesting to construct potentials defined by algebraic function in three variables. We will discuss a topic related with this question later.

4 Flat Structures Without Potentials

Some of the arguments so far go well without the existence of potentials. We are going to explain the idea based on [6, 7].

We start with introducing weighted homogeneous polynomials $g_1(x), g_2(x), \ldots, g_n(x)$ such that $Eg_j = (w_j + w_n)g_j \quad (j = 1, 2, \ldots, n)$ and that

$$g_j = \begin{cases} x_jx_n + g_j^{(0)}(x') \ (j = 1, 2, \ldots, n-1), \\ \frac{1}{2}x_n^2 + g_n^{(0)}(x') \ (j = n) \end{cases}$$

with polynomials $g_j^{(0)}(x')$ of $x' = (x_1, \ldots, x_{n-1})$. Using $g_j(x)$ $(j = 1, 2, \ldots, n)$, we define an $n \times n$ matrix C such that $C_{ij} = \partial_i g_j$. It is easy to see that $C_{nj} = x_j$ $(j = 1, 2, \ldots, n)$. We define matrices $\tilde{B}^{(p)} = \partial_p C$ $(p = 1, 2, \ldots, n)$ and $T = \sum_{j=1}^n w_jx_j\partial_j C = \sum_{j=1}^n w_jx_j\tilde{B}^{(j)}$. We denote by $b_{ij}^{(p)}$ the (i, j)-entry of $\tilde{B}^{(p)}$ and collect basic properties of $\tilde{B}^{(p)}$ $(p = 1, 2, \ldots, n)$:

1. $\partial_p\tilde{B}^{(q)} = \partial_q\tilde{B}^{(p)} \quad (\forall p, q)$,
2. $b_{pq}^{(r)} = b_{rq}^{(p)} \quad (\forall p, q, r)$,
3. $b_{nq}^{(p)} = \delta_{pq} \quad (\forall p, q)$,
4. $\tilde{B}^{(n)} = I_n$,
5. $\partial_n\tilde{B}^{(p)} = O \quad (p = 1, 2, \ldots, n-1)$,

where δ_{pq} is Kronecker's delta and I_n is the identity matrix.

Definition 4.1 If $\tilde{B}^{(p)}\tilde{B}^{(q)} = \tilde{B}^{(q)}\tilde{B}^{(p)}$ $(\forall p,\ q = 1,2,\ldots,n)$, then $\vec{g} = (g_1, g_2, \ldots, g_n)$ is called a potential vector field and $(x_1, x_2, \ldots, x_n)$ is a flat coordinate.

Remark 4.2 The potential vector field in Definition 4.1 is same as the local vector potential introduced in Manin [9]. The authors thank P. Lorenzoni for informing them of the paper [9].

Remark 4.3 Similar to the case of the existence of potentials, the potential vector field is a solution of a certain system of non-linear differential equations arising from the commutativity of matrices $\tilde{B}^{(p)}$ $(p = 1, 2, \cdots, n)$. In this sense, $\tilde{B}^{(p)}\tilde{B}^{(q)} = \tilde{B}^{(q)}\tilde{B}^{(p)}$ $(\forall p,\ q)$ is called an extended WDVV equation.

5 A Generalization of Ordinary Differential Equations of Okubo Type

In this section we freely use the notation in previous sections without any comment. The matrix T is defined by C and E. We always assume in this section that $(g_1, g_2, \ldots, g_n)$ is a potential vector field. Then by definition, $\tilde{B}^{(p)}\tilde{B}^{(q)} = \tilde{B}^{(q)}\tilde{B}^{(p)}$ for all $p,\ q$.

We introduce a diagonal matrix $B_\infty^{(n)}$ by $B_\infty^{(n)} = \mathrm{diag}(r + w_1, r + w_2, \ldots, r + w_n)$ for some constant $r \in \mathbf{C}$ and define $n \times n$ matrices $B^{(p)}$ $(p = 1, 2, \ldots, n)$ by

$$B^{(p)} = -T^{-1}\tilde{B}^{(p)}B_\infty^{(n)} \tag{2}$$

and a system of differential equations

$$\partial_p Y = B^{(p)}Y \quad (p = 1, 2, \ldots, n), \tag{3}$$

where $Y = {}^t(y_1, \cdots, y_n)$. The system (3) is rewritten by

$$dY = \Omega Y, \tag{4}$$

where Ω is the 1-form defined by $\Omega = \sum_{p=1}^{n} B^{(p)}dx_p$.

Theorem 5.1 *The system (4) is integrable.*

Remark 5.2 We put $T_0 = x_n I_n - \dfrac{1}{w_n}T$. Since $T - w_n x_n I_n$ does not depend on x_n and since $B^{(n)} = -T^{-1}B_\infty^{(n)}$, the differential equation

$$\partial_n Y = B^{(n)}Y \tag{5}$$

turns out to be

$$(x_n I_n - T_0)\partial_n Y = -\frac{1}{w_n} B_\infty^{(n)} Y. \tag{6}$$

Regarding (6) as an ordinary differential equation with respect to the variable x_n, (6) is an ordinary differential equation of Okubo type. In this sense, the system (4) [or (3)] is one of generalizations of Okubo type ordinary differential equation to several variables case.

6 Algebraic Solutions of Painlevé VI and Differential Equations of Okubo Type

In this section we mention a relationship between the system of differential equations (4) and algebraic solutions of Painlevé VI. Algebraic solutions of Painlevé VI are studied by many authors. For our purpose we assume that $n = 3$. Since $h = \det T$ is a cubic polynomial of x_3, let $z_j(x')$ $(j = 1, 2, 3)$ be defined by $h(x) = \prod_{i=1}^{3}(x_3 - z_i(x'))$ where $x' = (x_1, x_2)$. We consider $B^{(3)} = -T^{-1}\tilde{B}^{(3)}B_\infty^{(3)}$ as before. It follows from the definition that if $i \neq j$, the (i, j)-entry of $hB^{(3)}$ is a linear function of x_3. Noting this, we define $z_{ij}(x')$ $(i \neq j)$ by the condition that $x_3 = z_{ij}(x')$ is the zero of the (i, j)-entry of $hB^{(3)}$. If each of the diagonal entries of $B_\infty^{(3)}$ is not zero, that is, $r \neq -w_k$ $(k = 1, 2, 3)$, then by an easy computation, we have $z_{ij} = \left.\frac{\det(T)(T^{-1})_{ij}}{T_{ij}}\right|_{x_3=0}$. This formula holds for $j \neq k$ in the case $r = -w_k$. It can be shown that if the (k, k)-entry of the diagonal matrix $B_\infty^{(3)}$ is zero, namely, if $r = -w_k$ and $j \neq k$, then $w_{ij} = \frac{z_{ij}(x') - z_1(x')}{z_2(x') - z_1(x')}$ is an algebraic solution of Painlevé VI as a function of $t = \frac{z_3(x') - z_1(x')}{z_2(x') - z_1(x')}$.

We stop here the arguments on the relationship between the system of differential equations (4) and algebraic solutions of Painlevé VI because of space restriction and we focus our attention to the case treated by Dubrovin and Mazzocco [4] in the subsequent sections. Dubrovin [3] proved that three dimensional semisimple Frobenius manifolds correspond to one parameter family of the Painlevé VI equation. Algebraic solutions by Dubrovin and Mazzocco are included in this one parameter family. We shall give many examples of potential vector fields related with algebraic solutions to Painlevé VI equation elsewhere.

7 Frobenius Manifolds Treated by Dubrovin-Mazzocco [4]

We treat in this section three algebraic solutions of Painlevé VI related with the reflection group of type H_3 obtained by Dubrovin-Mazzocco [4]. The first one is called icosahedron solution. This solution is constructed from the polynomial potential introduced in Sect. 3. The second one and third one are called great icosahedron solution and great dodecahedron solution, respectively. We construct these two solutions by introducing potentials of algebraic functions. The argument in this section is mostly owing to [4] and [1]. Our contribution is, if exists, the construction of potentials of algebraic functions. In the last subsection, we discuss the relationship between these three solutions and free divisors in $\mathbf{C}^3$.

The Case Corresponding to Icosahedron Solution (H_3) in [4]

In this case, we start with the potential

$$F = \frac{t_1^{11}}{3960} + \frac{t_1^5 t_2^2}{20} + \frac{t_1^2 t_2^3}{6} + \frac{t_2^2 t_3}{2} + \frac{t_1 t_3^2}{2}.$$

The weights of the variables t_1, t_2, t_3 are given by

$$w(t_1) = \frac{1}{5},\ w(t_2) = \frac{3}{5},\ w(t_3) = 1.$$

The concrete form of the matrix C [cf. (1)] is then

$$C = \begin{pmatrix} t_3 & \frac{1}{2}t_1^4 t_2 + t_1 t_2^2 & \frac{1}{36}t_1^9 + t_1^3 t_2^2 + \frac{1}{3}t_2^3 \\ t_2 & \frac{1}{10}t_1^5 + t_1^2 t_2 + t_3 & \frac{1}{2}t_1^4 t_2 + t_1 t_2^2 \\ t_1 & t_2 & t_3 \end{pmatrix}.$$

In this case the Euler vector field E is given by

$$E = \frac{1}{5}t_1 \partial_{t_1} + \frac{3}{5}t_2 \partial_{t_2} + t_3 \partial_{t_3}$$

and $T = EC$ is given by

$$T = \frac{1}{5}\begin{pmatrix} 5t_3 & 7(\frac{1}{2}t_1^4 t_2 + t_1 t_2^2) & 9(\frac{1}{36}t_1^9 + t_1^3 t_2^2 + \frac{1}{3}t_2^3) \\ 3t_2 & 5(\frac{1}{10}t_1^5 + t_1^2 t_2 + t_3) & 7(\frac{1}{2}t_1^4 t_2 + t_1 t_2^2) \\ t_1 & 3t_2 & 5t_3 \end{pmatrix}.$$

Putting $B_\infty^{(3)} = \mathrm{diag}\left[-r - \frac{2}{5}, -r, -r + \frac{2}{5}\right]$, we define 3×3 matrices

$$B^{(p)} = -T^{-1}\tilde{B}^{(p)} B_\infty^{(3)}\ (p = 1, 2, 3)$$

and a system of differential equations

$$\partial_{t_p} Y = B^{(p)} Y \quad (p = 1, 2, 3). \tag{7}$$

The singularities of this system is contained in the hypersurface defined by $\det(T) = 0$. In this case

$$\begin{aligned} \det(T) = & -\tfrac{1}{1000} t_1^{15} - \tfrac{1}{100} t_1^{12} t_2 + \tfrac{2}{25} t_1^9 t_2^2 + \tfrac{1}{50} t_1^6 t_2^3 + \tfrac{23}{25} t_1^3 t_2^4 + \tfrac{27}{125} t_2^5 \\ & - \tfrac{1}{100} t_1^{10} t_3 - \tfrac{6}{5} t_1^4 t_2^2 t_3 - \tfrac{9}{5} t_1 t_2^3 t_3 + \tfrac{1}{10} t_1^5 t_3^2 + t_1^2 t_2 t_3^2 + t_3^3 \end{aligned} \tag{8}$$

Remark 7.1 The polynomial (8) is regarded as the discriminant of the reflection group of type H_3.

Substituting $t_2 = \frac{(1-23s^2+91s^4-5s^6)}{(1+3s^2)^3} t_1^3$, we find that

$$\det(T) = (t_3 - z_1)(t_3 - z_2)(t_3 - z_3),$$

where

$$\begin{aligned} z_1 &= -\tfrac{(25-585s^2+3530s^4-6690s^6-3955s^8+507s^{10})t_1^5}{10(1+3s^2)^5}, \\ z_2 &= \tfrac{(7-215s^2+1910s^4+4096s^5-5150s^6-20480s^7-6125s^8+357s^{10})t_1^5}{10(1+3s^2)^5}, \\ z_3 &= \tfrac{(7-215s^2+1910s^4-4096s^5-5150s^6+20480s^7-6125s^8+357s^{10})t_1^5}{10(1+3s^2)^5}. \end{aligned}$$

By direct computation, the $(3, 2)$-entry of the matrix $\det(T)T^{-1}$ is

$$\varphi_{32} = \frac{1}{50} t_2 (7t_1^5 + 14t_1^2 t_2 - 30t_3)$$

Regard $\varphi_{32} = 0$ as an equation of t_3. Then

$$w_{32} = \frac{7}{30}(t_1^5 + 2t_1^2 t_2) = \frac{7(3 - 37s^2 + 209s^4 + 17s^6)t_1^5}{30(1 + 3s^2)^3}$$

is its solution. Then

$$\begin{aligned} t &= \frac{z_3 - z_1}{z_2 - z_1} = \frac{(-1+s)^5(1+3s)^3(-1+4s+s^2)}{(1+s)^5(-1+3s)^3(-1-4s+s^2)}, \\ w &= \frac{w_{32} - z_1}{z_2 - z_1} = \frac{(-1+s)^2(1+3s)^2(3+s^2)}{3(1+s)^3(-1+3s)(-1-4s+s^2)}. \end{aligned}$$

As is explained in the previous section, we conclude that w is an algebraic solution to Painleve VI with respect to the variable t. The solution (t, w) is equivalent to the icosahedron solution obtained by Dubrovin-Mazzocco [4].

The Case Corresponding to Great Icosahedron Solution $(H_3)'$ *in [4]*

In this case, we start with introducing an algebraic function z of t_1, t_2 defined by

$$t_2 + t_1 z + z^4 = 0. \tag{9}$$

The weights of the variables t_1, t_2, t_3 and the algebraic function z are given by

$$w(t_1) = \frac{3}{5},\ w(t_2) = \frac{4}{5},\ w(t_3) = 1,\ w(z) = \frac{1}{5}.$$

Let

$$F = \frac{1}{2}(t_2^2 t_3 + t_1 t_3^2) - \frac{1}{18}t_1^4 z - \frac{7}{72}t_1^3 z^4 - \frac{17}{105}t_1^2 z^7 - \frac{2}{9}t_1 z^{10} - \frac{64}{585}z^{13}$$

be an algebraic function of t_1, t_2, t_3. In this case, the matrix C of (1) is given by

$$C = \begin{pmatrix} t_3 & \frac{1}{6}(t_1 - 2z^3)(t_1 + z^3) & -\frac{1}{84}z(28t_1^2 + 35t_1 z^3 + 16z^6) \\ t_2 & \frac{1}{10}(10t_3 - 5t_1 z^2 - 8z^5) & \frac{1}{6}(t_1 - 2z^3)(t_1 + z^3) \\ t_1 & t_2 & t_3 \end{pmatrix}$$

Noting that z is an algebraic function of t_1, t_2, we obtain

$$\tilde{B}_1(= \partial_{t_1} C) = \begin{pmatrix} 0 & \frac{1}{3}(t_1 + z^3) & -\frac{1}{12}z(4t_1 + z^3) \\ 0 & \frac{1}{2}z^2 & \frac{1}{3}(t_1 + z^3) \\ 1 & 0 & 0 \end{pmatrix},$$

$$\tilde{B}_2(= \partial_{t_2} C) = \begin{pmatrix} 0 & \frac{1}{2}z^2 & \frac{1}{3}(t_1 + z^3) \\ 1 & z & \frac{1}{2}z^2 \\ 0 & 1 & 0 \end{pmatrix},$$

$$\tilde{B}_3(= \partial_{t_3} C) = I_3.$$

Then a direct computation implies that $\tilde{B}_1\tilde{B}_2 = \tilde{B}_2\tilde{B}_1$, which means that F is a potential.

Since $T = \frac{3}{5}t_1\tilde{B}_1 + \frac{4}{5}t_2\tilde{B}_2 + t_3\tilde{B}_3$, it follows that

$$T = \frac{1}{5}\begin{pmatrix} 5t_3 & (t_1 - 2z^3)(t_1 + z^3) & -\frac{1}{12}z(28t_1^2 + 35t_1 z^3 + 16z^6) \\ 4t_2 & \frac{1}{2}(10t_3 - 5t_1 z^2 - 8z^5) & (t_1 - 2z^3)(t_1 + z^3) \\ 3t_1 & 4t_2 & 5t_3 \end{pmatrix}.$$

Putting $B^{(3)}_\infty = \operatorname{diag}\left[-r - \frac{1}{5}, -r, -r + \frac{1}{5}\right]$, we define 3×3 matrices

$$B^{(p)} = -T^{-1}\tilde{B}^{(p)}B^{(3)}_\infty \ (p = 1, 2, 3)$$

and a system of differential equations

$$\partial_{t_p} Y = B^{(p)} Y \quad (p = 1, 2, 3). \tag{10}$$

Then the singularities of this system is contained in the hypersurface defined by $\det(T) = 0$. In this case

$$\begin{aligned}\det(T) = {} & \tfrac{3}{125}t_1^5 + t_3^3 + \tfrac{3}{5}t_1^3 t_3 z - \tfrac{1}{2}t_1 t_3^2 z^2 - \tfrac{73}{50}t_1^4 z^3 + \tfrac{7}{20}t_1^2 t_3 z^4 - \tfrac{4}{5}t_3^2 z^5 \\ & -\tfrac{173}{120}t_1^3 z^6 - \tfrac{4}{5}t_1 t_3 z^7 - \tfrac{37}{25}t_1^2 z^9 - \tfrac{16}{25}t_3 z^{10} - \tfrac{56}{75}t_1 z^{12} - \tfrac{64}{375}z^{15}\end{aligned} \tag{11}$$

Remark 7.2 By a certain weight preserving change of variables, the polynomial above coincides with the $\det(T)$ in the case of icosahedron given in (8).

As in the previous case, by the substitution $t_1 = -\frac{4(1-15s^2+75s^4+3s^6)z^3}{(1+3s^2)^3}$, we find that

$$\det(T) = (t_3 - z_1)(t_3 - z_2)(t_3 - z_3),$$

where

$$\begin{aligned} z_1 &= \frac{6(5-125s^2+930s^4-2090s^6-1975s^8+183s^{10})z^5}{5(1+3s^2)^5}, \\ z_2 &= -\frac{6(3-75s^2+430s^4+1024s^5-870s^6-5120s^7-545s^8+33s^{10})z^5}{5(1+3s^2)^5}, \\ z_3 &= -\frac{6(3-75s^2+430s^4-1024s^5-870s^6+5120s^7-545s^8+33s^{10})z^5}{5(1+3s^2)^5}. \end{aligned}$$

By direct computation, the $(3, 2)$-entry of the matrix $\det(T)T^{-1}$ is

$$\varphi_{32} = 3t_1^3 - 20t_2 t_3 - 3t_1^2 z^3 - 6t_1 z^6.$$

Regard $\varphi_{32} = 0$ as an equation of t_3 and $t_3 = w_{32}$ is its solution. Then

$$w_{32} = \frac{3(t_1^3 - t_1^2 z^3 - 2t_1 z^6)}{20t_2} = -\frac{18(1 - 15s^2 + 75s^4 + 3s^6)(1 - 7s^2 + 59s^4 + 11s^6)z^5}{5(1 + 3s^2)^6}.$$

We define $t,\ w$ by

$$\begin{aligned} t &= \frac{z_3 - z_1}{z_2 - z_1} = \frac{(-1 + s)^5(1 + 3s)^3(-1 + 4s + s^2)}{(1 + s)^5(-1 + 3s)^3(-1 - 4s + s^2)}, \\ w &= \frac{w_{32} - z_1}{z_2 - z_1} = \frac{(-1 + s)^4(1 + 3s)^2}{(1 + s)(-1 + 3s)(-1 - 4s + s^2)(1 + 3s^2)}. \end{aligned}$$

Then w is an algebraic solution to Painlevé VI of the variable t equivalent to the great dodecahedron solution in [4].

To compare with the case (H_3), we rewrite the system (10) by using the variable (z, t_1, t_3). Using the relation (9), we define $C_M = C|_{t_2=-t_1z-z^4}$, $T_M = T|_{t_2=-t_1z-z^4}$. Then

$$C_M = \begin{pmatrix} t_3 & \frac{1}{6}(t_1-2z^3)(t_1+z^3) & -\frac{1}{84}z(28t_1^2+35t_1z^3+16z^6) \\ -z(t_1+z^3) & \frac{1}{10}(10t_3-5t_1z^2-8z^5) & \frac{1}{6}(t_1-2z^3)(t_1+z^3) \\ t_1 & -z(t_1+z^3) & t_3 \end{pmatrix}$$

$$T_M = \frac{1}{5}\begin{pmatrix} 5t_3 & (t_1-2z^3)(t_1+z^3) & -\frac{1}{12}z(28t_1^2+35t_1z^3+16z^6) \\ -4z(t_1+z^3) & \frac{1}{2}(10t_3-5t_1z^2-8z^5) & (t_1-2z^3)(t_1+z^3) \\ 3t_1 & -4z(t_1+z^3) & 5t_3 \end{pmatrix}.$$

As a consequence, we obtain a system of differential equations

$$\begin{cases} \partial_z\tilde{Y} = -T_M^{-1}(\partial_z C_M)B_\infty^{(3)}\tilde{Y}, \\ \partial_{t_1}\tilde{Y} = -T_M^{-1}(\partial_{t_1} C_M)B_\infty^{(3)}\tilde{Y}, \\ \partial_{t_3}\tilde{Y} = -T_M^{-1}(\partial_{t_3} C_M)B_\infty^{(3)}\tilde{Y}. \end{cases} \tag{12}$$

on the (z, t_1, t_3)-space.

Remark 7.3 The singularities of the system (12) is contained in the free divisor defined by $\det(T) = 0$, where $\det(T)$ is given by (11). It is underlined here that $\det(T)$ is regarded as the discriminant of the reflection group of type H_3 [cf. *Remark* 7.2].

Remark 7.4 B. Dubrovin informed the authors that his student Alejo Keuroghlanian computed the algebraic Frobenius manifold for the case of the great icosahedron in his master thesis "Varieta di Frobenius algebriche di dimensione 3" (2008).

The Case Corresponding to Great Dodecahedron Solution $(H_3)''$ in [4]

In this case, we start with introducing an algebraic function z of t_1, t_2 defined by

$$-t_1^2 + t_2 + z^2 = 0. \tag{13}$$

The weights of the variables t_1, t_2, t_3 and the algebraic function z are given by

$$w(t_1) = \frac{1}{3},\ w(t_2) = \frac{2}{3},\ w(t_3) = 1,\ w(z) = \frac{1}{3}$$

Let

$$F = \frac{4063}{1701}t_1^7 + \frac{1}{2}(t_2^2 t_3 + t_1 t_3^2) + \frac{19}{135}t_1^5 z^2 - \frac{73}{27}t_1^3 z^4 + \frac{11}{9}t_1 z^6 - \frac{16}{35}z^7.$$

be an algebraic function of t_1, t_2, t_3. In this case, the matrix C of (1) is given by

$$C = \begin{pmatrix} t_3 & C_{12} & C_{13} \\ t_2 & C_{22} & C_{23} \\ t_1 & t_2 & t_3 \end{pmatrix},$$

where

$$\begin{aligned} C_{12} = C_{23} &= \tfrac{1}{9}(t_1 + z)(91t_1^3 - 91t_1^2 z + 105t_1 z^2 - 33z^3), \\ C_{13} &= \tfrac{4}{405}(8281t_1^5 - 4410t_1^3 z^2 - 1620t_1^2 z^3 + 585t_1 z^4 - 324z^5), \\ C_{22} &= \tfrac{1}{27}(-146t_1^3 + 27t_3 + 198t_1 z^2 - 108z^3). \end{aligned}$$

Then by direct computation, we conclude that $\tilde{B}_1 = \partial_{t_1} C$, $\tilde{B}_2 = \partial_{t_2} C$, $\tilde{B}_3 = \partial_{t_3} C = I_3$ are commutative and therefore F is a potential.

Since $T = \frac{1}{3}t_1\tilde{B}_1 + \frac{2}{3}t_2\tilde{B}_2 + t_3\tilde{B}_3$, it follows that

$$T = \frac{1}{3}\begin{pmatrix} 3t_3 & 4C_{12} & 5C_{13} \\ 2t_2 & 3C_{22} & 4C_{23} \\ t_1 & 2t_2 & 3t_3 \end{pmatrix}.$$

Putting $B_\infty^3 = \text{diag}\left[-r - \frac{1}{3}, -r, -r + \frac{1}{3}\right]$, we define 3×3 matrices

$$B^{(p)} = -T^{-1}\tilde{B}^{(p)}B_\infty^{(3)} \ (p = 1, 2, 3)$$

and a system of differential equations

$$\partial_{t_p} Y = B^{(p)} Y \quad (p = 1, 2, 3). \tag{14}$$

Then the singularities of this system are contained in the hypersurface defined by $\det(T) = 0$. In this case

$$\det(T) = t_3^3 + K_1 t_3^2 + K_2 t_3 + K_3, \tag{15}$$

where

$$K_1 = -\tfrac{2}{27}(73t_1^3 - 99t_1 z^2 + 54z^3),$$

$$K_2 = -\tfrac{4}{729}(11557t_1^6 - 7182t_1^4 z^2 + 972t_1^3 z^3 - 1107t_1^2 z^4 - 2916t_1 z^5 + 1188z^6),$$

$$K_3 = \tfrac{8}{19683}\begin{pmatrix} 902629t_1^9 - 1473381t_1^7 z^2 + 535626t_1^6 z^3 + 693063t_1^5 z^4 \\ -12636t_1^4 z^5 - 169155t_1^3 z^6 - 39366t_1^2 z^7 + 12636t_1 z^8 - 5832z^9 \end{pmatrix}.$$

We introduce u by the relation

$$u^2 = s(8 - 11s + 8s^2). \tag{16}$$

Then substituting $t_1 = \frac{(-1-12s+12s^2-4s^3)z}{5(-1+2s)^2}$, we find that

$$\det(T) = (t_3 - z_1)(t_3 - z_2)(t_3 - z_3),$$

where

$$z_1 = \frac{32z^3\begin{pmatrix}724 - 6336s + 16704s^2 - 27276s^3 + 17514s^4 + 10026s^5 \\ -21489s^6 + 12996s^7 - 3924s^8 + 436s^9\end{pmatrix}}{3375(-1+2s)^6},$$

$$z_2 = -\frac{32z^3\begin{pmatrix}76 - 864s + 3096s^2 - 5424s^3 + 1161s^4 + 8649s^5 - 11211s^6 \\ +6804s^7 - 3276s^8 + 364s^9 + 675(-1+s)su^3\end{pmatrix}}{3375(-1+2s)^6},$$

$$z_3 = -\frac{32z^3\begin{pmatrix}76 - 864s + 3096s^2 - 5424s^3 + 1161s^4 + 8649s^5 - 11211s^6 \\ +6804s^7 - 3276s^8 + 364s^9 - 675(-1+s)su^3\end{pmatrix}}{3375(-1+2s)^6}.$$

By direct computation, the $(3,2)$-entry of the matrix $\det(T)T^{-1}$ is

$$\varphi_{32} = \frac{2}{81}(144t_1^5 + 104t_1^3t_2 - 66t_1t_2^2 - 27t_2t_3 + 144t_1^4z - 144t_1^2t_2z).$$

Regard $\varphi_{32} = 0$ as an equation of t_3 and $t_3 = w_{32}$ is its solution. Then

$$w_{32} = \frac{2(72t_1^5 + 52t_1^3t_2 - 33t_1t_2^2 + 72t_1^4z - 72t_1^2t_2z)}{27t_2}.$$

We define $t,\ w$ by

$$t = \frac{z_1 - z_2}{z_3 - z_2},\ w = \frac{w_{32} - z_2}{z_3 - z_2}.$$

Then

$$\begin{aligned} t &= \tfrac{1}{2} + \frac{(1+s)(32-320s+1112s^2-2420s^3+3167s^4-2420s^5+1112s^6-320s^7+32s^8)}{54(-1+s)su^3}, \\ w &= \tfrac{1}{2} - \frac{(8-28s+75s^2+31s^3-269s^4+318s^5-166s^6+56s^7)}{18(-1+s)s(3-4s+4s^2+2s^3)u}. \end{aligned} \tag{17}$$

The transformation $(s,u) \to (1/s_1, -u_1/s_1^2)$ leaves Eq. (16) invariant and write t, w by s_1, u_1, we find that t, w defined in (17) coincide with t, y of Theorem C in [1]. As a consequence, (t, w) is equivalent to the great dodecahedron solution of [4].

Remark 7.5 Historically, the great dodecahedron solution given in [4] was defined by a solution curve $F(t, y) = 0$ which took about ten pages to write down. Later

Boalch [1] obtained the solution (t, w) given in (17) which is equivalent to the one by Dubrovin-Mazzocco and is quite simple. Our interest is to construct a system of linear differential equations of three variables related with an algebraic solution to Painlevé VI. Starting from the solution given in Theorem C in [1], we accomplished this purpose and finally reached at the position of finding the algebraic potential given at the first place of this subsection.

Remarks on Fundamental Groups

In this section, we constructed three holonomic systems of (linear) differential equations (7), (10), (14) on $\mathbf{C}^3$ which correspond to icosahedron solution, great icosahedron solution and great dodecahedron solution, respectively. In particular we rewrite (10) to (12).

We now discuss a relationship between the monodromy groups of the systems (7), (12), (14) and the reflection group $W(H_3)$.

It is known that $W(H_3)$ is generated by three reflections on $\mathbf{R}^3$. One choice of generating reflections is

$$R_1 = \begin{pmatrix} -1 & 0 & 0 \\ 0 & 1 & 0 \\ 0 & 0 & 1 \end{pmatrix}, \ R_2 = \frac{1}{2}\begin{pmatrix} \bar{a} & 1 & -a \\ 1 & a & -\bar{a} \\ -a & -\bar{a} & 1 \end{pmatrix}, \ R_3 = \begin{pmatrix} 1 & 0 & 0 \\ 0 & 1 & 0 \\ 0 & 0 & -1 \end{pmatrix},$$

where $a = \frac{1+\sqrt{5}}{2}$, $\bar{a} = \frac{1-\sqrt{5}}{2}$. It is easy to show that

$$R_j^2 = I_3 \ (j = 1, 2, 3), \ (R_1R_2)^5 = I_3, \ (R_2R_3)^3 = I_3, \ (R_1R_3)^2 = I_3.$$

Then $G_A = \langle R_1, R_2, R_3 \rangle$ is a realization of $W(H_3)$ as a subgroup of $O(3)$. To show a relationship between G_A and the realization of $W(H_3)$ as the quotient group of a binary icosahedral group defined by F. Klein [8], we introduce a matrix $M = \begin{pmatrix} (\zeta-1)\zeta^2 & 0 & -(\zeta-1)\zeta^2 \\ \zeta+\zeta^4 & \sqrt{2} & -\zeta-\zeta^4 \\ -1 & \sqrt{2}(\zeta+\zeta^4) & 1 \end{pmatrix}$ and define $\tilde{R}_j = M^{-1}R_jM \ (j = 1, 2, 3)$ where $\zeta = e^{2\sqrt{-1}\pi/5}$. Then the group $G_K = \langle \tilde{R}_1, \tilde{R}_2, \tilde{R}_3 \rangle$ is a realization as the quotient group of the binary icosahedral group introduced by F. Klein [8, p. 41] (see also [5, p. 197]). It is underlined here that G_K has an action $\zeta \to \zeta^2$.

We return to the systems (7), (12). Each of the systems has a parameter r. The monodromy groups of (7), (12) depend on the parameter r. Consider the case $r = 1/2$. In this case, it can be shown that monodromy groups of the systems (7), (12) are finite. In particular, the monodromy group of (7) is isomorphic to the group G_K and the monodromy group of (12) is realized by changing $\tilde{R}_j$ to the matrix $\tilde{R}'_j = \tilde{R}_j|_{\zeta\to\zeta^3}$ $(j = 1, 2, 3)$.

The treatment of the remaining case (14) is different from (7), (12). We first note the relationship between the algebraic solution given in (17) and the solution 27 in [1]. Since the solution 27 case was discussed in [6, Sect. 8.1] in detail, we start with reviewing the argument there. In the solution 27 case, the weights of flat coordinates (t_1, t_2, t_3) are given by

$$w(t_1) = \frac{1}{15}, \; w(t_2) = \frac{1}{3}, \; w(t_3) = 1$$

and a potential vector field is defined by (g_1, g_2, g_3), where

$$\begin{aligned} g_1 &= -\tfrac{1}{33} t_1 (3t_1^{10} t_2 + 11t_2^3 - 33t_3), \\ g_2 &= \tfrac{1}{76}(-5t_1^{20} + 114t_1^{10}t_2^2 + 19t_2^4 + 76t_2t_3), \\ g_3 &= \tfrac{1}{870}(100t_1^{30} + 1740t_1^{20}t_2^2 - 5220t_1^{10}t_2^4 + 116t_2^6 + 435t_3^2). \end{aligned}$$

From g_1, g_2, g_3, we can introduce the matrices C, T. In this case we write C_{s27}, T_{s27}, respectively. Then $\det(T_{s27}) = 0$ defines a free divisor in $\mathbf{C}^3$ related with the solution 27. For our purpose, we change $t_1 = \xi_1^{1/10}, t_2 = \xi_2, t_3 = \xi_3$ and write $\det(T_{s27})$ by $\xi = (\xi_1, \xi_2, \xi_3)$ which we denote by $F_{s27}(\xi)$. The result is

$$\begin{aligned} F_{s27}(\xi) = \tfrac{1}{135}(&300\xi_1^4\xi_2 + 560\xi_1^3\xi_2^3 + 8328\xi_1^2\xi_2^5 - 4560\xi_1\xi_2^7 - 20\xi_2^9 \\ &+75\xi_1^3\xi_3 + 1395\xi_1^2\xi_2^2\xi_3 - 4455\xi_1\xi_2^4\xi_3 + 105\xi_2^6\xi_3 \\ &-270\xi_1\xi_2\xi_3^2 - 90\xi_2^3\xi_3^2 - 135\xi_3^3). \end{aligned} \tag{18}$$

We note here that the polynomial $F_{s27}(\xi)$ coincides with $F_{B,6}$ introduced in [15] by a certain coordinate change and that the fundamental group $\pi_1(\mathbf{C}^3 - \{F_{B,6} = 0\})$ is computed by Saito and Ishibe [13]. In fact, $\pi_1(\mathbf{C}^3 - \{F_{B,6} = 0\})$ is generated by three generators a, b, c with the relations

$$aba = bab, \; aca = bac, \; acaca = cacac.$$

On the other hand, by the substitution

$$\begin{aligned} \xi_1 &= -\tfrac{5}{4\cdot 3^{1/3}}(t_1 + z)^2, \\ \xi_2 &= -\tfrac{1}{2\cdot 3^{2/3}}(5t_1 - 3z), \\ \xi_3 &= \tfrac{1}{9}(-19t_1^3 + 9t_3 - 15t_1^2 z + 27t_1 z^2 - 9z^3) \end{aligned}$$

the polynomial $\det(T)$ in (15) coincides with $F_{s27}(\xi)$. This suggests that there is a relationship between $\pi_1(\mathbf{C}^3 - \{F_{B,6} = 0\})$ and $\pi_1(\mathbf{C}^3 - \{\det(T) = 0\})$. We now treat the system (14). It is a conjecture that in the case $r = 1/2$, the monodromy group of this system is finite and is isomorphic to $W(H_3)$. This implies, in particular, that there is a group homomorphism of $\pi_1(\mathbf{C}^3 - \{\det(T) = 0\})$ to $W(H_3)$, which matches the argument in [2, Sect. 6.1].

Remark 7.6 The system (7) is obtained in [5]. But it is not expressed in a flat coordinate system.

After the work was done, one of the authors of this chapter (Mitsuo Kato) proved the conjecture in page 397 affirmatively.

Acknowledgements This work was supported in part by a grant-in-aid from the Japan Society for the Promotion of Science (Grant Numbers 25800082 and 26400111).

References

1. P. Boalch, The fifty-two icosahedral solutions to Painlevé VI. J. Reine Angew. Math. **596**, 183–214 (2006)
2. P. Boalch, Towards a nonlinear Schwarz's list, in *The Many Facets of Geometry* (Oxford University Press, Oxford, 2010), pp. 210–236
3. B. Dubrovin, Geometry of 2D topological field theories, in *Integrable Systems and Quantum Groups. Montecatini Terme, 1993*, ed. by M. Francoviglia, S. Greco. Lecture Notes in Mathematics, vol. 1620 (Springer, Berlin/Heidelberg, 1996), pp.120–348
4. B. Dubrovin, M. Mazzocco, Monodromy of certain Painlevé VI transcendents and reflection groups. Invent. Math. **141**, 55–147 (2000)
5. M. Kato, J. Sekiguchi, Uniformization systems of equations with singularities along the discriminant sets of complex reflection groups of rank three. Kyushu J. Math. **68**, 181–221 (2014)
6. M. Kato, T. Mano, J. Sekiguchi, Flat structures without potentials. Revue Roumaine Math. Pures Appl. **60**(4), 481–505 (2015)
7. M. Kato, T. Mano, J. Sekiguchi, Flat structure on the space of isomonodromic deformations. Preprint `arXiv:1511.01608`
8. F.C. Klein, Vorlesungen über das Ikosaeder und die Auflösung der Gleichungen vom fünnften Grade, in *Herausgegeben mit einer Einführung und mit Kommentaren von Perer Slodowy* (Birkhäuser, Basel/Boston/Berlin, 1993)
9. Y.I. Manin, F-manifolds with flat structure and Dubrovin's duality. Adv. Math. **198**, 5–26 (2005)
10. C. Sabbah, *Isomonodromic Deformations and Frobenius Manifolds. An Introduction.* Universitext (Springer, London, 2008)
11. K. Saito, Theory of logarithmic differential forms and logarithmic vector fields. J. Fac. Sci. Univ. Tokyo Sect. IA Math. **27**, 265–291 (1980)
12. K. Saito, On a linear structure of the quotient variety by a finite reflexion group. Preprint RIMS-288 (1979); Publ. RIMS Kyoto Univ. **29**, 535–579 (1993)
13. K. Saito, T. Ishibe, Monoids in the fundamental groups of the complement of logarithmic free divisors in $\mathbf{C}^3$. J. Algebra **344**, 137–160 (2011)
14. K. Saito, T. Yano, J. Sekiguchi, On a certain generator system of the ring of invariants of a finite reflection group. Commun. Algebra **8**, 373–408 (1980)
15. J. Sekiguchi, A classification of weighted homogeneous Saito free divisors. J. Math. Soc. Jpn. **61**, 1071–1095 (2009)

Relation of Semi-Classical Orthogonal Polynomials to General Schlesinger Systems via Twistor Theory

Hironobu Kimura

Abstract We study the relation between semi-classical orthogonal polynomials and nonlinear differential equations coming from the isomonodromic deformation of linear system of differential equations on $\mathbb{P}^1$. There are many works establishing this kind of relations between the Painlevé equations and semi-orthogonal polynomials with the weight functions taking from the integrands for hypergeometric, Kummer, Bessel, Hermite, Airy integrals. Some extension of these results is obtained for the semi-classical orthogonal polynomials with the weight functions coming from the general hypergeometric integrals on the Grassmannian $G_{2,N}$. To establish the desired relations, we make use of the Atiyah-Ward Ansatz construction of particular solutions for the 2×2 Schlesinger system and its degenerated ones.

Keywords Isomonodromic deformation • Semi-classical orthogonal polynomial • Twistor theory

Mathematics Subject Classification (2000). Primary 34M56; Secondary 33C45

1 Introduction

In this note we discuss a relation of semi-classical orthogonal polynomials to the nonlinear systems of partial differential equations obtained from the theory of isomonodromic deformation of linear differential equations on the projective line $\mathbb{P}^1$.

First we explain our motivation. Let $w(t)$ be a positive weight function on some subset $I \subset \mathbb{R}$ and let $(f, g) = \int_I f(t)g(t)w(t)dt$ be the inner product for polynomials with respect to the measure $w(t)dt$. By the process of orthogonalization we have a series of monic orthogonal polynomials $p_n(t)$ of degree $n \in \mathbb{Z}_{\geq 0}$. One of the

H. Kimura (✉)
Department of Mathematics, Kumamoto University, 2-39-1 Kurokami, Kumamoto 860-8555, Japan
e-mail: hiro.kimu@gmail.com

G. Filipuk et al. (eds.), *Analytic, Algebraic and Geometric Aspects of Differential Equations*, Trends in Mathematics, DOI 10.1007/978-3-319-52842-7_12

important features of orthogonal polynomials is the three-term recurrence relations

$$tp_n(t) = p_{n+1}(t) + \alpha_n p_n(t) + \beta_n p_{n-1}(t).$$

It is important to know the coefficients α_n, β_n. These quantities can be expressed using the determinants

$$D_n = \det\left(\int_I t^{j+k} w(t)dt\right)_{j,k=0}^{n-1}$$

of the Hankel matrix whose (i,j) entry is the $i+j$ th moment of $w(t)$. It is known that β_n can be expressed as

$$\beta_n = \frac{D_{n-1}D_{n+1}}{D_n^2}$$

and α_n is also computable in terms of $\{D_n\}$. For the classical orthogonal polynomials, namely, Jacobi, Laguerre and Hermite polynomials, we take $w(t) = t^\alpha(1-t)^\beta$ on $[0, 1]$, $t^\alpha e^{-t}$ on $[0, \infty)$ and e^{-t^2} on $(-\infty, \infty)$ as the weight function, respectively. Evidently, we impose the condition $\alpha, \beta > -1$ so that the inner product can be defined for polynomials. In these cases, D_n are constants depending on the parameters α, β contained in the weight function. It should be noted here that, these weight functions are integrands of Beta, Gamma and Gaussian integrals, respectively:

$$B(\alpha+1, \beta+1) = \int_0^1 t^\alpha(1-t)^\beta dt,\ \Gamma(\alpha+1) = \int_0^\infty t^\alpha e^{-t} dt,\ \sqrt{\pi} = \int_{-\infty}^\infty e^{-t^2} dt.$$

Several semi-classical orthogonal polynomials are defined using the weight functions $w(t, x)$ depending on a parameter x. In these cases the Hankel determinants D_n depend on x, so we denote them as $D_n(x)$. A numerous works are devoted to clarify how $D_n(x)$ are related to the Painlevé equations P2,…,P6. For example, Dai and Zhang [3] considered the semi-classical orthogonal polynomials attached to the weight function $w(t, x) = t^\alpha(1-t)^\beta(t-x)^\gamma$ and showed that the function

$$H_n(x) := x(x-1)\frac{d}{dx}\log D_n(x) + d_1 x + d_2$$

with

$$d_1 = -n(n+\alpha+\beta+\gamma) - \frac{(\alpha+\beta)^2}{4},$$

$$d_2 = -\frac{1}{4}\left[2n(n+\alpha+\beta+\gamma) + \beta(\alpha+\beta) - \gamma(\alpha-\beta)\right]$$

satisfies the Okamoto's σ-form equation for the sixth Painlevé equation P6. This result indicates that $D_n(x)$ is the τ-function for some particular solution of P6.

Similar connection of semi-classical orthogonal polynomials of other type to the Painlevé equations was also discussed [1, 2, 4, 9]. The form of weight functions and the related Painlevé equations are listed in the following table.

$w(t, x)$	Painlevé	Special function
$t^{\alpha}(1-t)^{\beta}(t-x)^{\gamma}$	P6	Gauss
$t^{\alpha}(1-t)^{\beta}e^{-x/t}$	P5	Kummer
$(1+t)^{\alpha}(1-t)^{\beta}e^{-xt}$	"	"
$\|t-x\|^{\alpha}e^{-t^2}$	P4	Hermite-Weber
$t^{\alpha}e^{-t^2+xt}$	"	"
$\|t^{2\alpha+1}\|e^{-t^4+xt^2}$	"	"
$t^{\alpha}e^{-t-x/t}$	P3	Bessel
$e^{t^3/3+xt}$	P2	Airy

In the third column of the above table, we listed the name of special functions, where the weight function, or rather the measure $w(t,x)dt$, is essentially the integrand of the integral representation of the corresponding special function.

It is natural to ask if it is possible to extend the above story by taking an appropriate class of weight functions and a class of nonlinear differential equations. Here we take, as a weight function w, the integrand of integral representation of the general hypergeometric function (GHGF) on the Grassmannian manifold $G_{2,N}$ consisting of 2-dimensional subspaces in $\mathbb{C}^N$, see [6]. As is explained in Sect. 2, GHGF is defined as a Radon transform of a character of the universal covering group of a maximal abelian subgroup $H_{\lambda} \subset \mathrm{GL}_N(\mathbb{C})$ indexed by a partition λ of N. We observe that the Beta, Gamma and Gaussian integral are regarded as GHGF on $G_{2,3}$ for the partitions $\lambda = (1,1,1), (2,1), (3)$, respectively, and the special functions listed above, namely, Gauss, Kummer, Bessel, Hermite-Weber and Airy are GHGF on $G_{2,4}$ for the partitions $\lambda = (1,1,1,1), (2,1,1), (2,2), (3,1)$ and (4), respectively.

The nonlinear differential equations which we consider are those obtained from the isomonodromic deformation of systems of linear differential equations on $\mathbb{P}^1$ with regular and irregular singular points for 2×2 unknowns, so the nonlinear equations are equivalent to the Garnier system [5] and the systems of its confluent type. We call these systems as general Schlesinger system (GSS). The twistor theoretic approach is used to describe the isomonodromic deformation, where the deformation parameters live in the subspace $Z_{\lambda}$of $\mathrm{Mat}_{2,N}(\mathbb{C})$ whose quotient space $\mathrm{GL}_2(\mathbb{C}) \setminus Z_{\lambda}$ is a Zariski open subset of $G_{2,N}$, and the infinitesimal action of the group H_{λ} on Z_{λ} plays an important role.

A connection of the Hankel determinants to GSS is a consequence of the result due entirely to Shah and Woodhouse [10] on the construction of particular solutions, so called Ward ansatz solutions, for GSS.

This note is of expository nature and serves as a remark on the recognition of a possible extension of the connection between the theory of semi-classical orthogonal polynomials and nonlinear systems coming from isomonodromic deformation.

This note is organized as follows. In Sect. 2, we recall the definition of general hypergeometric functions (GHGF) on the Grassmannian manifold $G_{2,N}$. Then we review the twistor theoretic treatment of isomonodromic deformation in Sect. 3 following [10] and [8]. In Sect. 4, we explain the construction of Ward ansatz solution of the generalized anti-self-dual Yang-Mills equation (GYM) and of the related GSS in terms of general hypergeometric functions, which say that the determinant of the Hankel matrix, whose entries are moments of integrand of general hypergeometric integral on the Grassmannian, describes a particular solution of GSS. This establishes an extension of the results on the relation of semi-classical orthogonal polynomial theory to Painlevé equations.

2 Hypergeometric Function on $G_{2,N}$

Maximal Abelian Subgroup

We shall recall the definition of general hypergeometric functions (GHGF). Let N be a positive integer and $\lambda = (n_1, \dots, n_\ell)$ be a partition of N. For λ, we associate a maximal abelian subgroup of complex general linear group $\mathrm{GL}_N(\mathbb{C})$ defined by

$$H_\lambda := J(n_1) \times \cdots \times J(n_\ell),$$

where $J(n) \subset \mathrm{GL}_n(\mathbb{C})$ is the abelian subgroup obtained as a centralizer of the shift matrix $\Lambda = (\delta_{i+1,j})_{0 \le i,j < n}$ and is called the Jordan group of size n. Explicitly we have

$$J(n) = \left\{ h = h_0 I + h_1 \Lambda + \cdots + h_{n-1} \Lambda^{n-1} \mid h_0 \neq 0 \right\} \subset \mathrm{GL}_n(\mathbb{C}),$$

from which we can know the isomorphism $J(n) \simeq \mathbb{C}[X]/(X^n)$ as multiplicative groups, where $\mathbb{C}[X]$ is the ring of polynomials in X and (X^n) is the ideal generated by X^n. The Lie algebras for H_λ and $J(n)$ will be denoted by $\mathfrak{h}_\lambda$ and $\mathfrak{j}(n)$, respectively.

Character

Let $\tilde{H}_\lambda$ be the universal covering group of H_λ and consider a character of $\tilde{H}_\lambda$, namely a group homomorphism $\chi : \tilde{H}_\lambda \to \mathbb{C}^\times$. Explicit description is as follows. Let $\theta_m(x)$ $(m \ge 0)$ be the functions of $x = (x_0, x_1, \dots)$ defined by

$$\sum_{0 \le m < \infty} \theta_m(x) T^m = \log(x_0 + x_1 T + x_2 T^2 + \cdots). \tag{1}$$

$$= \log x_0 + \sum_{m=1}^{\infty} \frac{(-1)^{m-1}}{m} \left(\frac{x_1}{x_0} T + \frac{x_2}{x_0} T^2 + \cdots \right)^m . \tag{2}$$

Then we see that $\theta_0(x) = \log x_0$ and

$$\theta_1(x) = \frac{x_1}{x_0}$$
$$\theta_2(x) = \frac{x_2}{x_0} - \frac{1}{2}\left(\frac{x_1}{x_0}\right)^2$$
$$\theta_3(x) = \frac{x_3}{x_0} - \left(\frac{x_1}{x_0}\right)\left(\frac{x_2}{x_0}\right) + \frac{1}{3}\left(\frac{x_1}{x_0}\right)^3$$
$$\vdots$$

Since the correspondence $\tilde{J}(n) \to \mathfrak{j}(n)$, given by $h \mapsto (\theta_0(h), \theta_1(h), \dots, \theta_{n-1}(h))$, defines the identification of $\tilde{J}(n)$ with its Lie algebra $\mathfrak{j}(n)$, exponentiating a character of $\mathfrak{j}(n)$ and using this correspondence, we have a character $\chi_n : \tilde{J}(n) \to \mathbb{C}^\times$ as $\chi_n(h;\alpha) = \exp\left(\sum_{0\le i<n} \alpha_i\theta_i(h)\right)$ with a weight $\alpha = (\alpha_0, \alpha_1, \dots, \alpha_{n-1}) \in \mathbb{C}^n$. Since $\tilde{H}_\lambda$ is a direct product of $\tilde{J}(n_k)$, the characters χ of $\tilde{H}_\lambda$ are given by

$$\chi(h;\alpha) = \prod_{1\le k\le \ell} \chi_{n_k}(h^{(k)}, \alpha^{(k)}) = \prod_{1\le k\le \ell} \exp\left(\sum_{0\le i<n_k} \alpha_i^{(k)}\theta_i(h^{(k)})\right),$$

for $h = (h^{(1)}, \dots, h^{(\ell)}) \in \tilde{H}_\lambda$, $h^{(k)} \in \tilde{J}(n_k)$. Here $\alpha = (\alpha^{(1)}, \dots, \alpha^{(\ell)}) \in \mathbb{C}^N$, $\alpha^{(k)} = (\alpha_0^{(k)}, \dots, a_{n_k-1}^{(k)}) \in \mathbb{C}^{n_k}$ is a weight.

General Hypergeometric Function

The general hypergeometric function (GHGF) is defined as a Radon transform of the characters χ of $\tilde{H}_\lambda$ as follows. Let Z_λ be the open subset of $\mathrm{Mat}_{2,N}(\mathbb{C})$ consisting of matrices $z = (z^{(1)}, \dots, z^{(\ell)})$, $z^{(k)} = (z_0^{(k)}, \dots, z_{n_k-1}^{(k)}) \in \mathrm{Mat}_{2,n_k}(\mathbb{C})$ satisfying

$$\det(z_0^{(k)}, z_1^{(k)}) \neq 0, \ (\text{if } n_k \geq 2)$$
$$\det(z_0^{(k)}, z_0^{(l)}) \neq 0, \ (k \neq l).$$

Definition 2.1 Assume that the weight of a character χ of $\tilde{H}_\lambda$ satisfies the condition

$$\sum_{1\le k\le \ell} \alpha_0^{(k)} = -2. \tag{3}$$

Then the general hypergeometric function of type λ is defined by

$$F(z,\alpha) = \int_C \chi(\vec{t}z,\alpha)dt \qquad (z \in Z_\lambda) \tag{4}$$

where $\vec{t} = (1,t)$, $\vec{t}z = (\vec{t}z_0^{(1)},\dots,\vec{t}z_{n_1-1}^{(1)},\dots,\vec{t}z_0^{(\ell)},\dots,\vec{t}z_{n_\ell-1}^{(\ell)})$ and C is a one dimensional cycle in $\mathbb{C}$ of the homology group defined by the integrand. We do not enter in detailed explanation for the homology group.

On the space Z_λ, the groups $\mathrm{GL}_2(\mathbb{C})$ and H_λ act from left and right, respectively, by the matrix multiplication $\mathrm{GL}_2(\mathbb{C}) \times Z_\lambda \times H_\lambda \ni (g,z,h) \mapsto gzh \in Z_\lambda$. Then we know the following results.

Proposition 2.2 *$F(z,\alpha)$ satisfies*

$$F(gz,\alpha) = (\det g)^{-1}F(z,\alpha) \quad (g \in \mathrm{GL}_2(\mathbb{C})),$$

$$F(zh,\alpha) = \chi(h,\alpha)F(z,\alpha) \quad (h \in \tilde{H}_\lambda) \tag{5}$$

$$(\partial_{0i}\partial_{1j} - \partial_{1i}\partial_{0j})F(z,\alpha) = 0 \quad (\forall i,j) \tag{6}$$

Roughly speaking, the last equation (6) comes from the fact that GHGF is defined as a Radon transform of a function on $\tilde{H}_\lambda$.

Relation to the Classical Special Functions

We explain how the integral representation for the classical special functions is obtained as GHGF on the Grassmannian manifold $G_{2,4}$. We list up the following data:

1. the character of $\tilde{H}_\lambda$,
2. GHGF of type λ,
3. a subspace X_λ of Z_λ which is a realization of $\mathrm{GL}_2(\mathbb{C})\backslash Z_\lambda/H_\lambda$,
4. restriction of GHGF to X_λ with a normalization of parameters.

Gauss HGF($\lambda = (1,1,1,1)$)

1. $\chi(h) = h_1^{\alpha_1}\cdots h_4^{\alpha_4}$ with $\alpha_1+\alpha_2+\alpha_3+\alpha_4 = -2$,
2. $F(z,\alpha) = \int_C (z_{01}+z_{11}t)^{\alpha_1}\cdots(z_{04}+z_{14}t)^{\alpha_4}dt$,
3. $X_{(1,1,1,1)} = \left\{ \mathbf{x} = \begin{pmatrix} 1 & 0 & 1 & -x \\ 0 & 1 & -1 & 1 \end{pmatrix} \mid x \neq 0,1 \right\}$,
4. $F(\mathbf{x},\alpha) = \int_C t^{\alpha_2}(1-t)^{\alpha_3}(t-x)^{\alpha_4}dt$.

Kummer's Confluent HGF($\boldsymbol{\lambda} = (2, 1, 1)$)

1. $\chi(h) = h_1^{\alpha_1} \exp(\alpha_2 \frac{h_2}{h_1}) h_3^{\alpha_3} h_4^{\alpha_4}$ with $\alpha_1 + \alpha_3 + \alpha_4 = -2,\ \alpha_2 = -1$
2. $F(z, \alpha) = \int_C (z_{01} + z_{11}t)^{\alpha_1} \exp\left(\alpha_2 \frac{z_{02}+z_{12}t}{z_{01}+z_{11}t}\right) \prod_{i=3,4}(z_{0i} + z_{1i}t)^{\alpha_i} dt,$
3. $X_{(2,1,1)} = \left\{ \mathbf{x} = \begin{pmatrix} 1 & 0 & 0 & 1 \\ 0 & x & 1 & -1 \end{pmatrix} \mid x \neq 0 \right\},$
4. $F(\mathbf{x}, \alpha) = \int_C e^{-xt} t^{\alpha_3} (1-t)^{\alpha_4} dt.$

Hermite-Weber($\boldsymbol{\lambda} = (3, 1)$)

1. $\chi(h) = h_1^{\alpha_1} \exp\left(\alpha_2 \frac{h_2}{h_1} + \alpha_3 (\frac{h_3}{h_1} - \frac{1}{2}(\frac{h_2}{h_1})^2)\right) h_4^{\alpha_4}$ with $\alpha_1 + \alpha_4 = -2,\ \alpha_2 = 0,\ \alpha_3 = 1,$
2. $F(z, \alpha) = \int_C (z_{01} + z_{11}t)^{\alpha_1} \exp\left(\frac{z_{03}+z_{13}t}{z_{01}+z_{11}t} - \frac{1}{2}(\frac{z_{02}+z_{12}t}{z_{01}+z_{11}t})^2\right) (z_{04} + z_{14}t)^{\alpha_4} dt,$
3. $X_{(3,1)} = \left\{ \mathbf{x} = \begin{pmatrix} 1 & 0 & 0 & 0 \\ 0 & 1 & x & 1 \end{pmatrix} \mid x \in \mathbb{C} \right\},$
4. $F(\mathbf{x}, \alpha) = \int_C \exp(xt - \frac{1}{2}t^2) t^{\alpha_4} dt.$

Bessel($\boldsymbol{\lambda} = (2, 2)$)

1. $\chi(h) = h_1^{\alpha_1} \exp\left(\alpha_2 \frac{h_2}{h_1}\right) h_3^{\alpha_3} \exp(\alpha_4 \frac{h_4}{h_3})$ with $\alpha_1 + \alpha_3 = -2,\ \alpha_2 = 1,\ \alpha_3 = 1,$
2. $F(z, \alpha) = \int_C (z_{01} + z_{11}t)^{\alpha_1} \exp\left(\frac{z_{02}+z_{12}t}{z_{01}+z_{11}t}\right) (z_{03} + z_{13}t)^{\alpha_3} \exp\left(\frac{z_{04}+z_{14}t}{z_{03}+z_{13}t}\right) dt,$
3. $X_{(2,2)} = \left\{ \mathbf{x} = \begin{pmatrix} 1 & 0 & 0 & -x \\ 0 & 1 & 1 & 0 \end{pmatrix} \mid x \neq 0 \right\},$
4. $F(\mathbf{x}, \alpha) = \int_C \exp(t - \frac{x}{t}) t^{\alpha_3} dt.$

Airy($\boldsymbol{\lambda} = (4)$)

1. $\chi(h) = h_1^{\alpha_1} \exp\left(\alpha_2 \frac{h_2}{h_1} + \alpha_3 (\frac{h_3}{h_1} - \frac{1}{2}(\frac{h_2}{h_1})^2) + \alpha_4 (\frac{h_4}{h_1} - \left(\frac{h_2}{h_1}\right)\left(\frac{h_3}{h_1}\right) + \frac{1}{3}\left(\frac{h_2}{h_1}\right)^3)\right)$ with $\alpha_1 = -2,\ \alpha_2 = \alpha_3 = 0,\ \alpha_4 = 1,$
2. $F(z, \alpha) = \int_C (z_{01} + z_{11}t)^{\alpha_1} \exp\left(\frac{z_{04}+z_{14}t}{z_{01}+z_{11}t} - \frac{z_{02}+z_{12}t}{z_{01}+z_{11}t} \cdot \frac{z_{03}+z_{13}t}{z_{01}+z_{11}t} + \frac{1}{3}(\frac{z_{02}+z_{12}t}{z_{01}+z_{11}t})^3\right) dt,$
3. $X_{(4)} = \left\{ \mathbf{x} = \begin{pmatrix} 1 & 0 & 0 & 0 \\ 0 & 1 & 0 & x \end{pmatrix} \mid x \in \mathbb{C} \right\},$
4. $F(\mathbf{x}, \alpha) = \int_C \exp(xt + \frac{1}{3}t^3) dt.$

3 General Schlesinger System and the Result

Schlesinger System

Consider a family of linear differential equations on $\mathbb{P}^1$

$$\frac{dy}{d\zeta} = \left(\sum_{j=1}^{N-1} \frac{A_j(x)}{\zeta - x_j} \right) y, \quad A_j(x) \in sl_2(\mathbb{C}) \tag{7}$$

where $A_j(x)$ depends holomorphically on x in some open subset of $(\mathbb{P}^1)^{N-1}$. Assume that $A_N := -\sum_{j=1}^{N-1} A_j(x) = \mathrm{diag}(a, -a)$.

Definition 3.1 The Eq. (7) gives an isomonodromic family if there exists a fundamental system of solutions $Y(\zeta, x)$ such that the associated monodromy representation is invariant under the variation of x_js.

Assume here that, for each j, 2 eigenvalues of A_j do not differ by an integer. Then we know the following result.

Proposition 3.2 *The family of Eqs. (7) gives an isomonodromic family with a fundamental system of solutions* $Y(\zeta, x)$ *which has the form* $Y(\zeta, x) = (\sum_{m=0}^{\infty} Y_m \zeta^{-m}) \mathrm{diag}(\zeta^a, \zeta^{-a})$ *at* $\zeta = \infty$ *with* $Y_0 = I_2$, *if and only if (7) together with*

$$\frac{\partial y}{\partial x_j} = -\frac{A_j(x)}{\zeta - x_j} \quad (1 \le j < N)$$

form an integrable system. This condition can be written as the Schlesinger system:

$$dA_j = \sum_{i(\neq j)} [A_i, A_j] d \log(x_i - x_j) \qquad (1 \le j < N). \tag{8}$$

General Schlesinger System

For a given partition $\lambda = (n_1, \ldots, n_\ell)$ of N, let us consider the system of linear differential equations of the form

$$\frac{dy}{d\zeta} = \left(\sum_{k=1}^{\ell} \sum_{j=0}^{n_k - 1} A_j^{(k)}(z) \frac{d\theta_j(\vec{\zeta} z^{(k)})}{d\zeta} \right) y \tag{9}$$

where θ_j are those defined in (1), $z = (z^{(1)}, \dots, z^{(\ell)})$ varies in some open subset $U \subset Z_\lambda$ and $A_j^{(k)}(z) \in sl_2(\mathbb{C})$ depends holomorphically on $z \in U$ and satisfy $\sum_{k=0}^{\ell} A_0^{(k)}(z) = 0$. The equation has the singular points $\{x_1, \dots, x_\ell\}$, where $x_k = -\frac{z_{00}^{(k)}}{z_{10}^{(k)}}$ $(1 \le k \le \ell)$. Since $\frac{d\theta_j(\vec{\xi z}^{(k)})}{d\xi}$ has a pole of order n_k, the Eq. (9) has in general an irregular singular point at x_k when $n_k \ge 2$. Roughly speaking, the family (9) gives an isomonodromic family if there is a fundamental system of solutions such that the associated monodromy representation is independent of z and the connection matrices among canonical solutions at irregular singular points(including Stokes matrices) are also independent of z. We refer Sect. 4 of [10] for the detailed explanation for it. The nonlinear system for $A_j^{(k)}(z)$ governing the isomonodromic deformation is called the general Schlesinger system (GSS). We remark that when the partition of N is $\lambda = (1, \dots, 1)$, the GSS coincides with the Schlesinger system. Note also that the Schlesinger system is known to be completely integrable, but the integrability of GSS is checked in a particular case [7].

Result

Here we present a particular solution of GSS given in terms of the Hankel determinants of moments associated with the general hypergeometric function of type λ.

As above, take a partition $\lambda = (n_1, \dots, n_\ell)$ of N, a character $\chi(h, \alpha)$ of $\tilde{H}_\lambda$. For any fixed $n_0 \in \mathbb{Z}$, consider moments of $\chi(\vec{tz}, \alpha)$:

$$\phi_n(z) = \int_C t^{n+n_0} \chi(\vec{tz}, \alpha) dt, \tag{10}$$

where C is a cycle and the Hankel determinant $\tau_m^p(z) = \det(\phi_{i+j+p-m+1}(z))_{i,j=0}^{m-1}$, where m is a size of matrix and p denotes the index of moment which is arrayed in the main anti-diagonal entries. For example when $p = 0$,

$$\tau_m^0(z) = \begin{vmatrix} \phi_{1-m} & \phi_{2-m} & \cdots & & \phi_0 \\ \phi_{2-m} & & & \phi_0 & \\ \vdots & & \cdot^{\cdot^{\cdot}} & & \vdots \\ & \phi_0 & & & \phi_{m-2} \\ \phi_0 & & \cdots & \phi_{m-2} & \phi_{m-1} \end{vmatrix},$$

where ϕ_0 is arrayed along the main anti-diagonal line.

Put

$$f_0(z) = \frac{(-1)^m}{\tau_m^0} \begin{pmatrix} \tau_m^1 & \tau_{m+1}^0 \\ \tau_{m-1}^0 & \tau_m^{-1} \end{pmatrix} \tag{11}$$

We know that $f_0(z)$ belongs to $SL_2(\mathbb{C})$ by virtue of Sylvester's formula of determinants.

Theorem 3.3 *For any positive integer m, we have an isomonodromic family (9) of the form*

$$A_j^{(k)}(z) = -\frac{1}{2}\begin{pmatrix} \alpha_j^{(k)} & \\ & -\alpha_j^{(k)} \end{pmatrix} + \sum_{0 \le i} z_{1i}^{(k)} \frac{\partial f_0}{\partial z_{1,i+j}^{(k)}} \cdot f_0^{-1} \quad (1 \le k \le \ell, 0 \le j < n_k)$$

and hence a particular solution to GSS expressed in terms of Hankel determinants associated with GHGF of type λ.

This theorem is just a rephrase of the result due to Shah and Woodhouse [10]. It relies on the description of isomonodromic deformation via twistor theory and on the construction of particular solution of generalized anti-self-dual Yang-Mills equation on $G_{2,N}$ using Ward ansatz. It is not yet known that τ_m^0 is so called the τ-function for the isomonodromy problem. In the following sections, we explain how the above solution can be obtained.

4 Twistor Theory and Isomonodromic Deformation

In this section, we explain how the Schlesinger system and its confluent type systems can be obtained from the twistor theoretic point of view following [8, 10].

Generalized Yang-Mills Equation

Let $Z = \{z \in \mathrm{Mat}_{2,N}(\mathbb{C}) \mid rk\, z = 2\}$ with the coordinates $z = (z_{ij})_{0 \le i \le 1, 1 \le j \le N}$. Let $U \subset Z$ be an open set and consider a holomorphic connection D on the trivial bundle $U \times \mathbb{C}^2$ with connection matrices in $sl_2(\mathbb{C})$. Then D can be written as

$$D = d + \sum_{i,j} \Phi_{ij}(z) dz_{ij} = \sum_{ij} D_{ij} dz_{ij},$$

where

$$D_{ij} = \frac{\partial}{\partial z_{ij}} + \Phi_{ij}(z), \quad \Phi_{ij}(z) \in sl_2(\mathbb{C}).$$

Definition 4.1 A holomorphic connection D is called the generalized (anti-self-dual) Yang-Mills (GYM) connection, if we have

$$[\zeta D_{0j} - D_{1j}, \zeta D_{0k} - D_{1k}] = 0 \quad (j \neq k, \zeta \in \mathbb{C}). \tag{12}$$

Equivalently, we have the nonlinear equations for $\Phi_{ij}(z)$:

$$[D_{0j}, D_{0k}] = 0, \quad [D_{1j}, D_{1k}] = 0, \quad [D_{0j}, D_{1k}] + [D_{1j}, D_{0k}] = 0, \tag{13}$$

$1 \leq j, k \leq N$, which we call the generalized Yang-Mills equation.

Ward-Penrose Transform

The important feature in treating GYM equation is to encode its solutions in terms of some class of vector bundles on the twistor space $\mathbb{P}^{N-1}$. Consider a double fibration

$$\begin{array}{ccc} & \mathbb{P}\mathcal{C} & \\ \pi_1 \swarrow & & \searrow \pi_2 \\ \mathbb{P}^{N-1} & & Z \end{array}$$

where $\mathbb{P}\mathcal{C} := \{([t_0, t_1], z) \in \mathbb{P}^1 \times Z \mid (t_0, t_1) \neq (0, 0)\}$ with the homogeneous coordinates (t_0, t_1) of $\mathbb{P}^1$ and the maps π_1 and π_2 are defined by

$$\pi_1([t_0, t_1], z) = [t_0\vec{z}_0 + t_1\vec{z}_1], \quad z = \begin{pmatrix} \vec{z}_0 \\ \vec{z}_1 \end{pmatrix}$$

$$\pi_2([t_0, t_1], z) = z.$$

$\mathbb{P}\mathcal{C}$ is called the correspondence space and $\mathbb{P}^{N-1}$ the twistor space. The above double fibration gives the correspondence from Z to the twistor space $\mathbb{P}^{N-1}$ by

$$Z \ni z \mapsto \hat{z} := \pi_1(\pi_2^{-1}(z)) = \{[t_0\vec{z}_0 + t_1\vec{z}_1] \mid (t_0, t_1) \neq (0, 0)\} \subset \mathbb{P}^{N-1}.$$

Here $\hat{z}$ is a projective line in $\mathbb{P}^{N-1}$ joining two points $[\vec{z}_0], [\vec{z}_1]$ and is called the twistor line determined by z. On the other hand, it gives a correspondence from $\mathbb{P}^{N-1}$ to Z by

$$\mathbb{P}^{N-1} \ni p = [x] \mapsto \tilde{p} := \pi_2(\pi_1^{-1}([x])) = \{z \in Z \mid \vec{z}_0 \wedge \vec{z}_1 \wedge x = 0\},$$

where $\tilde{p}$ is a plane in Z of dim $= N - 1$ called the twistor surface.

It is known that a connection D on $U \times \mathbb{C}^2$ is GYM if and only if $D|_{\tilde{p}}$ is integrable for $\forall p \in \hat{U} = \pi_1(\pi_2^{-1}(U)) \subset \mathbb{P}^{N-1}$.

The Ward-Penrose transform is the correspondence between the following two sets

$$\left\{\begin{array}{l}\text{holomorphic } \mathrm{SL}_2(\mathbb{C})\text{-vector bundles}\\ E \to \hat{U}, \text{trivial on twistor lines } \hat{q}(q \in U)\end{array}\right\} \leftrightarrow \left\{\begin{array}{l}\text{solutions } D \text{ of}\\ \text{GYM on } U\end{array}\right\}$$

We call an element of the left-hand side a twistor bundle. We shall explain the Ward-Penrose transform from a twistor bundle to a solution of GYM.

Let $E \to \hat{U}$ be a $\mathrm{SL}_2(\mathbb{C})$-twistor bundle, and $\pi_1^* E$ be the lift of E to $\pi_1^{-1}(\hat{U}) = \mathbb{P}^1 \times U$. If $F \in \mathrm{SL}_2(\mathbb{C})$ is a patching function for E, then that for $\pi_1^* E$ is $F^* = F(\vec{z}_0 + \zeta \vec{z}_1)$, where $\zeta = t_1/t_0$ is the affine coordinates of $\mathbb{P}^1$. Let $\{V, \tilde{V}\}$ be an open covering of $\pi_1^{-1}(\hat{U})$ defined by

$$V = \{|\zeta| < r\} \times U, \ \tilde{V} = \{|\zeta| > \tilde{r}\} \times U, \quad \tilde{r} < r.$$

We may assume that F^* is defined on the intersection $V \cap \tilde{V}$. Since E is trivial on twistor lines, there is a Birkhoff decomposition $F^* = \tilde{f}^{-1} \cdot f$, where $f, \tilde{f} \in \mathrm{SL}_2(\mathbb{C})$ are holomorphic respectively on V and $\tilde{V}$ such that $\tilde{f}(\infty, z) = 1_2$. Combining this with the fact that F^* is a lift of F, we have $(\zeta\partial_{0j} - \partial_{1j})F^* = (\zeta\partial_{0j} - \partial_{1j})(\tilde{f}^{-1} \cdot f) = 0$ for any j. It implies

$$\zeta\partial_{0j}f \cdot f^{-1} - \partial_{1j}f \cdot f^{-1} = \zeta\partial_{0j}\tilde{f} \cdot \tilde{f}^{-1} - \partial_{1j}\tilde{f} \cdot \tilde{f}^{-1}. \tag{14}$$

The left-hand side and the right-hand side are defined on V and $\tilde{V}$, respectively. By Liouville Theorem, both sides define a polynomial function in ζ with a simple pole at $\zeta = \infty$. From $\tilde{f}(\infty, z) = 1_2$, we see that $\zeta = \infty$ is not a pole of both sides, and hence (14) defines a $sl_2(\mathbb{C})$ valued function depending only on z, which we denote as $\Phi_{1j}(z) \in sl_2(\mathbb{C})$. Then we have

$$[\zeta(\partial_{0j} + 0) - (\partial_{1j} + \Phi_{1j}(z))]f = 0, \quad [\zeta(\partial_{0j} + 0) - (\partial_{1j} + \Phi_{1j}(z))]\tilde{f} = 0.$$

This implies that the connection $\nabla = d + \sum_j \Phi_{1j}(z)dz_{1j}$ is a solution to GYM equation.

Note that, if $f = f_0(z) + f_1(z)\zeta + \cdots$, then Φ_{1j} can be determined only from f_0;

$$\Phi_{1j}(z) = -\partial_{1j}f_0 \cdot f_0^{-1} \quad (1 \le j \le N). \tag{15}$$

Isomonodromic Deformation

First we introduce some notation. Let $x = (x_1, \ldots, x_N)$ be the homogeneous coordinates of $\mathbb{P}^{N-1}$. We denote by $[x]$ (or sometimes by x) a point of $\mathbb{P}^{N-1}$. For $\xi \in \mathfrak{h}_\lambda$, define a vector field X_ξ on $\mathbb{P}^{N-1}$ and Y_ξ on $\mathbb{P}C$ by

$$X_\xi g := \frac{d}{ds} g([x \exp s\xi])|_{s=0}, \quad Y_\xi h := \frac{d}{ds} h(([t_0, t_1], z \exp s\xi))|_{s=0}.$$

Sometimes we use also the notation $x = (x^{(1)}, \ldots, x^{(\ell)})$, $x^{(k)} = (x_0^{(k)}, \ldots, x_{n_k-1}^{(k)}) \in \mathbb{C}^{n_k}$ for the homogeneous coordinates of $\mathbb{P}^{N-1}$.

Definition 4.2 $\mathrm{SL}_2(\mathbb{C})$-twistor bundle $E \to \hat{U}$ is said to be symmetric with respect to H_λ, if the infinitesimal action of H_λ on $\hat{U} \subset \mathbb{P}^{N-1}$ can be lifted to E, in other terms, if there is a Lie derivation $\mathcal{L}_\xi$ for any $\xi \in \mathfrak{h}_\lambda$, which acts on local sections of E, such that $\xi \mapsto \mathcal{L}_\xi$ is a Lie algebra homomorphism.

Note that the Lie derivation $\mathcal{L}_\xi$ can be written locally as $\mathcal{L}_\xi = X_\xi + B_\xi(x)$ with $B_\xi(x) \in sl_2(\mathbb{C})$. Since X_ξ spans a tangent space $T_x\mathbb{P}^{N-1}$at any points x such that $x_0^{(k)} \neq 0$ for any k, symmetry of the twistor bundle implies the integrable connection ∇ on $\hat{U}$ whose connection form can be written locally as

$$\sum_{k=1}^{\ell} \sum_{j=0}^{n_k-1} B_j^{(k)}(x) d\theta_j(x^{(k)}), \tag{16}$$

$B_j^{(k)} \in sl_2(\mathbb{C})$ is B_ξ for $\xi = E_j^{(k)} := 0 \oplus \cdots \oplus (\Lambda^{(k)})^j \oplus \cdots \oplus 0) \in \mathfrak{h}_\lambda$, where $\Lambda^{(k)} = (\delta_{i+1,j}) \in \mathfrak{j}(n_k)$ is the shift matrix of size n_k.

We want to get an isomonodromic family of linear differential equations on $\mathbb{P}^1$ by restricting the connection ∇ on twistor lines, or ∇^* on the lines $\mathbb{P}^1 \times \{z\} \subset \mathbb{P}\mathcal{C}$. To get a such family, we trivialize the twistor bundle, which is symmetric with respect to H_λ, on twistor lines. Let ∇^* be the lift of ∇ to the pullback bundle $\pi_1^* E \to \pi_1^{-1}(\hat{U}) \subset \mathbb{P}\mathcal{C}$ whose connection form on $V, \tilde{V}$ are of the forms

$$\omega = \sum_{k=1}^{\ell} \sum_{j=0}^{n_k-1} B_j^{(k)}(\zeta, z) d\theta_j(\vec{\xi} z^{(k)}), \quad \tilde{\omega} = \sum_{k=1}^{\ell} \sum_{j=0}^{n_k-1} \tilde{B}_j^{(k)}(\zeta, z) d\theta_j(\vec{\xi} z^{(k)})$$

and let F^* be the patching function for the bundle $\pi_1^* E$ on $V \cap \tilde{V}$. Then it is known that the symmetry is assured if F^* satisfies the condition

$$Y_\xi F^* = F^* B_\xi - \tilde{B}_\xi F^* \qquad (\forall \xi \in \mathfrak{h}_\lambda), \tag{17}$$

where B_ξ and $\tilde{B}_\xi$ are holomorphic on V and $\tilde{V}$ respectively which comes from the local form of $\mathcal{L}_\xi$ on V and $\tilde{V}$. Since $\pi_1^* E$ is trivial on $\mathbb{P}^1 \times \{z\}$ for any $z \in U$, we can find $f, \tilde{f} \in \mathrm{SL}_2(\mathbb{C})$, holomorphic respectively on V and $\tilde{V}$, such that $F^* = \tilde{f}^{-1} f$ with $\tilde{f}(\infty, z) = 1_2$. Then the integrable connection ∇^* on the product bundle:

$$\nabla^* = d + \Omega := d + f\omega f^{-1} - df \cdot f^{-1} = d + \tilde{f}\tilde{\omega}\tilde{f}^{-1} - d\tilde{f} \cdot \tilde{f}^{-1} \tag{18}$$

gives an isomonodromic deformation and the integrability condition $(\nabla^*)^2 = 0$ is a nonlinear system of differential equations. This nonlinear equation is the general Schlesinger system(GSS) of type λ. We want to express Ω more explicitly. Put $\Theta_\xi(z) := fB_\xi f^{-1} - Y_\xi f \cdot f^{-1} = \tilde{f}\tilde{B}_\xi \tilde{f}^{-1} - Y_\xi \tilde{f} \cdot \tilde{f}^{-1}$. Then we can show the following.

Proposition 4.3 *The connection form* Ω *for the connection* ∇^* *is given by*

$$\Omega = -\sum_{k=1}^{\ell}\sum_{j=0}^{n_k-1} A_j^{(k)}(z)d\theta_j(\vec{\zeta}z^{(k)}) + \Phi, \tag{19}$$

where $A_j^{(k)}(z) := \Theta_{E_j^{(k)}}(z) - i_{Y_{E_j^{(k)}}}\Phi$, $i_{Y_{E_j^{(k)}}}$ *being the interior product with respect to the vector field* $Y_{E_j^{(k)}}$, *and* $\Phi := \sum \Phi_{1j}(z)dz_{1j}$ *is a solution of GYM given by (15) corresponding to the twistor bundle E.*

The integrability of the connection $\nabla^* = d + \Omega$ describes the isomonodromic deformation of the linear differential equation (9).

5 Ward Ansatz Solution of GYM

We explain the construction of particular solutions of GSS of type λ following the description of [10], which gives Theorem 3.3.

Ward Ansatz Solution

Let $U \subset Z$ be an open set as in the previous section, and put $\hat{U} = \pi_1(\pi_2^{-1}(U))$. We set the following Ansatz:

(i) $SL_2(\mathbb{C})$-twistor bundle E on $\hat{U}$ corresponds to a solution of GYM equation,
(ii) the patching function F^* of π_1^*E has the form

$$F^* = \begin{pmatrix} \zeta^m & \phi(\zeta, z) \\ & \zeta^{-m} \end{pmatrix} \text{ on } V \cap \tilde{V}. \tag{20}$$

Let $\phi(\zeta, z) = \sum_{n=-\infty}^{\infty} \phi_n(z)\zeta^{-n}$ be the Laurent expansion with respect to ζ. Then we follow the process of Ward-Penrose transform explained above. We can construct the Birkhoff decomposition

$$F^* = \tilde{f}^{-1} \cdot f \tag{21}$$

uniquely under the condition $\tilde{f}(\infty, z) = 1_2$ using linear algebra. Especially the constant term f_0 in the Taylor expansion of f at $\zeta = 0$ can be given by (11). Hence if we can find the twistor bundle E such that the lifted patching function F^* satisfies the ansatz (i), (ii) and the condition (17) of symmetry with respect to H_λ, we can get a particular solution of GSS by the process of previous section.

First we consider the condition for $\phi(\zeta, z)$ so that F^* has the form (20). Since F^* is a lift of a transition F of E, we have $(\zeta\partial_{0j} - \partial_{1j})F^* = 0$, equivalently, $(\zeta\partial_{0j} - \partial_{1j})\phi = 0$ for any j. Substituting the Laurent expansion of $\phi(\zeta, z)$ with respect to ζ, we get

$$\partial_{0j}\phi_n = \partial_{1j}\phi_{n-1} \quad (1 \le j \le N, n \in \mathbb{Z}). \tag{22}$$

Notice that (22) implies

$$(\partial_{0j}\partial_{1k} - \partial_{0k}\partial_{1j})\phi_n = 0 \quad (j \ne k, n \in \mathbb{Z}), \tag{23}$$

which are just the Eqs. (6) used in characterizing the image of Radon transform.

Particular Solutions for GSS

In the construction of Ward ansatz solution described in the previous subsection, let us determine $\phi_n(z)$ so that the resulting twistor bundle becomes symmetric with respect to H_λ and as a result, it gives a particular solution to GSS of type λ.

Let $\chi : \tilde{H}_\lambda \to \mathbb{C}^\times$ be a character with a weight $\alpha \in \mathbb{C}^N$. Take a fixed $n_0 \in \mathbb{Z}$ and define

$$\phi_n(z) = \int_C t^{n+n_0}\chi(\vec{tz}, \alpha)dt, \tag{24}$$

where C is a cycle of the homology group associated with $\chi(\vec{tz}, \alpha)$. This choice of C assure the exchange of differentiation with respect to z and integration. We can check easily that $\phi_n(z)$ satisfies contiguous relation (22) and $\phi_n(zh) = \phi_n(z)\chi(h, \alpha)$. The last identity implies

$$Y_\xi\phi_n(z) = \langle \xi, \alpha\rangle\, \phi_n(z) \quad (\xi = \sum_{j,k} \xi_j^{(k)} E_j^{(k)} \in \mathfrak{h}_\lambda)$$

where $\langle \xi, \alpha\rangle = \sum_{j,k} \xi_j^{(k)}\alpha_j^{(k)}$. It follows that F^*, given by (20) and (24), satisfies the condition (17) for the symmetry of the twistor bundle with

$$B_\xi(z) = -\tilde{B}_\xi(z) = \frac{1}{2}\begin{pmatrix} \langle \xi, \alpha\rangle & \\ & -\langle \xi, \alpha\rangle \end{pmatrix}. \tag{25}$$

Finally, putting in $\Theta_\xi(z)$ the expression (25) for $\tilde{B}_\xi(z)$ and $\Phi = -\sum \partial_{1j}f_0 \cdot f_0^{-1}$ with f_0 given by (11) with (24), Proposition 4.3 produces a particular solution of GSS in terms of GHGF given in Theorem 3.3.

Acknowledgements The author is supported in part by JSPS grants-in-aid No.15K04903.

References

1. E. Basor, Y. Chen, T. Ehrhardt, Painlevé V and time-dependent Jacobi polynomials. J. Phys. A **43**(1), 015204, 25 pp (2010)
2. P.A. Clarkson, K. Jordaan, The relationship between semi-classical Laguerre polynomials and the fourth Painlevé equation. Constr. Approx. **39**(1), 223–254 (2014)
3. D. Dai, L. Zhang, Painlevé VI and Hankel determinants for the generalized Jacobi weight. J. Phys. A **43**(5), 055207, 14 pp (2010)
4. G. Filipuk, W. Van Assche, L. Zhang, The recurrence coefficients of semi-classical Laguerre polynomials and the fourth Painlevé equation. J. Phys. A **45**(20), 205201, 13 pp (2012)
5. K. Iwasaki, H. Kimura, S. Shimomura, M. Yoshida, *From Gauss to Painlevé* (Vieweg, Braunschweig, 1991)
6. H. Kimura, T. Koitabashi, Normalizer of maximal abelian subgroup of $GL(n)$ and general hypergeometric functions. Kumamoto J. Math. **9**, 13–43 (1996)
7. H. Kimura, Y. Nakamura, Analogue of Ward correspondence for a degenerated Schlesinger system. Kumamoto J. Math. **22**, 35–41 (2009)
8. H. Kimura, D. Tseveennamjil, General Schlesinger systems and their symmetry from the view point of twistor theory. J. Nonlinear Math. Phys. **20**(suppl. 1), 130–152 (2013)
9. A.P. Magnus, Painlevé-type differential equations for the recurrence coefficients of semi-classical orthogonal polynomials. J. Comput. Appl. Math. **57**(1–2), 215–237 (1995)
10. M.R. Shah, N.M.J. Woodhouse, Multivariate hypergeometric cascades, isomonodromy problems and Ward ansätze. J. Phys. A **43**(43), 434031, 16 pp (2010)

Some Notes on the Multi-Level Gevrey Solutions of Singularly Perturbed Linear Partial Differential Equations

Alberto Lastra and Stéphane Malek

Abstract This paper is a slightly modified, abridged version of the work (Lastra and Malek, Adv Differ Equ 21:767–800, 2016). It deals with some questions made to the authors during the conference Analytic, Algebraic and Geometric Aspects of Differential Equations, held in Będlewo (Poland) during the second week of September, 2015.

We study analytic and formal solutions related to a singularly perturbed partial differential equation and relate them by means of a multi-level Gevrey order asymptotic behavior, with respect to the perturbation parameter.

Keywords Borel-Laplace transform • Borel summability • Formal power series • Gevrey asymptotic expansion • Linear partial differential equations • Singular perturbation

Mathematics Subject Classification (2000). Primary 35C10; Secondary 35C20

1 Introduction

The main aim of the present work is to give answer to certain questions and fruitful mathematical discussions held with some participants of the conference Analytic, Algebraic and Geometric Aspects of Differential Equations (AAGADE), held in Będlewo (Poland) during the second week of September, 2015, where we presented the work [7]. For the sake of completeness and clarity, we provide an sketch of the results in that work.

A. Lastra (✉)
Departamento de Física y Matemáticas, University of Alcalá, Ap. de Correos 20, E-28871 Alcalá de Henares (Madrid), Spain
e-mail: alberto.lastra@uah.es

S. Malek
Laboratoire Paul Painlevé, University of Lille 1, 59655 Villeneuve d'Ascq Cedex, France
e-mail: Stephane.Malek@math.univ-lille1.fr

G. Filipuk et al. (eds.), *Analytic, Algebraic and Geometric Aspects of Differential Equations*, Trends in Mathematics, DOI 10.1007/978-3-319-52842-7_13

The main purpose in [7] is to study a family of singularly perturbed linear partial differential equations of the form

$$(\epsilon^{r_2}(t^{k+1}\partial_t)^{s_2} + a_2)(\epsilon^{r_1}(t^{k+1}\partial_t)^{s_1} + a_1)\partial_z^S X(t,z,\epsilon)$$

$$= \sum_{(s,\kappa_0,\kappa_1)\in\mathcal{S}} b_{\kappa_0\kappa_1}(z,\epsilon)t^s(\partial_t^{\kappa_0}\partial_z^{\kappa_1}X)(t,z,\epsilon), \quad (1)$$

for given initial conditions

$$(\partial_z^j X)(t,0,\epsilon) = \phi_j(t,\epsilon), \quad 0 \leq j \leq S-1, \quad (2)$$

where r_1 stands for a nonnegative integer (i. e. it belongs to $\mathbb{N} = \{0,1,\ldots\}$), and r_2, s_1, s_2 are positive integers. We also fix $a_1, a_2 \in \mathbb{C}^\star$. $\mathcal{S}$ consists of a finite subset of elements $(s,\kappa_0,\kappa_1) \in \mathbb{N}^3$. We assume that $S > \kappa_1$ for every $(s,\kappa_0,\kappa_1) \in \mathcal{S}$, and also that $b_{s,\kappa_0,\kappa_1}(z,\epsilon)$ belongs to the space of holomorphic functions in a neighborhood of the origin in $\mathbb{C}^2$, denoted by $\mathcal{O}\{z,\epsilon\}$.

The initial data consist of holomorphic functions defined in a product of finite sectors with vertex at the origin.

The case for complex perturbation parameter ϵ has also been studied when solving partial differential equations; in particular, when dealing with solutions belonging to spaces of analytic functions for singularly perturbed partial differential equations which exhibit several singularities of different nature. On this direction, one can cite the work by Canalis-Durand, Mozo-Fernández and Schäfke [2], Kamimoto [4], the second author [8, 9], and the first and the second author and Sanz [5]. In this last work, the appearance of both, irregular and fuchsian singularities in the problem causes that the Gevrey type concerning the asymptotic representation of the formal solution varies with respect to a problem in which only one type of such singularities appears.

The asymptotic behavior of the solution in the problem (1), (2) distinguish both singularly perturbed irregular operators located at the head of the main equation, in the sense that different Gevrey orders would appear relating asymptotically the analytic and the formal solution in the perturbation parameter ϵ. The main purpose of this work is to exhibit this interesting behavior of the asymptotics related. For this reason, we do not consider an Eq. (1) in which nonlinear terms have been taken into consideration. In our opinion, the relevant asymptotic phenomenon coming from the problem would not change, but computations would become tedious and unclear.

We construct actual holomorphic solutions $X(t,z,\epsilon)$ of (1), (2) which are represented by the formal solution

$$\hat{X}(t,z,\epsilon) = \sum_{\beta\geq 0} H_\beta(t,z)\frac{\epsilon^\beta}{\beta!} \in \mathbb{E}[[\epsilon]], \quad (3)$$

where $\mathbb{E}$ is an adequate complex Banach space. The solution is holomorphic in a domain of the form $\mathcal{T} \times \mathcal{U} \times \mathcal{E}$, where $\mathcal{T}$ and $\mathcal{E}$ are sectors of finite radius and vertex at the origin, and $\mathcal{U}$ is a neighborhood of the origin. In the asymptotic representation several Gevrey orders will appear.

In these notes, we also present some improvements with respect to the restrictions made on the coefficients appearing in the equation, and the geometry in which the problem rests. Moreover, we provide some details on the appearance of a higher number of operators appearing at the head of the equation and the asymptotic dependence on this data.

2 Summary of the Strategy Followed and Main Results

In this section, we present the main results in [7] giving only some detail on the crucial points for this notes. We refer to [7] for the complete details.

Let $S \geq 1$ be an integer. We also consider a nonnegative integer r_1 and positive integers r_2, s_1, s_2, k. Let $r := \frac{r_2}{s_2 k}$. We fix $a_1, a_2 \in \mathbb{C}^\star$ and a finite subset $\mathcal{S}$ of $\mathbb{N}^3$. For every $(s, \kappa_0, \kappa_1) \in \mathcal{S}$, let $b_{\kappa_0\kappa_1}(z, \epsilon)$ be a holomorphic and bounded function in a product of discs centered at the origin.

The problem (1) is studied for ϵ in each of the elements in a good covering in $\mathbb{C}^\star$.

Definition 2.1 Let $(\mathcal{E}_i)_{0 \leq i \leq \nu-1}$ be a finite family of open sectors such that $\mathcal{E}_i$ has its vertex at the origin and common finite radius $r_{\mathcal{E}_i} := r_{\mathcal{E}} > 0$ for every $0 \leq i \leq \nu - 1$. We say this family conforms a good covering in $\mathbb{C}^\star$ if $\mathcal{E}_i \cap \mathcal{E}_{i+1} \neq \emptyset$ for $0 \leq i \leq \nu - 1$ (we put $\mathcal{E}_\nu := \mathcal{E}_0$) and $\cup_{0 \leq i \leq \nu-1} \mathcal{E}_i = \mathcal{U} \setminus \{0\}$ for some neighborhood of the origin $\mathcal{U}$.

Definition 2.2 Let $(\mathcal{E}_i)_{0 \leq i \leq \nu-1}$ be a good covering in $\mathbb{C}^\star$. For every $0 \leq i \leq \nu - 1$, we assume

$$\mathcal{E}_i = \{\epsilon \in \mathbb{C}^\star : |\epsilon| < r_{\mathcal{E}}, \theta_{1,\mathcal{E}_i} < \arg(\epsilon) < \theta_{2,\mathcal{E}_i}\},$$

for some $r_{\mathcal{E}} > 0$ and $0 \leq \theta_{1,\mathcal{E}_i} < \theta_{2,\mathcal{E}_i} < 2\pi$. We write $d_{\mathcal{E}_i}$ for the bisecting direction of $\mathcal{E}_i$, $(\theta_{1,\mathcal{E}_i} + \theta_{2,\mathcal{E}_i})/2$. Let $\mathcal{T}$ be an open sector with vertex at 0 and finite radius, say $r_{\mathcal{T}} > 0$. We also fix a family of open sectors

$$S_{d_i,\theta,r_{\mathcal{E}}^r r_{\mathcal{T}}} = \left\{ t \in \mathbb{C}^\star : |t| \leq r_{\mathcal{E}}^r r_{\mathcal{T}}, |d_i - \arg(t)| < \frac{\theta}{2} \right\},$$

with $d_i \in [0, 2\pi)$ for $0 \leq i \leq \nu - 1$, and $\theta > \pi/k$, under the following properties:

1. **Assumption (A)**: one has $d_i \neq \frac{\pi(2j+1)+\arg(a_2)}{ks_2}$, for every $j = 0, \ldots, ks_2 - 1$.
2. **Assumption (B)**: one has $s_1 r_2 - s_2 r_1 > s_2 > 0$ and $|d_i - d_{\mathcal{E}_i j}| > \delta_{2i}$, for $j = 0, \ldots, ks_1 - 1$, where $\delta_{2i} := \frac{s_1 r_2 - s_2 r_1}{2ks_1 s_2}(\theta_{2,\mathcal{E}_i} - \theta_{1,\mathcal{E}_i})$, and $d_{\mathcal{E}_i j} = \frac{1}{ks_1}(\pi(2j+1) + \arg(a_1) + \frac{s_1 r_2 - s_2 r_1}{s_2}\left(\frac{\theta_{1,\mathcal{E}_i} + \theta_{2,\mathcal{E}_i}}{2}\right))$.
3. for every $0 \leq i \leq \nu - 1$, $t \in \mathcal{T}$ and $\epsilon \in \mathcal{E}_i$, one has $\epsilon^r t \in S_{d_i,\theta,r_{\mathcal{E}}^r r_{\mathcal{T}}}$.

Under the previous settings, we say the family $\{(S_{d_i,\theta,r_\mathcal{E}^r r_\mathcal{T}})_{0\leq i\leq \nu-1}, \mathcal{T}\}$ is associated to the good covering $(\mathcal{E}_i)_{0\leq i\leq \nu-1}$.

Assumption (A) in the previous definition is concerned with the existence of $d_i \in [0, 2\pi)$ such that the argument of every root of the polynomial $\tau \mapsto (k\tau^k)^{s_2} + a_2$ has positive distance to d_i, for every $0 \leq i \leq \nu - 1$.

The first part in Assumption (B) is motivated by the next

Assumption (C):

$$\theta_{2,\mathcal{E}_i} - \theta_{1,\mathcal{E}_i} < \frac{2\pi s_2}{s_1 r_2 - s_2 r_1},$$

which guarantees the existence of possible choices for directions $d_i \in [0, 2\pi)$ compatible with Assumption (B), in the sense that

$$d_i \notin \frac{1}{ks_1}\left[\pi(2j+1) + \arg(a_1) + \frac{s_1 r_2 - s_2 r_1}{s_2}\arg(\epsilon)\right],$$

for every $\tau \in S_{d_i,\theta,r_\mathcal{E}^r r_\mathcal{T}}$, $j = 0, \ldots, ks_1 - 1$, $\epsilon \in \mathcal{E}_i$ and $0 \leq i \leq \nu - 1$.

The second part in Assumption (B) is related to the existence of $d_i \in [0, 2\pi)$ such that the argument of every root of the polynomial $\tau \mapsto \epsilon^{r_1 - s_1 rk}(k\tau^k)^{s_1} + a_1$ has positive distance to d_i, for every $0 \leq i \leq \nu - 1$, independently of $\epsilon \in \mathcal{E}_i$.

We also make the further assumption that for every $(s, \kappa_0, \kappa_1) \in \mathcal{S}$, one has $\kappa_0 \geq 1$, $S > \kappa_1$, and there exists an integer $\delta_{\kappa_0,s} \geq k$ such that $s = \kappa_0(k+1) + \delta_{\kappa_0,s}$, and that $S > \left\lfloor b\left(\frac{\delta_{\kappa_0,s}}{k} + \kappa_0\right)\right\rfloor + 1$, for some $b > 1$. Observe that $S > \kappa_0$.

This last assumption allows to write the operators $T^s\partial_T^{\kappa_0}$ in such a way that the initial problem is transformed into an auxiliary equation via a slightly modified formal Borel transform (see [10] for the source of this idea and [6] for the properties held by this transformation).

Let $(\mathcal{E}_i)_{0\leq i\leq \nu-1}$ be a good covering, and let $\{(S_{d_i,\theta,r_\mathcal{E}^r r_\mathcal{T}})_{0\leq i\leq \nu-1}, \mathcal{T}\}$ be a family associated to that good covering. For every $0 \leq i \leq \nu - 1$, we study the Cauchy problem

$$\begin{aligned}(\epsilon^{r_2}(t^{k+1}\partial_t)^{s_2} + a_2)(\epsilon^{r_1}(t^{k+1}\partial_t)^{s_1} + a_1)\partial_z^S X_i(t, z, \epsilon)\\ = \sum_{(s,\kappa_0,\kappa_1)\in\mathcal{S}} b_{\kappa_0\kappa_1}(z, \epsilon)t^s(\partial_t^{\kappa_0}\partial_z^{\kappa_1}X_i)(t, z, \epsilon),\end{aligned} \tag{4}$$

for given initial conditions

$$(\partial_z^j X)(t, 0, \epsilon) = \phi_{i,j}(t, \epsilon), \quad 0 \leq j \leq S - 1, \tag{5}$$

where the functions $\phi_{i,j}$ are constructed as follows: for every $0 \leq i \leq \nu - 1$ and all $0 \leq j \leq S-1$, let $W_{i,j}(\tau, \epsilon) \in \mathcal{O}((S_{d_i} \cup D) \times \mathcal{E}_i)$, for some neighborhood of the origin

D, and $S_{d_i} = \{t \in \mathbb{C}^\star : |d_i - \arg(t)| < \theta/2\}$. Moreover, we make the assumption that

$$|W_{i,j}(\tau,\epsilon)| \le M_0 \left|\frac{\tau}{\epsilon^r}\right| \frac{1}{1+\left|\frac{\tau}{\epsilon^r}\right|^{2k}} \exp\left(\sigma\left|\frac{\tau}{\epsilon^r}\right|^k\right), \qquad (\tau,\epsilon) \in (S_{d_i} \cup D) \times \mathcal{E}_i, \tag{6}$$

for some $M_0, \sigma > 0$. Also, we assume $W_{i,j} \equiv W_{i+1\,j}$ in the domain $(S_{d_i} \cup D) \times (\mathcal{E}_i \cap \mathcal{E}_{i+1})$, for all $0 \le i \le \nu - 1$ and every $0 \le j \le S-1$. Let $L_{d_i} = [0,\infty)e^{\sqrt{-1}d_i}$. For every $0 \le i \le \nu - 1$ and all $0 \le j \le S-1$, we define

$$\phi_{i,j}(t,\epsilon) = k \int_{L_{d_i}} W_{i,j}(u,\epsilon) e^{-\left(\frac{u}{\epsilon^r t}\right)^k} \frac{du}{u},$$

for $(t,\epsilon) \in \mathcal{T} \times \mathcal{E}_i$. $\phi_{i,j}$ turns out to be a holomorphic function in $\mathcal{T} \times \mathcal{E}_i$.

Under these settings, one is able to construct the solution of (4) with initial conditions (5). We have $X_i(t,z,\epsilon) \in \mathcal{O}(\mathcal{T} \times D' \times \mathcal{E}_i)$, for some neighborhood of the origin D' in the form

$$X_i(t,z,\epsilon) = \sum_{\beta \ge 0} X_{i,\beta}(t,\epsilon) \frac{z^\beta}{\beta!}, \tag{7}$$

where

$$X_{i,\beta}(t,\epsilon) = k \int_{L_{d_i}} W_{i,\beta}(u,\epsilon) e^{-\left(\frac{u}{\epsilon^r t}\right)^k} \frac{du}{u}. \tag{8}$$

The elements $(W_{i,\beta}(\tau,\epsilon))_{\beta \ge 0}$ are constructed by a recurrence relation provided that the formal power series $W_i(\tau,z,\epsilon) = \sum_{\beta \ge 0} W_{i,\beta}(\tau,\epsilon) \frac{z^\beta}{\beta!}$ is a formal solution of

$$((k\tau^k)^{s_2} + a_2)(\epsilon^{r_1 - s_1 rk}(k\tau^k)^{s_1} + a_1)\partial_z^S W_i(\tau,z,\epsilon) \tag{9}$$

$$= \sum_{(s,\kappa_0,\kappa_1) \in \mathcal{S}} b_{\kappa_0\kappa_1}(z,\epsilon)\epsilon^{-r(s-\kappa_0)} \left[\frac{\tau^k}{\Gamma\left(\frac{\delta_{\kappa_0,s}}{k}\right)} \int_0^{\tau^k} (\tau^k - h)^{\frac{\delta_{\kappa_0,s}}{k} - 1} (kh)^{\kappa_0} \partial_z^{\kappa_1} W_i(h^{1/k},z,\epsilon) \frac{dh}{h} \right.$$

$$\left. + \sum_{1 \le p \le \kappa_0 - 1} A_{\kappa_0,p} \frac{\tau^k}{\Gamma\left(\frac{\delta_{\kappa_0,s} + k(\kappa_0 - p)}{k}\right)} \int_0^{\tau^k} (\tau^k - h)^{\frac{\delta_{\kappa_0,s} + k(\kappa_0 - p)}{k} - 1} (kh)^p \partial_z^{\kappa_1} W_i(h^{1/k},z,\epsilon) \frac{dh}{h} \right],$$

for given initial data

$$(\partial_z^j W_i)(\tau,0,\epsilon) = W_{i,j}(\tau,\epsilon), \quad 0 \le j \le S-1. \tag{10}$$

Here, $A_{\kappa_0,p} \in \mathbb{C}$. The previous equation is the result of applying formal Borel transform to both sides in (4), bearing in mind its properties, and taking into account Assumption (C) in order to rewrite the right-hand side of the main equation. More precisely, the form of formal Borel transform under use in this work is given by

$$\mathcal{B}(\hat{X})(\tau) = \sum_{n=1}^{\infty} \frac{a_n}{\Gamma\left(\frac{n}{k}\right)} \tau^n \in \tau\mathbb{E}[[\tau]],$$

for any $\hat{X}(T) = \sum_{n=1}^{\infty} a_n T^n \in T\mathbb{E}[[T]]$, where $\mathbb{E}$ is a complex Banach space. The concrete properties satisfied by this transformation are the following: let $\hat{f}(t) = \sum_{n\geq 1} f_n t^n \in \mathbb{E}[[t]]$, where $(\mathbb{E}, \|\cdot\|_{\mathbb{E}})$ is a Banach algebra. Let $k, m \geq 1$ be integers. One has

$$\mathcal{B}(t^{k+1}\partial_t \hat{f}(t))(\tau) = k\tau^k \mathcal{B}(\hat{f}(t))(\tau),$$

$$\mathcal{B}(t^m \hat{f}(t))(\tau) = \frac{\tau^k}{\Gamma\left(\frac{m}{k}\right)} \int_0^{\tau^k} (\tau^k - s)^{\frac{m}{k}-1} \mathcal{B}(\hat{f}(t))(s^{1/k}) \frac{ds}{s}.$$

One can observe from Eq. (9) that a small denominator phenomenon appears when calculating the coefficients $W_{i,\beta}(\tau, \epsilon)$. The domain of definition depends on ϵ and has to avoid the roots of the two polynomials at the head of the equation. This implies the domain of definition of the function $\tau \mapsto W_{i,\beta}(\tau, \epsilon)$ depends on $\epsilon \in \mathcal{E}_i$. Indeed, it is defined for $\tau \in \Omega(\epsilon) = S_{d_i} \cup (D \setminus \Omega_1(\epsilon))$, where $\Omega_1(\epsilon)$ turns out to be a finite collection of sets of the form $\{\tau \in \mathbb{C} : |\tau| > \rho(|\epsilon|), |\arg(\tau) - d_i| < \delta_2\}$, where $x \in (0, r_{\mathcal{E}}) \mapsto \rho(x)$ is a monotone increasing function with $\rho(x) \to 0$ when $x \to 0$ (see Fig. 1).

A fixed point technique allow us to conclude the existence of $M, Z_0 > 0$ such that

$$|W_{i,\beta}(\tau, \epsilon)| \leq M Z_0^{\beta} \beta! \left|\frac{\tau}{\epsilon^r}\right| \frac{1}{1 + \left|\frac{\tau}{\epsilon^r}\right|^{2k}} \exp\left(\sigma r_b(\beta) \left|\frac{\tau}{\epsilon^r}\right|^k\right), \qquad (\tau, \epsilon) \in (S_{d_i} \cup D) \times \mathcal{E}_i, \tag{11}$$

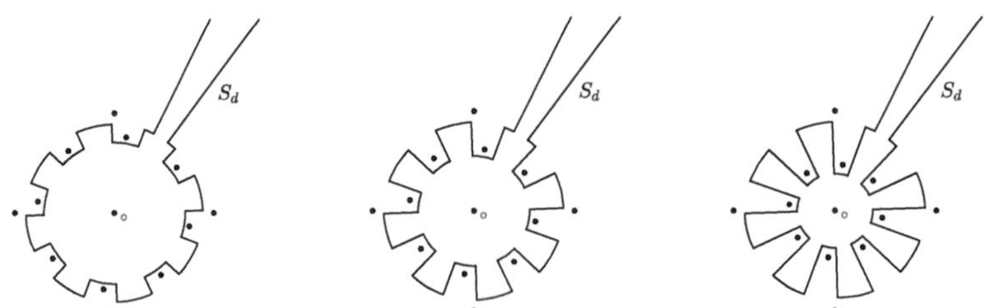

Fig. 1 Roots of the polynomials at the head of (9) and domain $\Omega(\epsilon)$

with $r_b(\beta) = \sum_{n=0}^{\beta} 1/(n+1)^b$. The estimates in (11) yield (8) is well-defined for $(t,\epsilon) \in \mathcal{T} \times \mathcal{E}_i$.

Theorem 2.3 *Under Assumptions (A)–(C), the hypotheses made on the elements involved in* (1), *and the construction of the initial conditions described through this section, there exist* $K, M > 0$ *(not depending on* ϵ*), such that*

$$\sup_{t\in\mathcal{T}, z\in D'} |X_{i+1}(t,z,\epsilon) - X_i(t,z,\epsilon)| \leq K \exp\left(-\frac{M}{|\epsilon|^{\hat{r}_i}}\right),$$

for every $\epsilon \in \mathcal{E}_i \cap \mathcal{E}_{i+1}$, *and some positive real number* $\hat{r}_i$ *which depends on i.*

Proof This result corresponds to Theorem 2 in [7]. We give some detail at certain steps of the proof. There are three different situations when estimating the difference of two solutions defined in consecutive elements in the good covering (Fig. 2).

1. If there are no singular directions $\frac{\pi(2j+1)+\arg(a_2)}{ks_2}$ for $j = 0, \dots, ks_2 - 1$ (we will refer to such directions as singular directions of first kind) nor $\tilde{d}$ with $|\tilde{d}_i - d_{\mathcal{E}_i,j}| \leq \delta_{2i}$ for $j = 0, \dots, ks_1$ (we will say these are singular directions of second kind) in between d_i and d_{i+1}, then one can deform the path $d_{\gamma_{i+1}} - d_{\gamma_i}$ to a point by means of Cauchy theorem so that the difference $X_{i+1} - X_i$ is null. In this case, one can reformulate the problem by considering a new good covering combining $\mathcal{E}_i$ and $\mathcal{E}_{i+1}$ in a unique sector.
2. If there exists at least a singular direction of first kind but no singular directions of second kind in between d_i and d_{i+1}, then the movable singularities depending on ϵ do not affect the geometry of the problem, whereas the path can only be deformed taking into account those singularities which do not depend on ϵ. In this case $\hat{r}_i := r_2/s_2$.
3. If there is at least a singular direction of second kind in between d_i and d_{i+1}, then the movable singularities depend on ϵ, and tend to zero. As a consequence, this affects the geometry of the problem, and the path deformation has to be made accordingly. In this case, $\hat{r}_i := r_1/s_1$.

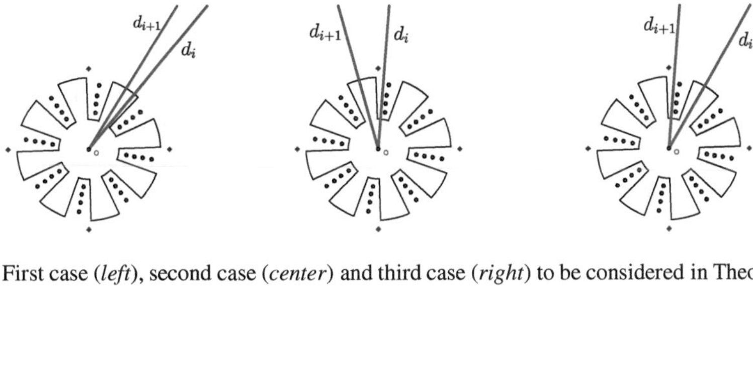

Fig. 2 First case (*left*), second case (*center*) and third case (*right*) to be considered in Theorem 2.3

Regarding the situation in which only singular directions of first kind appear, one can deform the integration path for the integrals along direction d_i and d_{i+1} in (8).

For every $\epsilon \in \mathcal{E}_i \cap \mathcal{E}_{i+1}$ and $t \in \mathcal{T}$ one has

$$X_{i+1,\beta}(t,\epsilon) - X_{i,\beta}(t,\epsilon) = k \int_{L_{\rho_0/2,d_{i+1}}} W_{i+1,\beta}(u,\epsilon) e^{-(\frac{u}{t\epsilon^r})^k} \frac{du}{u}$$

$$- k \int_{L_{\rho_0/2,d_i}} W_{i,\beta}(u,\epsilon) e^{-(\frac{u}{t\epsilon^r})^k} \frac{du}{u} + \int_{C(\rho_0/2,d_i,d_{i+1})} W_{i,i+1,\beta}(u,\epsilon) e^{-(\frac{u}{t\epsilon^r})^k} \frac{du}{u}.$$

Here, $\rho_0 > 0$ such that $\rho_0 \in D'$, $L_{\rho_0/2,d_{i+1}} := [\frac{\rho_0}{2}, +\infty)e^{\sqrt{-1}d_{i+1}}$, $L_{\rho_0/2,d_i} := [\frac{\rho_0}{2}, +\infty)e^{\sqrt{-1}d_i}$ and $C(\rho_0/2, d_i, d_{i+1})$ is an arc of circle with radius $\rho_0/2$ connecting $\rho_0/2e^{\sqrt{-1}d_{i+1}}$ and $\rho_0/2e^{\sqrt{-1}d_i}$ with a well chosen orientation. Moreover, $W_{i,i+1,\beta}$ denotes the function $W_{i,\beta}$ in an open domain which contains the closed path $(L_{d_{i+1}} \setminus L_{\rho_0/2,d_{i+1}}) - C(\rho_0/2, d_i, d_{i+1}) - (L_{d_i} \setminus L_{\rho_0/2,d_i})$, in which $W_{i,\beta}$ and $W_{i+1,\beta}$ coincide. This is a consequence of the construction of the initial data in our problem.

In the third situation, an analogous argument can be followed. One has to substitute ρ_0 by the function $\epsilon \mapsto \rho(|\epsilon|)$.

The result follows from here after usual estimates. □

The classical definition of Gevrey asymptotics on functions with values in complex Banach space are considered to describe the asymptotic behavior relating the analytic and the formal solution of the main problem under study.

Definition 2.4 Let $(\mathbb{E}, \|\cdot\|_{\mathbb{E}})$ be a complex Banach space and $\mathcal{E}$ be an open and bounded sector with vertex at 0. We also consider a positive real number α.

We say that a function $f : \mathcal{E} \to \mathbb{E}$, holomorphic on $\mathcal{E}$, admits a formal power series $\hat{f}(\epsilon) = \sum_{k\geq 0} a_k\epsilon^k \in \mathbb{E}[[\epsilon]]$ as its Gevrey asymptotic expansion with the order α if, for any closed proper subsector $\mathcal{W} \subseteq \mathcal{E}$ with vertex at the origin, there exist $C, M > 0$ such that

$$\left\| f(\epsilon) - \sum_{k=0}^{N-1} a_k\epsilon^k \right\|_{\mathbb{E}} \leq CM^N N!^{\alpha} |\epsilon|^N,$$

for every $N \geq 1$, and all $\epsilon \in \mathcal{W}$.

For the existence of a formal power series in ϵ and the asymptotic relation to the analytic solutions, we make use of a novel version of Ramis-Sibuya theorem in two levels and Theorem 2.3, in order to conclude with the main result in [7]. For a classical reference on this result, we provide [3] as a reference.

Theorem 2.5 *Under the previous assumptions, there exists a formal power series*

$$\hat{X}(t,z,\epsilon) = \sum_{\beta\geq 0} H_\beta(t,z) \frac{\epsilon^\beta}{\beta!} \in \mathbb{E}[[\epsilon]], \tag{12}$$

where $\mathbb{E}$ stands for the Banach space of holomorphic and bounded functions on the set $\mathcal{T} \times D'$ equipped with the supremum norm, which formally solves the equation

$$(\epsilon^{r_2}(t^{k+1}\partial_t)^{s_2} + a_2)(\epsilon^{r_1}(t^{k+1}\partial_t)^{s_1} + a_1)\partial_z^S \hat{X}(t,z,\epsilon)$$

$$= \sum_{(s,\kappa_0,\kappa_1)\in\mathcal{S}} b_{\kappa_0\kappa_1}(z,\epsilon)t^s(\partial_t^{\kappa_0}\partial_z^{\kappa_1}\hat{X})(t,z,\epsilon). \quad (13)$$

Moreover, $\hat{X}$ can be written in the form

$$\hat{X}(t,z,\epsilon) = a(t,z,\epsilon) + \hat{X}^1(t,z,\epsilon) + \hat{X}^2(t,z,\epsilon), \quad (14)$$

where $a(t,z,\epsilon) \in \mathbb{E}\{\epsilon\}$ is a convergent series on some neighborhood of $\epsilon = 0$ and $\hat{X}^1(t,z,\epsilon)$, $\hat{X}^2(t,z,\epsilon)$ are elements in $\mathbb{E}[[\epsilon]]$. Moreover, for every $0 \le i \le \nu - 1$, the $\mathbb{E}$-valued function $\epsilon \mapsto X_i(t,z,\epsilon)$ constructed in (7) *is of the form*

$$X_i(t,z,\epsilon) = a(t,z,\epsilon) + X_i^1(t,z,\epsilon) + X_i^2(t,z,\epsilon), \quad (15)$$

where $\epsilon \mapsto X_i^j(t,z,\epsilon)$ is a $\mathbb{E}$-valued function which admits $\hat{X}_i^j(t,z,\epsilon)$ as its $\hat{r}_i$-Gevrey asymptotic expansion on $\mathcal{E}_i$, for $j = 1, 2$.

Corollary 2.6 *Observe that $r_1/s_1 < r_2/s_2$. If one assumes the existence of $i_0 \in \{0, \dots, \nu-1\}$ such that $\mathcal{E}_{i_0}$ has opening larger than $\pi s_2/r_2$, such that every index in the set $I_{\delta_1,i,\delta_2} = \{i_0-\delta_1, \dots, i_0, \dots, i_0+\delta_2\}$ satisfies 2. in the proof of Theorem 2.3, for some $\delta_1, \delta_2 \ge 0$ and also*

$$\mathcal{E}_{i_0} \subseteq S_{\pi s_1/r_1} \subseteq \cup_{h\in I_{\delta_1,i,\delta_2}} \mathcal{E}_h,$$

where $S_{\pi s_1/r_1}$ stands for a sector with vertex at 0 and opening larger than $\pi s_1/r_1$, then the decomposition in (14) *and* (15) *is unique. In terms of [1], $\hat{X}(t,z,\epsilon)$, as a formal power series in ϵ, with coefficients in $\mathbb{E}$ is $(r_2/s_2, r_1/s_1)$-summable on $\mathcal{E}_{i_0}$, and its $(r_2/s_2, r_1, s_1)$-sum is the function $X_{i_0}(t,z,\epsilon)$ on $\mathcal{E}_{i_0}$.*

A practical situation has been considered in [7].

3 Some Additional Comments and Further Work

We focus our attention on Assumption (B), which is considered for geometric reasons, as we pointed out before. We now provide an alternative approach to avoid the assumption $s_1r_2 - s_2r_1 > s_2$, following different strategies.

Case 1: Assumption (B.1) $s_1r_2 - s_2r_1 < -s_1$.

Under Assumption (B.1), one can interchange the roles of the operators involved at the head of the main equation in (4), namely $\epsilon^{r_2}(t^{k+1}\partial_t)^{s_2} + a_2$ and $\epsilon^{r_1}(t^{k+1}\partial_t)^{s_1} +$

a_1. We consider $r := \frac{r_1}{ks_1}$ and put $T := \epsilon^r t$. After this change of variable, one rewrites the equation obtained by means of the idea in [10], as before. The operators $T^s \partial_t^{\kappa_0}$ can be rewritten so that the properties of formal Borel transform applied at both sides of the transformed equation lead to an auxiliary problem within the Borel plane. We omit all the details here because they follow analogous arguments as in the former construction. After this procedure, one gets the next problem, instead of (9):

$$(\epsilon^{r_2 - s_2 rk}(k\tau^k)^{s_2} + a_2)((k\tau^k)^{s_1} + a_1)\partial_z^S W_i(\tau, z, \epsilon) \tag{16}$$

$$= \sum_{(s,\kappa_0,\kappa_1)\in\mathcal{S}} b_{\kappa_0\kappa_1}(z,\epsilon)\epsilon^{-r(s-\kappa_0)} \left[\frac{\tau^k}{\Gamma\left(\frac{\delta_{\kappa_0,s}}{k}\right)} \int_0^{\tau^k} (\tau^k - h)^{\frac{\delta_{\kappa_0,s}}{k}-1}(kh)^{\kappa_0}\partial_z^{\kappa_1} W_i(h^{1/k}, z, \epsilon)\frac{dh}{h}\right.$$

$$\left. + \sum_{1\le p\le \kappa_0-1} A_{\kappa_0,p} \frac{\tau^k}{\Gamma\left(\frac{\delta_{\kappa_0,s}+k(\kappa_0-p)}{k}\right)} \int_0^{\tau^k} (\tau^k - h)^{\frac{\delta_{\kappa_0,s}+k(\kappa_0-p)}{k}-1}(kh)^p \partial_z^{\kappa_1} W_i(h^{1/k}, z, \epsilon)\frac{dh}{h}\right].$$

Regarding Assumption (B.1), parallel results to Lemmas 1 and 2 in [7] can be proved. More precisely, Lemma 2 in [7] reads as follows:

Lemma 3.1 *Let $0 \le i \le \nu - 1$ and $\epsilon \in \mathcal{E}_i$. Under Assumption (B.1), there exists a constant $C_2 > 0$, not depending on ϵ, such that*

$$\left| \frac{1}{\epsilon^{r_2 - s_2 rk}(k\tau^k)^{s_2} + a_2} \right| \le C_2,$$

for every $\tau \in \Omega(\epsilon)$.

Indeed, this lemma holds under the less restrictive condition $s_1 r_2 - s_2 r_1 < 0$. By means of a fixed point argument (analogous to that in Sect. 3 in [7]) we guarantee a formal solution of (16) under initial conditions (10) in the form $W_i(\tau, z, \epsilon) = \sum_{\beta\ge 0} W_{i,\beta}(\tau,\epsilon)\frac{z^\beta}{\beta!}$, and such that (11) holds.

In Theorem 2.3, the situations to handle differ. Indeed, the singularities of first kind and of second kind interchange their roles: singular directions $\frac{\pi(2j+1)+\arg(a_1)}{ks_1}$ for $j = 0, \dots, ks_1 - 1$ become fixed singular directions not depending on $\epsilon \in \mathcal{E}_i$ for any fixed i, i.e. of first kind; whilst directions $\tilde{d}_i \in [0, 2\pi)$ with $|\tilde{d}_i - d_{\mathcal{E}_i,j}| \le \delta_{2i}$ for $j = 0, \dots, ks_2$ turn into movable singular directions with respect to $\epsilon \in \mathcal{E}_i$. If there exist a singular direction of first kind but no singular directions of second kind in between d_i and d_{i+1}, we define $\hat{r}_i := r_1/s_1$. If there is at least a singular direction of second kind in between d_i and d_{i+1}, then we put $\hat{r}_i := r_2/s_2$.

Then, Theorem 2.5 holds under Assumption (B.1) with the same enunciate.

Case 2: Assumption (B.2) $s_1 r_2 - s_2 r_1 = 0$.

It is worth mentioning this particular case because under Assumption (B.2), the geometry of the problem changes. There is no longer a distinction between singularities depending on the perturbation parameter and fixed singularities, only remaining the fixed ones. Indeed, $r := \frac{r_2}{s_2 k} = \frac{r_1}{s_1 k}$. The same procedure leads to the auxiliary equation

$$((k\tau^k)^{s_2} + a_2)((k\tau^k)^{s_1} + a_1)\partial_z^S W_i(\tau, z, \epsilon) \tag{17}$$

$$= \sum_{(s,\kappa_0,\kappa_1)\in\mathcal{S}} b_{\kappa_0\kappa_1}(z,\epsilon)\epsilon^{-r(s-\kappa_0)} \left[\frac{\tau^k}{\Gamma\left(\frac{\delta_{\kappa_0,s}}{k}\right)} \int_0^{\tau^k} (\tau^k - h)^{\frac{\delta_{\kappa_0,s}}{k}-1} (kh)^{\kappa_0} \partial_z^{\kappa_1} W_i(h^{1/k}, z, \epsilon) \frac{dh}{h} \right.$$

$$\left. + \sum_{1\le p\le \kappa_0-1} A_{\kappa_0,p} \frac{\tau^k}{\Gamma\left(\frac{\delta_{\kappa_0,s}+k(\kappa_0-p)}{k}\right)} \int_0^{\tau^k} (\tau^k - h)^{\frac{\delta_{\kappa_0,s}+k(\kappa_0-p)}{k}-1} (kh)^p \partial_z^{\kappa_1} W_i(h^{1/k}, z, \epsilon) \frac{dh}{h} \right],$$

which can be solved by a fixed point theorem, leading to a unique Gevrey order appearing in the asymptotic representation of the solution of (4). More precisely, one has

Theorem 3.2 *There exists a formal power series $\hat{X}(t,z,\epsilon)$ in the form of* (12) *which formally solves* (13). *Moreover, for every $0 \le i \le \nu - 1$, the $\mathbb{E}$ valued function $\epsilon \mapsto X_i(t,z,\epsilon)$ constructed in* (7) *admits $\hat{X}(t,z,\epsilon)$ as its r_1/s_1-Gevrey asymptotic expansion on $\mathcal{E}_i$.*

Corollary 2.6 is reduced to the existence of an index $0 \le i_0 \le \nu - 1$ such that the opening of the sector $\mathcal{E}_{i_0}$ is larger than $\pi s_1/r_1$. In this case, $\hat{X}$, as a formal power series in ϵ with coefficients in $\mathbb{E}$ is r_1/s_1-summable in $\mathcal{E}_{i_0}$ by Watson's lemma.

Case 3: Assumption (B.3) $0 < s_1 r_2 - s_2 r_1 < s_2$.

As it has been pointed out, the condition $s_2 < s_1 r_2 - s_2 r_1$ in Assumption (B) is of geometric nature. It is imposed to guarantee the existence of rays from the origin which do not cross the movable singularities appearing at the head of the equation. One may substitute the good covering by any other consisting of sectors with small enough openings.

Case 4: Assumption (B.4) $-s_1 < s_1 r_2 - s_2 r_1 < 0$. Can be studied in the same way as Case 3.

Regarding the geometry of the problem involved, one can consider a more general problem under study, which can be solved analogously. A first approach could be to study the equation

$$(\epsilon^{r_2}(t^{k+1}\partial_t)^{s_2} + a_2)^{m_2}(\epsilon^{r_1}(t^{k+1}\partial_t)^{s_1} + a_1)^{m_1}\partial_z^S X(t,z,\epsilon)$$

$$= \sum_{(s,\kappa_0,\kappa_1)\in\mathcal{S}} b_{\kappa_0\kappa_1}(z,\epsilon) t^s (\partial_t^{\kappa_0}\partial_z^{\kappa_1} X)(t,z,\epsilon),$$

for any positive integers m_1, m_2.

This more general consideration does not change the configuration of the problem. Indeed, one can follow the same arguments to arrive at the auxiliary equation (9) in which the head of the equation has been substituted by

$$((k\tau^k)^{s_2} + a_2)^{m_2}(\epsilon^{r_1 - s_1 rk}(k\tau^k)^{s_1} + a_1)^{m_1}\partial_z^S W_i(\tau, z, \epsilon).$$

It is straight to check that no additional assumptions have to be added, because the roots of the polynomials $((k\tau^k)^{s_2} + a_2)^{m_2}$ coincide for any positive integer m_2. Also, the same holds for the polynomial $(\epsilon^{r_1 - s_1 rk}(k\tau^k)^{s_1} + a_1)^{m_1}$ for any positive integer m_1. The direction d_i at positive distance to the roots of both polynomials can be chosen independently of m_1 nor m_2. The problem can be solved following the same arguments as in [7]. The main result can be rewritten word by word.

A more general approach to this one could be to consider more than two singularly perturbed terms at the head of the equation. More precisely, one may consider the equation

$$(\epsilon^{r_h}(t^{k+1}\partial_t)^{s_h} + a_h)^{m_h}(\epsilon^{r_{h-1}}(t^{k+1}\partial_t)^{s_{h-1}} + a_{h-1})^{m_{h-1}} \dots (\epsilon^{r_1}(t^{k+1}\partial_t)^{s_1} + a_1)^{m_1}\partial_z^S X_i(t, z, \epsilon)$$
$$= \sum_{(s,\kappa_0,\kappa_1)\in\mathcal{S}} b_{\kappa_0\kappa_1}(z, \epsilon)t^s(\partial_t^{\kappa_0}\partial_z^{\kappa_1}X_i)(t, z, \epsilon),$$

for some integer $h \geq 2$, $a_j \in \mathbb{C}^\star$, and where r_j stands for a nonnegative integer whilst s_j, m_j are positive integers for every $j = 1, \dots, h$. Under this situation, one chooses the indices $\{h_1, \dots, h_\ell\} \subseteq \{1, \dots, h\}$ such that r_{h_μ}/s_{h_μ} coincide for every $\mu = 1, \dots, \ell$ and $r_{h_\mu}/s_{h_\mu} > r_p/s_p$ for every $p \in \{1, \dots, h\} \setminus \{h_1, \dots, h_\ell\}$. We write $\overline{r}/\overline{s} := r_{h_\mu}/s_{h_\mu}$ for any $\mu \in \{1, \dots, \ell\}$.

An analogous procedure can be followed in this situation. We do not enter into details for the sake of clarity, but it is worth mentioning that, under an appropriate geometry for the problem, several Gevrey orders appear in the asymptotic study of the equation. More precisely, the analytic solution can be split in several terms, in the shape of (15) and the formal solution can be written in the form of a sum of the same number of terms as the formal one. One of the terms in the analytic solution admits the corresponding one in the formal solution, as its Gevrey asymptotic expansion of order $\overline{r}/\overline{s}$ in each of the domains of definition of the perturbation parameter. The asymptotic expansions have to be considered as in Theorem 2.5, with coefficients of the formal power series, and functions with values in the Banach space $\mathbb{E}$. This term corresponds to the fixed singularity appearing in the auxiliary equation, in the Borel plane. The roots to be avoided are all the roots of the polynomials $(k\tau^k)^{s_{h_\mu}} + a_\mu = 0$, for $\mu = 1, \dots, \ell$.

Regarding the remaining terms at the head of the equation, corresponding to $(\epsilon^{r_p}(t^{k+1}\partial_t)^{s_p} + a_p)^{m_p}$, for $p \in \{1, \dots, h\} \setminus \{h_1, \dots, h_\ell\}$, one observes the phenomenon of movable singularities described in Theorem 2.5 at each term. The geometry becomes more complicated and one has to choose the direction d_i so that it avoids all singularities.

More precisely, Assumption (A) and Assumption (B) are substituted by the following ones.

Assumption (A): For every $0 \leq i \leq \nu - 1$ and $\mu \in \{1, \ldots, \ell\}$ one has $d_i \neq \frac{\pi(2j+1)+\arg(a_{h_\mu})}{ks_{h_\mu}}$ for every $j = 0, \ldots, ks_{h_\mu} - 1$.

Assumption (B): For every $p \in \{1, \ldots, h\} \setminus \{h_1, \ldots, h_\ell\}$ and $\mu = 1, \ldots, \ell$, one has $s_p r_{h_\mu} - s_{h_\mu} r_p > s_p > 0$ and $|d_i - d_{\mathcal{E}_{i,j,p,\mu}}| > \delta_{2,i,p,\mu}$ for $j = 0, \ldots, ks_p - 1$, where $\delta_{2,i,p,\mu} := \frac{s_p r_{h_\mu} - s_{h_\mu} r_p}{2ks_{h_\mu} s_p}(\theta_{2,\mathcal{E}_i} - \theta_{1,\mathcal{E}_i})$, and $d_{\mathcal{E}_{i,j,p,\mu}} = \frac{1}{ks_p}(\pi(2j+1) + \arg(a_p) + \frac{s_p r_{h_\mu} - s_{h_\mu} r_p}{s_{h_\mu}}\left(\frac{\theta_{1,\mathcal{E}_i} + \theta_{2,\mathcal{E}_i}}{2}\right))$.

Acknowledgements Alberto Lastra and Stéphane Malek are partially supported by the project MTM2012-31439 of Ministerio de Ciencia e Innovacion, Spain.

References

1. W. Balser, *Formal Power Series and Linear Systems of Meromorphic Ordinary Differential Equations*. Universitext (Springer, New York, 2000), xviii+299 pp.
2. M. Canalis-Durand, J. Mozo-Fernández, R. Schäfke, Monomial summability and doubly singular differential equations. J. Differ. Equ. **233**, 485–511 (2007)
3. P. Hsieh, Y. Sibuya, *Basic Theory of Ordinary Differential Equations*. Universitext (Springer, New York, 1999)
4. S. Kamimoto, On the exact WKB analysis of singularly perturbed ordinary differential equations at an irregular singular point. Preprint RIMS-1779 (2013)
5. A. Lastra, S. Malek, J. Sanz, On Gevrey solutions of threefold singular nonlinear partial differential equations. J. Differ. Equ. **255**, 3205–3232 (2013)
6. A. Lastra, S. Malek, On parametric Gevrey asymptotics for some nonlinear initial value Cauchy problems. Adv. Differ. Equ. **2015**, 200 (2015)
7. A. Lastra, S. Malek, Multi-level Gevrey solutions of singularly perturbed linear partial differential equations. Adv. Differ. Equ. **21**, 767–800 (2016)
8. S. Malek, On Gevrey functions solutions of partial differential equations with Fuchsian and irregular singularities. J. Dyn. Control Syst. **15**, 277–305 (2009)
9. S. Malek, On the summability of formal solutions for doubly nonlinear partial differential equations. J. Dyn. Control Syst. **18**, 45–82 (2012)
10. H. Tahara, H. Yamazawa, Multisummability of formal solutions to the Cauchy problem for some linear partial differential equations. J. Differ. Equ. **255**, 3592–3637 (2013)

Reducibility of Hypergeometric Equations

Toshio Oshima

Abstract We study a necessary and sufficient condition so that hypergeometric equations are reducible. Here the hypergeometric equations with one variable mean the rigid Fuchsian linear ordinary differential equations. If the equations with one variable have more than four singular points, they naturally define hypergeometric equations with several variables including Appell's hypergeometric equations. We also study the reducibility of such equations with several variables and we find a new kind of reducibility, which appears, for example, in a decomposition of Appell's F_4.

Keywords Hypergeometric function • Middle convolution • Monodromy representations • Pfaffian system

Mathematics Subject Classification (2000). Primary 34M03; Secondary 34A30, 34M35, 33C65

1 Introduction and Preliminary Results

The Gauss hypergeometric equation

$$x(1-x)u'' + \bigl(c-(a+b+1)x\bigr)u' - abu = 0 \tag{1}$$

is reducible if and only if at least one of the numbers

$$a,\ b,\ a-c,\ b-c \tag{2}$$

is an integer. An elementary proof of this result using neither an integral representation nor a connection formula of the solution is given in [9]. Here the linear ordinary differential equation with coefficients in rational functions is said to be reducible if and only if the equation has a non-zero solution satisfying a linear ordinary

T. Oshima (✉)
Faculty of Science, Josai University, 2-3-20, Hirakawacho, Chiyodaku, Tokyo 102-0093, Japan
e-mail: t-oshima@josai.ac.jp

G. Filipuk et al. (eds.), *Analytic, Algebraic and Geometric Aspects of Differential Equations*, Trends in Mathematics, DOI 10.1007/978-3-319-52842-7_14

differential equation with coefficients in rational functions whose order is lower than the original equation. Note that the reducibility of the equation is equivalent to the reducibility of the monodromy group of the solutions if the equation is Fuchsian.

The Riemann scheme of the Gauss hypergeometric equation is

$$\begin{Bmatrix} x=\infty & 0 & 1 \\ a & 0 & 0 \\ b & 1-c & c-a-b \end{Bmatrix}, \tag{3}$$

which is a table of singular points of (1) and the characteristic exponents at each singular point. The characteristic exponents 0 and $1-c$ at $x=0$ mean that (1) has solutions $u(x)$ satisfying $u(x) \sim x^0$ and $u(x) \sim x^{1-c}$, respectively, when $x \to 0$. If the parameters a, b and c are generic, the differential equation with this Riemann scheme is (1). Since the coefficients of the equation are polynomial functions of the parameters a, b and c, Eq. (1) is naturally and uniquely defined by this Riemann scheme for any values of parameters.

In [8] we define a (generalized) Riemann scheme

$$\begin{aligned} \{\lambda_{\mathbf{m}}\} &= \bigl([\lambda_{j,\nu}]_{(m_{j,\nu})}\bigr)_{\substack{0\le j\le p \\ 1\le \nu\le n_j}} \\ &= \begin{Bmatrix} x=c_0=\infty & c_1 & \cdots & c_p \\ [\lambda_{0,1}]_{(m_{0,1})} & [\lambda_{1,1}]_{(m_{1,1})} & \cdots & [\lambda_{p,1}]_{(m_{p,1})} \\ \vdots & \vdots & \vdots & \vdots \\ [\lambda_{0,n_0}]_{(m_{0,n_0})} & [\lambda_{1,n_1}]_{(m_{1,n_1})} & \cdots & [\lambda_{p,n_p}]_{(m_{p,n_p})} \end{Bmatrix} \end{aligned} \tag{4}$$

for a general Fuchsian ordinary differential equation $Pu = 0$ of order n. Here $c_0, c_1, \dots, c_p$ are singular points of the equation and the sets of characteristic exponents of the equation at $x = c_j$ are

$$\{\lambda_{j,\nu} + i \mid i = 0, 1, \dots, m_{j,\nu} - 1,\ \nu = 1, \dots, n_j\}, \tag{5}$$

respectively, and moreover the local monodromies of the solutions of the equation at $x = x_j$ are semisimple if $\lambda_{j,\nu} - \lambda_{j,\nu'} \notin \mathbb{Z}$ for $1 \le \nu < \nu' \le n_j$ and $j = 0, \dots, p$. We say that $\{\lambda_{\mathbf{m}}\}$ is the Riemann scheme of P and the $(p+1)$ tuples of partitions

$$\mathbf{m} = (m_{j,1}, \dots, m_{j,n_j})_{j=0,\dots,p} \tag{6}$$

of n is called the *spectral type* of the equation $Pu = 0$ or the operator P and we put $\operatorname{ord}\mathbf{m} = n$ which equals $m_{j,1} + \cdots + m_{j,n_j}$. If there is no confusion, $\mathbf{m}$ is shortly expressed by $m_{0,1}\cdots m_{0,n_0}, \cdots, m_{p,1}\cdots m_{p,n_p}$ and for example, the spectral type of Gauss hypergeometric equation is $11, 11, 11$.

The existence of such equation implies the Fuchs relation

$$|\{\lambda_{\mathbf{m}}\}| := \sum_{j=0}^{p}\sum_{\nu=1}^{n_j} m_{j,\nu}\lambda_{j,\nu} - \operatorname{ord}\mathbf{m} + \tfrac{1}{2}\operatorname{idx}\mathbf{m} = 0. \tag{7}$$

Here the *index of rigidity* of $\mathbf{m}$ is defined by Katz [7] as follows.

$$\operatorname{idx}\mathbf{m} := \sum_{j=0}^{p}\sum_{\nu=1}^{n_j} m_{j,\nu}^2 - (p-1)(\operatorname{ord}\mathbf{m})^2. \tag{8}$$

On the other hand, we say that $\mathbf{m}$ is *irreducibly realizable* if there exists an irreducible Fuchsian equation $Pu = 0$ with the Riemann scheme (4) for generic parameters $\lambda_{j,\nu}$ under the Fuchs relation. A characterization of irreducibly realizable spectral types $\mathbf{m}$ is given in [8]. Moreover if $\mathbf{m}$ is irreducibly realizable, then the equation $Pu = 0$ with the Riemann scheme (4) satisfying (7) has $(1 - \frac{1}{2}\operatorname{idx}\mathbf{m})$ accessory parameters and the differential operator P is polynomial functions of the parameters $\lambda_{j,\nu}$ and the accessory parameters, which is the *universal operator* in [8, Theorem 6.14]. If an irreducibly realizable spectral type $\mathbf{m}$ satisfies

$$\operatorname{idx}\mathbf{m} = 2, \tag{9}$$

then $\mathbf{m}$ is called *rigid*. In this case the Fuchsian differential equation with the Riemann scheme (4) has no accessory parameters and hence the equation is uniquely determined by local structure, namely, by characteristic exponents and conjugacy classes of local monodromies at the singular points.

The middle convolution for a Fuchsian system

$$\frac{du}{dx} = \sum_{j=1}^{p}\frac{A_j}{x-c_j}u \tag{10}$$

with constant square matrices $A_1, \dots, A_p$ of size n is introduced by Katz [7] and Dettweiler and Reiter [2] to analyze the rigid local system (cf. Sect. 3). Denoting $\partial = \frac{d}{dx}$ and $\vartheta = x\partial$, we define in [8] the middle convolution $\operatorname{mc}_\mu(P)$ with a complex number μ by

$$\operatorname{mc}_\mu(P) = \partial^{-k}\sum_{i,j} c_{i,j}x^i(\vartheta-\mu)^j, \tag{11}$$

$$\partial^N P = \sum_{i,j} c_{i,j}\partial^i\vartheta^j \tag{12}$$

for an element P of the ring $W[x]$ of linear ordinary differential operators with polynomial coefficients. Here N is a sufficiently large integer so that $\partial^N P$ is of the form (12) with $c_{i,j} \in \mathbb{C}$ and then k is the maximal integer so that $\mathrm{mc}_\mu(P) \in W[x]$.

Theorem 1.1 ([8]) *Suppose that the equation $Pu = 0$ with the Riemann scheme* (4) *is irreducible and the coefficients of P are polynomials without a common zero and moreover suppose*

$$\lambda_{j,1} - \lambda_{j,\nu} \notin \mathbb{Z} \text{ or } m_{j,1} \geq m_{j,\nu} \quad (\nu = 1, \dots, n_j,\ j = 0, \dots, p) \tag{13}$$

and

$$\lambda_{j,1} = 0 \quad (j = 1, \dots, p). \tag{14}$$

If $\lambda_{j,\nu}$ are generic (see [8, Theorem 5.2] for the precise condition) or the number

$$d_1(\mathbf{m}) := m_{0,1} + \cdots + m_{p,1} - (p-1)\operatorname{ord}\mathbf{m} = 2\operatorname{ord}\mathbf{m} - \sum_{j=0}^{p}\sum_{\nu=2}^{n_j} m_{j,\nu} \tag{15}$$

is positive, the Riemann scheme $\{\lambda'_{\mathbf{m}'}\}$ of $P' = \mathrm{mc}_{\lambda_{0,1}-1}(P)$ is determined by

$$\begin{cases} \lambda'_{j,\nu} = \begin{cases} 2 - \lambda_{0,1} & (j = 0, \qquad \nu = 1), \\ 0 & (j = 1, \dots, p,\ \nu = 1), \\ \lambda_{j,\nu} - \lambda_{0,1} + 1 & (j = 0, \qquad \nu = 2, \dots, n_0), \\ \lambda_{j,\nu} + \lambda_{0,1} - 1 & (j = 1, \dots, p,\ \nu = 2, \dots, n_j), \end{cases} \\ m'_{j,\nu} = \begin{cases} m_{j,1} - d_1(\mathbf{m}) & (j = 0, \dots, p,\ \nu = 1), \\ m_{j,\nu} & (j = 0, \dots, p,\ \nu = 2, \dots, n_j,) \end{cases} \end{cases} \tag{16}$$

and $P = \mathrm{mc}_{1-\lambda_{0,1}} \circ \mathrm{mc}_{\lambda_{0,1}-1}(P)$. Moreover $\mathrm{mc}_{\lambda_{0,1}-1}(P)$ is irreducible if $d_1(\mathbf{m}) \geq 0$. Here we allow that some $m_{j,1}$ are 0.

Remark 1.2

i) If $d_1(\mathbf{m}) > 0$, then $m_{j,1} > 0$ for $j = 0, \dots, p$ in the theorem.
ii) Suppose $P' = \mathrm{mc}_{\lambda_{0,1}-1}(P)$ is defined when the values of the parameters $\lambda_{j,\nu}$ are generic. Then we define P' for other values of the parameters by the analytic continuation of the parameters. In this case P and P' may be reducible for certain values of the parameters.

We define

$$\partial_1(\mathbf{m}) := \mathbf{m}' \tag{17}$$

by (15) and (16). Since

$$d_1(\mathbf{m}) \cdot \operatorname{ord} \mathbf{m} = \operatorname{idx} \mathbf{m} + \sum_{j=0}^{p} \sum_{\nu=1}^{n_j} (m_{j,1} - m_{j,\nu}) m_{j,\nu}, \tag{18}$$

we have $d_1(\mathbf{m}) > 0$ if $\operatorname{idx} \mathbf{m} = 2$ and moreover $\mathbf{m}$ is monotone, namely

$$m_{j,1} \geq m_{j,\nu} \quad (\nu = 1, \ldots, n_j,\ j = 0, \ldots, p). \tag{19}$$

The equation $Pu = 0$ is called rigid if it is irreducible and it has rigid spectral type. By the gauge transformation $u(x) \mapsto v(x) = \prod_{i=1}^{p} (x - c_i)^{\mu_i} u(x)$, the characteristic exponents of the Riemann scheme are changed from $\lambda_{j,\nu}$ to $\lambda_{0,\nu} - \sum_{i=1}^{p} \mu_i$ and $\lambda_{j,\nu} + \mu_j$ according to $j = 0$ and $j = 1, \ldots, p$, respectively. The corresponding transformation of P is called addition and we denote the transformation of P by $\mathrm{RAd}_{\mu_1,\ldots,\mu_p}(P)$. Here

$$\mathrm{RAd}_{\mu_1,\ldots,\mu_p}(P) \in W[x] \cap W(x) \prod_{j=1}^{p} (x - c_j)^{\mu_j} \cdot P \cdot \prod_{j=1}^{p} (x - c_j)^{-\mu_j}$$

and the coefficients of the differential operator $\mathrm{RAd}_{\mu_1,\ldots,\mu_p}(P)$ has no common zero. We put $\mathrm{R}(P) = \mathrm{RAd}_{0,\ldots,0}(P)$. The addition keeps the order of $P \in W[x]$.

Hence if the equation $Pu = 0$ is rigid and $\operatorname{ord} P > 1$, then we have $\operatorname{ord}(\mathrm{mc}_\mu \circ \mathrm{RAd}_{\mu_1,\ldots,\mu_p}(P)) < \operatorname{ord} P$ with suitable numbers $\mu, \mu_1, \ldots, \mu_p$. By a successive application of suitable additions and middle convolutions, the rigid equation $Pu = 0$ is transformed into the equation $\frac{du}{dx} = 0$. Since additions and middle convolutions are invertible, we can construct any rigid Fuchsian equation $Pu = 0$ by a successive application of suitable additions and middle convolutions to the equation $\frac{du}{dx} = 0$.

The combinatorial aspect of additions and middle convolutions are well interpreted by a star-shaped Kac-Moody root system (W, Π) as follows, which was introduced by Crawley-Boevey [1] to analyze irreducible Fuchsian systems. The set of simple roots is

$$\Pi = \{\alpha_0, \alpha_{j,\nu} \mid j = 0, 1, \ldots,\ \nu = 1, 2, \ldots\} = \{\alpha_i \mid i \in I\} \tag{20}$$

with

$$I := \{0, (j, \nu) \mid j = 0, 1, \ldots,\ \nu = 1, 2, \ldots\} \tag{21}$$

and the inner product of the roots are given by

$$
\begin{aligned}
&(\alpha_0|\alpha_0) = 2,\ (\alpha_0 \mid \alpha_{j,\nu}) = -\delta_{\nu,1},\\
&(\alpha_{j,\nu}|\alpha_{j',\nu'}) = \begin{cases} 2 & (j = j',\ \nu = \nu'),\\ -\delta_{j,j'} & (|\nu-\nu'| = 1),\\ 0 & (j \neq j' \text{ or } |\nu-\nu'| > 1). \end{cases}
\end{aligned}
\tag{22}
$$

The Weyl group W is generated by the simple reflection

$$
s_\alpha(x) = s_i(x) = x - (\alpha|x)\alpha \qquad (\alpha = \alpha_i \in \Pi,\ i \in I,\ x \in \sum_{\alpha\in\Pi} \mathbb{R}\alpha). \tag{23}
$$

The set of positive real roots Σ_+^{re} and the set of negative real roots Σ_-^{re} are

$$
\Sigma_+^{re} = \Big\{\sum_{\alpha\in\Pi} k_\alpha\alpha \in W\alpha_0 \mid k_\alpha \geq 0\Big\},\ \ \Sigma_-^{re} = \big\{-\alpha \mid \alpha \in \Sigma_+^{re}\big\} \tag{24}
$$

and then $W\Pi = \Sigma_+^{re} \cup \Sigma_-^{re}$. For a tuple of partition $\mathbf{m}$, [1] attached an element $\alpha_{\mathbf{m}}$ of the root lattice $\sum_{\alpha\in\Pi}\mathbb{Z}\alpha$ by

$$
\alpha_{\mathbf{m}} = \operatorname{ord}\mathbf{m}\cdot\alpha_0 + \sum_{j=0}^{p}\sum_{\nu=1}^{n_j}\sum_{s=\nu+1}^{n_j} m_{j,s}\alpha_{j,\nu}. \tag{25}
$$

Then $\operatorname{idx}\mathbf{m} = (\alpha_{\mathbf{m}}|\alpha_{\mathbf{m}})$, $s_{\alpha_0}(\alpha_{\mathbf{m}})$ corresponds to the middle convolution because $(\alpha_0|\alpha_{\mathbf{m}}) = d_1(\mathbf{m})$ and $s_{\alpha_{j,\nu}}(\alpha_{\mathbf{m}})$ corresponds to the transposition between $m_{j,\nu}$ and $m_{j,\nu+1}$. Moreover we have

Theorem 1.3 ([1, 8]) *The spectral type* $\mathbf{m}$ *is rigid if and only if* $\alpha_{\mathbf{m}} \in \Sigma_+^{re}$.

Remark 1.4

i) This theorem is given by Oshima [8, Chap. 7] and the corresponding theorem for the first order system of Schlesinger canonical form is proved by Crawley-Boevey [1].
ii) There exist positive real roots which do not correspond to the rigid spectral type. In fact we have

$$
\begin{aligned}
\Sigma_+^{re} &= \{\alpha_{\mathbf{m}} \mid \mathbf{m} \text{ are rigid spectral types}\}\\
&\quad \cup \{\alpha_{j,\nu} + \alpha_{j,\nu+1} + \cdots + \alpha_{j,\nu'} \mid 1 \leq \nu \leq \nu',\ j = 0, 1, \ldots\}.
\end{aligned}
\tag{26}
$$

We will examine the condition of irreducibility of the rigid equation $Pu = 0$ with the Riemann scheme (4). For an element $w \in W$, the expression $w = s_{\beta_1}s_{\beta_2}\cdots s_{\beta_\ell}$ with the reflections s_{β_i} with respect to simple roots $\beta_i \in \Pi$ is called minimal if ℓ is smallest among this product expression and in this case the number ℓ is called the length of w and denoted by $L(w)$. For a positive real root $\alpha \in \Sigma_+^{re}$, an element $w_\alpha \in W$ is uniquely determined by the conditions $w_\alpha \alpha = \alpha_0$ and $L(w_\alpha)$ is minimal. Moreover we put

$$\Delta(w) := \Sigma_+^{re} \cap w^{-1}\Sigma_-^{re} \text{ and } \Delta(\mathbf{m}) := \Delta(w_{\alpha_{\mathbf{m}}}) \tag{27}$$

for a rigid spectral type $\mathbf{m}$. Note that the number $|\Delta(w)|$ of elements of $\Delta(w)$ equals $L(w)$ and if $w = s_{\beta_1}s_{\beta_2}\cdots s_{\beta_L}$ is a minimal expression, we have

$$\Delta(w) = \{\beta_L, s_{\beta_L}\beta_{L-1}, \ldots, s_{\beta_L}\cdots s_{\beta_2}\beta_1\}, \tag{28}$$

$$\Delta(ws_\beta) = \begin{cases} s_\beta(\Delta(w)\setminus\{\beta\}) & (\beta \in \Delta(w)\cap\Pi), \\ s_\beta\Delta(w)\cup\{\beta\} & (\beta \notin \Delta(w)\cap\Pi). \end{cases} \tag{29}$$

For a rigid spectral type $\mathbf{m}$, the set of positive integers

$$[\Delta(\mathbf{m})] := \{(\alpha|\alpha_{\mathbf{m}}) \mid \alpha \in \Delta(\mathbf{m})\} \tag{30}$$

is a partition of the non-negative integer $h(\alpha_{\mathbf{m}})-1$ which is called the type of $\Delta(\mathbf{m})$, where

$$h(\alpha) := k_0 + \sum_{j\geq 0}\sum_{\nu\geq 1} k_{j,\nu} \text{ for } \alpha = k_0\alpha_0 + \sum_{j\geq 0}\sum_{\nu\geq 1} k_{j,\nu}\alpha_{j,\nu}. \tag{31}$$

Suppose $\mathbf{m}$ is rigid and monotone and $\operatorname{ord}\mathbf{m} > 1$. Let ν_j be the maximal integers satisfying $m_{j,1} - d_1(\mathbf{m}) < m_{j,\nu_j+1}$ for $j = 0, 1, \ldots$. Then [8, Proposition 7.9] shows

$$\begin{aligned} \Delta(\mathbf{m}) &= s_0\Bigl(\prod_{\substack{j\geq 0\\ \nu_j>0}} s_{j,1}\cdots s_{j,\nu_j}\Bigr)\Delta(s\partial_1\mathbf{m}) \cup \{\alpha_0\} \\ &\quad \cup \bigcup_{j=1}^{p}\{\alpha_0 + \alpha_{j,1} + \cdots + \alpha_{j,\nu} \mid 1 \leq \nu \leq \nu_j\}, \\ [\Delta(\mathbf{m})] &= [\Delta(s\partial_1\mathbf{m})] \cup \{d_1(\mathbf{m})\} \\ &\quad \cup \bigcup_{j=1}^{p}\{m_{j,\nu+1} - m_{j,1} + d_1(\mathbf{m}) \in \mathbb{Z}_{>0} \mid 1 \leq \nu \leq \nu_j\}. \end{aligned} \tag{32}$$

Here $\mathbb{Z}_{>0}$ is a set of positive integers and $s\mathbf{m}'$ is a monotone spectral type obtained by permutations of the sequences $(m'_{j,1}, \ldots, m'_{j,n'_j})$ of integers. The transformation s is realized by an element of a subgroup W' of W generated by $\{s_{j,\nu} \mid j = 0, 1, \ldots, \nu = 1, 2, \ldots\}$.

Theorem 1.5 ([8, Theorem 10.14]) *A rigid Fuchsian differential equation $Pu = 0$ with the Riemann scheme* (4) *is reducible if and only if at least one of the numbers*

$$\sum_{j=0}^{p}\sum_{\nu=1}^{n_j} \mathbf{m}(\beta)_{j,\nu}\lambda_{j,\nu} \quad (\beta \in \Delta(\mathbf{m})) \tag{33}$$

is an integer. Here

$$\mathbf{m}(\beta)_{j,\nu} = k_{j,\nu-1} - k_{j,\nu} \quad (\nu = 1,\dots,n_j,\ j = 0,\dots,p) \tag{34}$$

by putting

$$\begin{cases} \beta = k_0\alpha_0 + \sum_{j=0}^{p}\sum_{\nu=1}^{n_j-1} k_{j,\nu}\alpha_{j,\nu}, \\ k_{j,0} = k_0 \ \textit{and}\ k_{j,\nu} = 0 \ \textit{if}\ \nu \geq n_j. \end{cases} \tag{35}$$

For example, suppose $\mathbf{m} = 11, 11, 11$, namely, the spectral type of Gauss hypergeometric equation. Then $\alpha_{\mathbf{m}} = 2\alpha_0 + \alpha_{0,1} + \alpha_{1,1} + \alpha_{2,1}$ and we have

$$\begin{aligned} 2\alpha_0 + \alpha_{0,1} + \alpha_{1,1} + \alpha_{2,1} &\xrightarrow{s_0} \alpha_0 + \alpha_{0,1} + \alpha_{1,1} + \alpha_{2,1} \xrightarrow{s_{0,1}} \alpha_0 + \alpha_{1,1} + \alpha_{2,1} \\ &\xrightarrow{s_{1,1}} \alpha_0 + \alpha_{2,1} \xrightarrow{s_{2,1}} \alpha_0, \\ w_{\alpha_{\mathbf{m}}} &= s_{2,1}s_{1,1}s_{0,1}s_0, \\ \Delta(\mathbf{m}) &:= \{\alpha_0,\ s_0\alpha_{0,1},\ s_0s_{0,1}\alpha_{1,1},\ s_0s_{0,1}s_{1,1}\alpha_{2,1}\} \\ &= \{\alpha_0,\ \alpha_0 + \alpha_{0,1},\ \alpha_0 + \alpha_{1,1},\ \alpha_0 + \alpha_{2,1}\}. \end{aligned}$$

We rewrite the above in terms of tuples of partitions as follows.

$$\begin{array}{lllll} 11,11,11 \xrightarrow{\partial_1} & 01,01,01 \xrightarrow{s_{0,1}} & 10,01,01 \xrightarrow{s_{1,1}} & 10,10,01 \xrightarrow{s_{2,1}} & 10,10,10 \\ 10,10,10 \leftarrow & * & & & \\ 01,10,10 \leftarrow & -11,00,00 \leftarrow & * & & \\ 10,01,10 \leftarrow & 00,-11,00 \leftarrow & 00,-11,00 \leftarrow & * & \\ 10,10,01 \leftarrow & 00,00,-11 \leftarrow & 00,00,-11 \leftarrow & 00,00,-11 \leftarrow & * \end{array}$$

and

$$\{10,10,10,\ 01,10,10,\ 10,01,10,\ 10,10,01\} = \{\mathbf{m}(\beta) \mid \beta \in \Delta(11,11,11)\}.$$

Then for the Riemann scheme (3) we have

$$\left\{\sum_{j=0}^{2}\sum_{\nu=1}^{2} \mathbf{m}(\beta)_{j,\nu}\lambda_{j,\nu} \mid \beta \in \Delta(\mathbf{m})\right\} = \{a, b, 1 + a - c, c - b\}$$

and the condition for the reducibility of (1) by Theorem 1.5 [cf. (3)].

For a rigid spectral type $\mathbf{m}$ let $w_{\alpha_{\mathbf{m}}} = s_{i_1}s_{i_2}\cdots s_{i_L}$ be a minimal expression with respect to simple reflections. Put $w(j) = s_{i_1}s_{i_2}\cdots s_{i_j}$ and $\alpha(j) = w(j)^{-1}\alpha_0$ for $j = 1,\ldots,L$. Then we have the expressions of $\alpha(j)$ inductively as follows.

$$\begin{aligned}
\alpha(j) &= k_\nu\beta(j,\nu) + \gamma(j,\nu) \qquad (\nu = 1,\ldots,j),\\
\beta(j,\nu) &= s_{i_j}\beta(j-1,\nu),\ \gamma(j,\nu) = s_{i_j}\gamma(j-1,\nu), \quad (\nu = 1,2,\ldots,j-1),\\
\beta(j,j) &= \alpha_{i_j},\ \gamma(j,j) = \alpha(j-1),\\
k_j &= -(\alpha(j-1)|\alpha_{i_j}) = (\alpha(j)|\alpha_{i_j}) \in \mathbb{Z}_{>0},\\
\Delta(w(j)) &= \{\beta(j,\nu)|\nu = 1,\ldots,j\}.
\end{aligned} \tag{36}$$

Then we have

$$\Delta(\mathbf{m}) = \{\beta_\nu := \beta(L,\nu) \mid \nu = 1,\ldots,L\}. \tag{37}$$

A successive application of $s\partial_1$ to a rigid monotone spectral type $\mathbf{m}$, we have a sequence

$$\mathbf{m} \xrightarrow{s\partial_1} \mathbf{m}^{(1)} \xrightarrow{s\partial_1} \mathbf{m}^{(2)} \xrightarrow{s\partial_1} \cdots \xrightarrow{s\partial_1} \mathbf{m}^{(k)} = 10\cdots,10\cdots,10\cdots,\ldots \tag{38}$$

with a non-negative integer k. Then rewriting each transformation s as a product of transpositions defined by $s_{j,\nu}$, we have a diagram as above to get $\Delta(\mathbf{m})$. In the diagram each arrow corresponds to an element of $\Delta(\mathbf{m})$, which is expressed by

$$\begin{array}{ccccccc}
\mathbf{m} = \mathbf{m}^{(L)} \rightarrow \cdots \rightarrow & \mathbf{m}^{(j+1)} & \xrightarrow{s_{i_{j+1}}} & \mathbf{m}^{(j)} & \xrightarrow{s_{i_j}} & \mathbf{m}^{(j-1)} \rightarrow \\
k_\ell\mathbf{n}^{(L,j)} \quad \leftarrow \cdots \leftarrow & k_\ell\mathbf{n}^{(j+1,j)} & \leftarrow & k_\ell\mathbf{n}^{(j,j)} & \leftarrow & *
\end{array} \tag{39}$$

for $j = 1,\ldots,|\Delta(\mathbf{m})|$ and we have

$$\begin{aligned}
k_j\mathbf{n}^{(j,j)} &= \mathbf{m}^{(j)} - \mathbf{m}^{(j-1)}, \quad \alpha_{\mathbf{n}^{(j,j)}} \in \Pi, \quad k_j = (\alpha_{\mathbf{m}}|\alpha_{\mathbf{n}^{(L,j)}}),\\
\alpha_{\mathbf{n}^{(\ell+1,j)}} &= s_{i_{\ell+1}}\alpha_{\mathbf{n}^{(\ell,j)}},\\
\Delta(\mathbf{m}) &= \{\alpha_{\mathbf{n}^{(L,j)}} \mid j = 1,\ldots,|\Delta(\mathbf{m})|\},
\end{aligned} \tag{40}$$

and k_j is a greatest common divisor of $\mathbf{m}(j)_{i,\nu} - \mathbf{m}(j-1)_{i,\nu}$ with $i \geq 0$ and $\nu \geq 1$.

2 Condition for Reducibility

In this section we will further examine the condition of reducibility of Fuchsian ordinary differential equations with a rigid spectral type. First we show a general lemma assuring that if the rank of the equation does not depend on the parameter, the condition for the reducibility does not depend on the realization of the equation nor an integral shift of the parameter. Then we will classify the numbers (33) giving the condition for the reducibility into three types, which is related to the reducibility of a Pfaffian system studied in Sect. 3. In particular we show that we may omit some numbers among them. Lastly in Theorem 2.7 we give a condition that a reducible equation has a non-trivial quotient without an apparent singularity, which is a *ground state* for the reducibility in view of a shift operator.

Put $x = (x_1, \dots, x_n)$ and denote the ring of differential operators of x with polynomial coefficients by $W[x]$, which is called a Weyl algebra. The ring of differential operators with coefficients in the field $\mathbb{C}(x)$ of rational functions of x is denoted by $W(x)$. We identify a system of linear differential equation

$$\mathcal{M} : \sum_{j=1}^{N} P_{i,j} u_j = 0 \quad (i = 1, \dots, M) \tag{41}$$

with a left $W(x)$-module. Here $P_{i,j} \in W(x)$ and $u_1, \dots, u_N$ are generators of the left $W(x)$-module and (41) defines fundamental relations among the generators. The *rank* of $\mathcal{M}$ is the dimension of the vector space of $\mathcal{M}$ over the field $W(x)$ and we denote it by $\operatorname{rank} \mathcal{M}$. If $n = 1$, the rank of the equation $Pu = 0$ with $P \in W(x)$ equals $\operatorname{ord} P$, the order of P. Suppose $\operatorname{rank} \mathcal{M} < \infty$. Then $\mathcal{M}$ is said to be reducible if $\mathcal{M}$ has a quotient left $W(x)$-module $\mathcal{M}'$ satisfying $0 < \operatorname{rank} \mathcal{M}' < \operatorname{rank} \mathcal{M}$.

Lemma 2.1 *Let $\mathcal{M}_t$ and $\mathcal{N}_t$ are systems of linear differential equations with holomorphic parameter $t \in D := \{t \in \mathbb{C} \mid |t| < 1\}$. Suppose there exist a positive integer K such that* $\operatorname{rank} \mathcal{M}_t = \operatorname{rank} \mathcal{N}_t = K$ *for any $t \in D$ and a homomorphism ϕ_t of $\mathcal{M}_t$ to $\mathcal{N}_t$. Here it means a homomorphism between left $W(x)$-modules. Assume that ϕ_t is holomorphically depend on t and ϕ_t is an isomorphism if $t \neq 0$. Then $\mathcal{M}_0$ is reducible if and only if $\mathcal{N}_0$ is reducible.*

Proof Replacing ϕ_t by $t^m \phi_t$ with a suitable integer m, we may assume $\phi_0 \neq 0$. Then ϕ_0 is also a non-zero homomorphism of $\mathcal{M}_0$ to $\mathcal{N}_0$ by analytic continuation. If ϕ_0 is an isomorphism, the claim of the theorem is clear. If ϕ_0 is not bijective, then the kernel and the image of ϕ_0 is non-trivial proper invariant $W(x)$-submodules of $\mathcal{M}_0$ and $\mathcal{N}_0$, respectively, and hence we have the lemma. □

Hereafter in this section we examine the reducible condition for a Fuchsian ordinary differential equations with a rigid spectral type. Let

$$P(\lambda)u = 0 \tag{42}$$

be a differential equation with the Riemann scheme (4) satisfying (7). We assume that the spectral type $\mathbf{m}$ of the equation is rigid. Then we have the following remark, which also follows from Theorem 1.5.

Remark 2.2 Let $\epsilon_{j,\nu}$ be integers satisfying $\sum_{j,\nu} m_{j,\nu} \epsilon_{j,\nu} = 0$. Then [8, Theorem 11.2] shows that there exists a homomorphism ϕ_λ of the equation $P(\lambda)u = 0$ to the equation $P(\lambda+\epsilon)v=0$. Since ϕ_λ is holomorphically depend on λ, the theorem implies that $P(\lambda)u=0$ is reducible if and only if $P(\lambda+\epsilon)v=0$ is reducible.

Lemma 2.3 *Let $P(\lambda)$ be the universal operator with a rigid Riemann scheme $\{\lambda_{\mathbf{m}}\}$. Suppose that the characteristic exponents $\lambda_{j,\nu}$ holomorphically depend on $t\in D$. Assume that $P(\lambda(t))u=0$ is irreducible if $t\neq 0$ and*

$$d_1(\mathbf{m}) := m_{j,0}+m_{j,1}+\cdots+m_{j,p}-(p-1)\operatorname{ord}\mathbf{m}>0.$$

Put

$$\begin{aligned}\mu(t)&=\lambda_{0,1}(t)+\lambda_{1,1}(t)+\cdots+\lambda_{p,1}(t),\\ \tilde P(\lambda(t))&=\mathrm{RAd}_{\lambda_{1,1}(t),\ldots,\lambda_{p,1}(t)}\,P(\lambda(t)),\\ Q(\lambda(t))&=\mathrm{mc}_{1-\mu(t)}\tilde P(\lambda(t))\end{aligned}$$

and let $\mathbf{m}'$ be the spectral type of $Q(\lambda(t))$.

i) *If $\mu(0)=1$, $P(\lambda(0))u=0$ has a quotient* $\mathrm{RAd}_{-\lambda_{1,1}(0),\ldots,-\lambda_{p,1}(0)}\,Q(\lambda(0))u=0$.
ii) *Let d' be a positive integer satisfying $1\leq d'\leq d_1(\mathbf{m})$. If $\mu(0)=1-d'$, $P(\lambda(0))u=0$ has solutions $r(x)\prod_{j=1}^p(x-c_j)^{\lambda_{j,1}}$, where $r(x)$ are arbitrary polynomials of degree $<d'$ and in particular $\prod_{j=1}^p(x-c_j)^{\lambda_{j,1}}$ is a solution.*

Proof We may assume $Q(\lambda(t))u=0$ is irreducible if $t\neq 0$. Put $n=\operatorname{ord}\mathbf{m}$.

Note that $\operatorname{ord}Q(\lambda(t))=n-d_1(\mathbf{m})$ and the Riemann scheme of $\tilde P(\lambda(t))$ is

$$\begin{Bmatrix} x=c_0=\infty & c_j\ (j=1,\ldots,p)\\ [\sum_{i=0}^p\lambda_{i,1}]_{(m_{0,1})} & [0]_{(m_{j,\nu})}\\ [\lambda_{0,\nu}+\sum_{i=1}^p\lambda_{i,1}]_{(m_{0,\nu})}\ (\nu=1,\ldots,n_0) & [\lambda_{j,\nu}-\lambda_{j,1}]_{(m_{j,\nu})}\ (\nu=1,\ldots,n_j)\end{Bmatrix}.$$

If $\mu(0)=1$, the definition of $\mathrm{mc}_{1-\mu(t)}$ implies that $Q(\lambda(0))$ is a quotient of $\tilde P(\lambda(0))$, which proves i).

Put $f(x)=\prod_{j=1}^p(x-c_j)^{n-m_{j,1}}$. Then the Riemann scheme of $P^\vee u=0$ with $P^\vee=(f(x)^{-1}\tilde P)^*$ equals

$$\begin{Bmatrix} x=c_0=\infty & c_j\ (j=1,\ldots,p)\\ [2-n-m_{0,1}-\sum_{i=0}^p\lambda_{i,1}]_{(m_{0,1})} & [n-m_{j,1}]_{(m_{j,\nu})}\\ [2-n-m_{0,\nu}-\lambda_{0,\nu}-\sum_{i=1}^p\lambda_{i,1}]_{(m_{0,\nu})} & [n-m_{j,\nu}-\lambda_{j,\nu}+\lambda_{j,1}]_{(m_{j,\nu})}\end{Bmatrix}$$

as was given in [8, Theorem 4.19 ii)] and that of $\tilde P^*u=0$ equals

$$\begin{Bmatrix} x=c_0=\infty & c_j\ (j=1,\ldots,p)\\ [2-d_1-\sum_{i=0}^p\lambda_{i,1}]_{(m_{0,1})} & [0]_{(m_{j,\nu})}\\ [2-d_1-m_{0,\nu}+m_{0,1}-\lambda_{0,\nu}-\sum_{i=1}^p\lambda_{i,1}]_{(m_{0,\nu})} & [m_{j,1}-m_{j,\nu}-\lambda_{j,\nu}+\lambda_{j,1}]_{(m_{j,\nu})}.\end{Bmatrix}.$$

Here $d_1 = d_1(\mathbf{m})$. Note that $P(\lambda(t))^*$ is the universal operator with the above Riemann scheme and $2 - d_1 - \sum_{i=0}^{p} \lambda_{i,1}(0) = 1 + d' - d_1$. Then in the case $d' = d_1(\mathbf{m})$, the claim ii) follows from the definition of $P^\vee$ and $\mathrm{mc}_{1-\mu(t)}$.

We may split the characteristic exponent $[\lambda_{0,1}]_{(m_{0,1})}$ to

$$[\lambda_{0,1} + d_1 - d']_{(m_{0,1}-d_1+d')} \text{ and } [\lambda_{0,1}]_{(d_1-d')}.$$

Then $m_{0,1}$, $d_1(\mathbf{m})$ and n_0 changed into $m_{0,1} - d_1(\mathbf{m}) + d'$, d', $n_0 + 1$, respectively, and the same argument as above proves ii). □

Definition 2.4 The elements $\beta \in \Delta(\mathbf{m})$ are classified into three types

$$\begin{aligned} &(\text{Type 1}) && \operatorname{ord} s_\beta \alpha_{\mathbf{m}} > 0, \\ &(\text{Type 2}) && \operatorname{ord} s_\beta \alpha_{\mathbf{m}} = 0, \\ &(\text{Type 3}) && \operatorname{ord} s_\beta \alpha_{\mathbf{m}} < 0. \end{aligned}$$

Putting $\gamma := s_\beta \alpha_{\mathbf{m}}$, we have $\gamma = \alpha_{\mathbf{m}} - k_\beta \beta$ with $k_\beta = (\alpha_{\mathbf{m}}|\beta)$ and therefore

$$\alpha_{\mathbf{m}} = k_\beta \beta + \gamma. \tag{43}$$

Here we put $\operatorname{ord} \gamma = n_0$ if $\gamma = \sum_{i \in I} n_i \alpha_i$.

Proposition 2.5

i) *We have* $\operatorname{ord} \beta > 0$ *for any* $\beta \in \Delta(\mathbf{m})$.
ii) *If* $k_\beta = 1$, *then* β *is of Type 1.*
iii) *The numbers* (33) *for* $\beta \in \Delta(\mathbf{m})$ *of Type 3 may be omitted for Theorem 1.5.*

Proof Under the notation in the preceding section, we have $\beta = \beta(L, \nu_o)$ with a certain ν_o in (36).

i) Since $\mathbf{m}$ is monotone, if $\alpha \in \Sigma_+^{re}$ satisfies $(\alpha|\alpha_{\mathbf{m}}) > 0$, then $\operatorname{ord} \alpha > 0$.
ii) The claim is given in [8, Proposition 7.9 iv)].
iii) Note that the condition $\sum_{j,\nu} \mathbf{m}(\beta)_{j,\nu} \lambda_{j,\nu} \in \mathbb{Z}$ implies $\sum_{j,\nu} \mathbf{m}(\gamma)_{j,\nu} \lambda_{j,\nu} \in \mathbb{Z}$ because of the Fuchs relation. Since $\gamma(\nu_o, \nu_o) \in \Sigma_+^{re}$, there exists $j > \nu_o$ satisfying $\beta(j,j) = \alpha_{i_j} = \gamma(j-1, \nu_o) = -\gamma(j, \nu_o) \in \Pi$. Then $\beta(L,j) = -\gamma(L, \nu_o) \in \Delta(\mathbf{m})$. Hence the condition for the number (33) with $\beta(L, \nu_o)$ can be omitted for Theorem 1.5.

Put $\beta_\nu = \beta(L, \nu)$, $\gamma_\nu = \gamma(L, \nu)$ and $k_\nu = k_{\beta_\nu}$ for simplicity. Then

$$\alpha_{\mathbf{m}} = k_{\nu_o} \beta_{\nu_o} + \gamma_{\nu_o} \text{ with } k_{\nu_o} = (\alpha_{\mathbf{m}}|\beta_{\nu_o})$$

and

$$k_j = (\alpha_{\mathbf{m}}|\beta_j) = (\alpha_{\mathbf{m}}|k_{\nu_o}\beta_{\nu_o} - \alpha_{\mathbf{m}}) = k_{\nu_o}^2 - 2 > 1.$$

If $\beta(L,j)$ is of Type 3, we repeat the same way and this procedure ends in finite steps because $\nu_o < j \leq L$.

Remark 2.6

i) There always exists $\beta \in \Delta(\mathbf{m})$ of Type 1 and $k_\beta = 1$ in (43) if $\operatorname{ord}\mathbf{m} > 1$ (cf. [8, Proposition 10.7]). In this case we have
$$|\{\lambda_{\mathbf{m}(\beta)}\}| + |\{\lambda_{\mathbf{m}(\gamma)}\}| = 1$$
and the condition $\sum \mathbf{m}(\beta)_{j,\nu}\lambda_{j,\nu} \in \mathbb{Z}$ is equivalent to $\sum \mathbf{m}(\gamma)_{j,\nu}\lambda_{j,\nu} \in \mathbb{Z}$.

ii) Type 2 appears only in the case when there exist j_o and $1 \leq \nu_o < \nu_o' \leq n$ such that $m_{j_o,\nu_o} + m_{j_o,\nu_o'}$ and $m_{j,\nu}$ with $(j,\nu) \notin \{(j_o,\nu_o),(j_o,\nu_o')\}$ are divisible by a common integer k larger than 1. There is an example of Type 2 in Sect. 3.

iii) There is an example of Type 3 in Sect. 3.

Theorem 2.7 *Let $P(\lambda)$ be the universal operator with the Riemann scheme* (4) *with rigid spectral type* $\mathbf{m}$. *Let $\beta \in \Delta(\mathbf{m})$. Then if*
$$1 \leq d' := \operatorname{ord}\beta - \sum_{j=0}^{p}\sum_{\nu=1}^{n_j} \mathbf{m}(\beta)_{j,\nu}\lambda_{j,\nu} \leq (\alpha_{\mathbf{m}}|\beta), \tag{44}$$
$P(\lambda) \in W(x)Q(\lambda)$. Here $Q(\lambda)$ is the universal operator with the spectral type $\mathbf{m}(\beta)$ and if $d' = 1$, the Riemann scheme of $Q(\lambda)$ equals $\{\lambda_{\mathbf{m}(\beta)}\}$. Here $0 < \operatorname{ord}Q(\lambda) = \operatorname{ord}\beta < \operatorname{ord}P$.

If $(\alpha_{\mathbf{m}}|\beta) = 1$, then replacing β by $\gamma = \alpha_{\mathbf{m}} - \beta$, the above statement also holds.

Proof First note that the condition $d' = 1$ for β is the Fuchs relation of $Q(\lambda)$ written by $|\{\lambda_{\mathbf{m}(\beta)}\}| = 0$ and hence this number d' is invariant under the action of W.

We will prove the theorem by the induction with respect to the number k in (38). Suppose $\beta = \alpha_0$ or $\alpha_0 + \alpha_{j,1} + \cdots + \alpha_{j,\nu}$ in (32). Applying Lemma 2.3 ii) with $d' = 1$ to P after the transposition of the indices $(j, \nu+1)$ and $(j,1)$ if necessary, we have the claim of the theorem for β. We have also the claim of the theorem for γ by Lemma 2.3 i).

Put $w' = (\prod_{\substack{j\geq 0\\ \nu_j>0}} s_{j,1}\cdots s_{j,\nu_j})s_0$ in (32) and let $\beta \in w'^{-1}\Delta(s\partial_1\mathbf{m})$. Then β corresponds to an element $\beta' \in \Delta(s\partial_1\mathbf{m})$ with $\beta' = w'\beta$ and $w' \in W$ corresponds to the map $s\partial_1$, which corresponds to $\psi = \mathrm{RAd}_{\mu_1,\ldots,\mu_p} \circ \mathrm{mc}_\mu$, the combination of a middle convolution and an addition.

Here we note that $\beta, \beta' \in \Sigma_+^{re}$, $\operatorname{ord}\beta > 0$ and $\operatorname{ord}\beta' > 0$.

Then the statement for β is obtained by applying ψ^{-1} to the operator $\psi(P)$ and the operator corresponding to $\beta' \in \Delta(s\partial_1\mathbf{m})$ whose existence is assured by the hypothesis of the induction.

If β is of Type 1, then $\gamma' := w'\gamma \in \Sigma_+^{re}$ and $\operatorname{ord}\gamma' > 0$, we have also the last claim. □

3 Hypergeometric Equations with Several Variables

Any four points t_0, t_1, t_∞, t_x in the Riemann sphere $\mathbb{P}^1_{\mathbb{C}}$ can be transformed to 0, 1, ∞, x by the fractional transformation defined by

$$x = \frac{(t_x - t_0)(t_\infty - t_1)}{(t_1 - t_0)(t_\infty - t_x)}.$$

Hence Gauss hypergeometric function $F(a, b, c; x)$ is naturally considered as a hypergeometric function on the configuration space of four points in $P^1_{\mathbb{C}}$ by

$$F(a, b, c; \tfrac{t_x - t_0}{t_1 - t_0} \tfrac{t_\infty - t_1}{t_\infty - t_x}).$$

Then the Riemann scheme of this function is

$$\begin{Bmatrix} t_x = t_0 & t_x = t_1 & t_x = t_\infty & t_0 = t_1 & t_0 = t_\infty & t_1 = t_\infty \\ 0 & 0 & a & a & 0 & 0 \\ 1 - c & 1 - a - b & b & b & 1 - a - b & 1 - c \end{Bmatrix}.$$

The solutions of the universal rigid Fuchsian equations $Pu = 0$ with a rigid spectral type $\mathbf{m}$ have natural integral representations (cf. [8]). When $\mathbf{m}$ is a $(p + 1)$ tuples of partitions, P has $(p + 1)$ singular points. We may specialize the points as 0, 1, ∞, y_1, ..., y_{p-2} and then the solutions $u(x, y_1, \dots, y_{p-2})$ has $(p - 1)$ variables by the integral representation. These functions are a kind of hypergeometric functions with several variables and also they can be considered hypergeometric functions on the configuration space of $(p + 2)$ points in $\mathbb{P}^1_{\mathbb{C}}$.

For these hypergeometric functions with several variables, it will be convenient to use the differential equations of Pfaffian form satisfied by the functions. They have been studied by Dettweiler and Reiter [2] for the case of single variable and by Haraoka [3] for several variables. We will shortly explain it in the case of two variables x and y as an example. Then the Pfaffian system is

$$\mathcal{M}_\lambda : du = \left(A_1 \frac{dx}{x} + A_2 \frac{d(x - y)}{x - y} + A_3 \frac{d(x - 1)}{x - 1} + A_4 \frac{dy}{y} + A_5 \frac{d(y - 1)}{y - 1} \right) u. \tag{45}$$

Here A_j are square matrices of size n. They are called the residue matrices along the corresponding hypersurfaces. If the eigenvalues of A_j are $\lambda_{j,\nu}$ with multiplicity $m_{j,\nu}$ for $\nu = 1, \dots, n_j$ and A_j are semisimple, the set of characteristic exponents corresponding to A_j is defined by $\{[\lambda_{j,1}]_{m_{j,1}}, \dots, [\lambda_{j,n_j}]_{m_{j,n_j}}\}$, and we can define the generalized Riemann scheme for this Pfaffian system. For example, the set of characteristic exponents at $x = 0$, $x = y$ and $x = 1$ are these sets for $j = 1$, $j = 2$ and $j = 3$, respectively, and the set of characteristic exponents at $x = \infty$ is define by the matrix $A_0 = -(A_1 + A_2 + A_3)$.

The addition with parameters $\lambda_1, \lambda_2, \lambda_3$ (for x-variable) is defined by

$$\mathrm{Ad}_{\lambda_2,\lambda_2,\lambda_3}(A_1, A_2, A_3, A_4, A_5) \mapsto (A_1 + \lambda_1, A_2 + \lambda_2, A_3 + \lambda_3, A_4, A_5).$$

The middle convolution mc_μ to these matrices is defined by

$$\mathrm{mc}_\mu(A_j) := \tilde{A}_j \mod \mathcal{K}_\mu := \begin{pmatrix} \ker A_1 \\ \ker A_2 \\ \ker A_3 \end{pmatrix} \oplus \ker(\tilde{A}_1 + \tilde{A}_2 + \tilde{A}_3) \tag{46}$$

with

$$\tilde{A}_1 = \begin{pmatrix} A_1 + \mu & A_2 & A_3 \\ 0 & 0 & 0 \\ 0 & 0 & 0 \end{pmatrix}, \tilde{A}_2 = \begin{pmatrix} 0 & 0 & 0 \\ A_1 & A_2 + \mu & A_3 \\ 0 & 0 & 0 \end{pmatrix}, \tilde{A}_3 = \begin{pmatrix} 0 & 0 & 0 \\ 0 & 0 & 0 \\ A_1 & A_2 & A_3 + \mu \end{pmatrix},$$

$$\tilde{A}_4 = \begin{pmatrix} A_4 + A_2 & -A_2 & 0 \\ -A_1 & A_4 + A_1 & 0 \\ 0 & 0 & A_4 \end{pmatrix} \text{ and } \tilde{A}_5 = \begin{pmatrix} A_5 & 0 & 0 \\ 0 & A_5 + A_3 & -A_3 \\ 0 & -A_2 & A_5 + A_2 \end{pmatrix}.$$

Here $\tilde{A}_j$ are considered to be linear maps on the vector space $\mathbb{C}^{3n}$ and the space $\mathcal{K}_\mu$ are invariant by these maps and then we represent A_j as square matrices of size $3n - \dim \mathcal{K}_\mu$.

Considering y as a parameter and forgetting A_4 and A_5, the above definition is due to [2]. In this case, $x = 0,\ 1,\ y,\ \infty$ are regular singular points of the differential equation with the variable x. The above operation for A_4 and A_5 is defined and studied by Haraoka [3], where x and y are equally considered as variables.

When we regard y as a parameter, the structure of operations of additions and middle convolutions are compatible to the corresponding operations of Fuchsian ordinary differential equations, which are briefly explained in the previous sections. Moreover any rigid monodromy group is equally realized by solutions of both type of equations. Assume that the spectral type $\mathbf{m}$ is rigid. Then there is a homomorphism ψ_λ of the universal Fuchsian differential equation $P(\lambda)u = 0$ to the rigid Pfaffian system $\mathcal{M}_\lambda$, where ψ_λ is meromorphically depend on λ and it defines an onto isomorphism between two equations for generic λ, which follows from the fact that they are constructed from the trivial equation by successive applications of additions and middle convolutions.

Table 1 is the number of rigid spectral types of order at most 15 whose numbers of singular points are smaller than 7.

It is well-known that the integral representation of the solution of Jordan-Pochhammer equation, which is characterized by rigid spectral type $21, 21, 21, 21$, gives a solution of Appell's F_1 (in general, Lauricella's F_D with more variables). As is given in the table, the rigid Pfaffian system corresponding to the spectral type $31, 31, 22, 211$ corresponds to Appell's F_2 and F_3 and that to $31, 22, 22, 22$ corresponds to Appell's F_4 as was shown in [3]. Hence we have a plenty of generalizations of Appell's hypergeometric functions.

Table 1 Hypergeometric equations with less than six variables

Order	2	3	4	5	6	7	8	9	10	11	12	13	14	15
One variable	1	1	3	5	13	20	45	74	142	212	421	588	1004	1481
Two variables		1	2	4	11	16	35	58	109	156	299	402	685	924
Three variables			1	1	3	5	12	17	43	52	104	135	263	327
Four variables				1	0	1	3	5	8	14	24	39	60	79
Five variables					1	0	0	2	3	4	6	6	14	20

Table 2 Hierarchy of rigid quartets (cf. [8])

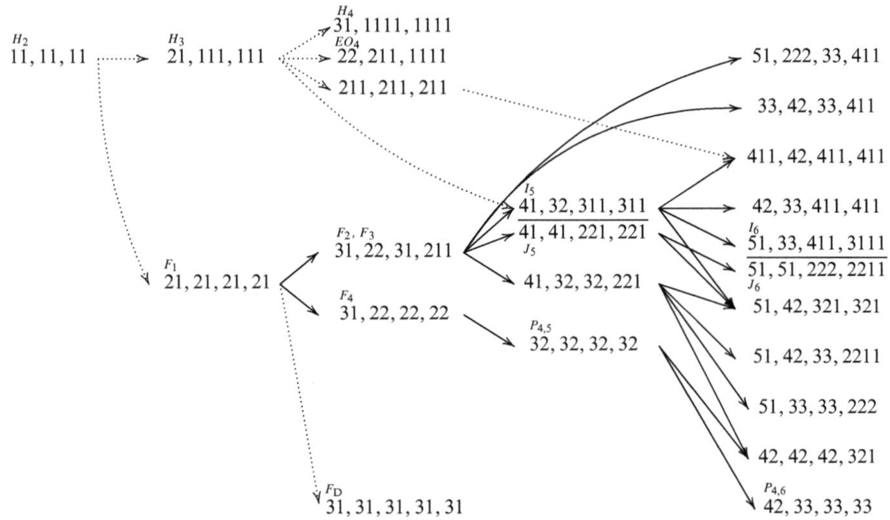

Remark 3.1

i) The number of parameters of the equation with a rigid spectral type after a suitable addition is given by

$$\#\text{parameters} = \sum \big(\#\text{ blocks at singular points} - 1\big) = \sum_{j=0}^{p}(n_j - 1) \tag{47}$$

as in the case of the Gauss hypergeometric function.

ii) In Table 2, the arrow shows that two spectral types are connected by an addition and a middle convolution. Moreover in Table 2 we see that F_2 and F_3 are obtained by restrictions of a Pfaffian system to different complex lines and the pair of I_5 and J_5 has the same property.

iii) It follows from [8, Theorem 11.2] that there exists a non-zero homomorphism realizing any integral shift of characteristic exponents in the Pfaffian system corresponding to a rigid Fuchsian ordinary differential equation. Lemma 2.1 assures that the irreducibility of the system is invariant under the integral shift.

iv) There exists a universal Fuchsian equation with a rigid spectral type as is stated in Sect. 1. But in the case of a Pfaffian system the extension to a special value of λ which corresponds to a reducible monodromy group is not unique. Consider the Gauss hypergeometric system with the Riemann scheme $\begin{Bmatrix} \lambda_{0,1} & \lambda_{1,1} & \lambda_{2,1} \\ \lambda_{0,2} & \lambda_{1,2} & \lambda_{2,2} \end{Bmatrix}$ for generic values of parameters under the condition $\lambda_{0,j}+\lambda_{1,j}+\lambda_{2,j} = 0$ for $j = 1$ and 2. Then the Riemann scheme of its irreducible quotient is $\{\lambda_{0,1}\ \lambda_{1,1}\ \lambda_{2,1}\}$ or $\{\lambda_{0,2}\ \lambda_{1,2}\ \lambda_{2,2}\}$. But there is no natural way to determine it and it depends on its construction using additions and middle convolutions.

Theorem 3.2 *Let*

$$du = \Bigl(\sum_{i=1}^{n}\sum_{k=1}^{q} A_{i,k}\frac{dy_i}{y_i - c_k} + \sum_{1\le i<j\le n} B_{i,j}\frac{d(y_i - y_j)}{y_i - y_j}\Bigr)u \tag{48}$$

be a completely integrable Pfaffian system with variables $y_1,\dots,y_n$, where $A_{i,k}$ and $B_{i,j}$ are constant $N \times N$ matrices and $c_1,\dots,c_q$ are mutually different complex numbers. Suppose

$$\frac{du}{dx} = \Bigl(\sum_{k=1}^{q}\frac{A_{1,k}}{x - c_k} + \sum_{j=2}^{n}\frac{B_{1,j}}{x - y_j}\Bigr)u \tag{49}$$

is an irreducible rigid Fuchsian system with mutually different complex numbers $c_1,\dots,c_q,y_2,\dots,y_n$. Then (48) is irreducible and it is obtained by a successive application of additions and middle convolutions extended by Haraoka for the variable $x = y_1$ to a Pfaffian differential equation

$$dv = \Bigl(\sum_{i=2}^{n}\sum_{k=1}^{q} \alpha_{i,k}\frac{dy_i}{y_i - c_k} + \sum_{2\le i<j\le n} \beta_{i,j}\frac{d(y_i - y_j)}{y_i - y_j}\Bigr)v \tag{50}$$

of the first order. Here $\alpha_{i,k}$ and $\beta_{i,j}$ are complex numbers. Note that the solution of (50) is a constant multiple of $\prod_{i=2}^{n}\prod_{k=1}^{q}(y_i - c_k)^{\alpha_{i,k}} \cdot \prod_{2\le i<j\le n}(y_i - y_j)^{\beta_{i,j}}$.

Proof This theorem is not clearly stated in [3] but it is essentially obtained in or easily obtained by Haraoka [3].

Since (49) is rigid, it is reduced to the trivial equation $du = 0$ by a successive application of additions and middle convolutions. Then we get an equation of the form (50) by the successive application of the corresponding operations for y_1 variable to (48). Since these operations are invertible, we obtain (48) from (50) by these operations.

Since (49) is irreducible, the monodromy group of the solutions of (48) is irreducible even if $y_2,\dots,y_2$ are fixed. Hence Eq. (48) is irreducible. □

Theorem 3.3 *Let* $\mathbf{m}$ *be a rigid spectral type and let* (49) *be the Fuchsian system with the rigid spectral type* $\mathbf{m}$. *Let* $\left\{[\lambda_{j,\nu}]_{m_{j,\nu}}\right\}_{\substack{0\le j\le p\\1\le\nu\le n_j}}$ *be the Riemann scheme of the Fuchsian system* (49) *with* $p = q + n - 1$. *Let* (48) *be the corresponding completely integrable Pfaffian system of rank* $N = \operatorname{ord}\mathbf{m}$.

i) *Suppose any one of the numbers* (33) *is not an integer, the Pfaffian system* (48) *is irreducible.*
ii) *Suppose there exists* $\beta \in \Delta(\mathbf{m})$ *such that*

$$\sum_{j=0}^{p}\sum_{\nu=1}^{n_j}\mathbf{m}(\beta)_{j,\nu}\lambda_{j,\nu} \in \mathbb{Z}. \tag{51}$$

If β *is of Type 1 or Type 3 (cf. Definition 2.4), then* (48) *is reducible.*

Suppose (48) *is irreducible. Fix a suitable base of local solutions of* (48) *in a neighborhood of a generic point* $y^o = (y_1^0, \dots, y_n^o)$ *of* $\mathbb{C}^n$. *Then*

$$\tilde{A}_\nu = \tilde{A}'_\nu \otimes I_r \quad (1 \le \nu \le p) \tag{52}$$

are the corresponding monodromy matrices along the closed loops γ_ν *starting from* y^o *in the* y_1*-plane and* r *is an integer larger than* 1. *Here putting* $c_{q+j-1} = y_j$ *for* $2 \le j \le n$, *we denote by* γ_ν *the closed loops starting from* y^o *in the* y_1*-plane which satisfy* $\frac{1}{2\pi\sqrt{-1}}\int_{\gamma_\nu}\frac{dy_1}{y_1 - c_{\nu'}} = \delta_{\nu,\nu'}$ *with* $\nu,\ \nu' \in \{1, 2, \dots, p\}$.

Proof The claim i) follows from Theorem 3.2.

Fix $\beta \in \Delta(\mathbf{m})$. We may fix generic $\lambda_{j,\nu}$ satisfying (51) and the Fuchs relation. Then (49) is reducible. Let $u(y)$ be a local solution of (48) in a neighborhood of y^o such that a component of $u(x, y_2, \dots, y_n)$ is a solution of an irreducible Fuchsian ordinary differential equation with the variable x so that the order R of the equation is smaller than N. Since the coefficients of the equation are rational functions of (x, y), the analytic continuations of the component satisfy the same equation.

Suppose (48) is irreducible. Then the dimension of the linear span V of the local solutions obtained by analytic continuations of $u(y)$ with respect to the variables $y_1, \dots, y_n$ equals N. Hence there exist $u_{\ell,1}(y), \dots, u_{\ell,R}(y)$ in V for $1 \le \ell \le r$ with $N = rR$ such that the spaces spanned by $u_{\ell,1}(y), \dots, u_{\ell,R}(y)$ are stable under the analytic continuation along the loops γ_ν and moreover the dimension of the space $\sum_{\ell=1}^{r}\sum_{i=1}^{R}\mathbb{C}u_{\ell,i}(y)$ equals N. We may moreover assume that the monodromies along γ_ν with respect to the base $\{u_{\ell,1}(y), \dots, u_{\ell,R}(y)\}$ do not depend on ℓ. Then the multiplicity of the eigenvalues of local monodromy matrices at each singular points of (49) is divisible by r. This never happens if β is of Type 1 nor Type 3 because of the genericity condition for the values $\lambda_{j,\nu}$. Hence β is of Type 2 and we have the theorem. □

Remark 3.4 Under the notation in Theorems 1.5, 3.2 and 3.3 the ordinary differential equation (49) is reducible if and only if at least one of the numbers (33) for $\beta \in \Delta(\mathbf{m})$ of Type 1 or Type 2 is an integer. But it seems that the system (48) is reducible if and only if at least one of the numbers (33) for $\beta \in \Delta(\mathbf{m})$ of Type 1 is an integer (cf. Sect. 3). We note that for most rigid spectral types $\mathbf{m}$ there is no $\beta \in \Delta(\mathbf{m})$ of Type 2 [cf. Remark 2.6 ii)] and in this case the system (48) is reducible if and only if so is Eq. (49).

We will examine all of our Pfaffian systems with two variables corresponding to a rigid spectral type $\mathbf{m}$ whose ranks are at most 5 (cf. [10]). The Pfaffian systems satisfied by Appell's hypergeometric functions are described in [4].

Computer Algebra

A computer algebra `Risa/Asir` with the library [11] calculates $\Delta(\mathbf{m})$ and mc_μ in (11) and (46) and in particular the examples given in this section.

Appell's F_1

$\mathbf{m} = 21, 21, 21, 21$: rank $= 3$ with four parameters and $[\Delta(\mathbf{m})] = 1^4 \cdot 2^1$

$21, 21, 21, 21 \rightarrow 01, 01, 01, 01 \quad H_2 : 11, 11, 11, 20$ (by a middle convolution)
$= 10, 10, 10, 01 \oplus 11, 11, 11, 20$ (reducibility: 4 cases by the symmetry)
$= 2(10, 10, 10, 10) \oplus 01, 01, 01, 01$ (1 case)

The generalized Riemann scheme for the two variables (x, y) is

$$\begin{Bmatrix} x=0 & x=1 & x=y & x=\infty & y=0 & y=1 & y=\infty \\ [0]_2 & [0]_2 & [0]_2 & [e]_2 & [0]_2 & [0]_2 & [-c-e]_2 \\ a & b & c & d & a+c+2e & b+c+2e & d \end{Bmatrix}$$

with the following Fuchs relation.

$a + b + c + d + 2e = 2, 0$ (Fuchs rel.)
$10, 10, 10, 01 : d \notin \mathbb{Z}$ (irred. cond.)
$10, 10, 01, 10 : c + e \notin \mathbb{Z}$
$10, 01, 10, 10 : b + e \notin \mathbb{Z}$
$01, 10, 10, 10 : a + e \notin \mathbb{Z}$
$10, 10, 10, 10 : e \notin \mathbb{Z}$

	t_∞	t_0	t_y	t_1	t_x	idx
t_∞		21	21	21	21	2
t_0	21		21	21	21	2
t_y	21	21		21	21	2
t_1	21	21	21		21	2
t_x	21	21	21	21		2

In the Fuchs relation, "$= 2$" is valid for a single Fuchsian differential equation and "$= 0$" is valid for the corresponding Pfaffian system.

The set of five points $\{t_0, t_1, t_\infty, t_x, t_y\}$ in $\mathbb{P}^1_{\mathbb{C}}$ is transformed into $\{0, 1, \infty, x, y\}$. Hence a Pfaffian system with five variables $t_0, t_1, t_\infty, t_x, t_y$ is defined through this map and the spectral type at each singular hypersurface is given in the above.

The above five conditions for the irreducibility are obtained by the following procedure, which is explained in Sect. 1.

$$\underline{2}1,\underline{2}1,\underline{2}1,\underline{2}1 \overset{-2}{\underset{2}{\Rightarrow}} \underline{0}1,\underline{0}1,\underline{0}1,\underline{0}1 \underset{1}{\to} \underline{10},01,01,01 \underset{1}{\to} 10,\underline{10},01,01 \underset{1}{\to} 10,10,\underline{10},01 \underset{1}{\searrow} 10,10,10,10$$

$$2(10,10,10,10) \overset{+2}{\Leftarrow} *$$

$$01,10,10,10 \overset{+1}{\Leftarrow} \underline{-1}1,\underline{0}0,\underline{0}0,\underline{0}0 \leftarrow *$$

$$10,01,10,10 \overset{+1}{\Leftarrow} \underline{0}0,\underline{-1}1,\underline{0}0,\underline{0}0 \leftarrow \underline{00},-11,00,00 \leftarrow *$$

$$10,10,01,10 \overset{+1}{\Leftarrow} \underline{0}0,\underline{0}0,\underline{-1}1,\underline{0}0 \leftarrow \underline{00},00,-11,00 \leftarrow 00,\underline{00},-11,00 \leftarrow *$$

$$10,10,10,01 \overset{+1}{\Leftarrow} \underline{0}0,\underline{0}0,\underline{0}0,\underline{-1}1 \leftarrow \underline{00},00,00,-11 \leftarrow 00,\underline{00},00,-11 \leftarrow 00,00,\underline{00},-11 \leftarrow *$$

The Pfaffian system corresponds to the hypergeometric function

$$F_1(\alpha;\beta,\beta';\gamma;x,y) = \sum_{m,n=0}^{\infty} \frac{(\alpha)_{m+n}\,(\beta)_m\,(\beta')_n}{(\gamma)_{m+n}\,m!\,n!} x^m y^n$$

with

$$a = \beta' - \gamma + 1, \quad b = \gamma - \alpha - \beta - 1, \quad c = -\beta - \beta', \quad d = \alpha, \quad e = \beta$$

and then the necessary and sufficient condition for the irreducibility is

$$\{\alpha,\ \beta,\ \beta',\ \alpha - \gamma,\ \beta + \beta' - \gamma\} \cap \mathbb{Z} = \emptyset. \tag{53}$$

Appell's F_2, F_3

$211,22,31,31$: rank $= 4$, 5 parameters, $(1^6 \cdot 2^2)$
$\to\ F_1 : 201,21,21,21 \quad H_2 : 011,02,11,11$
$= 010,01,10,10 \oplus 201,21,21,21 \quad (4)$
$= 101,11,11,20 \oplus 110,11,20,11 \quad (2)$
$= 2(100,01,10,10) \oplus 011,20,11,11 \quad (2)$
These are of Type 1.

	t_∞	t_0	t_y	t_1	t_x	idx
t_∞		211	211	211	211	-8
t_0	211		31	31	22	2
t_y	211	31		22	31	2
t_1	211	31	22		31	2
t_x	211	22	31	31		2

$2a + b + c + 2d + e + f = 3,\ 0$

$$\left\{\begin{matrix} x=0 & x=y & x=1 & x=\infty & y=0 & y=1 & y=\infty & x=y=\infty \\ [0]_2 & [0]_3 & [0]_3 & [d]_2 & [0]_3 & [0]_2 & [-a-b-e-d]_2 & [f]_3 \\ [a]_2 & b & c & e & a+b+2d & [e-f]_2 & f & -a-b-e \\ & & & f & & & a+f & \end{matrix}\right\}$$

$$\underline{2}11,\underline{2}2,\underline{3}1,\underline{3}1 \overset{-2}{\underset{2}{\Rightarrow}} \underline{01}1,02,11,11 \to 1\underline{01},02,11,11 \to 110,\underline{02},11,11 \to 110,20,11,11 \Rightarrow$$

$$2(100,10,10,10) \overset{+2}{\Leftarrow} *$$

$$010,10,10,10 \overset{+1}{\Leftarrow} \underline{-1}10,\underline{0}0,\underline{0}0,\underline{0}0 \leftarrow *$$

$$001,10,10,10 \overset{+1}{\Leftarrow} \underline{-1}01,\underline{0}0,\underline{0}0,\underline{0}0 \leftarrow \underline{0-1}1,\underline{0}0,\underline{0}0,\underline{0}0 \leftarrow *$$

$$
\begin{array}{l}
2(100,01,10,10) \overset{+1}{\Leftarrow} 2(\underline{0}00,\underline{-1}1,\underline{0}0,\underline{0}0) \leftarrow 2(\underline{00}0,-11,00,00) \leftarrow 2(0\underline{00},-11,00,00) \\
010,01,10,10 \overset{+0}{\Leftarrow} \underline{0}10,\underline{0}1,\underline{1}0,\underline{1}0 \leftarrow \underline{10}0,01,10,10 \leftarrow 1\underline{00},01,10,10 \leftarrow 100,\underline{10},10,10 \\
001,01,10,10 \overset{+0}{\Leftarrow} \underline{0}01,\underline{0}1,\underline{1}0,\underline{1}0 \leftarrow \underline{00}1,01,10,10 \leftarrow 1\underline{10},01,10,10 \leftarrow 010,\underline{10},10,10 \\
110,11,11,20 \overset{+1}{\Leftarrow} \underline{0}10,\underline{0}1,\underline{0}1,\underline{1}0 \leftarrow \underline{10}0,01,01,10 \leftarrow 1\underline{00},01,01,10 \leftarrow 100,\underline{10},01,10 \\
110,11,20,11 \overset{+1}{\Leftarrow} \underline{0}10,\underline{0}1,\underline{1}0,\underline{0}1 \leftarrow \underline{10}0,01,10,01 \leftarrow 1\underline{00},01,10,01 \leftarrow 100,\underline{10},10,01
\end{array}
$$

The condition for the irreducibility is

$$\{d,\ e,\ a+d,\ a+e,\ a+b+d+e,\ a+c+d+e, f,\ a+f\} \cap \mathbb{Z} = \emptyset. \tag{54}$$

This system corresponds to the equation satisfied by $F_2(\alpha;\beta,\beta';\gamma,\gamma';x,1-y)$ with

$$F_2(\alpha;\beta,\beta';\gamma,\gamma';x,y) = \sum_{m,n=0}^{\infty} \frac{(\alpha)_{m+n}\,(\beta)_m\,(\beta')_n}{(\gamma)_m\,(\gamma')_n\,m!\,n!} x^m y^n,$$

$$
\begin{aligned}
&a = 1-\gamma, && b = \gamma+\gamma'-\alpha-\beta-\beta'-2, && c = \gamma-\alpha-\beta+\beta'-1,\\
&d = \beta, && e = \alpha-\gamma'+1, && f = \alpha
\end{aligned}
$$

and (54) equals

$$\{\alpha,\ \beta,\ \beta',\ \alpha-\gamma,\ \alpha-\gamma',\ \beta-\gamma,\ \beta'-\gamma',\ \alpha-\gamma-\gamma'\} \cap \mathbb{Z} = \emptyset. \tag{55}$$

Appell's F_4

$22,22,31,22$: rank $=4$, 4 parameters,
$(1^8\cdot 2^1)\ \rightarrow\ F_1 : 12,12,21,12$
$= 01,01,10,01 \oplus 21,21,21,21\quad (8)$
$= 2(11,11,20,11)\oplus 00,00,(-1)1,00\quad (1)$
The last one is of Type 2.
$a+2b+2c+2d+2e = 3,\ 0$

	t_∞	t_0	t_y	t_1	t_x	idx
t_∞		211	22	211	22	−4
t_0	211		22	211	22	−4
t_y	22	22		22	31	2
t_1	211	211	22		22	−4
t_x	22	22	31	22		2

$$
\begin{Bmatrix}
x=0 & x=1 & x=y & y=0 & y=1 & x=\infty & y=\infty & x=y=\infty \\
[0]_2 & [0]_2 & [0]_3 & [0]_2 & [0]_2 & [d]_2 & [b+c+d]_2 & b+c+2d \\
[b]_2 & [c]_2 & a & [-b]_2 & [-c]_2 & [e]_2 & [b+c+e]_2 & b+c+2e \\
 & & & & & & & [0]_2
\end{Bmatrix}
$$

The addition corresponding to $u \mapsto y^b(1-y)^c u$ changes the above into

$$\left\{\begin{matrix} x=0 & x=1 & x=y & y=0 & y=1 & x=\infty & y=\infty & x=y=\infty \\ [0]_2 & [0]_2 & [0]_3 & [0]_2 & [0]_2 & [d]_2 & [d]_2 & 2d \\ [b]_2 & [c]_2 & a & [b]_2 & [c]_2 & [e]_2 & [e]_2 & 2e \\ & & & & & & & [-b-c]_2 \end{matrix}\right\}.$$

$$\begin{array}{l}
\underline{3}1, \underline{2}2, \underline{2}2, \underline{2}2 \overset{-1}{\underset{1}{\Rightarrow}} 21, \underline{12}, 12, 12 \rightarrow 21, 21, \underline{12}, 12 \rightarrow 21, 21, 21, \underline{12} \rightarrow 21, 21, 21, 21 \Rightarrow \\
10, 10, 10, 10 \overset{+1}{\Leftarrow} * \\
10, 01, 10, 10 \overset{+1}{\Leftarrow} \underline{0}0, \underline{-1}1, \underline{0}0, \underline{0}0 \leftarrow * \\
10, 10, 01, 10 \overset{+1}{\Leftarrow} \underline{00}, 00, -11, 00 \leftarrow 00, \underline{00}, -11, 00 \leftarrow * \\
10, 10, 10, 01 \overset{+1}{\Leftarrow} \underline{0}0, \underline{0}0, \underline{0}0, \underline{-1}1 \leftarrow 00, \underline{00}, 00, -11 \leftarrow 00, 00, \underline{00}, -11 \leftarrow * \\
2(20, 11, 11, 11) \overset{+1}{\Leftarrow} 2(\underline{1}0, \underline{0}1, \underline{0}1, \underline{0}1) \leftarrow 2(10, \underline{10}, 01, 01) \leftarrow 2(10, 10, \underline{10}, 01) \leftarrow 2(10, 10, 10, \underline{10}) \\
21, 21, 21, 21 \overset{+2}{\Leftarrow} \underline{0}1, \underline{0}1, \underline{0}1, \underline{0}1 \leftarrow 01, \underline{01}, 01, 01 \leftarrow 01, 01, \underline{10}, 01 \leftarrow 01, 10, 10, \underline{10} \\
10, 10, 01, 01 \overset{+0}{\Leftarrow} \underline{1}0, \underline{1}0, \underline{0}1, \underline{0}1 \leftarrow 10, \underline{01}, 01, 01 \leftarrow 10, 01, \underline{10}, 01 \leftarrow 10, 01, 10, \underline{10} \\
10, 01, 10, 01 \overset{+0}{\Leftarrow} \underline{1}0, \underline{0}1, \underline{1}0, \underline{0}1 \leftarrow 10, \underline{10}, 10, 01 \leftarrow 10, 10, \underline{01}, 01 \leftarrow 10, 10, 01, \underline{10} \\
10, 01, 01, 10 \overset{+0}{\Leftarrow} \underline{1}0, \underline{0}1, \underline{0}1, \underline{1}0 \leftarrow 10, \underline{10}, 01, 10 \leftarrow 10, 01, \underline{10}, 10 \leftarrow 10, 10, 10, \underline{01}
\end{array}$$

Kato [5, 6] gives the equation

$$\begin{cases} x(1-x)\dfrac{\partial^2 u}{\partial x^2} + \bigl(\gamma - (\alpha+\beta+1)x\bigr)\dfrac{\partial u}{\partial x} - \alpha\beta u + \epsilon\dfrac{y-1}{x-y}\Bigl(x\dfrac{\partial u}{\partial x} - y\dfrac{\partial u}{\partial y}\Bigr) = 0, \\ y(1-y)\dfrac{\partial^2 u}{\partial y^2} + \bigl(\gamma - (\alpha+\beta+1)y\bigr)\dfrac{\partial u}{\partial y} - \alpha\beta u + \epsilon\dfrac{x-1}{y-x}\Bigl(y\dfrac{\partial u}{\partial y} - x\dfrac{\partial u}{\partial x}\Bigr) = 0 \end{cases}$$

satisfied by $u(x,y) = F_4\bigl(\alpha, \beta; \gamma, \gamma'; xy, (1-x)(1-y)\bigr)$ with

$$F_4(\alpha, \beta; \gamma, \gamma'; x, y) = \sum_{m=n=0}^{\infty} \frac{(\alpha)_{m+n}\,(\beta)_{m+n}}{(\gamma)_m\,(\gamma')_n\,m!\,n!} x^m y^n,$$

$$\epsilon = \gamma + \gamma' - \alpha - \beta - 1. \tag{56}$$

We note that when $\epsilon = 0$, the above equation has the solution

$$u(\alpha, \beta, \gamma; x, y) = F(\alpha, \beta, \gamma; x) \cdot F(\alpha, \beta, \gamma; y). \tag{57}$$

Hence the monodromy group defined by the space of solutions is irreducible when $\epsilon = 0$ (generally $2\epsilon \in \mathbb{Z}$ by Lemma 2.1 or by the Riemann scheme) and α, β and γ are generic. The corresponding Pfaffian system (45) is given by

$$A_1 = \begin{pmatrix} 0 & 1 & 0 & 0 \\ 0 & 1-\gamma & 0 & 0 \\ 0 & \epsilon & 0 & 1 \\ 0 & 0 & 0 & 1-\gamma \end{pmatrix}, \ A_4 = \begin{pmatrix} 0 & 0 & 1 & 0 \\ 0 & 0 & \epsilon & 1 \\ 0 & 0 & 1-\gamma & 0 \\ 0 & 0 & 0 & 1-\gamma \end{pmatrix}, \ A_2 = \begin{pmatrix} 0 & 0 & 0 & 0 \\ 0 & \epsilon & -\epsilon & 0 \\ 0 & -\epsilon & \epsilon & 0 \\ 0 & 0 & 0 & 0 \end{pmatrix},$$

$$A_3 = \begin{pmatrix} 0 & 0 & 0 & 0 \\ -\alpha\beta & -\gamma' & \epsilon & 0 \\ 0 & 0 & 0 & 0 \\ 0 & 0 & -(\alpha+\epsilon)(\beta+\epsilon) & -\gamma' \end{pmatrix}, \ A_5 = \begin{pmatrix} 0 & 0 & 0 & 0 \\ 0 & 0 & 0 & 0 \\ -\alpha\beta & \epsilon & -\gamma' & 0 \\ 0 & -(\alpha+\epsilon)(\beta+\epsilon) & 0 & -\gamma' \end{pmatrix},$$

with

$$a = 2\epsilon, \quad b = 1-\gamma, \quad c = -\gamma', \quad d = \alpha, \quad e = \beta.$$

Then the condition for the irreducibility is

$$\{d,\ e,\ b+d,\ b+e,\ c+d,\ c+e,\ b+c+d,\ b+c+e\} \cap \mathbb{Z} = \emptyset \tag{58}$$

or equivalently

$$\{\alpha,\ \beta,\ \alpha-\gamma,\ \beta-\gamma,\ \alpha-\gamma',\ \beta-\gamma',\ \alpha-\gamma-\gamma',\ \beta-\gamma-\gamma'\} \cap \mathbb{Z} = \emptyset. \tag{59}$$

Note that Theorem 1.5 says that under the condition (58) the differential equation $\frac{du}{dx} = \left(\frac{A_1}{x} + \frac{A_2}{x-y} + \frac{A_3}{x-1}\right)u$ is irreducible if and only if $a \notin \mathbb{Z}$.

Rank 5 with Six Parameters

I_5 $41,32,311,311$, J_5 $41,41,221,221$: $(1^6 \cdot 2^4)$

$41,41,221,221 \ \rightarrow\ F_1 : 21,21,021,021$

$41,32,311,311 \ \rightarrow\ H_2 : 11,02,011,011$

$41,41,221,221$

$= 10,10,001,010 \oplus 31,31,220,211$ (4)

$= 20,11,110,110 \oplus 21,30,111,111$ (2)

$= 2(10,10,100,100) \oplus 21,21,021,021$ (4)

	t_∞	t_0	t_y	t_1	t_x	idx
t_∞		311	311	221	221	−10
t_0	311		32	311	41	2
t_y	311	32		311	41	2
t_1	221	311	311		221	−10
t_x	221	41	41	221		2

Rank 5 with Five Parameters

$41, 32, 32, 221 : (1^7 \cdot 2^3)$
$\to\ 31, 22, 22, 220 \quad F_2 : 31, 22, 31, 121$
$F_1 : 21, 12, 12, 021$
$= 10, 10, 10, 001 \oplus 31, 22, 22, 220 \quad (1)$
$= 10, 01, 10, 010 \oplus 31, 31, 22, 211 \quad (4)$
$= 20, 11, 11, 101 \oplus 21, 21, 21, 120 \quad (2)$
$= 2(10, 10, 10, 100) \oplus 21, 12, 12, 021 \quad (2)$
$= 2(20, 11, 11, 110) \oplus 01, 10, 10, 001 \quad (1)$

	t_∞	t_0	t_y	t_1	t_x	idx
t_∞		311	32	2111	32	−6
t_0	311		32	2111	32	−6
t_y	32	32		221	41	2
t_1	2111	2111	221		221	−18
t_x	32	32	41	221		2

Rank 5 with Four Parameters

$P_{4,5}\ \ 32, 32, 32, 32 : (1^8 \cdot 2^2)$
$\to F_4 : 22, 22, 22, 31 \quad F_1 : 12, 12, 12, 12$
$= 10, 10, 10, 01 \oplus 22, 22, 22, 31 \quad (4)$
$= 21, 21, 21, 12 \oplus 11, 11, 11, 20 \quad (4)$
$= 2(10, 10, 10, 10) \oplus 12, 12, 12, 12 \quad (1)$
$= 2(21, 21, 21, 21) \oplus -(10, 10, 10, 10)\ (1)$

	t_∞	t_0	t_y	t_1	t_x	idx
t_∞		221	221	211	32	−10
t_0	221		221	221	32	−10
t_y	221	221		221	33	−10
t_1	221	221	221		32	−10
t_x	32	32	32	32		2

The last one is of Type 3 and then $10, 10, 10, 10$ appears in the preceding decomposition [cf. Proposition 2.5 iii)].

Acknowledgements The author thanks the referee for several improvements of the paper, which makes this paper more readable.
The author is supported by Grant-in-Aid for Scientific Researches (B), No. 25287017, Japan Society of Promotion of Science.

References

1. W. Crawley-Boevey, On matrices in prescribed conjugacy classes with no common invariant subspaces and sum zero. Duke Math. J. **118**, 339–352 (2003)
2. M. Dettweiler, S. Reiter, An algorithm of Katz and its applications to the inverse Galois problems. J. Symb. Comput. **30**, 761–798 (2000)
3. Y. Haraoka, Middle convolution for completely integrable systems with logarithmic singularities along hyperplane arrangements. Adv. Stud. Pure Math. **62**, 109–136 (2012)
4. Y. Haraoka, T. Kikukawa, Rigidity of monodromies for Appell's hypergeometric functions. Opuscula Math. **5**, 567–594 (2015)
5. M. Kato, A Pfaffian system of Appell's F4. Bull. College Educ. Univ. Ryukyus **33**, 331–334 (1988)
6. M. Kato, Connection formulas for Appell's system F4 and some applications. Funkcial, Ekvac. **38**, 243–266 (1995)

7. H.M. Katz, *Rigid Local Systems*. Annals of Mathematics Studies vol. 139 (Princeton University Press, Princeton, 1995)
8. T. Oshima, *Fractional Calculus of Weyl Algebra and Fuchsian Differential Equations*. MSJ Memoirs, vol. 28 (Mathematical Society of Japan, Tokyo, 2012)
9. T. Oshima, *An Elementary Approach to the Gauss Hypergeometric Function*. Josai Mathematical Monographs, vol. 6 (2013), pp. 2–23
10. T. Oshima, *Transformations of KZ type equations*. RIMS Kôkyûroku Bessatsu **B61**, 141–162 (2017)
11. T. Oshima, `os_muldif.rr`. `ftp://akagi.ms.u-tokyo.ac.jp/pub/math/muldif`, 2008–2016

Parametric Borel Summability of Partial Differential Equations of Irregular Singular Type

Masafumi Yoshino

Abstract The parametric Borel summability of formal series solutions of a system of ordinary differential equations is fairly well understood due to Balser, Schäfke, Mozo-Fernández and so on. In this paper, we shall study the parametric Borel summability of formal solutions of a semilinear system of partial differential equations of irregular singular type.

Keywords Irregular singular • Parametric Borel summability • Semilinear system

Mathematics Subject Classification (2000). Primary 35C10; Secondary 45E10, 35Q15

1 Introduction

In [1, 2] and [3], the parametric Borel summability of a formal power series solution of a system of ordinary differential equations was extensively studied. In our preceding paper [4], we showed the parametric Borel summability for a semilinear system of partial differential equations of Fuchsian type. In this chapter, we extend the result in [4] to the equations containing irregular singular and Fuchsian variables.

We first construct two different formal power series solutions expanded in the independent variable x or in the perturbative parameter η. We then observe that they coincide with each other as a formal power series of x and η. After having shown the Gevrey estimate of the formal power series, we discuss the parametric Borel summability. When proving the summability, we make use of the method of characteristics in order to show the analytic continuation of the formal Borel transform of the formal series. Finally we remark that no Diophantine phenomena enter in the present situation because we assume Poincaré condition in our argument.

M. Yoshino (✉)
Department of Mathematics, Hiroshima University, 1-3-1 Kagamiyama, Higashi-Hiroshima, Hiroshima 739-8526, Japan
e-mail: yoshinom@hiroshima-u.ac.jp

G. Filipuk et al. (eds.), *Analytic, Algebraic and Geometric Aspects of Differential Equations*, Trends in Mathematics, DOI 10.1007/978-3-319-52842-7_15

This paper is organized as follows. In Sect. 2, we state the main theorem, Theorem 2.1. In Sect. 3, we construct two different types of formal series solutions and show the Gevrey estimate. In Sect. 4, we prepare lemmas of the convolution. In Sect. 5, we give the proof of our main result, Theorem 2.1 after the preparation of lemmas.

2 Statement of Results

Let $x = (x_1, \dots, x_n) \in \mathbb{C}^n$, $n \geq 1$ be the variable in $\mathbb{C}^n$. For an integer m with $1 \leq m < n$, let $s_j \in \mathbb{Z}$ $(1 \leq j \leq n)$ be integers such that $s_j \geq 2$ if $1 \leq j \leq m$, and $s_j = 1$ if $j > m$. Let $\lambda_j \in \mathbb{C}$, $\lambda_j \neq 0$ $(j = 1, 2, \dots, n)$ and $N \geq 1$ be an integer.

We shall study the solution u of the equation

$$\eta \mathcal{L} u = f(x, u, \eta), \tag{1}$$

such that $u(0) = 0$, where $\eta \in \mathbb{C}$ is a complex parameter,

$$\mathcal{L} := \sum_{j=1}^{n} \lambda_j x_j^{s_j} \frac{\partial}{\partial x_j} \tag{2}$$

and $u = (u^{(1)}, \dots, u^{(N)})$. Here $f(x, u, \eta) = (f^{(1)}(x, u, \eta), \dots, f^{(N)}(x, u, \eta))$ is a holomorphic vector function in some neighborhood of the origin of $x \in \mathbb{C}^n$, $\eta \in \mathbb{C}$ and $u \in \mathbb{C}^N$.

We write $x = (x', x'')$, $x' = (x_1, \dots, x_m)$, $x'' = (x_{m+1}, \dots, x_n)$, $\alpha = (\alpha', \alpha'')$, $\alpha' = (\alpha_1, \dots, \alpha_m)$, $\alpha'' = (\alpha_{m+1}, \dots, \alpha_n)$. We assume

$$f(x, 0, \eta)|_{x''=0} \equiv 0, \text{ for all } x', \eta, \quad \det(\nabla_u f(0, 0, 0)) \neq 0 \tag{3}$$

where $\nabla_u f(0, 0, 0)$ denotes the Jacobi matrix of $f(x, u, \eta)$ with respect to u at the point $x = 0, u = 0, \eta = 0$. Moreover, we assume

$$\nabla_u f(0, 0, \eta) \equiv \nabla_u f(0, 0, 0), \quad \text{for all } \eta, \tag{4}$$

$$\nabla_u f(x, 0, 0) \text{ is a diagonal matrix.} \tag{5}$$

Hence one may assume that $\nabla_u f(0, 0, 0)$ is the diagonal matrix with diagonal components given by $\mu_1, \dots, \mu_N$ in this order. Moreover, we assume

$$\lambda_j > 0, \ \Re \mu_k > 0, \quad (j = m + 1, \dots, n, \ k = 1, \dots, N), \tag{6}$$

where $\Re \mu_k$ means the real part of μ_k.

First we construct a formal power series solution of (1) in the space variable. We consider

$$u^{(k)}(x,\eta) = \sum_{\alpha\in\mathbb{Z}^n_+,\alpha\neq 0} u^{(k)}_\alpha(\eta)x^\alpha = \sum_{\alpha''\in\mathbb{Z}^{n-m}_+,\alpha\neq 0} u^{(k)}_{\alpha''}(x',\eta)(x'')^{\alpha''}, \quad k=1,2,\ldots,N. \tag{7}$$

If we assume the nonresonance condition

$$\eta\sum_{j=m+1}^{n}\lambda_j\alpha_j - \mu_k \neq 0, \qquad \text{for all } \alpha=(\alpha_{m+1},\ldots,\alpha_n)\in\mathbb{Z}^{n-m}_+, 1\le k\le N, \tag{8}$$

then we can determine the formal series solution $u(x,\eta)$. Moreover, every coefficient $u^{(k)}_\alpha(\eta)$ in $u^{(k)}(x,\eta)$ is analytic at $\eta=0$. By inserting the expansion of $u^{(k)}_\alpha(\eta)$ into (7) we obtain the formal power series expansion of $u(x,\eta)$ in x and η. For simplicity we denote it with $u(x,\eta)$ if there is no fear of confusion.

Next, we construct the solution $v(x,\eta)$ of (1) in the form

$$v(x,\eta) = \sum_{\nu=0}^{\infty}\eta^\nu v_\nu(x) = v_0(x)+\eta v_1(x)+\cdots, \tag{9}$$

where the series is a formal power series of η with coefficients $v_\nu(x)$'s being holomorphic vector functions of x in some open set independent of ν. We set $v_\nu(x)\equiv v_\nu = (v_\nu^{(1)},\ldots,v_\nu^{(N)})$. We denote by Ω_0 the neighborhood of the origin on which every coefficient $v_\nu(x)$ is defined. (cf. Sect. 3). Expand $v_\nu(x)$ into the power series of x and insert it into $v(x,\eta)$. Then we obtain the formal power series expansion of $v(x,\eta)$ in x and η. For simplicity we denote the formal expansion with the same $v(x,\eta)$. By Lemma 3.2 which follows we see that $u(x,\eta)=v(x,\eta)$ as a formal power series of x and η.

Therefore, we shall consider the Borel summability of $v(x,\eta)$ with respect to η. In order to state our results we recall some definitions from the summability theory. (cf. [1]). The formal Borel transform of $v(x,\eta)$ is defined by

$$B(v)(x,y) := \sum_{\nu=0}^{\infty} v_\nu(x)y^\nu/\Gamma(\nu+1), \tag{10}$$

where $\Gamma(z)$ is the Gamma function. For an opening $\theta>0$ and the bisecting direction ξ, define the sector $S_{\theta,\xi}$ by

$$S_{\theta,\xi} = \{z\in\mathbb{C};\ |\arg z-\xi|<\theta/2\}. \tag{11}$$

We say that $v(x,\eta)$ is 1-summable in the direction ξ with respect to η if $B(v)(x,y)$ converges in some neighborhood of $(x,y)=(0,0)$, and there exist a neighborhood U of $x=0$ and a $\theta>0$ such that $B(v)(x,y)$ can be analytically continued to the set

$\{(x,y) \in U \times S_{\theta,\xi}\}$ and of exponential growth of order 1 with respect to y in $S_{\theta,\xi}$. For the sake of simplicity we denote the analytic continuation with the same notation $B(v)(x,y)$. The Borel sum $V(x,\eta)$ of $v(x,\eta)$ is, then, given by the Laplace transform

$$V(x,\eta) := \eta^{-1} \int_0^{\infty e^{i\xi}} e^{-y\eta^{-1}} B(v)(x,y)dy. \tag{12}$$

Let C_0 be the convex closed positive cone with vertex at the origin containing μ_k $(k = 1,\ldots,N)$. Write

$$C_0 = \{z \in \mathbb{C}; -\theta_2 \le \arg z \le \theta_1\} \tag{13}$$

for some $0 \le \theta_1 < \pi/2$ and $0 \le \theta_2 < \pi/2$. Define $\xi = \pi + \frac{\theta_1-\theta_2}{2}$ and $\theta = \pi - \theta_1 - \theta_2$. We observe that $S_{\pi+\theta,\xi}$ is equal to $\mathbb{C} \setminus C_0$. Then we have

Theorem 2.1 *Suppose* (3)–(6). *Then there exists a neighborhood U of $x = 0$ such that $v(x,\eta)$ is 1-summable in the direction* $\arg\eta$ *with $\eta \in S_{\theta,\xi}$ when $x \in U$. Moreover, there exists a neighborhood W of $\eta = 0$ in $\mathbb{C}$ such that $V(x,\eta)$ is holomorphic and satisfies* (1) *when $(x,\eta) \in U \times (W \cap (\mathbb{C} \setminus C_0))$.*

Remark 2.2 We briefly state the similar results in the case $s_k \ge 2$ for $k = 1,2,\ldots,n$ in (2). The following theorem can be proved by the similar argument as in the proof of Theorem 2.1.

Let $\omega = (\omega_1,\ldots,\omega_n) \in \mathbb{R}^n$ be given and assume $0 < \omega_0 < \pi$. We define the multi-sector $\Sigma_{\omega_0,\omega}$ by $\Sigma_{\omega_0,\omega} := \prod_{j=1}^n S_{\omega_0,\omega_j}$, where S_{ω_0,ω_j} is given by (11). We assume (6) except for the condition $\lambda_j > 0$. Let ξ and θ be the ones given in Theorem 2.1. Then we have

Theorem 2.3 *Suppose* (3)–(6). *Then there exist a neighborhood U of $x = 0$ and $\omega_0 > 0$ such that $v(x,\eta)$ is 1-summable in the direction* $\arg\eta$ *with $\eta \in S_{\theta,\xi}$ when $x \in U \cap \Sigma_{\omega_0,\omega}$. Moreover, there exists a neighborhood W of $\eta = 0$ in $\mathbb{C}$ such that $V(x,\eta)$ is holomorphic and it satisfies* (1) *when $(x,\eta) \in (U \cap \Sigma_{\omega_0,\omega}) \times (W \cap (\mathbb{C} \setminus C_0))$.*

3 Formal Power Series Solution and Gevrey Estimates

As we stated in Sect. 2 there are two formal solutions $u(x,\eta)$ and $v(x,\eta)$ of (1). In this section we study their properties and relations.

Expansion in the space variable. By substituting (7) into (1) we obtain recurrence relations for $u_{\alpha''}^{(k)}$'s. First we consider the relation for $u_0^{(k)}$ with $\alpha'' = 0$

$$\eta \sum_{j=1}^m \lambda_j x_j^{s_j} (\partial/\partial x_j) u_0^{(k)} - \mu_k u_0^{(k)} = \tilde{f}^{(k)}(x', u_0, \eta) - \mu_k u_0^{(k)}, \tag{14}$$

where $\tilde{f}^{(k)} = f^{(k)}|_{x''=0}$ and $f^{(k)}$ is the k-th component of f. We can see that the solution of (14) such that $u_0^{(k)}(0,\eta) = 0$ vanishes identically. Indeed, we have $\tilde{f}^{(k)}(x',0,\eta) \equiv 0$ by (3). Recalling that $\mu_k = \nabla_u f^{(k)}(0,0,\eta)$ we have

$$\tilde{f}^{(k)}(x',u_0,\eta) - \mu_k u_0^{(k)} = \left(\nabla_u \tilde{f}^{(k)}(x',0,\eta) - \nabla_u f^{(k)}(0,0,\eta)\right) u_0^{(k)} + O(|u_0|^2).$$

We expand $u_0^{(k)}$ in the power series of x'. In the right-hand side of (14) there appears no term containing the linear term of the expansion of $u_0^{(k)}$. On the other hand, the first term of the right-hand side of (14) consists of higher order terms because $s_j \geq 2$. Hence the linear term in the expansion of u_0 vanishes because $\mu_k \neq 0$ by (8). In the same way, by calculating higher order terms recurrently one can show that $u_0 = 0$.

Next, we consider the recurrence relation for $u_{\alpha''}^{(k)}$, $(\alpha'' \neq 0)$

$$\begin{aligned} &\eta \sum_{j=1}^{m} \lambda_j x_j^{s_j} (\partial/\partial x_j) u_{\alpha''}^{(k)} + \left(\eta \sum_{j=m+1}^{n} \lambda_j \alpha_j - \mu_k\right) u_{\alpha''}^{(k)} \\ &= R(x, u_{\gamma''}^{(\nu)}; |\gamma''| < |\alpha''|, 1 \leq \nu \leq N), \quad 1 \leq k \leq N. \end{aligned} \tag{15}$$

Here the term in the right-hand side of (15) is a known term when we determine $u_{\alpha''}^{(k)}$ inductively in α''. Every coefficient $u_\alpha^{(k)}(\eta)$ in $u^{(k)}(x,\eta)$ is analytic at $\eta = 0$. By inserting the expansion of $u_\alpha^{(k)}(\eta)$ into (7) we obtain the formal power series expansion of $u(x,\eta)$ in x and η.

Expansion in the parameter η. We substitute (9) into (1) with $u = v$. The left-hand side is given by

$$\eta \mathcal{L} v = \sum_{\nu=0}^{\infty} \mathcal{L} v_\nu(x) \eta^{\nu+1}. \tag{16}$$

By the partial Taylor expansion of f with respect to η, $f(x,u,\eta) = \sum_{\ell=0}^{\infty} f_\ell(x,u)\eta^\ell$, the right-hand side of (1) is written as

$$\begin{aligned} f(x,v,\eta) &= \sum_{\ell=0}^{\infty} \eta^\ell f_\ell(x, v_0 + v_1\eta + v_2\eta^2 + \cdots) \\ &= f_0(x,v_0) + \eta(\nabla_u f_0)(x,v_0)v_1 + \eta f_1(x,v_0) + O(\eta^2). \end{aligned} \tag{17}$$

By comparing the coefficients of η, we obtain for $\eta^0 = 1$

$$f_0(x, v_0(x)) = 0 \tag{18}$$

and for η

$$\mathcal{L}v_0 - f_1(x, v_0) = (\nabla_u f_0)(x, v_0)v_1. \tag{19}$$

We can solve (18) with the condition $v_0(0) = 0$ on some neighborhood Ω_0 in terms of the implicit function theorem by the assumption (3) and $f_0(0,0) = f(0,0,0)$. Next we solve v_1 from (19) on Ω_0 where we may assume $\det(\nabla_u f_0(x, v_0(x))) \neq 0$ on Ω_0 since $\det(\nabla_u f(0,0,0)) \neq 0$ by shrinking Ω_0 if necessary.

In order to determine $v_\nu(x)$ ($\nu \geq 2$) we compare the coefficients of η^ν in both sides of (1). Indeed, we differentiate (17) $(\nu - 1)$-times with respect to η and we put $\eta = 0$. Then we obtain

$$\mathcal{L}v_{\nu-1} = (\nabla_u f_0)(x, v_0)v_\nu + (\text{terms consisting of } v_k, k < \nu). \tag{20}$$

We observe that the second term in the right-hand side appear from the terms $\eta^\ell f_\ell(x, v_0 + v_1\eta + v_2\eta^2 + \cdots)$ for $\ell \geq 0$ in (17), which are products of terms of the form $v_{i_j}\eta^{i_j}$ and η^ℓ such that

$$i_1 + i_2 + \cdots + i_k + \ell = \nu, \ i_1 \geq 0, \ldots, i_k \geq 0, i_j \neq 0$$

for some $k \geq 2$ and $j \leq k$. It follows that all terms in the second term have the form v_j for some $j < \nu$. Therefore one can write (20) in the following way

$$\nabla_u f_0(x, v_0)v_\nu = H_\nu(x, v_0, v_1, \ldots, v_{\nu-1}) \quad \text{for all } \nu \geq 2.$$

Since $\det(\nabla_u f(x, v_0(x), 0)) \neq 0$ on Ω_0, one can inductively determine v_ν. The next theorem gives the existence of a formal solution.

Proposition 3.1 *Assume* (3). *Then there exists a neighborhood of the origin* $x = 0$ *such that every coefficient of* (9) *is uniquely determined as a holomorphic function on* Ω_0.

We shall study the relation between $u(x, \eta)$ and $v(x, \eta)$.

Lemma 3.2 *Assume that* (8) *holds. Then we have* $u(x, \eta) = v(x, \eta)$ *as a formal power series of* x *and* η.

Proof Because every $u_\alpha(\eta)$ is analytic at $\eta = 0$, the limit $\lim_{\eta\to 0} u_\alpha(\eta)$ exists for every α. We define $\lim_{\eta\to 0} u(x, \eta) := \sum_\alpha \lim_{\eta\to 0} u_\alpha(\eta)x^\alpha$. The right-hand side is a well-defined formal power series of x. Set $u_0(x) := \lim_{\eta\to 0} u(x, \eta)$. By letting $\eta \to 0$ in (1) we have $f(x, u_0, 0) = 0$, which is identical with (18). Because the solution of (18) such that $v_0(0) = 0$ is unique, it follows that $u_0 = v_0$. Hence we have $u(x, 0) = v(x, 0)$. Next we define $u_1 = \lim_{\eta\to 0}(\partial/\partial\eta)u(x, \eta)|_{\eta=0}$. The right-hand side exists as a formal power series. Differentiate (1) by η and set $\eta = 0$. One can verify that u_1 satisfies (19) with $v_1 = u_1$. By the uniqueness of the analytic solution of (19) we see that $u_1 = v_1$. In the same way, by differentiating (1) appropriately many times in η and by setting $\eta = 0$ we have the assertion of the lemma. □

Gevrey estimate of formal series solutions. Let σ_j $(j = 0, 1, 2, \ldots, n)$ be nonnegative numbers. Set $\sigma := (\sigma_0, \sigma_1, \ldots, \sigma_n)$. We define the formal Borel transform $B_\sigma(v)(\zeta, y)$ of a power series $v := \sum_{\nu,\alpha\geq 0,\alpha\neq 0} v_{\nu,\alpha}\eta^\nu x^\alpha$ by

$$\sum_{\nu,\alpha\geq 0,\alpha\neq 0} v_{\nu,\alpha}\frac{y^\nu \zeta^\alpha}{\Gamma(1+\sigma_0\nu+\sigma_1\alpha_1+\cdots+\sigma_n\alpha_n)},$$

where Γ denotes the Gamma function and y and ζ are dual variables of η and x, respectively. Similarly we define the partial Borel transform by putting some σ_j to be zero. Then we have

Proposition 3.3 *Let $\sigma_0 = 1$ and $\sigma_j = 0$ for $j = 1, 2, \ldots, n$. Then, there exist a neighborhood U of $x = 0$ and a neighborhood W of $y = 0$ in $\mathbb{C}$ such that the partial Borel transform $B_\sigma(v)(x, y)$ converges in $(x, y) \in U \times W$.*

This proposition is proved by the same argument as the proof of Proposition 3.2 in [4]. As for the estimate of the series containing space variables we have

Proposition 3.4 *Suppose that $s_j \geq 2$ for $j = 1, 2, \ldots, n$. Let $\sigma_0 = 1$, $\sigma_j = 1/(s_j-1)$ for $j = 1, 2, \ldots, n$. Then, there exist a neighborhood U of $\zeta = 0$ and a neighborhood W of $y = 0$ in $\mathbb{C}$ such that $B_\sigma(v)(\zeta, y)$ converges in $(\zeta, y) \in U \times W$.*

4 Convolution Estimate

For $0 < \theta < \pi$ and $r_0 > 0$ let Ω be an open set containing the sector $S_{\theta,\pi}$ and the disk $\{|z| < r_0\}$ such that if $z \in \Omega$, then $\Re z < r_0$, $z - t \in \Omega$ for every $t \geq 0$ and $z\sigma \in \Omega$ for every $0 \leq \sigma \leq 1$. For $c > 0$, we define the space $\mathcal{H}_c(\Omega)$ as the set of those $h \in H(\Omega)$ such that there exists $K \geq 0$ for which

$$|h(z)| \leq Ke^{-c\Re z}(1+|z|)^{-2} \quad \text{for all } z \in \Omega, \tag{21}$$

where $H(\Omega)$ is the set of holomorphic functions in Ω. Obviously, $\mathcal{H}_c(\Omega)$ is the Banach space with the norm

$$\|h\|_{\Omega,c} := \sup_{z\in\Omega}|h(z)|(1+|z|)^2 e^{c\Re z}. \tag{22}$$

The convolution $f * g$ $(f, g \in \mathcal{H}_c(\Omega))$ is defined by

$$(f*g)(z) := \frac{d}{dz}\int_0^z f(z-t)g(t)dt = \frac{d}{dz}\int_0^z f(t)g(z-t)dt. \tag{23}$$

Write $f'(z) = (df/dz)(z)$. Then we have the next proposition. (cf. Proposition 4.2 of [4]).

Proposition 4.1 *For every $f, g \in \mathcal{H}_c(\Omega)$ such that $f(0) = g(0) = 0$ and $f', g' \in \mathcal{H}_c(\Omega)$ we have $f * g \in \mathcal{H}_c(\Omega)$ with the estimate*

$$\|f * g\|_{\Omega,c} \leq 8\|f'\|_{\Omega,c}\|g\|_{\Omega,c}, \quad \|f * g\|_{\Omega,c} \leq 8\|f\|_{\Omega,c}\|g'\|_{\Omega,c}. \tag{24}$$

For $f \in \mathcal{H}_c(\Omega)$ we define $D_z^{-1}f(z) := \int_0^z f(t)dt$, where the integration is taken on the straight line connecting the origin and z. Then we have

Lemma 4.2 *D_z^{-1} is a continuous linear operator defined on $\mathcal{H}_c(\Omega)$ into $\mathcal{H}_c(\Omega)$.*

Proof Let $f \in \mathcal{H}_c(\Omega)$. By definition we have

$$\begin{aligned} e^{c\Re z}(1+|z|)^2|D_z^{-1}f(z)| &\leq \int_0^z |f(t)|e^{c\Re z}(1+|z|)^2|dt| \\ &\leq \|f\|_c \int_0^z e^{-c\Re(t-z)}\frac{(1+|z|)^2}{(1+|t|)^2}|dt|. \end{aligned} \tag{25}$$

If $\Re t \geq (1-\epsilon)\Re z$, then we have $\Re(t-z) \geq -\epsilon\Re z$. Hence we have $e^{-c\Re(t-z)}(1+|z|)^2 \leq e^{c\epsilon\Re z}(1+|z|)^2$. Because $\Re z < r_0$ for $z \in \Omega$ we see that the right-hand side is bounded by some constant K_0 independent of z and f. Moreover, the integral $\int_{\Re t\geq(1-\epsilon)\Re z}(1+|t|)^{-2}|dt|$ is bounded by $\int_{-\infty}^{\infty}(1+|t|)^{-2}|dt|$. Hence it is bounded by some constant K_1. It follows that

$$\int_{\Re t\geq(1-\epsilon)\Re z} e^{-c\Re(t-z)}\frac{(1+|z|)^2}{(1+|t|)^2}|dt| \leq K_0K_1. \tag{26}$$

Next we shall estimate

$$\int_{\Re z\leq\Re t\leq(1-\epsilon)\Re z} e^{-c\Re(t-z)}\frac{(1+|z|)^2}{(1+|t|)^2}|dt|. \tag{27}$$

Because (27) is bounded if z moves in a bounded set in Ω, we may assume that z lies in the left-half plane and $|z|$ is sufficiently large. It follows that there exists $0 < c_0 < 1$ such that $|\Re z| \geq c_0|z|$ for such z. Because $|\Re t| \geq (1-\epsilon)|\Re z|$, we have

$$(1+|t|)^2 \geq (1+|\Re t|)^2 \geq (1+(1-\epsilon)c_0|z|)^2 \geq (1-\epsilon)^2c_0^2(1+|z|)^2.$$

Therefore (27) is bounded by

$$(1-\epsilon)^{-2}c_0^{-2}\int_{\Re z\leq\Re t\leq(1-\epsilon)\Re z} e^{-c\Re(t-z)}|dt| = (1-\epsilon)^{-2}c_0^{-2}\int_{0\leq\Re s\leq-\epsilon\Re z} e^{-c\Re s}|ds|,$$

where we set $s = t-z$. Therefore, by (26) there exists $K_3 > 0$ independent of f such that the right-hand side of (25) is bounded by $K_3\|f\|_c$. □

5 Proof of Theorem 2.1

First we define a function space. Let D and Ω be the open connected set in a neighborhood of the origin of $\mathbb{C}^n$ and the set given in (21), respectively. Let $H(D, \Omega)$ be the set of holomorphic functions in $(x, y) \in D \times \Omega$. Then we define $\mathcal{H}_c(D, \Omega)$ as the set of those $h \equiv h(x, y) \in H(D, \Omega)$ such that there exists $K_0 \geq 0$ for which

$$\sup_{x \in D} |h(x, y)| \leq K_0 e^{-c\Re y}(1 + |y|)^{-2} \quad \text{for all } y \in \Omega. \tag{28}$$

The space $\mathcal{H}_c(D, \Omega)$ is a Banach space with the norm $\|h\|_c = \inf K_0$ where K_0 is given in (28).

Proof of Theorem 2.1. We first show the summability of $v(x, \eta)$ in the direction $\arg \eta = \pi$ when $x \in U$, where U is given in Proposition 3.3. One may assume $\lambda_n = 1$ without loss of generality by dividing the equation with $\lambda_n \neq 0$. In terms of (1) with u replaced by $v_0 + u$, $f_0(x, v_0) = 0$ and

$$f_\ell(x, v) = f_\ell(x, v_0) + \nabla_u f_\ell(x, v_0) \cdot u + \sum_{|\beta| \geq 2} r_{\beta,\ell}(x, v_0) u^\beta, \quad \ell = 0, 1, 2, \ldots$$

we obtain

$$\mathcal{L}u = -\mathcal{L}v_0 + \eta^{-1} \nabla_u f_0(x, v_0) u + \eta^{-1} \sum_{|\beta| \geq 2} r_{\beta,0}(x, v_0) u^\beta \tag{29}$$

$$+ \sum_{\ell \geq 1} \eta^{\ell-1} \left(f_\ell(x, v_0) + \nabla_u f_\ell(x, v_0) \cdot u + \sum_{|\beta| \geq 2} r_{\beta,\ell}(x, v_0) u^\beta \right).$$

Let $\hat{u}(y) := \mathcal{B}(u)$ be the Borel transform of u with respect to η, where y is the dual variable of η. By the Borel transform of (29) and by recalling that η^{-1} corresponds to $\partial/\partial y$, we obtain

$$\mathcal{L}\hat{u} = -\mathcal{L}v_0 + \nabla_u f_0(x, v_0)(\partial \hat{u}/\partial y) + (\partial/\partial y) \sum_{|\beta| \geq 2} r_{\beta,0}(x, v_0)(\hat{u})^{*\beta} \tag{30}$$

$$+ \sum_{\ell \geq 1} D_y^{1-\ell} \left(f_\ell(x, v_0) + \nabla_u f_\ell(x, v_0) \cdot \hat{u} + \sum_{|\beta| \geq 2} r_{\beta,\ell}(x, v_0)(\hat{u})^{*\beta} \right),$$

where $(\hat{u})^{*\beta} = (\hat{u}_1)^{*\beta_1} \cdots (\hat{u}_N)^{*\beta_N}$, $\beta = (\beta_1, \ldots, \beta_N)$, and $(\hat{u}_j)^{*\beta_j}$ is the β_j-convolution product, $(\hat{u}_j)^{*\beta_j} = \hat{u}_j * \cdots * \hat{u}_j$. Here $D_y^{1-\ell} = (D_y^{-1})^{\ell-1}$ and D_y^{-1} is the integration, $D_y^{-1} g(y) = \int_0^y g(t)dt$.

Let v be the formal solution given by Proposition 3.1 and let $B(v)$ be the formal Borel transform of v. Define $\hat{u}(x, y) := B(v) - v_0$. Then $\hat{u}(x, y)$ is analytic when $(x, y) \in U \times W$, and is the solution of (30) in some neighborhood of $y = 0$ such that $\hat{u}(x, 0) \equiv 0$ in x. We see that every solution of (30) being analytic at $y = 0$ and satisfying $\hat{u}(x, 0) \equiv 0$ is uniquely determined from the recurrence relation in the expansion of y because $\nabla_u f_0(x, v_0)$ is invertible. Therefore, if we can show the existence of the solution of (30) being analytic in $(x, y) \in U \times W$ and of exponential growth with respect to y in Ω, then we have the analytic continuation of the formal Borel transform of v which is of exponential growth in $y \in \Omega$. Hence we have the summability of v. Therefore, it is sufficient to prove the following theorem.

Theorem 5.1 *There exist $c > 0$, a neighborhood D of the origin $x = 0$, and Ω as in* (22) *such that* (30) *has a solution $\hat{u}$ in $\mathcal{H}_c(D, \Omega)$.*

For the proof of Theorem 5.1 we prepare six lemmas. Let $c > 0$, D and Ω be given. We may assume that D is contained in an open ball centered at the origin. In order to prove the solvability of (30) when x is in a neighborhood of the origin and $y \in \Omega$ we first solve

$$\mathcal{L}w - (\nabla_u f_0)(x, 0)(\partial w/\partial y) = g(x, y), \tag{31}$$

where $w = (w_1, \ldots, w_N)$ and $g = g(x, y) = (g_1, \ldots, g_N)$, $g_j \in \mathcal{H}_c(D, \Omega)$ is a given vector function and $f_0(x, u) = f(x, u, \eta)|_{\eta=0}$.

In view of the assumption (5), for every j, $1 \le j \le N$ we denote the j-th diagonal component of $(\nabla_u f_0)(x, 0)$ by $(\nabla_u f_0)_j(x, 0)$. We use the method of characteristics in order to solve (31). Namely, we consider

$$\frac{d\zeta}{\zeta} = \frac{dx_k}{\lambda_k x_k^{s_k}} = -\frac{dy}{(\nabla_u f_0)_j(x, 0)}, \quad k = 1, 2, \ldots, n-1. \tag{32}$$

Let $b \in \mathbb{C}$, $b \neq 0$ be sufficiently small and $y_0 \in \Omega$ be given. By integrating (32) we have

$$x_k = x_k(\zeta) = ((s_k - 1)(c_k - \lambda_k \log \zeta))^{1/(1-s_k)} \ (k = 1, 2, \cdots, m) \tag{33}$$

$$x_k = x_k(\zeta) = c_k \zeta^{\lambda_k} \ (k = m+1, \ldots, n-1), \ y = y_0 - \Phi_j(\zeta, b),$$

where

$$\Phi_j(\zeta, b) = \int_b^\zeta (\nabla_u f_0)_j(x_1(s), \cdots, x_{n-1}(s), s, 0) s^{-1} ds, \tag{34}$$

and the integral is taken along the non self-intersecting curve which does not encircle the origin. Then we make analytic continuation around the origin. Here $y_0 := y(b) \in \Omega$ is the initial value of $y = y(\zeta)$ at $\zeta = b$ and c_k's are chosen so that the initial point $x^{(0)} := (x_1(b), \ldots, x_{n-1}(b), b)$ lies in D and $\Re\,(c_k/\lambda_k) > 0$ if

$1 \le k \le m$. The last condition implies that $c_k - \lambda_k \log \zeta \neq 0$ when ζ is in a small neighborhood of the origin. Define $\Phi(\zeta, b) := (\Phi_1(\zeta, b), \ldots, \Phi_N(\zeta, b))$. Then we have

Lemma 5.2 *Let $\zeta_0 \in D \setminus \{0\}$. Then, for every j, $1 \le j \le N$ there exists a curve $\gamma_{\zeta_0, j}$ which passes ζ_0 and tends to the origin such that $\Im\, \Phi_j(\zeta, b) = \Im\, \Phi_j(\zeta_0, b)$ for every $\zeta \in \gamma_{\zeta_0, j}$, where $\Im\, \Phi_j$ denotes the imaginary part of Φ_j.*

Proof The condition $\Im\, \Phi_j(\zeta, b) = \Im\, \Phi_j(\zeta_0, b)$ is equivalent to $\Im\, \Phi_j(\zeta, \zeta_0) = 0$. We shall look for the curve $\gamma_{\zeta_0, j}$ satisfying the latter condition. By the assumption (5) there exists a neighborhood of the origin $x = 0$, Ω_0 such that $(\nabla_u f_0)_j(x, 0) = \mu_j + O(|x|)$ when $x \in \Omega_0$. Then we have

$$\Phi_j(\zeta, \zeta_0) = \int_{\zeta_0}^{\zeta} s^{-1} (\nabla_u f_0)_j(x, 0) ds = \mu_j \log(\zeta/\zeta_0) + R(\zeta), \tag{35}$$

where $x = x(s)$ is given by (33) and $R(\zeta)$ corresponds to the $O(|x|)$ term. Clearly, $R(\zeta)$ is a multi-valued analytic function of ζ in some neighborhood of the origin except for $\zeta = 0$. We shall study the behavior of $R(\zeta)$ modulo bounded functions when $\zeta \to 0$ because $R(\zeta)$ is not bounded in general. In view of (33) with $\zeta = s$, $R(\zeta)$ is given by the sum of constant times of the integral

$$\int_{\zeta_0}^{\zeta} s^{-1} \prod_{j=1}^{m} (c_j - \lambda_j \log s)^{\alpha_j/(1-s_j)} \prod_{j=m+1}^{n} s^{\alpha_j \lambda_j} ds \tag{36}$$

where $\lambda_n = 1$ and $\alpha_j \ge 0$ is an integer. If we assume that s tends to the origin from some sector with vertex at the origin, then we may assume that $\Re \log s \to -\infty$ and $\Im \log s$ is bounded as s tends to the origin. Because we take c_j such that $\Re\,(c_j/\lambda_j) > 0$, it follows that $c_j - \lambda_j \log s$ does not vanish and tends to infinity when $s \to 0$ from some sector. It follows from the assumption on λ_j that, if $\alpha_j > 0$ holds for some $j > m$, then the integral (36) is bounded when $\zeta \to 0$. Therefore, in order to estimate the growth of $R(\zeta)$ when $\zeta \to 0$ we assume that $\alpha_j = 0$ for $j > m$. If there exists $j, 1 \le j \le m$ such that $\rho \equiv \alpha_j/(s_j - 1) > 1$, then we consider the integral $\int_{\zeta_0}^{\zeta} (c_j - \lambda_j \log s)^{-\rho} s^{-1} ds$, because the other factors in the integrand of (36) are bounded by some constant. By setting $\sigma = \log s$, we may consider the integral $\int_{\log \zeta_0}^{\log \zeta} (c_j - \lambda_j \sigma)^{-\rho} d\sigma$. Clearly the term is bounded when $\zeta \to 0$ because $\rho > 1$. Therefore we find that unbounded terms in $R(\zeta)$ could appear from

$$\int_{\zeta_0}^{\zeta} s^{-1} \prod_{j=1}^{m} (c_j - \lambda_j \log s)^{\alpha_j/(1-s_j)} ds \tag{37}$$

where

$$0 \le \frac{\alpha_j}{s_j - 1} \le 1,\ \alpha_j \in \mathbb{Z}_+\ 1 \le j \le m,\ \alpha_1 + \cdots + \alpha_m \neq 0. \tag{38}$$

We see that there exist only finite number of pairs $(\alpha_1, \cdots, \alpha_m)$ which satisfy (38). Because we will use the estimate in the later argument, we shall calculate the integral $I \equiv \int_{\log \zeta_0}^{\log \zeta}(c_j - \lambda_j\sigma)^{-\rho}d\sigma$. By simple computations it is equal to

$$\frac{1}{\lambda_j(\rho-1)}\left((c_j - \lambda_j \log \zeta)^{1-\rho} - (c_j - \lambda_j \log \zeta_0)^{1-\rho}\right) = \frac{f(1) - f(0)}{\lambda_j(\rho-1)}, \tag{39}$$

where

$$f(t) := \left(c_j - \lambda_j(t \log \zeta + (1-t)\log \zeta_0)\right)^{1-\rho}. \tag{40}$$

The right-hand side of (39) is equal to

$$(\log \zeta - \log \zeta_0)\int_0^1 \left(c_j - \lambda_j(t \log \zeta + (1-t)\log \zeta_0)\right)^{-\rho} dt. \tag{41}$$

If ζ and ζ_0 move in some sector at the origin, then there exists $\rho > 0$ such that the growth of I when $\zeta \to 0$ from a sector is bounded by

$$\log(\zeta/\zeta_0)O((|\log \zeta| + |\log \zeta_0|)^{-\rho}). \tag{42}$$

Therefore (36) and $R(\zeta)$ also have the same estimate like (42).

We shall prove the existence of the curve $\gamma_{\zeta_0,j}$. By (35) the relation $\Im\, \Phi_j(\zeta, \zeta_0) = 0$ is written as

$$\Im(\mu_j \log \zeta) + \Im R(\zeta) - \Im(\mu_j \log \zeta_0) = 0. \tag{43}$$

By setting $\tilde{\xi} := \mu_j \log \zeta$, $\tilde{\xi}_0 := \mu_j \log \zeta_0$, and $\tilde{R}(\tilde{\xi}) := R(e^{\tilde{\xi}/\mu_j})$ (43) is written as

$$\Im\, \tilde{\xi} + \Im \tilde{R}(\tilde{\xi}) - \Im\, \tilde{\xi}_0 = 0. \tag{44}$$

Set $\tilde{\xi} = \tilde{x} + i\tilde{y}$ and $\tilde{\xi}_0 = \tilde{x}_0 + i\tilde{y}_0$. We shall determine $\tilde{y} = \tilde{y}(\tilde{x})$ from (44) such that $\tilde{y}(\tilde{x}_0) = \tilde{y}_0$. We remark that (44) holds if $\tilde{\xi} = \tilde{\xi}_0$. On the other hand, we have $\tilde{R}(\tilde{\xi}) = O(\tilde{\xi}^{1-\rho})$ by (39). Hence, if $|\tilde{x}_0|$ is sufficiently large and $\tilde{x} \le \tilde{x}_0 \le 0$, then the derivative with respect to $\tilde{y}$ of (44) does not vanish. Hence, by the implicit function theorem one can determine $\tilde{y} = \tilde{y}(\tilde{x})$ as an analytic function of $\tilde{x}$, if $|\tilde{x}_0|$ is sufficiently large and $\tilde{x} \le \tilde{x}_0 \le 0$. We denote the curve by $\tilde{\gamma}_{\tilde{\xi}_0,j}$.

We transform $\tilde{\gamma}_{\tilde{\xi}_0,j}$ in the $\tilde{\xi}$-space to $\gamma_{\zeta_0,j}$ in the ζ-space by expressing $\tilde{\xi}$ in terms of ζ . Set $\mu_j = \mu_{j,0} + i\mu_{j,1}$ with $\mu_{j,0} > 0$. Then we have

$$\tilde{\xi}\mu_j^{-1} = |\mu_j|^{-2}\left(\tilde{x}\mu_{j,0} + \tilde{y}\mu_{j,1} + i(\tilde{y}\mu_{j,0} - \tilde{x}\mu_{j,1})\right). \tag{45}$$

We recall that $\tilde{y}(\tilde{x})$ is a bounded function of $\tilde{x}$ and its derivative tends to zero as $\tilde{x} \to -\infty$ by definition. Because $\mu_{j,0} > 0$ and $\tilde{x} \to -\infty$, we see that $\zeta = \exp(\tilde{\zeta}/\mu_j) \to 0$. □

Lemma 5.3 *Let $c \neq 0$ and $\zeta_0 \neq 0$ be in some neighborhood of the origin. Then, for every j, $1 \leq j \leq N$, $\Re\, \Phi_j(\zeta, c)$ is monotone decreasing when ζ approaches to the origin along the curve $\gamma_{\zeta_0 j}$.*

Proof Because the curves $\gamma_{\zeta_0 j}$ and $\tilde{\gamma}_{\tilde{\zeta}_0 j}$ have a one-to-one correspondence by the relation $\zeta = \exp(\tilde{\zeta}/\mu_j)$ it is sufficient to show that $\Re\, \Phi_j(\zeta, \zeta_0)$ is monotone decreasing when $\tilde{\zeta}$ tends to $-\infty$ along $\tilde{\gamma}_{\tilde{\zeta}_0 j}$. In terms of (35) we have

$$\Re\, \Phi_j(\zeta, \zeta_0) + \Re\, (\mu_j \log \zeta_0) = \Re\, \tilde{\zeta} + \Re\, \tilde{R}(\tilde{\zeta}) = \tilde{x} + \Re\, \tilde{R}(\tilde{\zeta}). \tag{46}$$

By virtue of (39) we have $\tilde{R}(\tilde{\zeta}) = R(\zeta) = O(\zeta^{1-\rho})$ for some $\rho > 0$. Thus, with $\tilde{\zeta} = \tilde{x} + i\tilde{y}$ and $\tilde{y} = \tilde{y}(\tilde{x})$, we see that $\left|(d/d\tilde{x})\Re\, \tilde{R}(\tilde{\zeta})\right|$ is arbitrarily small if $\tilde{x} \leq \tilde{x}_0 \leq 0$ for sufficiently small $\tilde{x}_0$. Therefore we have $(d/d\tilde{x})\Re\, \Phi_j(\zeta, \zeta_0) > 0$.

Lemma 5.4 *Let $g = g(x, y) = (g_1, \ldots, g_N)$, $g_j \in \mathcal{H}_c(D, \Omega)$. Then the solution of* (31) *is given by*

$$w = P_0 g := (P_{0,1} g_1, \ldots, P_{0,N} g_N). \tag{47}$$

Here, for every j, $1 \leq j \leq N$ and $\zeta \neq 0$ in a neighborhood of the origin we take ζ_0 such that $\zeta \in \gamma_{\zeta_0 j}$ and $P_{0,j}$ is given by

$$P_{0,j} g_j := \int_{\zeta_0}^{\zeta} g_j(x_1(s), \cdots, x_{n-1}(s), s; y_0 - \Phi_j(s, b)) s^{-1} ds, \tag{48}$$

where the integral is taken along the curve $\gamma_{\zeta_0 j}$ from ζ_0 to $\zeta \in \gamma_{\zeta_0 j}$ and $x_j(s)$ is given by (33) *with $\zeta = s$. The independent variables x_k and y in* (48) *are related to c_k and y_0 via* (33).

Proof We show that the integrand in (48) is well defined. By (33) and (34) we have

$$y_0 - \Phi_j(s, b) = y - \Phi_j(s, b) + \Phi_j(\zeta, b) = y + \Phi_j(\zeta, s). \tag{49}$$

By Lemma 5.2 we see that $\Im \Phi_j(\zeta, s) = 0$ if $s \in \gamma_{\zeta_0 j}$ because $\zeta \in \gamma_{\zeta_0 j}$. On the other hand, by Lemma 5.3 we see that $\Re\, \Phi_j(\zeta, s)$ is a monotone decreasing function of $\zeta \in \gamma_{\zeta_0 j}$ when ζ approaches the origin. Hence we have $\Re\, \Phi_j(\zeta, s) \leq 0$ on $\gamma_{\zeta_0 j}$. In view of the assumption on Ω we have $y + \Re\, \Phi_j(\zeta, s) \in \Omega$ for every $y \in \Omega$.

Next we take the neighborhood U_0 of the origin such that the formal solution is holomorphic in U_0. Consider the substitution $x_k = x_k(s)$ into the integrand of (48) where $x_k(s)$ is given by (33) with $\zeta = s$. We want to show that this is possible for s which is on the segment of $\gamma_{\zeta_0 j}$ between ζ_0 and ζ. Indeed, we first observe that $s^{\lambda_k} c_k$

can be made arbitrarily small if c_k is taken sufficiently small. We observe

$$s^{\lambda_j} = \exp\left(\lambda_j(\log|s| + i\arg s)\right). \tag{50}$$

Because $\lambda_j > 0$, the absolute value of the right-hand side of (50) is monotone decreasing when $r = |s|$ tends to zero, namely s tends to the origin along $\gamma_{\zeta_0,j}$. This proves the assertion. Next we consider the case $x_k(s) = ((s_k - 1)(c_k - \lambda_k \log s))^{1/(1-s_k)}$. Clearly, the right-hand side is small if s tends to zero because $\Re\,(c_k/\lambda_k) > 0$ by assumption. Hence the right-hand side of (48) is well defined.

Next we shall show that $w_j := P_{0,j}g_j$ $(j = 1, 2, \ldots, N)$ satisfies Eq. (31). Indeed, by (32) and (33) we have

$$g_j(x,y)x_n^{-1} = \frac{dw_j}{d\zeta} = \sum_{k=1}^{n} \frac{\partial x_k}{\partial \zeta}\frac{\partial w_j}{\partial x_k} + \frac{\partial y}{\partial \zeta}\frac{\partial w_j}{\partial y} \tag{51}$$

$$= \sum_{k=1}^{n} \frac{\lambda_k x_k^{s_k}}{\zeta}\frac{\partial w_j}{\partial x_k} - \frac{(\nabla_u f_0)_j(x,0)}{\zeta}\frac{\partial w_j}{\partial y}.$$

Multiplying both sides with ζ and setting $\zeta = x_n$ we have the assertion. □

Let ζ_0 satisfy $|\zeta_0| = r_0 > 0$. Let $\varepsilon_0 > 0$ be a given small constant. In the following we take D such that $|\zeta|/|\zeta_0| \geq \varepsilon_0$ for ζ and ζ_0 in D, where ζ and ζ_0 are related by $\zeta \in \gamma_{\zeta_0,j}$.

Lemma 5.5 *There exists a constant c_1 such that, for every $1 \leq j \leq N$, $g_j \in \mathcal{H}(D,\Omega)$, we have*

$$\|P_{0,j}g_j\|_c \leq c_1\|g_j\|_c, \quad \left\|(\partial/\partial y)(P_{0,j}g_j)\right\|_c \leq c_1\|g_j\|_c. \tag{52}$$

The constant c_1 is independent of ζ_0, $|\zeta_0| = r_0 > 0$.

Proof We shall prove the estimate of $P_{0,j}g_j$ in (52). The latter one can be proved similarly. Let $\zeta \in \gamma_{\zeta_0,j}$ and consider the integral (48). Noting that $y_0 - \Phi_j(s,b) = y + \Phi_j(\zeta,s)$ we make the change of the variable $\sigma = y + \Phi_j(\zeta,s)$ in (48) from s to σ. We have $d\sigma = -(\nabla_u f)_j s^{-1}ds$. Observe that the right-hand side is independent of y. We have $\sigma = y$ for $s = \zeta$ and $\sigma = y + \tilde{\zeta}_0$ for $s = \zeta_0$, where $\tilde{\zeta}_0 = \Phi_j(\zeta,\zeta_0)$. Clearly, $s \in \gamma_{\zeta_0,j}$ is expressed as $\sigma \in y + \gamma_{\tilde{\zeta}_0,j}$, where $\gamma_{\tilde{\zeta}_0,j}$ is the straight line connecting 0 and $\tilde{\zeta}_0$. Then (48) is written as

$$w = -\int_{y+\gamma_{\tilde{\zeta}_0,j}} g(x_1(s),\cdots,x_{n-1}(s),s;\sigma)\frac{d\sigma}{(\nabla_u f_0)_j}, \tag{53}$$

where $(\nabla_u f_0)_j$ is bounded from the below by the assumption (3).

We next estimate the growth of $y_0 - \Phi_j(s, b)$. In terms of (49) we have

$$\exp\left(-c\Re\left(y_0 - \Phi_j(s, b)\right)\right) = \exp\left(-c\Re\left(y + \Phi_j(\zeta, s)\right)\right). \tag{54}$$

By Lemma 5.3 $\Re\,\Phi_j(\zeta, s)$ is decreasing in ζ as ζ tends to the origin along $\gamma_{\zeta_0 j}$. It follows that $\Re\,\Phi_j(\zeta, s) \le \Re\,\Phi_j(s, s) = 0$. Hence we need to estimate $e^{-c\Re\,\Phi_j(\zeta,s)}$, namely we shall estimate $\Re\,\Phi_j(\zeta, s)$ from the below.

Because $R(\zeta)$ in (35) has the estimate (42) $\Phi_j(\zeta, s)$ has the behavior

$$\Phi_j(\zeta, s) = \mu_j \log(\zeta/s)\left(1 + O((|\log \zeta| + |\log \zeta_0|)^{-\rho})\right), \tag{55}$$

for every $s \in \gamma_{\zeta_0, j}$ if $|\log \zeta_0|$ is sufficiently large. Set $\log(\zeta/s) = x + iy$ and $\mu_j = \alpha + i\beta$ with $\alpha > 0$. Then we have $\Re\,(\mu_j \log(\zeta/s)) = \alpha x - \beta y$ and $\Im\,(\mu_j \log(\zeta/s)) = \alpha y + \beta x$. It follows that $0 \ge \Re\,\Phi_j(\zeta, s) \ge \gamma_0(\alpha x - \beta y)$ for some $\gamma_0 > 0$ if $|\log \zeta_0|$ is sufficiently large. The condition holds if ζ and ζ_0 are in a sector in a neighborhood of the origin. Thus we have $e^{-c\Re\,\Phi_j(\zeta,s)} \le e^{-c\gamma_0(\alpha x - \beta y)}$.

By (55) there exist $\gamma_1 > 0$ and $\gamma_2 > 0$ such that

$$\gamma_1(\alpha y + \beta x) \le \Im\,\Phi_j(\zeta, s) \le \gamma_2(\alpha y + \beta x), \tag{56}$$

if $|\log \zeta_0|$ is sufficiently large. Because $\Im\,\Phi_j(\zeta, s)$ is constant when $\zeta \in \gamma_{\zeta_0 j}$, there exist $\tilde{\gamma}_1 > 0$ and $\tilde{\gamma}_2 > 0$ such that $\tilde{\gamma}_2 \le \alpha y + \beta x \le \tilde{\gamma}_1$, if $\zeta \in \gamma_{\zeta_0 j}$. Suppose that $\beta > 0$. Then, by (56) we have $-\beta y \ge \beta^2\alpha^{-1}x - \beta\alpha^{-1}\tilde{\gamma}_1$. we have

$$\alpha x - \beta y \ge (\alpha + \beta^2\alpha^{-1})x - \tilde{\gamma}_1\beta\alpha^{-1}.$$

Noting that $x = \log(|\zeta|/|s|) > \log(|\zeta|/|\zeta_0|) > \log \varepsilon_0$, we have

$$\begin{aligned}\exp(-c\gamma_0(\alpha x - \beta y)) &\le \exp(-c\gamma_0(\alpha + \beta^2\alpha^{-1})x + c\gamma_0\tilde{\gamma}_1\beta\alpha^{-1}) \\ &\le \exp\left(c\gamma_0(\alpha + \beta^2\alpha^{-1})\log \varepsilon_0^{-1} + c\gamma_0\tilde{\gamma}_1\beta\alpha^{-1}\right) =: K_0.\end{aligned}$$

This proves

$$\exp\left(-c\Re\left(y_0 - \Phi_j(s, b)\right)\right) \le K_0 \exp\left(-c\Re\, y\right). \tag{57}$$

We have the same estimate (57) in case $\beta \le 0$ by (56).

We shall estimate $|y_0 - \Phi_j(s, b)| = |y + \Phi_j(\zeta, s)|$ from the below. Because $\Im\,\Phi_j(\zeta, s) = 0$ and $\Re\,\Phi_j(\zeta, s) \le 0$ on $\gamma_{\zeta_0 j}$, there exists $C_1 > 0$ independent of ζ and s such that

$$(1 + |y_0 - \Phi_j(s, b)|)^{-2} \le C_1(1 + |y|)^{-2} \quad \text{for all } y \in \Omega. \tag{58}$$

Therefore, by (57) and (58) there exist $C_2 > 0$ and $C_3 > 0$ such that

$$
\begin{aligned}
\|w_j\|_c &\leq \\
&\leq \sup \left((1+|y|)^2 \exp(c\Re\, y) \int \|g_j\|_c \frac{\exp\left(-c\Re\,(y_0 - \Phi_j(s,b))\right)}{(1+|y_0-\Phi_j(s,b)|)^2} |d\sigma| \right) \\
&\leq C_2 \|g_j\|_c \int |d\sigma| \leq C_3 \|g_j\|_c.
\end{aligned}
\tag{59}
$$

□

We shall solve (30) in $\mathcal{H}_c(D,\Omega)$. First we note

$$
\nabla_u f_0(x, v_0)\frac{\partial \hat{u}}{\partial y} = \nabla_u f_0(x,0)\frac{\partial \hat{u}}{\partial y} + (\nabla_u f_0(x,v_0) - \nabla_u f_0(x,0))\frac{\partial \hat{u}}{\partial y}. \tag{60}
$$

We note $\|\nabla_u f_0(x,v_0) - \nabla_u f_0(x,0)\| = O(\|v_0\|)$ when $\|v_0\| \to 0$. We denote by $R(x,\hat{u})$ the fourth term of the right-hand side of (30). Then we define the approximate sequence $\hat{u}_k$ $(k = 0,1,2,\ldots)$ by $\hat{u}_0 = 0$ and

$$
\hat{u}_1 = -P_0 \mathcal{L} v_0 \tag{61}
$$

$$
\begin{aligned}
\hat{u}_2 = P_0 \sum_{|\beta|\geq 2} r_\beta(x,v_0)\frac{\partial}{\partial y}(\hat{u}_1)_*^\beta - P_0\mathcal{L}v_0 + P_0 R(x,\hat{u}_1) \\
+ P_0(\nabla_u f_0(x,v_0) - \nabla_u f_0(x,0))\frac{\partial \hat{u}_1}{\partial y},
\end{aligned}
\tag{62}
$$

$$\vdots$$

$$
\begin{aligned}
\hat{u}_{k+1} = P_0 \sum_{|\beta|\geq 2} r_\beta(x,v_0)\frac{\partial}{\partial y}(\hat{u}_k)_*^\beta - P_0\mathcal{L}v_0 + P_0 R(x,\hat{u}_k) \\
+ P_0(\nabla_u f_0(x,v_0) - \nabla_u f_0(x,0))\frac{\partial \hat{u}_k}{\partial y},
\end{aligned}
\tag{63}
$$

where $k = 1,2,\ldots$ We note that the term $P_0 R(x,\hat{u}_k)$ can be estimated by Lemmas 4.2 and 5.5. Then we have

Lemma 5.6 *Let D be as in Lemma* 5.5. *Then there exists a constant $K_3 > 0$ independent of k such that*

$$
\|\hat{u}_k\|_c \leq C\varepsilon K_3, \quad \|(\hat{u}_k)_y\|_c \leq C\varepsilon K_3, \; k = 0,1,2,\ldots \tag{64}
$$

Lemma 5.7 *Under the same assumptions as in Lemma 5.6 $\hat{u}_k$ $(k = 1,2,\ldots)$ converges in $\mathcal{H}_c(D,\Omega)$.*

Lemmas 5.6 and 5.7 are proved by exactly the same argument as in the proofs of Lemmas 5.6 and 5.7 of [4], respectively. We observe that Lemma 5.7 implies the solvability of (30) in $\mathcal{H}_c(D, \Omega)$.

Proof of Theorem 5.1. If one proves Lemmas 5.2, 5.3, 5.4, 5.5, 5.6 and 5.7 in the above, then one can prove Theorem 5.1 by the argument as in the proof of Theorem 5.1 in [4].

End of the Proof of Theorem 2.1. We shall prove the summability in the direction $\eta \in S_{\theta,\xi}$. By multiplying (1) with $e^{-i\theta}$ we see that η, λ_k, μ_j are replaced by $\eta e^{-i\theta}$, λ_k and $\mu_j e^{-i\theta}$, respectively. Since (6) holds for $0 \le \theta < \pi/2 - \theta_2$, the summability follows for $\eta = e^{i(\pi+\theta)}$ with $0 \le \theta < \pi/2 - \theta_2$. Hence the summability holds for $\pi \le \arg \eta < 3\pi/2 - \theta_2$.

Next we set $u = ve^{i\theta}$, and consider the equation of v. Clearly, η and μ_k are replaced by $\eta e^{i\theta}$ and $\mu_k e^{i\theta}$, respectively. λ_k does not change. On the other hand, we see that (6) is satisfied for the new equation when $0 \le \theta < \pi/2 - \theta_1$. Hence the summability holds for $\pi/2 + \theta_1 < \arg \eta \le \pi$. Therefore, the summability follows for $\pi/2 + \theta_1 < \arg \eta < 3\pi/2 - \theta_2$. In view of the definition of Borel sum we have the latter half of the assertion. This ends the proof of Theorem 2.1.

Acknowledgements The author would like to thank anonymous referee for useful suggestions. The author is partially supported by Grant-in-Aid for Scientific Research (No. 11640183), Ministry of Education, Science and Culture, Japan.

References

1. W. Balser, V. Kostov, Singular perturbation of linear systems with a regular singularity. J. Dyn. Control. Syst. **8**(3), 313–322 (2002)
2. W. Balser, J. Mozo-Fernández, Multisummability of formal solutions of singular perturbation problems. J. Differ. Equ. **183**(2), 526–545 (2002)
3. M. Canalis-Durand, J. Mozo-Fernández, R. Schäfke, Monomial summability and doubly singular differential equations. J. Differ. Equ. **233**(2), 485–511 (2007)
4. H. Yamazawa, M. Yoshino, Borel summability of some semilinear system of partial differential equations. Opusc. Math. **35**(6), 825–845 (2015)

MIX
Papier aus verantwortungsvollen Quellen
Paper from responsible sources
FSC® C105338

FSC
www.fsc.org

If you have any concerns about our products,
you can contact us on
ProductSafety@springernature.com

In case Publisher is established outside the EU,
the EU authorized representative is:
Springer Nature Customer Service Center GmbH
Europaplatz 3, 69115 Heidelberg, Germany

Printed by Libri Plureos GmbH
in Hamburg, Germany